Air Conditioning Engineering

Air Conditioning Engineering

W. P. Jones

M.Sc., C Eng., F.Inst.F., F.I.H.V.E.

EDWARD ARNOLD

First published 1967
by Edward Arnold (Publishers) Limited
25 Hill Street, London W1X 8LL
Reprinted 1969
Second Edition 1973
Reprinted 1975, 1977
ISBN: 0 7131 3312 0

Printed in Great Britain by
Richard Clay (The Chaucer Press) Ltd, Bungay, Suffolk

Preface to the Second Edition

The change to the use of the Système International d'Unités (SI) has meant a revision of all worked examples and exercises in this volume. In addition, alteration to some of the standards adopted and the methods of calculation has changed the basis of some examples; new values in whole numbers have been used for many design parameters because a simple translation from Imperial to metric would have yielded unworkable fractions. In most cases, the units in the text are pure SI, it being impossible at this early stage in the life of the new system to forecast future development and decide which units will be discarded as impractical and which retained. Only evolution will show this, but where it has seemed likely that alternative forms may become popular (e.g. litre/s instead of dm^3/s) they have been adopted on occasion.

Some errors in the original text have been corrected and several new sections added. In particular, additional text has been provided to deal with the modern views on psychrometry which form the basis of the latest I.H.V.E. tables. The chapters dealing with heat gains and automatic controls have been partly revised and the opportunity has been taken to add sections on the increasingly popular variable air volume system and the interesting development of integrated environmental design.

W.P.J.

Preface to the First Edition

Air conditioning (of which refrigeration is an inseparable part) has its origins in the fundamental work on thermodynamics which was done by Boyle, Carnot and others in the seventeenth and eighteenth centuries, but air conditioning as a science applied to practical engineering owes much to the ideas and work of Carrier, in the United States of America, at the beginning of this century. An important stepping stone in the path of progress which has led to modern methods of air conditioning was the development of the psychrometric chart, first by Carrier in 1906 and then by Mollier in 1923, and by others since.

The summer climate in North America has provided a stimulus in the evolution of air conditioning and refrigeration which has put that semi-continent in a leading position amongst the other countries in the world. Naturally enough, engineering enterprise in this direction has produced a considerable literature on air conditioning and allied subjects. The *Guide and Data Book* published by the American Society of Heating, Refrigeration and Air Conditioning Engineers has, through the years, been a foremost work of reference but, not least, the *Guide to Current Practice* of the Institution of Heating and Ventilating Engineers has become of increasing value, particularly of course in this country. Unfortunately, although there exists a wealth of technical literature in textbook form which is expressed in American terminology and is most useful for application to American conditions, there is an almost total absence of textbooks on air conditioning couched in terms of British practice. It is hoped that this book will make good the deficiency.

The text has been written with the object of appealing to a dual readership, comprising both the student studying for the associate membership examinations of the Institution of Heating and Ventilating Engineers and the practising engineer, with perhaps a 75 per cent emphasis being laid upon the needs of the former. To this end, the presentation follows the sequence which has been adopted by the author during the last few years in lecturing to students at the Polytechnic of the South Bank. In particular, wherever a new idea or technique is introduced, it is illustrated immediately by means of a worked example, when this is possible. It is intended that the text should cover those parts of the syllabus for the corporate membership examination that are relevant to air conditioning.

Inevitably some aspects of air conditioning have been omitted (the author particularly regrets the exclusion of a section on economics). Unfortunately, the need to keep the book within manageable bounds and the desire to avoid a really prohibitive price left no choice in the matter.

W.P.J.

ACKNOWLEDGEMENTS

Originally this book was conceived as a joint work, in co-authorship with Mr. L. C. Bull. Unfortunately, owing to other commitments, he was compelled largely to forego his interest. However, Chapters 9 and 14 (on the fundamentals of vapour compression and vapour-absorption refrigeration) are entirely his work. The author wishes to make this special acknowledgement to Mr. Bull for writing these chapters and also to thank him for his continued interest, advice and encouragement. Equations (7.30) and (7.35) are derived from his work.

The helpful comment of Mr. E. Woodcock is also appreciated.

The author is additionally grateful to the following for giving their kind permission to reproduce copyright material which appears in the text.

The Institution of Heating and Ventilating Engineers for various questions from external examination papers and also Table 7.1 and Figure 7.16 from the Guide to Current Practice.

The Polytechnic of the South Bank for various questions from internal examination papers.

Buffalo Forge Company, Buffalo, N.Y. for equations (4.1) and (4.2), from Fan Engineering.

H.M. Stationery Office for equation (4.3), from War Memorandum No. 17, Environmental Warmth and its Measurement by T. Bedford.

H. K. Lewis & Co. Ltd. for equation (4.4), from Basic Principles of Ventilation and Heating by T. Bedford.

Haden Young Ltd. for Tables 4.1 and 7.8.

The American Society of Heating, Refrigeration and Air Conditioning Engineers for Table 7.3.

The Steam and Heating Engineer for Table 7.11.

Industrial Press Inc., 200 Madison Avenue, New York, for Tables 10.1, 10.2 and 10.3, from articles by M. A. Ramsey in Air Conditioning, Heating and Ventilating.

John Wiley & Sons Inc., New York, for Figure 13.8, from Automatic Process Control by D. P. Eckman.

McGraw-Hill Book Company for Table 7.12.

Contents

Notation

Symbol	*Description*	*Unit*
Chapter 2		
A	surface area of a water droplet	m^2
$A_{ik}(T)$	second virial coefficient, molecules considered two at a time	m^3/kg
$A_{ijk}(T)$	third virial coefficient, molecules considered three at a time	m^3/kg
A_{aa}	second virial coefficient of dry air	m^3/kg
A_{ww}	second virial coefficient of water vapour	m^3/kg
A_{www}	third virial coefficient of water vapour	m^3/kg
A_{aw}	interaction coefficient	m^3/kg
A, B, C, D	constants	
A', B', etc.	constants	
a	constant	
c	specific heat of moist air	J/kg °C
c_a	specific heat of dry air	J/kg °C
c_s	specific heat of water vapour (steam)	J/kg °C
f_s	dimensionless coefficient	
g	moisture content	kg/kg dry air
g_{ss}	moisture content of saturated air	kg/kg dry air
H	enthalpy	kJ
h	specific enthalpy of moist air	kJ/kg dry air
h_a	enthalpy of 1 kg of dry air	kJ/kg
h_g	latent heat plus sensible heat in the water vapour associated with 1 kg of dry air	kJ/kg
h_s	enthalpy of 1 kg of water vapour at a temperature t, the temperature of the dry bulb	kJ/kg
h_c	coefficient of heat transfer through the gas film surrounding a water droplet, by convection	W/m^2 °C

Symbol	*Description*	*Unit*
h_r	coefficient of heat transfer to a water droplet by radiation from the surrounding surfaces	W/m² °C
h	$= h_c + h_r$	W/m² °C
h_{fg}	latent heat of evaporation	kJ/kg
(Le)	Lewis number	
M	molecular mass	
M_a	molecular mass of dry air	
M_s	molecular mass of water vapour (steam)	
m	mass of a gas	kg
m_a	mass of dry air	kg
m_s	mass of water vapour (steam)	kg
n_a	number of molecules of constituent a	
n_w	number of molecules of constituent w	
p	gas pressure	N/m²
p_a	pressure of dry air	N/m²
p_{at}	atmospheric pressure (barometric pressure)	N/m²
p_s	pressure of water vapour (steam)	N/m²
p_{ss}	saturation vapour pressure	N/m²
p'_{ss}	saturation vapour pressure at the wet-bulb temperature	N/m²
R_o	universal gas constant	J/kmol K
R	particular gas constant	J/kg K
R_a	particular gas constant of dry air	J/kg K
R_s	particular gas constant of water vapour (steam)	J/kg K
R_{ss}	particular gas constant of dry saturated steam	J/kg K
T	absolute temperature	K
t	temperature	°C
t_c	critical temperature	°C
t_d	dew-point temperature	°C
t_{ss}	saturation temperature	°C
t'	wet-bulb temperature	°C
t^*	temperature of adiabatic saturation	°C
U	internal energy	kJ
v	specific volume	m³/kg
V	gas volume	m³
V_a	volume of dry air	m³
V_m	volume of 1kmol	m³
V_s	volume of steam	m³
x	mole fraction	
α	coefficient of diffusion	kg/N

Symbol	*Description*	*Unit*
μ	percentage saturation	
ϕ	relative humidity	

Chapter 3

c_a	specific heat of dry air	J/kg °C or kJ/kg °C
E	effectiveness of an air washer or spray chamber	
g	moisture content	kg/kg dry air or g/kg dry air
h	enthalpy of most air	J/kg dry air
m_a	mass or mass flow of dry air	kg or kg/s
m_w	mass of feed water evaporated	kg/s
t	dry-bulb temperature	°C
t_d	dew-point temperature (or the dry-bulb temperature at a state D)	°C
t_w	feed-water temperature	°C
β	contact factor	
η	humidifying efficiency	

Chapter 4

a, b, T	constants	
C	rate of heat loss or gain by convection	W or kW
E	rate of heat loss by evaporation	W or kW
H	cooling power of a Kata thermometer	
M	metabolic rate	W or kW
p_s	partial pressure of the water vapour in the air	N/m^2
p_w	vapour pressure of water	N/m^2
R	rate of heat loss or gain by radiation	W or kW
S	rate at which heat is stored in the body	W or kW
t	dry-bulb temperature	°C
T_a	air dry-bulb temperature	K
T_g	reading of a globe thermometer	K
T_{rm}	mean radiant temperature	K
v	relative velocity of air flow	m/s
V	air velocity measured by a Kata thermometer	m/s
W	rate of useful working	W or kW

Chapter 6

A	surface area	m^2

Symbol	*Description*	*Unit*
c	specific heat of air	kJ/kg °C or J/kg °C
g	moisture content	g/kg dry air
g_a	moisture content at the apparatus dew point	g/kg
g_o	moisture content of outside air	g/kg
g_r	room moisture content	g/kg
g_s	supply air moisture content	g/kg
h_a	enthalpy of air at the apparatus dew point	kJ/kg
h_o	enthalpy of the outside air	kJ/kg
h_w	enthalpy of the air leaving the cooler coil	kJ/kg
h_{fg}	latent heat of evaporation	kJ/kg
m	mass of air supplied	kg or kg/s
m_a	mass flow rate of dry air	kg/s
n	air change rate per hour	
Q	rate of production of heat	kW
T	dry-bulb temperature	K
t_o	outside environmental dry-bulb temperature	°C
t_r	room dry-bulb temperature	°C
t_s	supply air temperature	°C
t_w	dry-bulb temperature of air leaving the cooler coil	°C
U	overall thermal transmittance	kW/m² °C
V	volume	m³
v	specific volume	m³/kg dry air
β	contact factor	
ρ	density of air	kg/m³
η	total fan efficiency	%

Chapter 7

A	area	m²
A, B	internal duct dimensions	m
A_f	floor area	m²
A_w	window area	m²
a	altitude of the sun	
a'	altitude of the sun at noon	
a_n	harmonic phase angle	
C	dimensionless constant	
c	specific heat	kJ/kg °C
D	internal duct diameter	m
d	declination of the sun	

Symbol	*Description*	*Unit*
F_g	angle factor for the ground	
F_s	angle factor for the sky	
f	decrement factor	
h	hour-angle, stack height in metres or coefficient of heat transfer according to the context	
h_{si}	inside surface film coefficient of heat transfer	W/m² °C
h_{so}	outside surface film coefficient of heat transfer	W/m² °C
I	intensity of direct solar radiation on a plane at right angles to the path of the rays of the sun	W/m²
I_h	component of I which is normally incident upon a horizontal surface	W/m²
I_r	intensity of radiation reflected from surrounding surfaces	W/m²
I_s	intensity of diffuse radiation normally incident on a surface	W/m²
I_t	intensity of total radiation on a surface	W/m²
I_v	component of I which is normally incident upon a vertical surface	W/m²
I_δ	component of I which is normally incident upon a surface which is tilted at an angle δ to the horizontal	W/m²
i	angle of incidence	
k	thermal conductivity of a wall or of duct insulation	W/m °C
L	wall thickness, or duct length	m
	latitude of a place on the surface of the earth	
l	thickness of insulation on a duct	m
n	solar-wall azimuth	
P	external duct perimeter	m
Q	rate of heat flow	W or W/m²
Q_m	mean rate of heat flow into a room	W
Q'	rate of heat flow into the outer surface of a wall	W/m²
Q_θ	rate of heat flow into a room at time θ	W
$Q_{(\theta+\varphi)}$	rate of heat flow into a room at time $(\theta+\varphi)$	W
q	rate of airflow	m³/s

Symbol	*Description*	*Unit*
q_i	instantaneous solar gain	W/m^2
q_{max}	maximum solar gain	W/m^2
R	depth of a window recess, in metres or millimetres as appropriate, or a remainder term, according to the context	
r_{si}	thermal resistance of the air film inside a duct	m^2 °C/W
r_{so}	thermal resistance of the air film outside a duct	m^2 °C/W
T	sun-time	h
t	time or temperature	h or °C
t_e	sol-air temperature	°C
t_{em}	mean sol-air temperature, over 24 hours	°C
$t_{e\theta}$	sol-air temperature at time θ	°C
t_g	mean glass temperature	°C
t_n	harmonic temperature coefficient	
t_o	outside air temperature	°C
t_r	room air temperature	°C
t_{si}	inside surface temperature	°C
t_{sm}	mean inside surface temperature	°C
t_{so}	outside surface temperature	°C
v	mean velocity of airflow	m/s
U	overall thermal transmittance	W/m^2 °C
V	air velocity in a duct	m/s
x, y	co-ordinates	m or mm
x	thickness of a floor slab	m or mm
z	azimuth of the sun	
α	coefficient of absorption for direct solar radiation	
α'	coefficient of absorption for diffuse solar radiation	
Δp	pressure difference	N/m^2
Δt	equivalent temperature differential $(t_o - t_r)$	°C
θ	time or temperature difference	h or °C
λ_e	equivalent decrement factor	
λ_1	fundamental decrement factor	
λ_n	decrement factor for the *n*th harmonic	
ρ	density	kg/m^3
τ	coefficient of transmissivity for direct solar radiation	
τ'	coefficient of transmissivity for diffuse solar radiation	

Symbol	*Description*	*Units*
φ	time lag or phase constant	h
φ_1	fundamental lag angle, in degrees	
φ_n	harmonic lag angle, in degrees; hence the time lag in hours is $\varphi_n/15$	
ω	angular velocity of the sun	h^{-1}
$1/K$	time constant	h^{-1}

Chapter 8

Symbol	*Description*	*Units*
c	humid specific heat of air	kJ/kg °C
g	moisture content of air	kg/kg or g/kg
h	enthalpy of air	kJ/kg
m	mass flow of air	kg/s
t	dry-bulb temperature of air	°C
Q_L	latent heat gain	W or kW
Q_S	sensible heat gain	W or kW
L	sensible heat gain due to electric lighting	W or kW
P	sensible heat gain due to people	W or kW
S	sensible heat gain due to solar radiation	W or kW
T	sensible heat gain due to transmission by virtue of air-to-air temperature difference	W
β	contact factor	

Chapter 9

Symbol	*Description*	*Units*
COP	coefficient of performance	
c	constant in equation $pv^n = c$ (see §9.4)	
c_1	specific heat of liquid	kJ/kg °C
c_p	specific heat of gas at constant pressure	kJ/kg °C
c_v	specific heat of gas at constant volume	kJ/kg °C
f	mass in kg of refrigerant which vaporises during throttling, per kg circulated	
f_1	mass in kg of refrigerant which vaporises during expansion in an engine, per kg circulated	
f_2	mass in kg of refrigerant which vaporises during compression, per kg circulated	
h_j	heat transferred to the cylinder jacket during compression	kJ/kg
h_{1c}	enthalpy of saturated liquid at condensing pressure	kJ/kg
h_{1e}	enthalpy of saturated liquid at evaporating pressure	kJ/kg

Symbol	*Description*	*Unit*
h_{m1}	enthalpy of mixture of liquid and vapour entering evaporator	kJ/kg
h_{m2}	enthalpy of mixture of liquid and vapour entering compressor	kJ/k
h_{vc}	enthalpy of saturated vapour at condensing pressure	kJ/kg
h_{vd}	enthalpy of superheated vapour at discharge condition	kJ/kg
h_{ve}	enthalpy of saturated vapour at evaporating pressure	kJ/kg
h_{fg}	latent heat of vaporisation at evaporating temperature	kJ/kg
m	mass of refrigerant circulated	kg/s
n	exponent in equation $pV^n = C$(see §9.4)	
p	pressure	N/m^2
p_1	pressure at the beginning of a process	N/m^2
p_2	pressure at the end of a process	N/m^2
p_c	condensing pressure	N/m^2
p_d	discharge pressure	N/m^2
p_e	evaporating pressure	N/m^2
Q	quantity of heat per unit mass	kJ/kg
R	particular gas constant	kJ/kg K
s	entropy	kJ/kg K
s_1	entropy at the beginning of a process	kJ/kg K
s_2	entropy at the end of a process	kJ/kg K
s_{AB}	entropy during process depicted by line *AB*	kJ/kg K
s_{CD}	entropy during process depicted by line *CD*	kJ/kg K
s_{1c}	entropy of saturated liquid at condensing pressure	kJ/kg K
s_{1e}	entropy of saturated liquid at evaporating pressure	kJ/kg K
s_{vc}	entropy of saturated vapour at condensing pressure	kJ/kg K
s_{ve}	entropy of saturated vapour at evaporating pressure	kJ/kg K
T	absolute temperature	K
T_1	temperature at beginning of process	K
T_2	temperature at end of process	K
T_c	condensing temperature	K
T_d	discharge temperature	K
T_e	evaporating temperature	K
TR	tons of refrigeration (*see* §9.2)	

Symbol	*Description*	*Unit*
V	volume per unit mass	m^3/kg
V_1	volume at the beginning of a process	m^3/kg
V_2	volume at the end of a process	m^3/kg
V_d	volume at discharge condition	m^3/kg
V_e	volume of saturated vapour at evaporating pressure	m^3/kg
W_c	work of compression	kW
X	cooling capacity or load, in kW of refrigeration	
γ	$= c_p/c_v$	

Chapter 10

Symbol	*Description*	*Unit*
A_f	face area of the cooler coil	m^2
A_r	total external surface area per row of the cooler coil	m^2
A_t	total external surface area of the cooler coil	m^2
c	specific heat of humid air	kJ/kg °C
h	enthalpy of humid air	kJ/kg °C
h_a	coefficient of heat transfer on the air-side of the cooler coil	kJ/kg °C
k	constant	
LMTD	logarithmic mean temperature difference	°C
m_a	mass flow of dry air	kg/s
m_w	mass flow of water inside the cooler coil tubes in kg water/kg dry air flowing over the outside of the cooler coil	
Q_s	rate of sensible heat transfer	kW
Q_t	rate of total heat transfer	kW
R_a	thermal resistance of the surface film on the air-side of a cooler coil	m^2 °C/W
R_m	thermal resistance of the metal of the cooler coil (wall + fins)	m^2 °C/W
R_r	thermal resistance of the surface film on the refrigerant side of a cooler coil	m^2 °C/W
R_w	thermal resistance of the surface film on the water-side of the cooler coil	m^2 °C/W
r	the number of rows in the cooler coil	
S	sensible heat removed by the coil/total heat removed by the coil	
t, t_a	dry-bulb temperature of the air flowing over a cooler coil	°C
t'	wet-bulb temperature	°C

Symbol	*Description*	*Unit*
t_d	dew-point temperature of the air	°C
t_w	chilled water temperature	°C
t_{wa}, t_{wb}	entering, leaving chilled water temperature	°C
t_{w1}	mean surface temperature for the first row of a cooler coil	°C
t_{w2}	mean coil surface temperature for the second row of a cooler coil	°C
t_{sm}	mean coil surface temperature for the whole cooler coil	°C
U_t	U-value for the cooler coil	W/m² °C
v_f	velocity of airflow over the face of the cooler coil	m/s
β	contact factor	

Chapter 11

a	cross-sectional area of the tower	m²
h_a	enthalpy of humid air kJ/kg of dry air	
k	coefficient of vapour diffusion for a unit value of Δh_m	kg/s m²
ks	volume transfer coefficient	
m_a	rate of mass flow of air	kg/s
m_w	rate of mass flow of water	kg/s
s	wetted surface area per unit volume of packing	m⁻¹
t	air temperature	°C
t_w	cooling water temperature	°C
Z	height of a cooling tower	m
Δh_m	mean driving force	kJ/kg

Chapter 13

A_v	flow coefficient	
a	cross-sectional area of a valve opening	m²
d	diameter of the pipe or duct	m
f	dimensionless coefficient of friction (Fanning)	
f_m	dimensionless coefficient of friction (Moody) = $4f$	
g	acceleration due to gravity	m/s²
H_1, H_0	head in a reservoir or sink, in m of fluid	
h_l	head lost along a pipe, in m of fluid	
h_v	head lost across the valve	m

Symbol	*Description*	*Unit*
h	static head, in m of fluid	
K, K_1	constants of proportionality	
k_d	derivative control factor	
k_i	integral control factor	
k_p	proportional control factor	
l	length	m
q	volumetric flow rate	m^3/s
q_0, z_0	maximum flow rate and maximum value lift	m^3/s and m
v	mean velocity of fluid flow	m/s
z	position of a valve stem	m
α	valve authority $= 1/(1+\beta z^2)$	
$\beta =$	$16 f_m l K_1^2/\pi^2 d^5$	
ϕ	potential correction	
θ	deviation	°C

Chapter 14

C_a	concentration of LiBr, per cent by weight, in the absorber	
C_g	concentration of LiBr, per cent by weight, in the generator	
COP	coefficient of performance	
c_1	specific heat of the liquid refrigerant	kJ/kg °C
H_a	heat removed at the absorber	kW/kW of refrigeration
H_c	heat removed at the condenser	kW/kW of refrigeration
H_g	heat added at the generator	kW/kW of refrigeration
h_a	enthalpy of the solution leaving the absorber	kJ/kg
h_g	enthalpy of the solution leaving the generator	kJ/kg
h_{1c}	enthalpy of the refrigerant liquid leaving the condenser	°C
h_{ve}	enthalpy of the refrigerant vapour leaving the evaporator	°C
h_{vg}	enthalpy of the vapour leaving the generator	kJ/kg
m_r	mass of refrigerant circulated	kg/s kW
m_{sa}	mass of solution leaving the absorber	kg/s kW
m_{sg}	mass of solution leaving the generator	kg/s kW
p_c	condensing pressure of the refrigerant	kN/m^2

Symbol	*Description*	*Unit*
p_e	evaporating pressure of the refrigerant	kN/m^2
Q_g	heat supplied at the generator, in kJ/kg of refrigerant	
Q_r	refrigerating effect, in kJ/kg of refrigerant	
T_c	$273+t_c$	
T_g	$273+t_g$	
t_c	temperature of the refrigerant leaving the condenser	°C
t_e	temperature of the refrigerant leaving the evaporator	°C
t_a	temperature of the refrigerant in the absorber	°C
t_g	temperature of the refrigerant in the generator	°C
W	work done per kg of refrigerant	kW/kg

Chapter 15

A	cross-sectional area of a duct	m^2
A'	reduced area available for airflow through the vena-contracta	m^2
a, b	duct dimensions	m or mm
a_1, a_2	cross-sectional areas	m^2
C	constant, or correction factor, or curve ratio, according to the context	
C_A	coefficient of area	
C_E	coefficient of entry	
C_V	coefficient of velocity	
d	internal duct diameter	m or mm
E	correction factor	
FP	fan power	kW
FTP	fan total pressure	N/m^2
f	dimensionless coefficient of friction	
g	acceleration due to gravity	m/s^2
H	head lost, in m of fluid flowing	
h	height	m
h_s	static suction set up in the plane of the vena-contracta	N/m^2
K	R_0/R_{n+1}	
k	loss coefficient	
k_s	absolute surface roughness	m
l	length of a duct	m
m	hydraulic mean gradient	

Symbol	*Description*	*Unit*
N_1	0.30842 ρ	
N_2	0.98567 μ	
n	integer, or speed of rotation of a fan, or exponent	
P	duct perimeter	m or mm
p	pressure	N/m^2
Q	rate of airflow	m^3/s
Q'	theoretical rate of airflow, in the absence of losses	m^3/s
R_e	Reynolds number	
R_c	centre-line radius	m or mm
R_t	throat radius	m or mm
R_0	throat radius of a bend without splitters	m or mm
R_{n+1}	heel radius of a bend without splitters	m or mm
s	spacing between turning vanes	m or mm
t	temperature or time	°C
TP	total pressure	N/m^2
SP	static pressure	N/m^2
VP	velocity pressure	N/m^2
V, v	mean velocity of airflow	m/s
$\bar{v}$	mean velocity of airflow	m/s
V'	theoretical maximum velocity of airflow	m/s
V_c	velocity in the vena-contracta	m/s
W	width of the cross-section of a duct	m or mm
Δp	pressure drop	N/m^2
η	total fan efficiency	
μ	absolute viscosity	kg/ms
ν	kinematic viscosity	m^2/s
ρ	density	kg/m^3

Chapter 16

Symbol	*Description*	*Unit*
A/T	ratio of primary air supplied to transmission heat loss	
f	boost factor	
g_r	moisture content of room air	g/kg
H	heat loss from building or humidification load	kW
H_0	design heat loss	kW
h_m	enthalpy of air (mixture) entering the primary air cooler coil	kJ/kg
K^{-1}	time constant	h
L	sensible heat liberated by light	W

Symbol	*Description*	*Unit*
M	mass flow rate	kg/s
P	sensible heat liberated by people	W
p_o	period of occupancy	h
p_b	boost period	h
p_c	cooling period	h
p_L	running period at full capacity	h
T	transmission heat loss	W
t	time	h
t_c	temperature of the air in the cold duct	°C
t_h	temperature of the air in the hot duct	°C
t_o	outside air temperature	°C
t_p	primary air temperature	°C
t_r	room air temperature	°C
t_s	supply air temperature	°C
t_w	air temperature leaving the primary cooler coil	°C
U	thermal transmittance coefficient	W/m² °C
V	ventilation load	W

Chapter 17

Symbol	Description	Unit
A	constant	
B	constant	
c	concentration of a contaminant, as a fraction, at any time	
c_a	concentration of CO_2 in the outside air, as a fraction	
c_o	initial concentration, at time zero	
G_a	rate of mass flow of air	kg/s
g_r	moisture content of room air	g/kg
H	enthalpy after time t	kJ
$H(t)$	enthalpy gain as a function of time	kJ/kg
H_0	initial enthalpy	kJ
h_a	enthalpy of air	kJ/kg
M	mass of air	kg
n	number of changes in a given time	
Q	volumetric rate of airflow	m^3/s
Q'	rate of supply of fresh air	m^3/s person
t	time	s
V	volume	m^3
V'	volume of a room	m^3/person
V_c	rate of production of CO_2 through respiration	m^3/s person
Δc	small change of concentration	

Symbol	*Description*	*Unit*
Δq	small volume of air	m^3
Δt	small interval of time	s

Chapter 18

Symbol	*Description*	*Unit*
C	constant	
I	intensity of light passing through a soiled filter paper	
I_o	intensity of light passing through a clean filter paper	
L	$\log_e(I_0/I)$ for the stain on the filter paper	
L_0	$\log_e(I_0/I)$ for the clean filter paper	
V	measured volume of an air-dust mixture	m^3

1

The Need for Air Conditioning

1.1 The meaning of air conditioning

Full air conditioning implies the automatic control of an atmospheric environment either for the comfort of human beings or animals or for the proper performance of some industrial or scientific process. The adjective 'full' demands that the purity, movement, temperature and relative humidity of the air be controlled, within the limits imposed by the design specification. (It is possible that, for certain applications, the pressure of the air in the environment may also have to be controlled.) Air conditioning is often misused as a term and is loosely and wrongly adopted to describe a system of simple ventilation. It is really correct to talk of air conditioning only when a cooling and dehumidification function is intended, in addition to other aims. This means that air conditioning is always associated with refrigeration and it accounts for the high cost of air conditioning. Refrigeration plant is precision-built machinery and is the major item of cost in an air conditioning installation, thus the expense of air conditioning a building is some four times greater than that of only heating it. Full control over relative humidity is not always exercised, hence for this reason a good deal of partial air conditioning is carried out; it is still referred to as air conditioning because it does contain refrigeration plant and is therefore capable of cooling and dehumidifying.

The ability to counter sensible and latent heat gains is, then, the essential feature of an air conditioning system and, by common usage, the term 'air conditioning' means that refrigeration is involved.

1.2 Comfort conditioning

Human beings are born into a hostile environment, but the degree of hostility varies with the season of the year and with the geographical locality. This suggests that the arguments for air conditioning might be based solely on climatic considerations, but although these may be valid in tropical and sub-tropical areas, they are not for temperate climates with industrialised social structures and rising standards of living.

Briefly, air conditioning is necessary for the following reasons. Heat gains from sunlight and electric lighting, in particular, may cause unpleasantly high temperatures in rooms, unless windows are opened. If windows are opened, then even moderate wind speeds cause excessive draughts, becoming worse on the upper floors of tall buildings. Further, if windows are opened, noise and dirt enter and are objectionable, becoming worse on the lower floors of buildings, particularly in urban districts and industrial areas. In any case, the relief provided by natural airflow through open windows is only effective for a depth of about 6 metres inward from the glazing. It follows that the inner areas of deep buildings will not really benefit at all from opened windows. Coupled with the need for high intensity continuous electric lighting in these core areas, the lack of adequate ventilation means a good deal of discomfort for the occupants. Mechanical ventilation without refrigeration is only a partial solution. It is true that it provides a controlled and uniform means of air distribution, in place of the unsatisfactory results obtained with opened windows (the vagaries of wind and stack effect, again particularly with tall buildings, produce discontinuous natural ventilation), but tolerable internal temperatures will prevail only during winter months. For much of the spring and autumn, as well as the summer, the internal room temperature will be several degrees higher than that outside, and it will be necessary to open windows in order to augment the mechanical ventilation.

The design specification for a comfort conditioning system is intended to be the framework for providing a comfortable environment for human beings throughout the year, in the presence of sensible heat gains in summer and sensible heat losses in winter. Dehumidification would be achieved in summer but the relative humidity in the conditioned space would be allowed to diminish as winter approached. There are two reasons why this is acceptable: first, human beings are comfortable within a fairly large range of humidities, from about 60 per cent to about 30 per cent and, secondly, the use of single glazing will cause the inner surfaces of windows to stream with condensed moisture if it is attempted to maintain too high a humidity in winter.

Thus, a system might be specified as capable of maintaining an internal condition of 22°C dry-bulb, with 50 per cent saturation, in the presence of an external summer state of 28°C dry-bulb, with 20°C wet-bulb, declining to an inside condition of 20°C dry-bulb, with an unspecified relative humidity, in the presence of an external state of −2°C saturated.

The essential feature of comfort conditioning is that it aims to produce an environment which is comfortable to the majority of the occupants. The ultimate in comfort can never be achieved, but the use of individual automatic control for individual rooms helps considerably in satisfying most people.

1.3 Industrial conditioning

Here the picture is quite different. An industrial or scientific process may perhaps, be performed properly only if it is carried out in an environment that has values of temperature and humidity lying within well defined limits. A departure from these limits may spoil the work being done. It follows that a choice of the inside design condition is not to be based on a statistical survey of the feelings of human beings but on a clearly defined statement of what is wanted.

Thus, a system might be specified to hold 21°C ± 0·5°C, with 50 per cent saturation ± 2½ per cent, provided that the outside state lay between 29·5°C dry-bulb, with 21°C wet-bulb and −4°C saturated.

2

Fundamental Properties of Air and Water Vapour Mixtures

2.1 The basis for rationalisation

Perhaps the most important thing for the student of psychrometry to appreciate from the outset is that the working fluid under study is a mixture of two different gaseous substances. One of these, dry air, is itself a mixture of gases, and the other, water vapour, is steam in the saturated or superheated condition. An understanding of this fact is important because in a simple analysis one applies the Ideal Gas Laws to each of these two substances separately, just as though one were not mixed with the other. The purpose of doing this is to establish equations which will express the physical properties of air and water vapour mixtures in a simple way. That is to say, the equations could be solved and the solutions used to compile tables of psychrometric data or to construct a psychrometric chart.

The justification for considering the air and the water vapour separately in this simplified treatment is provided by Dalton's *laws of partial pressure* and the starting point in the case of each physical property considered is its definition.

It must be acknowledged that the *ideal gas laws* are not strictly accurate, particularly at higher pressures. Although their use yields answers which have been adequately accurate in the past, they do not give a true picture of gas behaviour, since they ignore intermolecular forces. The most up-to-date psychrometric tables (I.H.V.E. 1970) are based on a fuller treatment, discussed in section 2.19. However, the Ideal Gas Laws may still be used for establishing psychrometric data at non-standard barometric pressures, with sufficient accuracy for most practical purposes.

2.2 The composition of dry air

Dry air is a mixture of five main component gases together with traces of a number of other gases. It is reasonable to consider all these as one homogeneous substance but to deal separately with the water vapour present because the latter is condensable at everyday pressures and temperatures whereas the associated dry gases are not.

One method of distinguishing between gases and vapours is to regard vapours as capable of liquefaction by the application of pressure alone but to consider gases as incapable of being liquefied unless their temperatures are reduced to below certain critical values. Each gas has its own unique

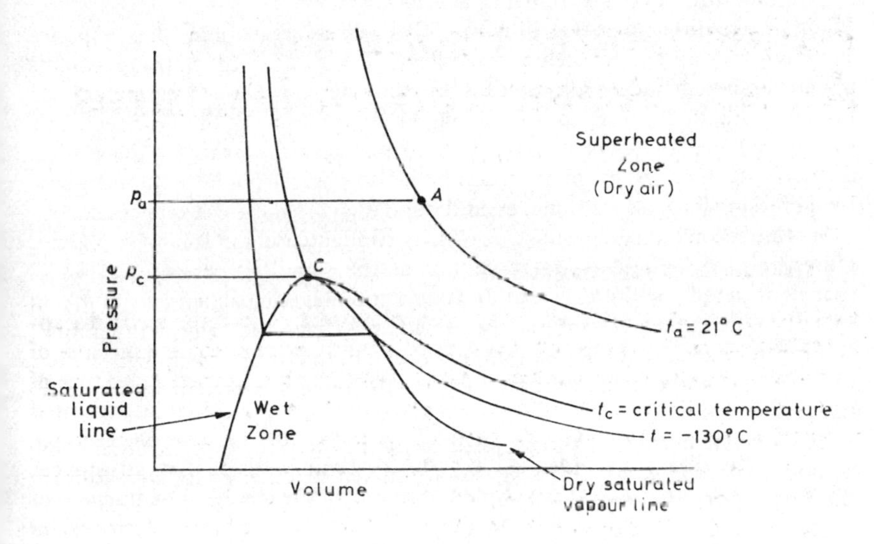

Fig. 2.1(*a*)

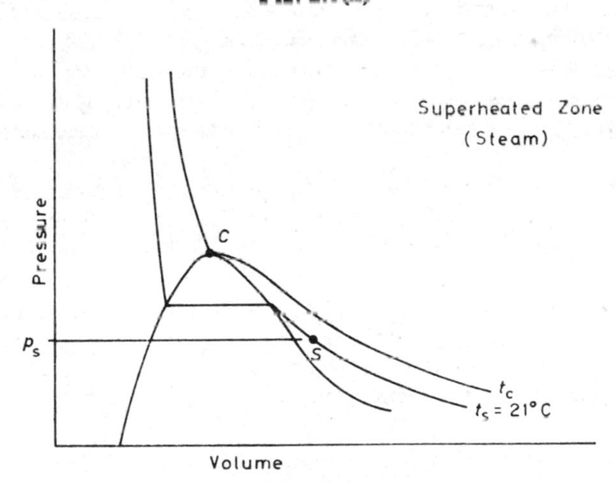

Fig. 2.1(*b*)

critical temperature, and it so happens that the critical temperatures of nitrogen and oxygen, the major constituents of dry air, are very much below the temperatures dealt with in air conditioning. On the other hand, the critical temperature of steam (374·2°C) is very much higher than these values and, consequently, the water vapour mixed with the dry air in the atmosphere may change its phase from gas to liquid if its pressure is increased, without any reduction in temperature. While this is occurring, the phase of the dry air will, of course, remain gaseous.

Figures 2.1(*a*) and 2.1(*b*) illustrate this. Pressure-volume diagrams are shown for dry air and for steam, separately. Point *A* in Fig. 2.1(*a*) represents a state of dry air at 21°C. It can be seen that no amount of increase of pressure will cause the air to pass through the liquid phase, but if its temperature is reduced to −130°C, say, a value just less than that of the critical isotherm, t_c, then the air may be compelled to pass through the liquid phase by increasing its pressure alone, even though its temperature is kept constant.

In the second diagram, Fig. 2.1(*b*), a similar case for steam is shown. Here, point *S* represents water vapour at the same temperature, 21°C, as that considered for the dry air. It is evident that atmospheric dry air and steam, because they are intimately mixed, will have the same temperature. But it can be seen that the steam is superheated, that it is far below its critical temperature, and that an increase of pressure alone is sufficient for its liquefaction.

The dry air portion of the atmosphere, then, may be thought of as being composed of true gases. These gases are mixed together as follows, to form the major part of the working fluid:

Gas	*Proportion* (%)	*Molecular mass*
Nitrogen	78·03	28·02
Oxygen	20·99	32·00
Carbon dioxide	0·03	44·00
Hydrogen	0·01	2·02
Argon	0·94	39·91

From the above, one may compute a value for the mean molecular mass of dry air:

$$\begin{aligned} M &= 28{\cdot}02\times 0{\cdot}7803 + 32\times 0{\cdot}2099 + 44\times 0{\cdot}0003 + 2{\cdot}02\times 0{\cdot}0001 \\ &\quad + 39{\cdot}91\times 0{\cdot}0094 \\ &= 28{\cdot}97 \end{aligned}$$

As will be seen shortly, this is used in establishing the value of the particular gas constant for dry air, prior to making use of the General Gas Law. In a similar connection, it is necessary to know the value of the particular gas constant for water vapour; it is therefore of use at this juncture to calculate the value of the mean molecular mass of steam.

Since steam is not a mixture of separate substances but a chemical compound in its own right, we do not use the proportioning technique adopted above. Instead, all that is needed is to add the masses of the constituent elements in a manner indicated by the chemical formula:

$$M = 2 \times 1{\cdot}01 + 1 \times 16$$
$$= 18{\cdot}02$$

2.3 Standards adopted

In general, those standards which have been used by the Institution of Heating and Ventilating Engineers in the 1970 edition of their *Guide to Current Practice*, are used here.

The more important values are:

Density of Air	1·293 kg/m^3 for dry air at 101325 N/m^2 and 0°C.
Density of Water	1000 kg/m^3 at 4°C and 998·23 kg/m^3 at 20°C.
Barometric Pressure	101325 N/m^2 (1013·25 mbar).

Normal temperature and pressure is 0°C and 101325 N/m^2, and the specific force due to gravity is taken as 9·807 N/kg (9·80665 N/kg exactly).

2.4 Boyle's law

This states that, for a true gas, the product of pressure and volume at constant temperature has a fixed value.

As an equation then, one can write

$$pV = \text{a constant.} \tag{2.1}$$

Graphically, this is a family of rectangular hyperbolas, each curve of which shows how the pressure and volume of a gas varies at a given temperature. Early experiment produced this concept of gas behaviour and subsequent theoretical study seemed to verify it. This theoretical approach is expressed in the *kinetic theory of gases*, the basis of which is to regard a gas as consisting of an assembly of spherically shaped molecules. These are taken to be perfectly elastic and to be moving in a random fashion. There are several other restricting assumptions, the purpose of which is to simplify the treatment of the problem. By considering that the energy of the moving molecules is a measure of the energy content of the gas and, that the change of momentum suffered by a molecule upon collision with the wall of the vessel containing the gas is an indication of the pressure of the gas, an equation identical with Boyle's law can be obtained.

However simple Boyle's law may be to use, the fact remains that it does not represent the manner in which a real gas behaves. Consequently one speaks of gases which are assumed to obey Boyle's law as being ideal gases. There are several other simple laws, namely, Charles' law, Dalton's laws of partial pressures, Avogadro's law, Joule's law and Gay Lussac's law, which are not strictly true but which are in common use. All these are known as the *ideal gas laws*.

Several attempts have been made to deal with the difficulty of expressing exactly the behaviour of a gas. It now seems clear that it is impossible to show the way in which pressure-volume changes occur at constant temperature by means of a simple algebraic equation. The expression which, in preference to Boyle's law, is today regarded as giving the most exact answer is in the form of a convergent infinite series:

$$pV = A(1+B/V+C/V^2+D/V^3+\ldots) \quad (2.2)$$

The constants A, B, C, D, etc., are termed the *virial coefficients* and they have different values at different temperatures.

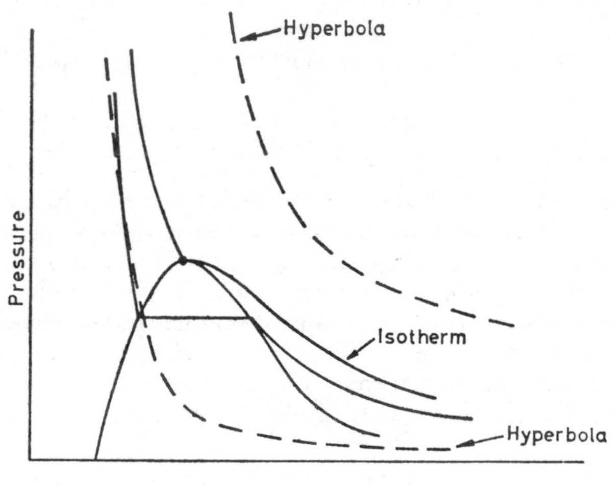

Fig. 2.2

It is sometimes more convenient to express the series in a slightly different way:

$$pV = A'+B'p+C'p^2+D'p^3+\ldots \quad (2.3)$$

At very low pressures the second and all subsequent terms on the right-hand side of the equation become progressively smaller and, consequently, the expression tends to become the same as Boyle's law. Hence, one may use Boyle's law without sensible error, provided the pressures are sufficiently small.

The second virial coefficient, B, is the most important. It has been found that, for a given gas, B has a value which changes from a large negative number at very low temperatures, to a positive one at higher temperatures, passing through zero on the way. The temperature at which B equals zero is called the Boyle temperature and, at this temperature, the gas obeys Boyle's law up to quite high pressures. For nitrogen, the main constituent of the atmosphere, the Boyle temperature is about 50°C. It seems that at this temperature, the gas obeys Boyle's law to an accuracy of better than 0·1 per cent for pressures up to about 1·9 MN/m². On the other hand, it seems that at 0°C, the departure from Boyle's law is 0·1 per cent for pressures up to 0·2 MN/m^2.

We conclude that it is justifiable to use Boyle's law for the expression of the physical properties of the atmosphere which are of interest to air conditioning engineers, in many cases.

In a very general sort of way, Fig. 2.2 shows what is meant by adopting Boyle's law for this purpose. It can be seen that the hyperbola of Boyle's law may have a shape similar to the curve for the true behaviour of the gas, provided the pressure is small. It also seems that if one considers a state sufficiently far into the superheated region, a similarity of curvature persists. However, it is to be expected that near to the dry saturated vapour curve, and also within the wet zone, behaviour is not ideal.

2.5 Charles' law

It is evident from Boyle's law that, for a given gas, the product pV could be used as an indication of its temperature and, in fact, this is the basis of a scale of temperature. It can be shown that for an ideal gas, at constant pressure, the volume is related to the temperature in a linear fashion. Experimental results support this, and reference to Fig. 2.3 shows just how this could be so. Suppose that experimental results allow a straight line to be drawn between two points A and B, as a graph of volume against temperature. If the line is extended to cut the abscissa at a point P, having a temperature of −273·15°C, it is clear that shifting the origin of the co-ordinate system to the left by 273·15°C will give an equation for the straight line, of the form

$$V = aT, \qquad (2.4)$$

where T is the temperature on the new scale and a is a constant representing the slope of the line.

Obviously

$$T = 273{\cdot}15^\circ + t \qquad (2.5)$$

This graphical representation of Charles' law shows that a direct proportionality exists between the volume of a gas and its temperature, as expressed on the new abscissa scale. It also shows that a new scale of temperature may be used. This new scale is an absolute one, so termed since it is possible to argue that all molecular movement has ceased at its zero, hence the internal energy of the gas is zero and, hence also, its temperature is at an absolute zero. Absolute temperature is expressed in kelvin, denoted by K, and the symbol T is used instead of t, to distinguish it from relative temperature on the Celsius scale.

EXAMPLE 2.1. 15 m^3/s of air at a temperature of 27°C passes over a cooler coil which reduces its temperature to 13°C. The air is then handled by a fan, blown over a reheater, which increases its temperature to 18°C, and finally supplied to a room.

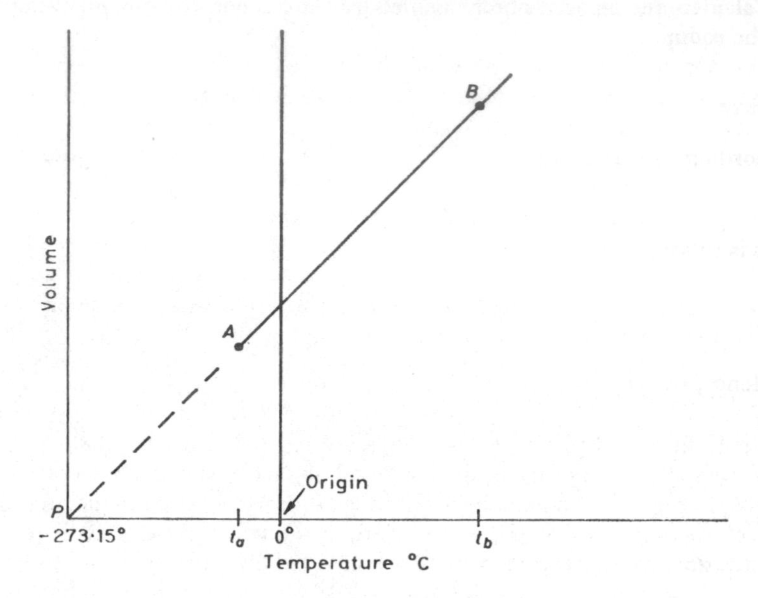

Fig. 2.3(*a*)

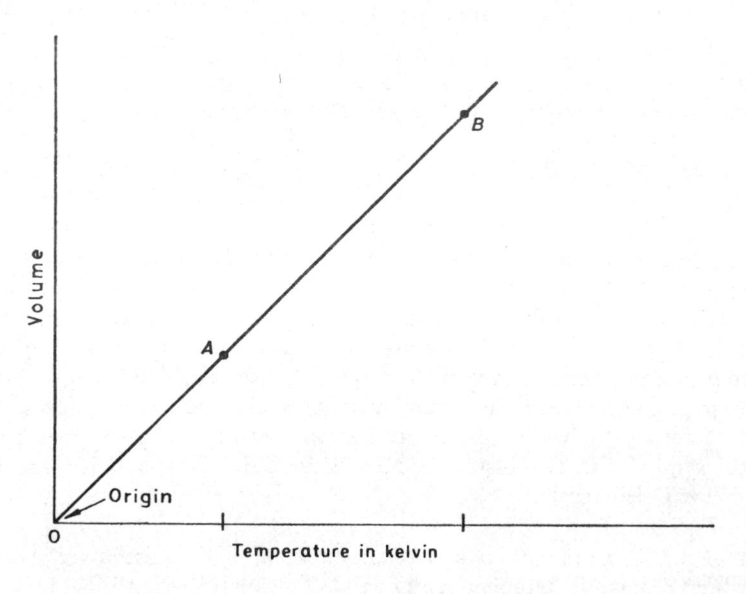

Fig. 2.3(*b*)

Calculate the amount of air handled by the fan and the quantity supplied to the room.

Answer

According to Charles' law:

$$V = aT,$$

that is to say,

$$V_2 = V_1 \frac{T_2}{T_1}$$

Hence, the air quantity handled by the fan

$$= 15 \frac{(273+13)}{(273+27)}$$

$$= 14.3 \text{ m}^3/\text{s}$$

and the air quantity supplied to the room

$$= 15 \frac{(273+18)}{(273+27)}$$

$$= 14.55 \text{ m}^3/\text{s}.$$

One further comment, it is clearly fallacious to suppose that the volume of a gas is directly proportional to its temperature right down to absolute zero; the gas liquefies before this temperature is attained.

2.6 The general gas law

It is possible to combine Boyle's and Charles' laws as one equation:

$$pV = mRT \tag{2.6}$$

where p = the pressure of the gas in N/m^2,
V = the volume of the gas in m^3,
m = the mass of the gas in kg,
R = a constant of proportionality,
T = the absolute temperature of the gas in K.

Avogadro's hypothesis argues that equal volumes of all gases at the same temperature and pressure contain the same number of molecules. Accepting this and taking as the unit of mass the kilomole (kmol), a mass in kilograms numerically equal to the molecular mass of the gas, a value for the universal gas constant can be established:

$$pV_m = R_o T \tag{2.7}$$

where V_m is the volume in m^3 of 1 k mol and is the same for all gases having the same values of p and T. Using the values $p = 101325 \text{ N/m}^2$ and $T = 273.15$ K, it has been experimentally determined that V_m equals 22.4145 m^3/k mol. Hence

$$R_o = \frac{pV_m}{T} = \frac{101325 \times 22.4145}{273.15} = 8314.66 \text{ J/kmol K}$$

Dividing both sides of eqn. (2.7) by the molecular mass, M, of any gas in question allows the determination of the particular gas constant, R, for the gas. This may then be used for a mass of 1 kg in eqn. (2.6) and we can write

$$pv = \frac{R_o T}{M} = RT \left(\text{whence } R = \frac{8314.66}{M}\right)$$

where v is the volume of 1 kg.

If a larger mass, m kg, is used, this expression becomes eqn. (2.6)

$$pV = mRT \tag{2.6}$$

where V is the volume of m kg and R has units of J/kg K.

$$\text{For dry air, } R = \frac{8314.66}{28.97} = 287 \text{ J/kg K}$$

$$\text{For steam, } R = \frac{8314.66}{18.02} = 461 \text{ J/kg K}$$

A suitable transposition of the general gas law yields expressions for density, pressure and volume.

EXAMPLE 2.2. Calculate the density of a sample of dry air which is at a pressure of 101325 N/m^2 and at a temperature of 20°C.

Answer

$$\text{Density} = \frac{\text{mass of the gas}}{\text{volume of the gas}}$$

$$= \frac{m}{V}$$

$$= \frac{p}{RT}$$

$$= \frac{101325}{287 \times (273+20)}$$

$$= 1{\cdot}204 \text{ kg/m}^3$$

We may compare this answer with that obtained by referring to the I.H.V.E. tables of psychrometric data, which quote the volume of air at 20°C dry bulb and 0 per cent saturation as 0·8301 m^3/kg of dry air.

The reciprocal of density is specific volume, hence the density of the air quoted by the tables is given by the reciprocal of 0·8301 m^3/kg and is 1·2047 kg/m^3.

2.7 Dalton's law of partial pressure

This may be stated as follows:

If a mixture of gases occupies a given volume at a given temperature, the total pressure exerted by the mixture equals the sum of the pressures of the constituents, each being considered at the same volume and temperature.

It is possible to show that if Dalton's law holds, each component of the mixture obeys the general gas law. As a consequence, it is sometimes more convenient to re-express the law in two parts:

(i) the pressure exerted by each gas in a mixture of gases is independent of the presence of the other gases, and
(ii) the total pressure exerted by a mixture of gases equals the sum of the partial pressures.

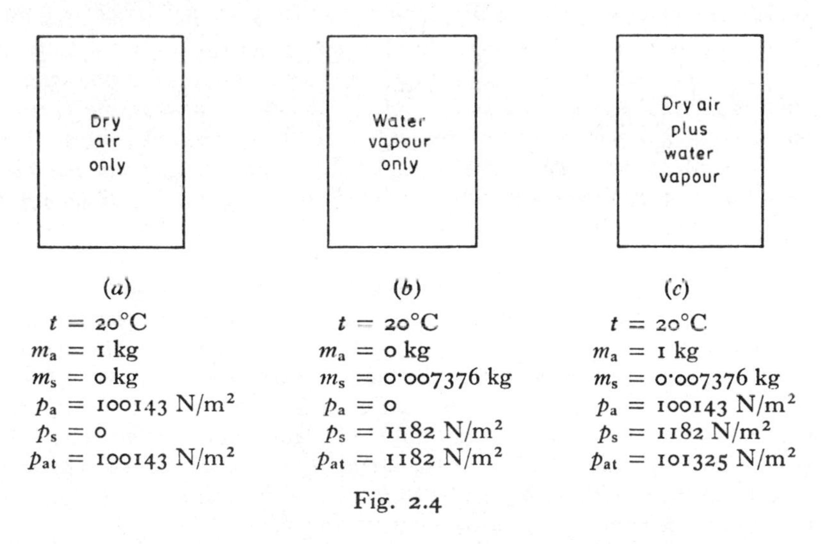

Fig. 2.4

Figure 2.4 illustrates this. It shows an air-tight container under three different conditions, from which it can be seen that the partial pressures exerted by the water vapour in (*b*) and (*c*) are equal, as are those exerted by the dry air in (*a*) and (*c*) and, that in (*a*), (*b*) and (*c*), the total pressure equals the sum of the partial pressures.

As in the two gas laws already considered, Dalton's law agrees with the results achieved by the kinetic theory of gases and, to some extent, finds substantiation in experiment.

It is now necessary to turn attention to the behaviour of water vapour at the state of saturation and to consider its partial pressure when it is in the superheated state and mixed with dry air.

2.8 Saturation vapour pressure

There are two requirements for the evaporation of liquid water to occur.

(i) Thermal energy must be supplied to the water.
(ii) The vapour pressure of the liquid must be greater than that of the steam in the environment.

These statements need some explanation.

Molecules in the liquid state are comparatively close to one another. They are nearer to one another than are the molecules in a gas and are less strongly bound together than those in a solid. The three states of matter are further distinguished by the extent to which an individual molecule may move. At a given temperature, a gas consists of molecules which have high individual velocities and which are arranged in a random fashion. A liquid at the same temperature is composed of molecules, the freedom of movement of which is much less, owing to the restraining effect which neighbouring molecules have on one another, by virtue of their comparative proximity. An individual molecule, therefore, has less kinetic energy if it is in the liquid state than it does if in the gaseous state. Modern thought is that the arrangement of molecules in a liquid is not entirely random as in a gas, but that it is not as regular as it is in most, if not all, solids. However, this is by the way.

It is evident that if the individual molecular kinetic energies are greater in the gaseous state, then energy must be given to a liquid if it is to change to the gaseous phase. This explains the first stated requirement for evaporation.

As regards the second requirement, the situation is clarified if one considers the boundary between a vapour and its liquid. Only at this boundary can a transfer of molecules between the liquid and the gas occur. Molecules at the surface have a kinetic energy which has a value related to the temperature of the liquid. Molecules within the body of the gas also have a kinetic energy which is a function of the temperature of the gas. Those gaseous molecules near the surface of the liquid will, from time to time, tend to hit the surface of the liquid, some of them staying there. Molecules within the liquid and near to its surface will, from time to time, also tend to leave the liquid and enter the gas, some of them staying there.

It is found experimentally that, in due course, an equilibrium condition arises for which the gas and the parent liquid both have the same temperature. For this state of equilibrium the number of molecules leaving the liquid is the same as the number of molecules entering it from the gas, on average.

Such a state of thermal equilibrium is exemplified by a closed insulated container which has within it a sample of liquid water. After a sufficient period of time, the space above the liquid surface, initially a vacuum, contains steam at the same temperature as the remaining sample of liquid water. The steam under these conditions is said to be saturated.

Before this state of equilibrium was reached the liquid must have been losing molecules quicker than it was receiving them. Another way of saying this is to state that the vapour pressure of the liquid exceeded that of the steam above it.

One point emerges from this fanciful example: since the loss of molecules from the liquid represents a loss of kinetic energy, and since the kinetic energy of the molecules in the liquid is an indication of the temperature of the liquid, then the temperature of the liquid must fall during the period preceding its attainment of thermal equilibrium.

It has been found that water in an ambient gas which is not pure steam but a mixture of dry air and steam, behaves in a similar fashion, and that for most practical purposes the relationship between saturation temperature and saturation pressure is the same for liquid water in contact only with steam. One concludes from this a very important fact: saturation vapour pressure depends solely upon temperature.

If we take the results of experiment and plot saturation vapour pressure against saturation temperature, we obtain a curve which has the appearance of the line on the psychrometric chart for 100 per cent saturation. The data on which this particular line is based can be found in tables of psychrometric information. Referring, for instance, to those tables published by the Institution of Heating and Ventilating Engineers, we can read off the saturation vapour pressure at, say 20°C, by looking at the value of the vapour pressure at 100 per cent saturation and a dry-bulb temperature of 20°C. It is important to note that the term 'dry-bulb' has a meaning only when we are speaking of a mixture of a condensable vapour and a gas. In this particular context the mixture is of steam and dry air, but we could have a mixture of, say, alcohol and dry air, which would have its own set of properties of dry- and wet-bulb temperatures.

According to the National Engineering Laboratory Steam Tables, 1964, the following equation may be used for the vapour pressure of steam over water up to 100°C:

$$\log p = 28{\cdot}59051 - 8{\cdot}2\log(t+273{\cdot}16) + 0{\cdot}0024804\,(t+273{\cdot}16) - \frac{3142{\cdot}31}{(t+273{\cdot}16)} \qquad (2.8)$$

where t is temperature in °C and p is pressure in bars.

Over ice, the equation to be used is the following, taken from National Bureau of Standards Circular No. 564:

$$\log p = 10{\cdot}5380997 - \frac{2663{\cdot}91}{(273{\cdot}15+t)} \tag{2.9}$$

2.9 The vapour pressure of steam in moist air

It is worth pausing a moment to consider the validity of the ideal gas laws as they are applied to the mixture of gases which comprises moist air.

Kinetic theory, which supports the ideal gas laws, fails to take account of the fact that intermolecular forces of attraction exist. In a mixture such forces occur between both like molecules and unlike molecules. That is to say, between molecules of dry air, between molecules of steam and, between molecules of steam and dry air. The virial coefficients mentioned in section 2.4 attempt to deal with the source of error resulting from attractive forces between like molecules. To offset the error accruing from the forces between unlike molecules, a further set of coefficients, termed 'interaction coefficients', is adopted.

An explanation of the modern basis of psychrometry, taking these forces into account, is given in section 2.19. For the moment, and for most practical purposes, we can take it that the saturation vapour pressure in humid air depends on temperature alone; that is, it is uninfluenced by barometric pressure.

EXAMPLE 2.3. Determine the saturation vapour pressure of moist air (*a*), at 15°C and a barometric pressure of 101325 N/m^2 and (*b*) at 15°C and a barometric pressure of 95000 N/m^2.

Answer

(*a*) From I.H.V.E. tables of psychrometric data, at 15°C dry-bulb and 100 per cent saturation, the saturation vapour pressure is 1704 N/m^2.

(*b*) From the same source exactly, we determine that the saturation vapour pressure is 1704 N/m^2 at 15°C dry-bulb and 100 per cent relative humidity. We can, of course, use the I.H.V.E. tables of psychrometric data for determining this saturation vapour pressure, even though the question speaks of 95000 N/m^2, since saturation vapour pressure does not depend on barometric pressure. On the other hand, it should be noted that at all relative humidities less than 100 per cent the vapour pressures quoted in these tables are valid only for the total or barometric pressure for which the tables are published, namely, 101325 N/m^2.

The question arises: how do we determine the vapour pressure for relative humidities other than 100 per cent? An empirical equation exists which answers this question:

$$p_s = p'_{ss} - p_{at}A(t - t') \qquad (2.10)$$

where

p_s = the vapour pressure required;
p'_{ss} = the saturation vapour pressure at a temperature numerically equal to the wet-bulb temperature of moist air being considered;
p_{at} = the barometric pressure;
t = the dry-bulb temperature of the moist air, in °C;
t' = the wet-bulb temperature of the moist air, in °C;
A = a constant which has values as follows:

	wet-bulb $\geq$ 0°C	wet-bulb < 0°C
Screen	$7{\cdot}99 \times 10^{-4}$ °C^{-1}	$7{\cdot}20 \times 10^{-4}$ °C^{-1}
Sling	$6{\cdot}66 \times 10^{-4}$ °C^{-1}	$5{\cdot}94 \times 10^{-4}$ °C^{-1}

EXAMPLE 2.4. Calculate the vapour pressure of moist air at a state of 20°C dry-bulb, 15°C wet-bulb (sling) and 950 mbar barometric pressure.

Answer

The saturation vapour pressure for 15°C is found from I.H.V.E. tables of psychrometric data for 1013·25 mbar to be 17·04 mbar. We can insert this as the value for p'_{ss} in eqn. (2.10) and write down:

$$p_s = 17{\cdot}04 - 0{\cdot}666 \times 0{\cdot}95 \times (20° - 15°)$$
$$= 13{\cdot}88 \text{ mbar}$$

By further reference to I.H.V.E. tables, we observe that, at a barometric pressure of 1013·25 mbar, the vapour pressure of moist air at the same values of dry- and wet-bulb temperature is 13·69 mbar. This may not seem to be much different from the value calculated for 950 mbar, but the difference becomes increasingly large and increasingly significant, the further one departs from 1013·25 mbar.

To illustrate the distinction between saturated vapour pressure and superheated vapour pressure, consider a sample of liquid water within a closed vessel. On the application of heat evaporation occurs, and for every temperature through which the liquid passes there is an equilibrium pressure, as has already been discussed. Fig. 2.5(*a*) shows a curve *A*, *B*, *B'* representing the relationship between saturation vapour pressure and absolute temperature. If heat is applied to the vessel beyond the instant when the last of the liquid water turns to saturated steam, the change of state of the steam can no longer be represented by the curve. The point *B* represents the state of the contents of the vessel at the instant when the last of the liquid has just evaporated. The

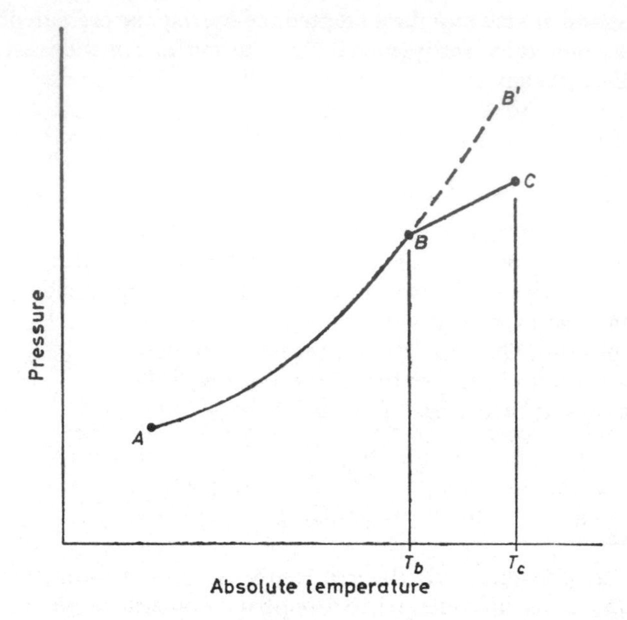

Fig. 2.5(a)

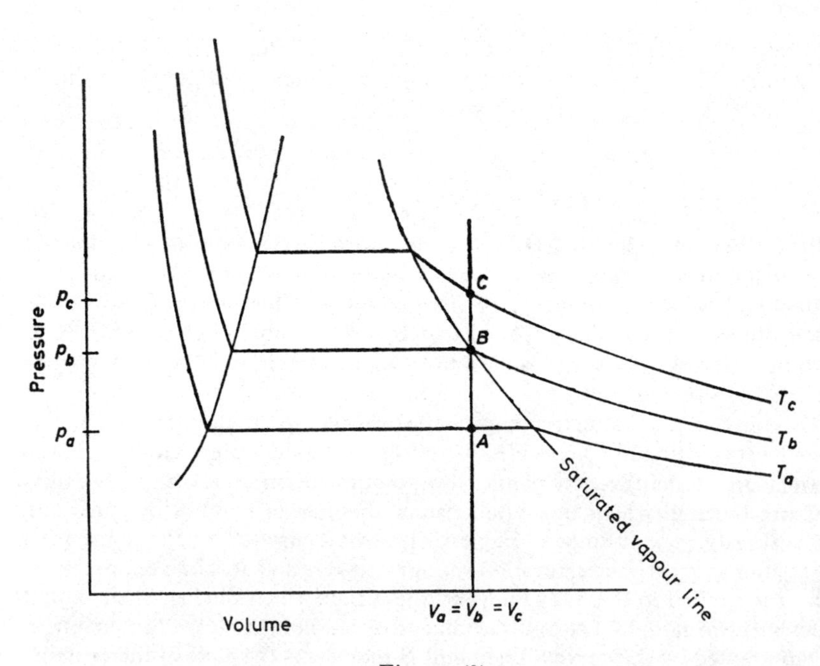

Fig. 2.5(b)

vessel contains dry saturated steam but, unlike the case so far, no liquid is present. By our earlier assumptions then, the contents of the vessel approximate an ideal gas and, therefore, may be taken to obey Charles' law for any further heating at constant volume. Equation 2.4 states this law, and further changes of state of the steam in the closed vessel may be represented by a straight line. This is shown in Fig. 2.5(*a*) by the line *BC*.

The changes can also be shown on another sort of diagram, Fig. 2.5(*b*), where pressure and volume are used as co-ordinates. The total volume of the liquid and vapour has remained constant throughout the application of all the heat, hence changes on the $p-V$ diagram must occur along a line of constant volume, for this example. At condition *A* the vessel contains saturated liquid and saturated vapour. Accordingly, on the $p-V$ diagram state *A* must lie within the wet zone. On the other hand, at point *B* the contents of the vessel are saturated steam only, hence *B* lies on the saturated vapour line. It can be seen that the change of state into the superheated zone at *C* is an extension of the vertical line *AB* as far as *C*.

2.10 Relative humidity

This is defined as the ratio of the partial pressure of the water vapour in moist air ***at a given temperature t***, to the partial pressure of the water vapour in saturated air, ***at the same temperature t***.

Emphasis is placed on the fact that the partial pressures in both the numerator and the denominator are at the same temperature.

The meaning of relative humidity is illustrated in Fig. 2.6, which shows pressure-volume changes for steam alone. That is to say, by making use of Dalton's law, the water vapour content of moist air is considered separately from the dry air content. The line *ABCD* is an isotherm for a value of absolute temperature denoted by *T*. Suppose that moist air at a relative humidity of less than 100 per cent contains steam with a partial pressure of p_a and that this steam is superheated so that it can then be represented on the $p-V$ diagram by a point *A*, in the superheated zone. Saturated steam is represented by point *B* on the same diagram. At state *B* the steam has a partial pressure of $p_b (= p_c)$, which is greater than p_a. Hence one can write down the relative humidity of the moist air being considered by the equation:

$$\phi = p_a/p_b \times 100 \text{ per cent.}$$

EXAMPLE 2.5. Calculate the relative humidity of moist air at 20°C dry-bulb, 15°C wet-bulb (sling) and 950 mbar barometric pressure.

Answer

By definition, relative humidity $= p_s/p_{ss}$, both pressures being expressed at the same temperature. The vapour pressure of the steam in the given state of air has already been calculated (*see* example 2.4). It is 13·88 mbar. Reference to I.H.V.E. tables of psychrometric data provides the information that, at 20°C, the saturation vapour pressure (p_{ss}) is 23·37 mbar. Hence, we can write

$$\phi = \frac{13{\cdot}88}{23{\cdot}37} \times 100$$

$$= 59{\cdot}5 \text{ per cent.}$$

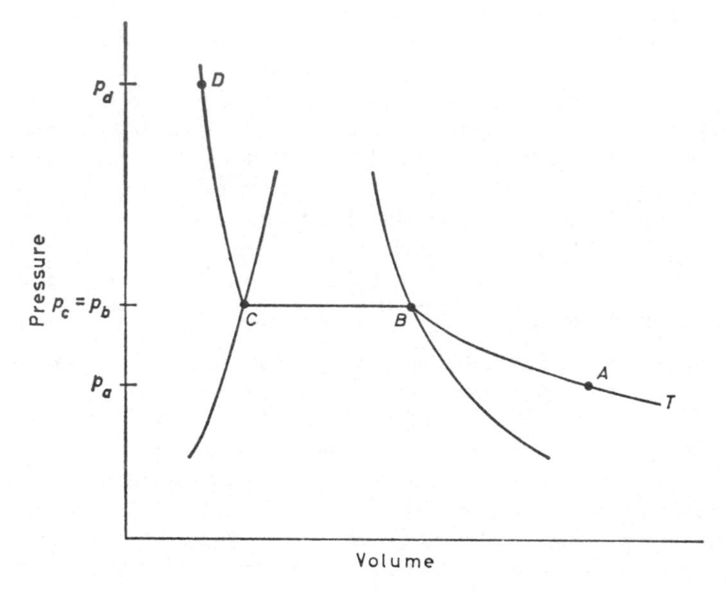

Fig. 2.6

Further reference to I.H.V.E. tables shows that at 1013·25 mbar, the relative humidity is 58·57 per cent at the same conditions of dry- and wet-bulb temperature. If, for moist air, a plot was made of the partial pressure of the water vapour content against temperature, lines of constant relative humidity would appear as curves the positions of which would depend on the barometric pressure of the moist air under consideration.

2.11 Moisture content and humidity ratio

Moisture content is defined as the mass of water vapour in kilograms which is associated with one kilogram of dry air in an air-water vapour mixture. It is sometimes called specific humidity or humidity ratio.

Starting with the definition, we can write—

$$\text{moisture content} = \text{mass of water vapour per unit mass of dry air}$$
$$= m_s/m_a$$

By using Dalton's law we can now apply the general gas law to each of the two constituents of moist air, just as though the other did not exist:

$$pV = mRT \text{ in general}$$

hence

$$p_s V_s = m_s R_s T_s \text{ for the water vapour}$$

and

$$p_a V_a = m_a R_a T_a \text{ for the dry air.}$$

The general gas law may be rearranged so that mass is expressed in terms of the other variables:

$$m = \frac{pV}{RT}$$

By transposing in the equations referring to water vapour and dry air, we can obtain an expression for moisture content based on its definition:

$$\text{moisture content} = \frac{p_s V_s R_a T_a}{R_s T_s p_a V_a}$$

$$= \frac{R_a p_s}{R_s p_a}$$

since the water vapour and the dry air have the same temperature and volume

The ratio of R_a to R_s is termed the relative density of water vapour with respect to dry air and, as already seen, it depends on the ratio of the molecular mass of water vapour to that of dry air:

$$\frac{R_a}{R_s} = \frac{M_s}{M_a} = \frac{18{\cdot}02}{28{\cdot}97} = 0{\cdot}622.$$

Since we are interested in calculating moisture content at different barometric pressures and, since eqn. (2.10) already provides a means of determining the vapour pressure of superheated steam, then it is convenient and simple to rearrange the expression for moisture content so that the partial pressure of the dry air is given in terms of barometric pressure and the partial pressure of the water vapour.

Hence, we may write

$$\text{moisture content} = 0{\cdot}622 \frac{p_s}{p_a}$$

that is,

$$g = 0{\cdot}622 \frac{p_s}{(p_{at} - p_s)} \text{ kg/kg of dry air} \qquad (2.11)$$

EXAMPLE 2.6 Calculate the moisture content of moist air at 20°C dry-bulb, 15°C wet-bulb (sling) and 950 mbar barometric pressure.

Answer

In Example 2.4 it was calculated that the vapour pressure of the water vapour in the given state of moist air was 13.88 mbar. Hence, using eqn. (2.11),

$$g = 0{\cdot}622 \frac{13{\cdot}88}{(950-13{\cdot}88)}$$

$$= 0{\cdot}00923 \text{ kg/kg of dry air.}$$

Although saturation vapour pressure is independent of barometric pressure, the moisture content at a condition of 100 per cent relative humidity is not.

EXAMPLE 2.7. Calculate the moisture content of saturated air at a state of 20°C and 950 mbar barometric pressure.

Answer

The saturation vapour pressure at 20°C is 23·37 mbar (from I.H.V.E. tables). Hence, using eqn. (2.11), and assuming $R_s = R_{ss}$ (see §2.12),

$$g = 0{\cdot}622 \frac{23{\cdot}37}{(950-23{\cdot}37)}$$

$$= 0{\cdot}0157 \text{ kg/kg of dry air.}$$

It may be verified that, at 20°C and 1013·25 mbar, saturated air has a moisture content of 0·01475 kg/kg of dry air.

The above example shows that the curve for 100 per cent relative humidity is in different positions for different barometric pressures. Figure 2.7 illustrates this and shows that the curve is higher up for lower barometric pressures.

2.12 Percentage saturation

This is not the same as relative humidity but is sometimes confused with it. For saturated air and for dry air the two are identical and within the range of states used for comfort conditioning they are virtually indistinguishable.

Percentage saturation is defined as the ratio of the moisture content of moist air at a given temperature, t, to the moisture content of saturated air at the same temperature t.

It is also known as the degree of saturation.

Applying the general gas law to the steam present in moist air which is not saturated, we may write

$$g = \frac{m_s}{m_a} = \frac{p_s V_s}{R_s T_s} \cdot \frac{R_a T_a}{p_a V_a}$$

$$= \frac{p_s}{p_a} \cdot \frac{R_a}{R_s}$$

since the steam and dry air occupy the same volume and are at the same temperature, being in intimate contact with one another.

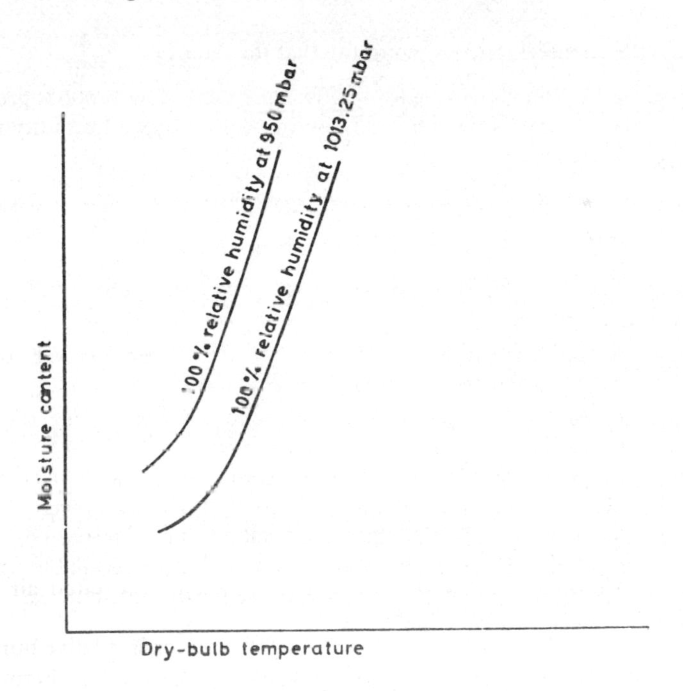

Fig. 2.7

We may then write

$$g = \frac{p_s}{(p_{at} - p_s)} \cdot \frac{R_a}{R_s}$$

for the unsaturated moist air.

Similarly, for saturated air the moisture content is given by

$$g_{ss} = \frac{p_{ss}}{(p_{at} - p_{ss})} \cdot \frac{R_a}{R_{ss}}$$

From the definition of percentage saturation we can then write

$$\mu = \frac{g}{g_{ss}} \cdot 100$$

$$= \frac{p_s}{p_{ss}} \cdot \frac{(p_{at} - p_{ss})}{(p_{at} - p_s)} \cdot \frac{R_{ss}}{R_s} \cdot 100$$

$$= \phi \cdot \frac{(p_{at} - p_{ss})}{(p_{at} - p_s)}$$

From what has been assumed earlier, it is clear that we must take R_{ss} as equal to R_s.

From I.H.V.E. tables, we may compute that the factor

$$\frac{(p_{at} - p_{ss})}{(p_{at} - p_s)}$$

has values:

0·9964 at 0°C dry-bulb and 50 per cent saturation,

0·9885 at 20°C dry-bulb and 50 per cent saturation.

It is seen that the difference between percentage saturation and relative humidity is quite small.

2.13 Dew point

This is defined as the temperature of saturated air which has the same vapour pressure as the moist air under consideration.

It is not possible to express this definition in the form of a simple equation by means of the ideal gas laws. This is evident if we refer to eqn. (2.8). It is more convenient to refer to tabulated values when saturated vapour pressure is required. However, provided such tables are available, we can determine the dew point of air at a given psychrometric state and barometric pressure.

EXAMPLE 2.8. Calculate the dew point of moist air at 20°C dry-bulb, 15°C wet-bulb (sling) and 950 mbar barometric pressure.

Answer

The vapour pressure of moist air at this state has already been calculated as 13·88 mbar. At its dew point temperature, the moist air must have a saturation vapour pressure equal to this value. Reference to I.H.V.E. tables of psychrometric data shows that at 12°C dry-bulb and 100 per cent saturation, the vapour pressure is 14·02 mbar whilst at 11°C saturated the vapour pressure is 13·12 mbar. By interpolation we establish that the dew point is 11·57°C.

Further reference to the tables shows, for purposes of comparison, that at the same dry-bulb and wet-bulb temperatures, but at a barometric pressure of 1013·25 mbar, the dew point is 11·6°C.

One last point: eqn. (2.11) showed that moisture content was a function of vapour pressure. It follows that an alternative definition of dew point may be considered in which moisture content replaces vapour pressure.

2.14 Specific volume

This is the volume in cubic metres of one kilogram of dry air together with the mass of water vapour associated with it. In the mixture the steam occupies the same volume as the dry air but each of these two constituents is at its own partial pressure. By Dalton's law the sum of these two partial pressures is the total pressure of the mixture. Thus, it is possible to use the ideal gas law to determine the humid volume in three ways:

(i) making use of the mass and partial pressure of the dry air,
(ii) making use of the mass and partial pressure of the water vapour and,
(iii) making use of the mass and total pressure of the mixture.

Since we know the particular gas constants for dry air and steam it is most convenient to adopt method (i) or, sometimes, method (ii).

EXAMPLE 2.9. Calculate the specific volume of air at 20°C dry-bulb, 15°C wet-bulb (sling) and 950 mbar barometric pressure.

Answer

Apply the ideal gas law to the dry air alone:

$$V_a = \frac{m_a R_a T_a}{p_a}$$

From an earlier example it is known that the vapour pressure is 13·88 mbar, hence the partial pressure of the dry air is

$$\begin{aligned} p_a &= 950 - 13{\cdot}88 \\ &= 936{\cdot}12 \text{ mbar} \\ &= 93612 \text{ N/m}^2 \end{aligned}$$

Hence

$$\begin{aligned} V_a &= \frac{1 \times 287 \times (273+20)}{93\,612} \\ &= 0{\cdot}898 \text{ m}^3 \end{aligned}$$

This is the volume of 1 kg of dry air, at its partial pressure.

Alternatively, we could consider the moisture mixed with the dry air.

$$V_s = \frac{m_s R_s T_s}{p_a}$$

From an earlier example it is known that the moisture content is 0·00923 kg/kg of dry air. Hence

$$V_s = \frac{0{\cdot}00923 \times 461 \times (273+20)}{1388}$$

$$= 0{\cdot}898\ \text{m}^3$$

This is the volume of 0·0088 kg of steam at its partial pressure of 13·88 mbar.

The example has shown that the humid volume may be calculated by one of two ways and that both of these yield the same answer.

2.15 Enthalpy: thermodynamic background

The *first law of thermodynamics* may be considered as a statement of the principle of the conservation of energy. A consequence of this is that a concept termed ' internal energy ' must be introduced, if the behaviour of a gas is to be explained with reasonable exactness during processes of heat transfer. Internal energy is the energy stored in the molecular and atomic structure of the gas and it may be thought of as being a function of two independent variables, the pressure and the temperature of the gas.

We can consider heat being supplied to a gas in either one of two ways: at constant volume or at constant pressure. Since the work done by a gas or on a gas, during a process of expansion or contraction, is expressed by the equation: work done = $\int p\, dV$, it follows that if heat is supplied to a gas at constant volume, no work will be done by the gas on its environment. Consequently the heat supplied to the gas serves only to increase its internal energy, U. If a heat exchange occurs at constant pressure, as well as a change in internal energy taking place, work may be done.

This leads to a definition of enthalpy, H:

$$H = U + pV \tag{2.12}$$

The equation is strictly true for a pure gas of mass m, pressure p, and volume V. However, it may be applied without appreciable error to the mixtures of gases associated with air conditioning.

Because of the way in which enthalpy is defined by eqn. (2.12), it is desirable that the expression ' heat content ' should not be used. This, and the other common synonym for enthalpy, ' total heat ', suggest that only the internal energy of the gas is being taken into account. Because of this, both terms are a little misleading and, in consequence, throughout the rest of this book the term enthalpy will be used.

2.16 Enthalpy in practice

It is not possible to give an absolute value to enthalpy since no assessment is possible of the absolute value of the internal energy of a gas. The reason for this is that, particularly at the atomic level, the contribution electrical, magnetic and other forces make is not capable of determination. The expression, mentioned earlier, of internal energy as a function of pressure and temperature, is a simplification. Fortunately, air conditioning involves only a calculation of changes in enthalpy. It follows that such changes may be readily determined if a datum level of enthalpy is adopted for its expression. Thus, we are really always dealing in relative enthalpy, although we do not so term it.

The enthalpy, h, used in psychrometry is defined by the equation:

$$h = h_a + gh_g \tag{2.13}$$

where h_a is the enthalpy of dry air, h_g is the enthalpy of water vapour, both expressed in kJ/kg, and g is the moisture content in kg/kg dry air.

The value of temperature chosen for the zero of enthalpy is 0°C for both dry air and water. The relationship between the enthalpy of dry air and its temperature is not quite linear and values taken from N.B.S. Circular No. 564, for the standard atmospheric pressure of 1013·25 mbar and suitably modified for the chosen zero, form the basis of the I.H.V.E. tables of the properties of humid air. An approximate equation for the enthalpy of dry air over the range 0°C to 60°C is, however

$$h_a = 1{\cdot}007\,t - 0{\cdot}026 \tag{2.14}$$

and for lower temperatures, down to −10°C, the approximate equation is

$$h_a = 1{\cdot}005\,t \tag{2.15}$$

Values of h_g for the enthalpy of vapour over water have been taken from N.E.L. steam tables, slightly increased to take account of the influence of barometric pressure and modified to fit the zero datum. The enthalpy of vapour over ice, on the other hand, is based on information in the A.S.H.R.A.E. Handbook of Fundamentals (1967).

For purposes of approximate calculation, without recourse to the I.H.V.E. psychrometric tables, we may assume that, in the range 0°C to 60°C, the vapour is generated from water at 0°C and that the specific heat of superheated steam is a constant. The following equation can then be used for the enthalpy of water vapour:

$$h_g = 2501 + 1{\cdot}84\,t \tag{2.16}$$

Equations (2.14) and (2.16) can now be combined, as typified by eqn. (2.13), to give us an approximate expression for the enthalpy of humid air at a barometric pressure of 1013·25 mbar:

$$h = (1{\cdot}007\,t - 0{\cdot}026) + g(2501 + 1{\cdot}84\,t) \tag{2.17}$$

EXAMPLE 2.10. Calculate the approximate enthalpy of an air-water vapour mixture at a state of 20°C dry-bulb, 15°C wet-bulb and 1013·25 mbar. Use I.H.V.E. tables of psychrometric data to establish the moisture content.

Answer

From tables, $g = 0{\cdot}008556$ kg/kg

Using eqn. (2.17):

$$h = (1{\cdot}007 \times 20 - 0{\cdot}026) + 0{\cdot}008556\,(2501 + 1{\cdot}84 \times 20)$$

$$= 41{\cdot}828 \text{ kJ/kg.}$$

The I.H.V.E. tables quote a value of 41·83 kJ/kg and are intentionally rounded off to four significant figures.

For the range of temperatures from −10°C to 0°C eqn. (2.16) is also approximately correct for the enthalpy of water vapour over ice. Using eqns. (2.15) and (2.16) the combined approximate equation for the enthalpy of humid air becomes

$$h = 1{\cdot}005\,t + g\,(2501 + 1{\cdot}84\,t) \tag{2.18}$$

EXAMPLE 2.11. Calculate the approximate enthalpy of humid air at −10°C dry-bulb, −12°C wet-bulb and 1013·25 mbar. Use I.H.V.E. tables to establish the moisture content.

Answer

From tables, $g = 0{\cdot}000611$ kg/kg

Using eqn. (2.18):

$$h = 1{\cdot}005 \times (-10) + 0{\cdot}000611\,(2501 + 1{\cdot}84 \times (-10))$$

$$= -8{\cdot}533 \text{ kJ/kg}$$

The I.H.V.E. tables quote −8·538 kJ/kg.

2.17 Wet-bulb temperature

A distinction must be drawn between measured wet-bulb temperature and the temperature of adiabatic saturation, otherwise sometimes known as the thermodynamic wet-bulb temperature. The wet-bulb temperature is a value indicated on an ordinary thermometer, the bulb of which has been wrapped round with a wick, moistened in water. The initial temperature of the water used to wet the wick, and the heat exchange between the wick and its environment by radiation, are both important factors that influence the temperature indicated by the wet-bulb thermometer. On the other hand, the temperature of adiabatic saturation is that obtained purely from an equation representing an adiabatic heat exchange. It is somewhat unfortunate, in air and water-vapour mixtures, that the two are almost numerically identical at normal temperatures and pressures, which is not so in mixtures of other gases and vapours.

Wet-bulb temperature is not a property only of mixtures of dry air and water vapour. Any mixture of a non-condensable gas and a condensable vapour will have a wet-bulb temperature. It will also have a temperature of adiabatic saturation. Consider a droplet of water suspended in an environment of moist air. Suppose that the temperature of the droplet is t_w and that its corresponding vapour pressure is p_w. The ambient moist air has a temperature t and a vapour pressure of p_s.

Provided that p_w exceeds p_s, evaporation will take place and, to effect this, heat will flow from the environment into the droplet by convection and radiation. If the initial value of t_w is greater than that of t, then, initially, some heat will flow from the drop itself to assist in the evaporation. Assuming that the original temperature of the water is less than the wet-bulb temperature of the ambient air, some of the heat gain to the drop from its surroundings will serve to raise the temperature of the drop itself, as well as providing for the evaporation. In due course, a state of equilibrium will be reached in which the sensible heat gain to the water exactly equals the latent heat loss from it, and the water itself will have taken up a steady temperature, t', which is termed the wet-bulb temperature of the moist air surrounding the droplet.

The condition can be expressed by means of an equation:

$$(h_c+h_r)\times A\times(t-t') = \alpha\times A\times h_{fg}\times(p'_{ss}-p_s) \qquad (2.19)$$

where

h_c = the coefficient of heat transfer through the gas film around the drop, by convection,

h_r = the coefficient of heat transfer to the droplet by radiation from the surrounding surfaces,

A = the surface area of the droplet,

h_{fg} = the latent heat of evaporation of water at the equilibrium temperature attained,

α = the coefficient of diffusion for the molecules of steam as they travel from the parent body of liquid through the film of vapour and non-condensable gas surrounding it.

The heat transfer coefficients (h_c+h_r) are commonly written, in other contexts, simply as h.

Equation (2.19) can be re-arranged:

$$(p'_{ss}-p_s) = \frac{(h_c+h_r)}{\alpha h_{fg}}\cdot(t-t') \qquad (2.20)$$

or

$$p_s = p'_{ss}-\left(\frac{h_c+h_r}{\alpha h_{fg}}\right)(t-t')$$

which is the basis of eqn. (2.10), the constant

$$\left(\frac{h_c+h_r}{\alpha h_{fg}}\right)$$

being a function of barometric pressure and temperature.

The radiation component of the sensible heat gain to the droplet is really a complicating factor since it depends on the absolute temperatures of the surrounding surfaces and their emissivities, and the transfer of heat by radiation is independent of the amount of water vapour mixed with the air, in the case considered. The radiation is also independent of the velocity of airflow over the surface of the droplet. This fact can be taken advantage of to minimise the intrusive effect of h_r. If the air velocity is made sufficiently large it has been found experimentally that

$$\frac{(h_c+h_r)}{h_c}$$

can be made to approach unity. In fact, in an ambient air temperature of about 10°C its value decreases from about 1·04 to about 1·02 as the air velocity is increased from 4 m/s to 32 m/s.

At this juncture it is worth observing that increasing the air velocity does not (contrary to what might be expected) lower the equilibrium temperature t'. As more pounds of air flow over the droplet each second there is an increase in the transfer of sensible heat to the water, but this is offset by an increase in the latent heat loss from the droplet. Thus t' will be uncharged. What will change, of course, is the time taken for the droplet to attain the equilibrium state. The evaporation rate is increased for an increase in air velocity and so the water more rapidly assumes the wet-bulb temperature of the surrounding air.

In general,

$$g = \frac{M_s}{M_a}.\frac{p_s}{p_a}$$

(see eqn. (2.11)) and so

$$(g'_{ss}-g) = \frac{M_s}{M_a}.\frac{(p'_{ss}-p_s)}{p_a},$$

it being assumed that p_a, the partial pressure of the non-condensable dry air surrounding the droplet, is not much changed by variations in p_s, for a given barometric pressure. The term g'_{ss} is the moisture content of saturated air at t'. Equation (2.20) can be re-arranged as

$$\frac{(g'_{ss}-g)}{(t-t')} = \frac{M_s}{M_a}.\frac{(h_c+h_r)}{p_a h_{fg}\alpha}.$$

Multiplying above and below by c, the specific heat of humid air, the equation reaches the desired form

$$\frac{(g'_{ss}-g)}{(t-t')} = (Le)\ \frac{c}{h_{fg}} \tag{2.21}$$

where (Le) is a dimensionless quantity termed the Lewis number and equal to

$$\frac{M_s}{M_a} \cdot \frac{(h_c+h_r)}{p_a \alpha c}.$$

In moist air at a dry-bulb temperature of about 21°C and a barometric pressure of 1013·25 mbar, the value of

$$\frac{M_s}{M_a} \cdot \frac{h_c}{p_a \alpha c}$$

is 0·945 for airflow velocities from 3·8 to 10·1 m/s and the value of

$$\frac{(h_c+h_r)}{h_c}$$

is 1·058 at 5·1 m/s. So for these sort of values of velocity, temperature and pressure the Lewis number is unity.

As will be seen in the next section, this coincidence, that the Lewis number should equal unity for moist air at a normal condition of temperature and pressure, leads to the similarity between wet-bulb temperature and the temperature of adiabatic saturation.

2.18 Temperature of adiabatic saturation

An adiabatic process is one in which no heat is supplied to or rejected from the gas during the change of state it undergoes.

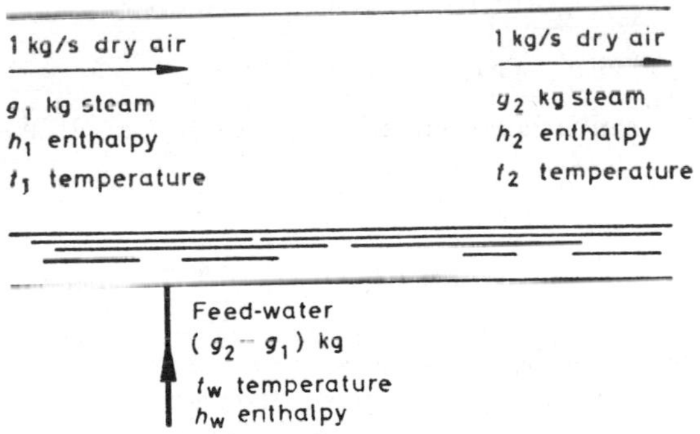

Fig. 2.8

Consider the flow of a mixture of air and water vapour through a perfectly insulated chamber. The chamber contains a large wetted surface, and water is evaporated from the surface to the stream of moist air flowing over it, provided the moist air is not already saturated. Feed-water is supplied to the chamber to make good that evaporated. Figure 2.8 shows such a situation.

A heat balance may be established:

$$c_a(t_1-t_2)+c_s g_1(t_1-t_2) = (g_2-g_1)\{(t_2-t_w)+h_{fg}\}$$

This equation expresses the physical changes that have taken place:

(*a*) g_2-g_1 kg of feed-water are sensibly heated from a temperature of t_w to a temperature t_2.

(*b*) g_2-g_1 kg of water are evaporated from the wetted surface within the chamber, at a temperature t_2. The latent heat required for this is h_{fg} kJ/kg at temperature t_2.

(*c*) 1 kg of dry air is cooled from a temperature t_1 to a temperature t_2.

(*d*) g_1 kg of steam are cooled from t_1 to t_2.

The left-hand side of the equation may be simplified by using a combined specific heat, c, for moist air:

$$c(t_1-t_2) = (g_2-g_1)\{(t_2-t_w)+h_{fg}\}.$$

In due course, if the chamber is infinitely long, the moist air will become saturated and will have a moisture content g_{ss} and a temperature t_{ss}.
The equation then becomes

$$c(t_1-t_{ss}) = (g_{ss}-g_1)\{(t_{ss}-t_w)+h_{fg}\}.$$

If the feed-water is supplied to the chamber at a temperature t_{ss}, then a further simplification results:

$$c(t_1-t_{ss}) = (g_{ss}-g_1)h_{fg}$$

From this we can write an equation representing an adiabatic saturation process:

$$\frac{(g_{ss}-g_1)}{(t_1-t_{ss})} = \frac{c}{h_{fg}} \tag{2.22}$$

This equation should be compared with eqn. (2.21). Both give a value for the slope of a line in a moisture content-temperature, co-ordinate system, which is shown in Fig. 2.9. It can also be seen that if the air is saturated by passing it through an adiabatic saturation chamber, its wet-bulb temperature will fall, if, as illustrated, the value of the Lewis number is greater than one. At the saturated condition, the temperature of adiabatic saturation, t^*, the wet-bulb temperature, t'_2, and the dry-bulb temperature, t_2, are all equal and may be denoted by t_{ss}.
An illustration of a Lewis number being greater than one is that of moist air, initially at 21°C dry bulb, 15°C wet bulb (sling) and 1013·25 mbar barometric

pressure which is flowing over a wet-bulb thermometer or through an adiabatic saturation chamber at a velocity of 1 m/s. The Lewis number is 1·05. If the velocity exceeded about 1 m/s, then the Lewis number would become less than unity and the wet-bulb temperature would increase as the moist air was humified by a process of adiabatic saturation.

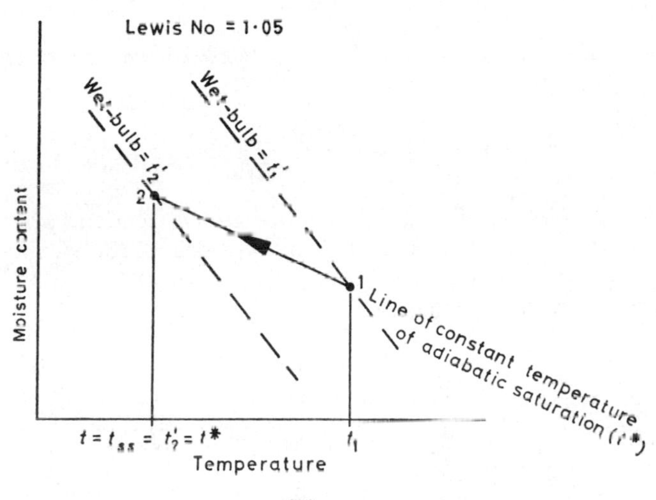

Fig. 2.9

Outside the field of air conditioning, if toluene is evaporated into dry air by a process of adiabatic saturation, it can be observed that the wet-bulb temperature is very much more than the temperature of adiabatic saturation. This is because, at the temperatures and pressures considered here, the Lewis number equals about 1·8, instead of unity.

2.19 Non-ideal Behaviour

The application of the ideal gas laws to mixtures of gases and vapours assumes the validity of Dalton's Law. To quote Goff (1949): "By present-day standards, Dalton's Law must be regarded as an inaccurate conjecture based upon an unwarranted faith in the ultimate simplicity of nature." The ideal laws and the kinetic theory of gases ignore the effects of inter-molecular forces. The more up-to-date approach of statistical mechanics however, gives us a theory of thermodynamics which takes account of forces between like molecules, in terms of virial coefficients, and between unlike molecules in terms of interaction coefficients.

The methods of statistical mechanics, in conjunction with the best experimental results, yield accurate evaluations of the properties of dry air and water vapour, published in N.B.S. Circular No. 564 and in the N.E.L. Steam Tables (1964), respectively.

The association of dry air and water vapour in a mixture is expressed by the moisture content on a mass basis and so percentage saturation is particularly relevant, since it defines the fraction of saturated water vapour present in a mixture, for a particular temperature. It is typified by the equation:

$$g = \frac{\mu g_{ss}}{100} \tag{2.23}$$

and the moisture content as saturation, g_{ss}, is defined more precisely than is possible with the ideal gas laws by

$$g = \frac{0{\cdot}62197 f_s p_{ss}}{p_{at} - f_s p_{ss}} \tag{2.24}$$

The dimensionless coefficient f_s is a function of barometric pressure and temperature, having an approximate value of 1·004 in the vicinity of 0°C and 1013·25 mbar. Using this value an approximate equation for the moisture content at saturation is

$$g_{ss} = \frac{0{\cdot}624\, p_{ss}}{p_{at} - 1{\cdot}004\, p_{ss}} \tag{2.25}$$

At other temperatures and pressures f_s assumes the following values:

TABLE 2.1

mbar	0°C	10°C	20°C	30°C	40°C	50°C	60°C
1013·25	1·0044	1·0044	1·0044	1·0047	1·0051	1·0055	1·0059
1000	1·0044	1·0043	1·0044	1·0047	1·0051	1·0055	1·0059
975	1·0043	1·0043	1·0043	1·0046	1·0050	1·0054	1·0058
950	1·0042	1·0042	1·0043	1·0045	1·0049	1·0053	1·0057
925	1·0041	1·0041	1·0042	1·0045	1·0049	1·0053	1·0056
900	1·0040	1·0040	1·0041	1·0044	1·0048	1·0052	1·0056
875	1·0039	1·0039	1·0040	1·0043	1·0047	1·0051	1·0055
850	1·0038	1·0038	1·0040	1·0042	1·0046	1·0050	1·0054
825	1·0037	1·0037	1·0039	1·0042	1·0045	1·0049	1·0053
800	1·0036	1·0036	1·0038	1·0041	1·0045	1·0049	1·0052

The relationships are not linear and are based on data published by Goff.

At sufficiently low pressures the behaviour of a mixture of gases can be accurately described by

$$pv = RT - p \sum_{ik} x_i x_k A_{ik}(T) - p^2 \sum_{ijk} x_i x_j x_k A_{ijk}(T) \tag{2.26}$$

The term $A_{ik}(T)$ is a function of temperature and represents the second virial coefficient if $i = k$, but an interaction coefficient if $i \neq k$, the molecules being considered two at a time. The function $A_{ijk}(T)$ then refers to the molecules considered three at a time. The terms in x are the mole fractions of the constituents, the mole fraction being defined by

$$x_a = \frac{n_a}{n_a + n_w} \tag{2.27}$$

where n_a is the number of moles of constituent a and n_w is the number of moles of constituent w. Thus if x_a is the mole fraction of constituent a, $(1-x_a)$ is the mole fraction of constituent w, eqn. (2.26) can then be re-phrased as

$$\begin{aligned} pv = RT - [x_a^2 A_{aa} + x_a(1-x_a)\, 2A_{aw} + (1-x_a)^2\, A_{ww}]p \\ -[(1-x_a)^3\, A_{www}]p^2 \end{aligned} \tag{2.28}$$

where the subscripts a and w refer to dry air and water vapour, respectively. The third virial coefficient, A_{www}, is a function of the reciprocal of absolute temperature and is insignificant for temperatures below 60°C. Other third order terms are ignored.

If eqn. (2.28) is to be used to determine the specific volume of dry air then p_a and R_a replace p and R. It then becomes

$$\begin{aligned} v &= \frac{R_a T}{p_a} - [x_a^2 A_{aa} + x_a(1-x_a)\, 2A_{aw} + (1-x_a)^2 A_{ww}] \\ &= \frac{82{\cdot}0567 \times 1013{\cdot}25}{28\,966} \cdot \frac{T}{(p_{at} - p_s)} - [x_a^2 A_{aa} + x_a(1-x_a)2A_{aw} \\ &\quad + (1-x_a)^2\, A_{ww}] \end{aligned} \tag{2.29}$$

For water vapour mixed with dry air

$$x_w = \frac{n_w}{n_a + n_w}$$

by eqn. (2.27) but

$$g = \frac{n_w M_w}{n_a\, M_a} = \frac{0{\cdot}62197\, n_w}{n_a}$$

therefore

$$x_w = \frac{g}{0{\cdot}62197 + g} \tag{2.30}$$

TABLE 2.2 *Virial and interaction coefficients for moist air.*

t (°C)	−10°	0°	10°	20°	30°	40°	50°	60°
A_{aa}	5·42 × 10⁻⁴	4·56 × 10⁻⁴	3·76 × 10⁻⁴	3·04 × 10⁻⁴	2·37 × 10⁻⁴	1·76 × 10⁻⁴	1·19 × 10⁻⁴	0·659 × 10⁻⁴
A_{ww}	7·94 × 10⁻²	6·318 × 10⁻²	5·213 × 10⁻²	4·35 × 10⁻²	3·708 × 10⁻²	3·19 × 10⁻²	2·772 × 10⁻²	2·434 × 10⁻²
A_{aw}	1·55 × 10⁻³	1·45 × 10⁻³	1·36 × 10⁻³	1·27 × 10⁻³	1·19 × 10⁻³	1·12 × 10⁻³	1·05 × 10⁻³	0·984 × 10⁻³

All the coefficients are in m³/kg.

The enthalpy of dry air increases very slightly as the pressure falls at constant temperature but, for most practical purposes, it can be regarded as a constant value at any given temperature from one standard atmosphere down to 800 mbar. It follows that the most significant effect of a change in barometric pressure on the enthalpy of moist air lies in the alteration of the moisture content. A revised expression for the enthalpy of moist air is

$$h = h_a + \left[\frac{0.62197 f_s p_{ss}}{p_{at} - f_s p_{ss}}\right] h_g \tag{2.31}$$

EXAMPLE 2.12. Calculate the enthalpy of moist air at 20°C and 825 mbar (*a*) when saturated, and (*b*) at 50 per cent saturation.

Answer

(*a*) From I.H.V.E. tables, $h_a = 20.11$ kJ/kg and $p_{ss} = 23.37$ mbar at 20°C. From N.E.L. steam tables $h_g = 2537.5$ kJ/kg at 20°C saturated. From Table 2.1 $f_s = 1.0039$ at 20°C and 825 mbar. If it is assumed that the liquid from which the steam was evaporated into the atmosphere was under a pressure of one standard atmosphere then an addition of 0·08 kJ/kg should be made to the N.E.L. value for h_g giving a new figure of 2537·58 kJ/kg. Then:

$$h = 20.11 + \frac{(0.62197 \times 1.0039 \times 23.37 \times 2537.58)}{(825 - 1.0039 \times 23.37)}$$

$$= 20.11 + 46.19$$

$$= 66.30 \text{ kJ/kg}$$

The I.H.V.E. Guide gives a table of additive values to be applied to enthalpies read from the psychrometric tables for 1013·25 mbar at any value of the temperature of adiabatic saturation. From this table the additive value at 20°C adiabatic saturation temperature and 825 mbar is 8·77 kJ/kg. The I.H.V.E. psychrometric tables for 1013·25 mbar and 20°C saturated quote 57·55 kJ/kg. Hence the value to be compared with our result is 66·32 kJ/kg.

(*b*) From eqn. (2.23) and eqn. (2.13)

$$h = h_a + \frac{\mu g_{ss}}{100} h_g$$

$$= 20{\cdot}11 + 0{\cdot}5 \times 46{\cdot}19$$

$$= 43{\cdot}21 \text{ kJ/kg.}$$

EXERCISES

1. (*a*) Air at a condition of 30°C dry-bulb, 17°C wet-bulb and a barometric pressure of 1050 mbar enters a piece of equipment where it undergoes a process of adiabatic saturation, the air leaving with a moisture content of 5 g/kg higher than entering. Calculate, using the data below, and the ideal gas laws,

(i) the moisture content of the air entering the equipment,

(ii) the dry-bulb temperature and enthalpy of the air leaving the equipment.

(*b*) Using the data below, calculate the moisture content of air at 17°C dry-bulb and 40 per cent saturation where the barometric pressure is 950 mbar.

Sat. vapour pressure at 17°C = 19.36 mbars
Specific heat of dry air = 1.015 kJ/kg °C
Specific heat of water vapour = 1.890 kJ/kg °C
Latent heat of evaporation at 17°C = 2459 kJ/kg
Latent heat of evaporation at 0°C = 2500 kJ/kg
Constant for psychrometric equation
for wet-bulb temperatures above 0°C = 0.666×10^{-3}

Note: Psychrometric tables are not to be used for this question.

Answers (*a*) (i) 6.15 g/kg, (ii) 18°C, 46.52 kJ/kg.

(*b*) 5.18 g/kg.

2. (*a*) Define the following psychrometric terms

(i) Vapour pressure.

(ii) Relative humidity.

(iii) Humid volume.

(iv) Dew point.

(*b*) Name two distinct types of instrument which are used to determine the relative humidity of atmospheric air and briefly explain the principle in each case.

3. (*a*) Calculate from first principles the enthalpy of air at 28°C dry-bulb and 19·26 mbar vapour pressure, at barometric pressure, using the ideal gas laws.

(*b*) Explain how the value would alter if the barometric pressure only were reduced.

Answer: 59·03 kJ/kg dry air.

4. (*a*) Calculate the dew point of air at 28°C dry-bulb, 21°C wet-bulb and 877 mbar barometric pressure. Use I.H.V.E. tables where necessary to determine saturation vapour pressure.

(*b*) Calculate its enthalpy.

Answers: (*a*) 18·11°C, (*b*) 66·70 kJ/kg dry air.

5. (*a*) Distinguish between saturation vapour pressure and the vapour pressure of moist air at an unsaturated state.

(*b*) Illustrate relative humidity by means of a $p-V$ diagram for steam.

(*c*) What happens if air at an unsaturated state is cooled to a temperature below its dew point?

(*d*) Calculate the relative humidity and percentage saturation of air at 1013·25 mbar, 21°C dry-bulb and 14·5°C wet-bulb (sling). You may use the I.H.V.E. tables of psychrometric data only to determine saturation vapour pressures.

Answers: (*a*) 48·7%, 48·2%.

BIBLIOGRAPHY

1. J. L. Threlkeld. *Thermal Environmental Engineering*. Prentice-Hall, 1962.
2. J. A. Goff. Standardisation of thermodynamic properties of moist air, *Trans. A.S.H.V.E.*, 1949, **55**, 463.
3. J. A. Goff and S. Gratch. Thermodynamic properties of moist air, *Trans. A.S.H.V.E.*, 1945, **51**, 125.
4. W. H. McAdams. *Heat Transmission*. McGraw-Hill, 1942.
5. D. G. Kern. *Process Heat Transfer*. McGraw-Hill, 1950.
6. W. H. Walker, W. K. Lewis, W. H. McAdams and E. R. Gilliland. *Principles of Chemical Engineering*. McGraw-Hill, 1937.
7. W. Goodman. *Air Conditioning Analysis with Psychrometric Charts and Tables*. The Macmillan Company, New York, 1943.
8. W. P. Jones. The Psychrometric Chart in S.I. Units, *J. Instn. Heat. Vent. Engrs.*, 1970, **38**, 93.

3

The Psychrometry of Air Conditioning Processes

3.1 The psychrometric chart

This provides a picture of the way in which the state of moist air alters as an air conditioning process takes place or a physical change occurs. Familiarity with the psychrometric chart is essential for a proper understanding of air conditioning.

Any point on the chart is termed a *state point*, the location of which, at a given barometric pressure, is fixed by any two of the psychrometric properties discussed in Chapter 2. It is customary and convenient to design charts at a constant barometric pressure because barometric pressure does not alter greatly over much of the inhabited surface of the earth. When the barometric pressure is significantly different from the standard adopted for the chart or psychrometric tables to hand, then the required properties can be calculated using the equations derived earlier.

The British standard is that adopted by the Institution of Heating and Ventilating Engineers for their Tables of Psychrometric Data and for their psychrometric chart. It is 1013·25 millibars. The American standard is also 1013·25 millibars and this value is used by the American Society of Refrigeration, Heating and Air Conditioning Engineers. It is also, incidentally, the value adopted by the Meteorological Office in Great Britain.

It is worth noting though, that there are a few differences of expression in the British and American charts. The most important of these is, of course, in the datum used for the enthalpy of dry air. Two other minor points of difference are that the American chart expresses the temperature of adiabatic saturation and specific humidity, whereas the British chart uses wet-bulb temperature (sling) and moisture content.

The psychrometric chart published by the I.H.V.E. uses two fundamental properties, mass and energy, in the form of moisture content and enthalpy, as co-ordinates. As a result, mixture states lie on the straight line which joins the state points of the two constituents. Lines of constant dry-bulb temperature are virtually straight but divergent, only the isotherm for 30°C being vertical. The reason for this is that to preserve the usual appearance of a psychrometric chart, in spite of choosing the two fundamental properties as co-ordinates, the co-ordinate axes are oblique, not rectangular. Hence, lines of constant enthalpy are both straight and parallel, as are lines of constant moisture content. Since both these properties are taken as linear, the lines of constant enthalpy are equally spaced as are, also, the lines of constant moisture content. This is not true of the lines of constant humid volume and constant wet-bulb temperature, which are slightly curved and

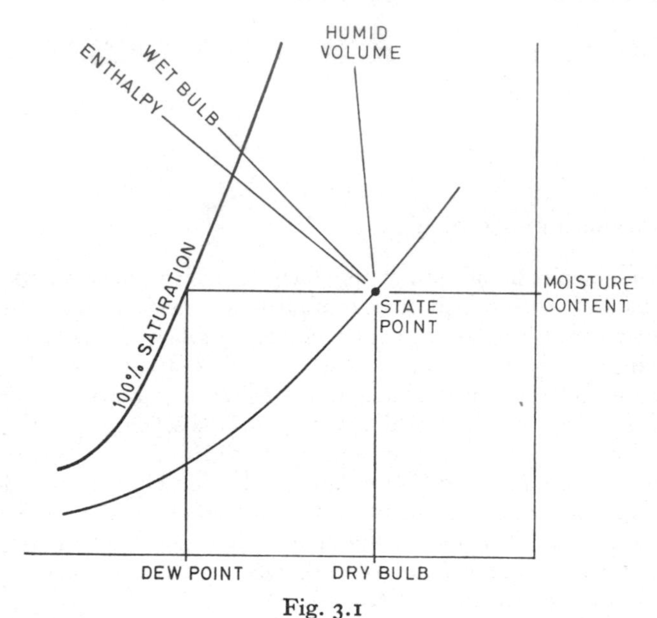

Fig. 3.1

divergent. Since their curvature is only slight in the region of practical use on the chart, they can be regarded as straight without sensible error resulting. In the sketches of psychrometric charts used throughout this text to illustrate changes of state, only lines of percentage saturation are shown curved. All others are shown straight, and isotherms are shown as vertical, for convenience.

The chart also has a protractor which allows the value of the ratio of the sensible heat gain to the total heat gain to be plotted on the chart. This ratio is an indication of the slope of the room ratio line (*see* § 6.4) and is of value in determining the correct supply state of the air that must be delivered to a conditioned space. Because the enthalpy of the added vapour depends on the temperature at which evaporation takes place, the zero value for the ratio is parallel to the isotherm for 30°C, it being assumed that most of the latent heat gain to the air in a conditioned room is by evaporation from the skin of the occupants and that their skin surface temperature is about 30°C.

Figure 3.1 shows a state point on a psychrometric chart.

3.2 Mixtures

Figure 3.2 shows what happens when two airstreams meet and mix adiabatically. Moist air at state 1 mixes with moist air at state 2, forming a mixture at state 3. The principle of the conservation of mass allows two mass balance equations to be written:

$$m_{a1} + m_{a2} = m_{a3} \text{ for the dry air and}$$

$$g_1 m_{a1} + g_2 m_{a2} = g_3 m_{a3} \text{ for the associated water vapour.}$$

$$\text{Hence } (g_1 - g_3) m_{a1} = (g_3 - g_2) m_{a2}$$

$$\text{Therefore } \frac{g_1 - g_3}{g_3 - g_2} = \frac{m_{a2}}{m_{a1}}$$

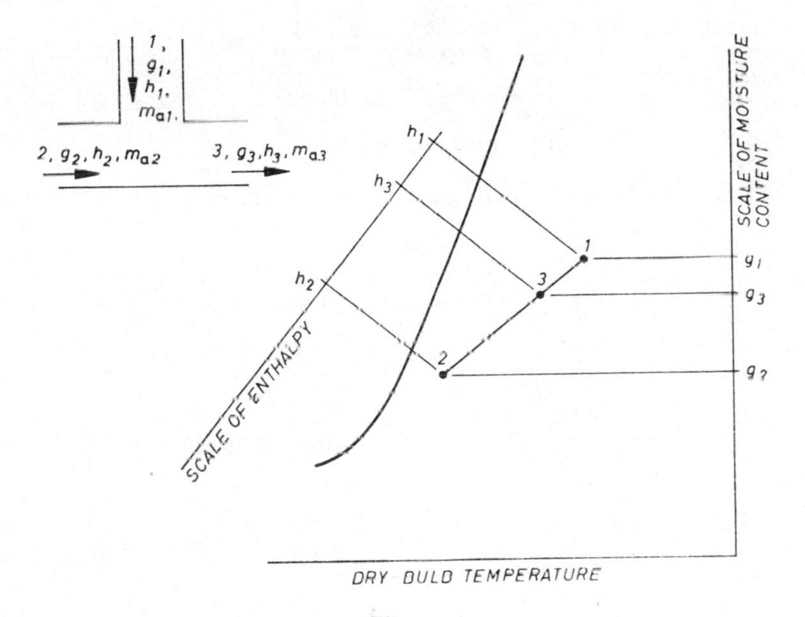

Fig. 3.2

Similarly, making use of the principle of the conservation of energy,

$$\frac{h_1 - h_3}{h_3 - h_2} = \frac{m_{a2}}{m_{a1}}$$

From this it follows that the three state points must lie on a straight line in a mass-energy co-ordinate system. The psychrometric chart published by the I.H.V.E. is such a system—with oblique co-ordinates. For this chart then, a principle can be stated for the expression of mixture states. When two airstreams mix adiabatically, the mixture state lies on the straight line which joins the state points of the constituents, and the position of the mixture state point is such that the line is divided inversely as the ratio of the masses of dry air in the constituent airstreams.

EXAMPLE 3.1. Moist air at a state of 60°C dry-bulb, 32·1°C wet-bulb (sling) and 1013·25 mbar barometric pressure mixes adiabatically with moist air at 5°C dry-bulb, 0·5°C wet-bulb (sling) and 1013·25 mbar barometric pressure. If the masses of dry air are 3 kg and 2 kg, respectively, calculate the enthalpy, specific humidity and dry-bulb temperature of the mixture.

Answer

From I.H.V.E. tables of psychrometric data,

$$g_1 = 18{\cdot}400 \text{ g/kg dry air}$$

$$g_2 = 2{\cdot}061 \text{ g/kg dry air}$$

$$h_1 = 108{\cdot}40 \text{ kJ/kg dry air}$$

$$h_2 = 10{\cdot}20 \text{ kJ/kg dry air.}$$

The principle of the conservation of mass demands that

$$g_1 m_{a1} + g_2 m_{a2} = g_3 m_{a3} = g_3(m_{a1} + m_{a2})$$

hence

$$g_3 = \frac{g_1 m_{a1} + g_2 m_{a2}}{m_{a1} + m_{a2}} = \frac{18{\cdot}4 \times 3 + 2{\cdot}061 \times 2}{3+2} = \frac{59{\cdot}322}{5} = 11{\cdot}864 \text{ g/kg dry air.}$$

Similarly, by the principle of the conservation of energy,

$$h_3 = \frac{h_1 m_{a1} + h_2 m_{a2}}{m_{a1} + m_{a2}} = \frac{108{\cdot}40 \times 3 + 10{\cdot}20 \times 2}{(3+2)} = 69{\cdot}12 \text{ kJ/kg dry air.}$$

To determine the dry-bulb temperature, the following practical equation must be used

$$h = (1{\cdot}007t - 0{\cdot}026) + g(2501 + 1{\cdot}84t) \tag{2.17}$$

Substituting the values calculated for moisture content and enthalpy, this equation can be solved for temperature:

$$h = 69{\cdot}12 = (1{\cdot}007t - 0{\cdot}026) + 0{\cdot}01186(2501 + 1{\cdot}84t)$$

$$t = \frac{39{\cdot}47}{1{\cdot}029} = 38{\cdot}4^\circ\text{C}$$

On the other hand, if the temperature were calculated by proportion, according to the masses of the dry air in the two mixing airstreams, a slightly different answer results:

$$t = \frac{3\times 60 + 2\times 5}{5}$$

$$= 38^\circ\text{C}$$

This is clearly the wrong answer, both numerically and by the method of its calculation. However, the error is small, considering that the values chosen for the two mixing states spanned almost the full range of the psychrometric chart. The error is even less for states likely to be encountered in everyday practice and is well within the acceptable limits of accuracy. To illustrate this, consider an example involving the mixture of two airstreams at states more representative of common practice.

EXAMPLE 3.2. A stream of moist air at a state of 21°C dry-bulb and 14·5°C wet-bulb (sling) mixes with another stream of moist air at a state of 28°C dry-bulb and 20·2°C wet bulb (sling), the respective masses of the associated dry air being 3 kg and 1 kg. With the aid of I.H.V.E. tables of psychrometric data calculate accurately the dry-bulb temperature of the mixture (*a*) using the principles of the conservation of energy and of mass and, (*b*), using a direct proportionality between temperature and mass.

Answer

The reader is left to go through the steps in the arithmetic, as an exercise, using the procedure adopted in Example 3.1.

The answers are (*a*) 22·76°C, dry-bulb, (*b*) 22·75°C dry-bulb.

The conclusion to be drawn is that the method used to obtain answer (*b*) is quite accurate enough for all practical purposes.

3.3 Sensible heating and cooling

Sensible heat transfer occurs when moist air flows across a heater battery or over the coils of a sensible cooler. In the heater, the temperature of the medium used to provide the heat is not critical. The sole requirement for heat transfer is that the temperature shall exceed the final air temperature. In sensible cooling there is a further restriction: the lowest water temperature must not be so low that moisture starts to condense on the cooler coils. If such condensation does occur, through a poor choice of chilled water temperature, then the process will no longer be one of sensible cooling since dehumidification will also be taking place. This complication will not be discussed further here but is dealt with in sections 3.4 and 10.6.

Figure 3.3 shows the changes of state which occur, sketched upon a psychrometric chart. The essence of both processes is that the change of state must occur along a line of constant moisture content. The variations in the physical properties of the moist air, for the two cases, are summarised below:

	Sensible heating	*Sensible cooling*
Dry bulb	increases	decreases
Enthalpy	increases	decreases
Humid volume	increases	decreases
Wet bulb	increases	decreases
Percentage saturation	decreases	increases
Moisture content	constant	constant
Dew point	constant	constant
Vapour pressure	constant	constant

EXAMPLE 3.3. Calculate the load on a battery which heats 1·5 m^3/s of moist air, initially at a state of 21°C dry-bulb, 15°C wet-bulb (sling) and 1013·25 mbar barometric pressure, by 20 degrees. If low pressure hot water at 85°C flow and 75°C return, is used to achieve this, calculate the flow rate necessary, in kilograms of water per second.

Answer

Heating load = (mass flow of moist air expressed in kg/s of associated dry air) × (increase in enthalpy of moist air expressed in kJ/kg of associated dry air.)

From I.H.V.E. tables of psychrometric data, the initial enthalpy is found to be 41·88 kJ/kg of dry air, and the initial humid volume to be 0·8439 m^3/kg of dry air. Since the air is being heated by 20 degrees, reference must now be made to tables in order to determine the enthalpy at the same moisture

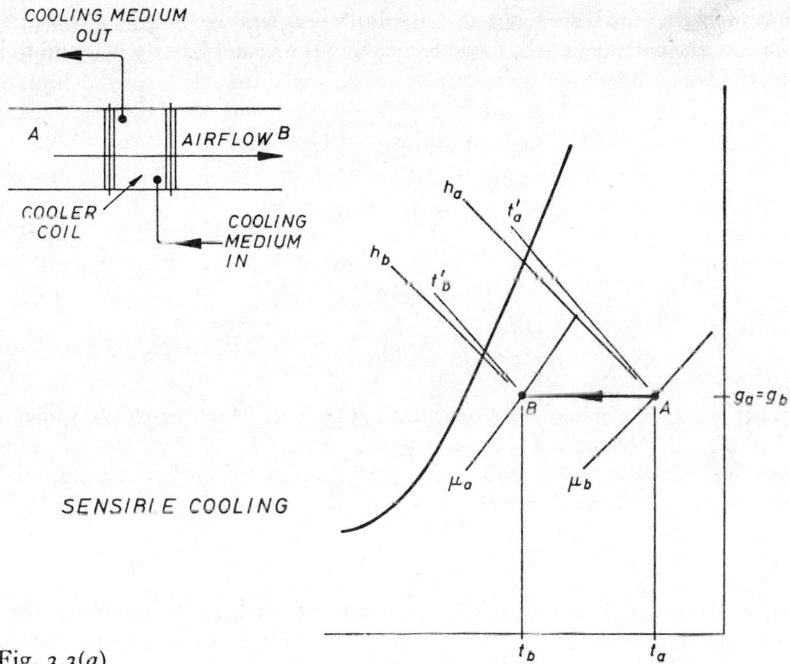

Fig. 3.3(*a*)

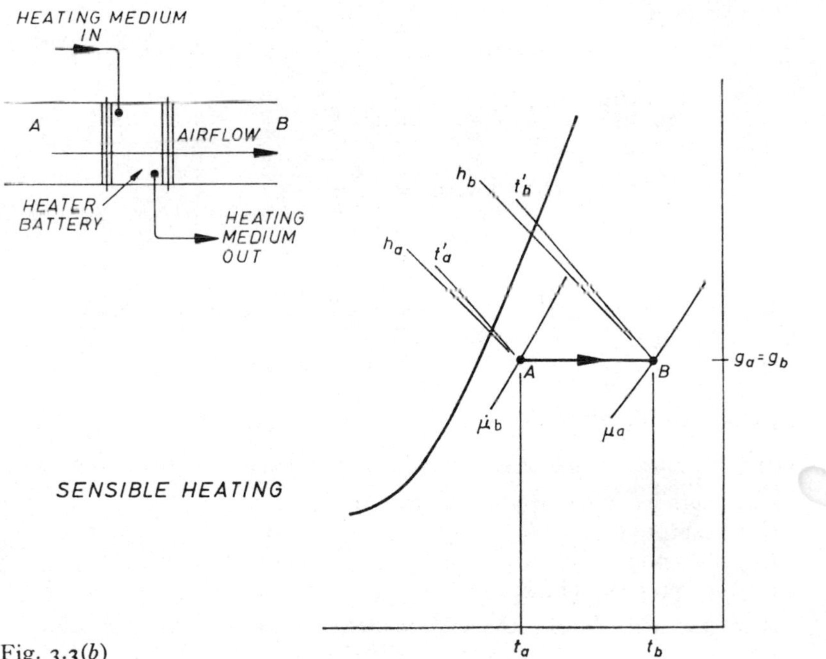

Fig. 3.3(*b*)

content as the initial state but at a dry-bulb temperature of 41°C. By interpolation, the enthalpy of the moist air leaving the heater battery is found to be 62·31 kJ/kg of dry air.

$$\text{Heating load} = \left(\frac{1{\cdot}5}{0{\cdot}8439}\right) \times (62{\cdot}31 - 41{\cdot}88)$$

$$= 36{\cdot}3 \text{ kW}$$

$$\text{Flow rate of L.P.H.W.} = \frac{36{\cdot}3}{(85^\circ - 75^\circ) \times 4{\cdot}2}$$

$$= 0{\cdot}864 \text{ kg/s}$$

EXAMPLE 3.4. Calculate the load on a cooler coil which cools the moist air mentioned in Example 3.3 by 5 degrees. What is the flow rate of chilled water necessary to effect this cooling if flow and return temperatures of 10°C and 15°C are satisfactory?

Answer

The initial enthalpy and humid volume are the same as in the first example. The final dry-bulb temperature of the moist air is 16°C but its moisture content is still 8·171 g/kg of dry air. At this state, its enthalpy is found from tables to be 36·77 kJ/kg of dry air.

$$\text{Cooling load} = \left(\frac{1{\cdot}5}{0{\cdot}8439}\right) \times (41{\cdot}88 - 36{\cdot}77)$$

$$= 9{\cdot}1 \text{ kW}$$

$$\text{Flow rate of chilled water} = \frac{9{\cdot}1}{5 \times 4{\cdot}2}$$

$$= 0{\cdot}433 \text{ kg/s}$$

It is to be noted that the selection of chilled water temperatures is a matter demanding some care.

3.4 Dehumidification

There are four principal methods whereby moist air can be dehumidified:

(i) cooling to a temperature below the dew point,
(ii) adsorption,
(iii) absorption,
(iv) compression followed by cooling.

The first method forms the subject matter of this section.

Cooling to a temperature below the dew point is done by passing the moist air over a cooler coil or through an air washer.

Figure 3.4 shows on a sketch of a psychrometric chart what happens when moist air is cooled and dehumidified in this fashion. Since dehumidification

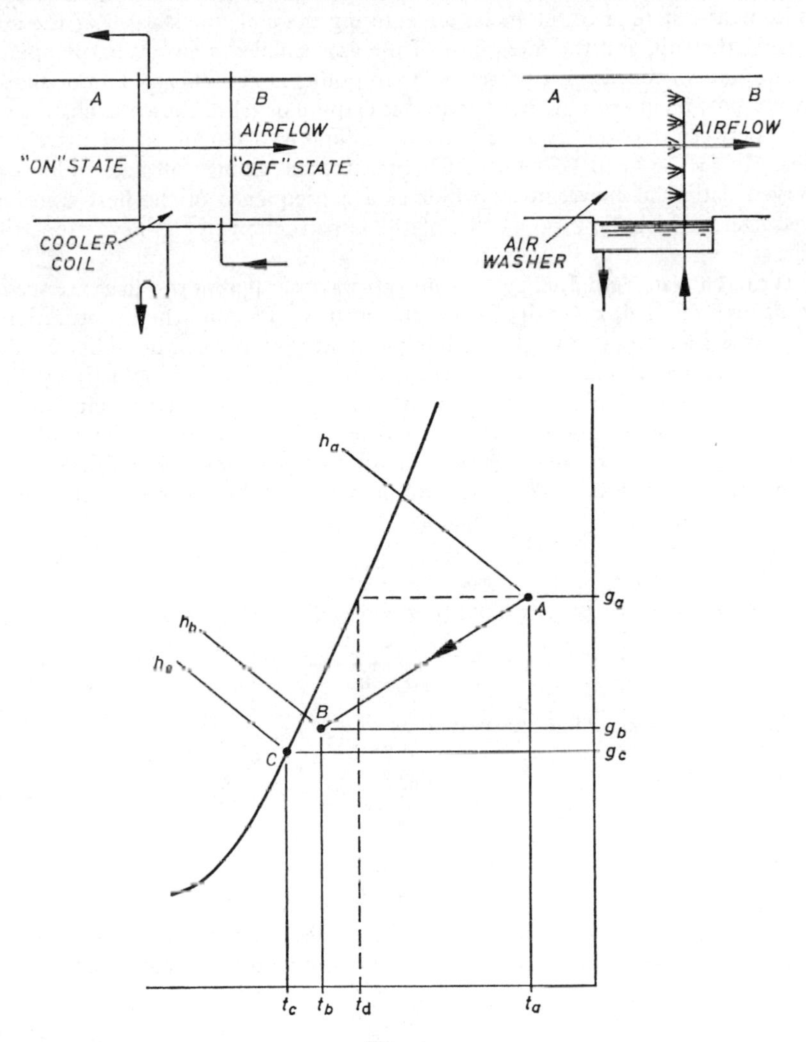

Fig. 3.4

is the aim, some of the spray water or some part of the cooler coil, must be at a temperature less than the dew point of the air entering the equipment. In the figure, t_d is the dew point of the moist air 'on' the coil or washer. The temperature t_c, corresponding to the point C on the saturation curve, is termed the *apparatus dew point.* This term is in use for both coils and washers but, in the case of cooler coils alone, t_c is also sometimes termed the *mean coil surface temperature*. The justification for this latter terminology is offered in Chapter 10.

For purposes of carrying out air conditioning calculations, it is sufficient to know the state A of the moist air entering the coil, the state B of the air leaving the coil, and the mass flow of the associated dry air. What happens to the state of the air as it passes between points A and B is seldom of more than academic interest. Consequently, it is quite usual to show the change of state between the 'on' and the 'off' conditions as occurring along a straight line. In fact, as will be seen in Chapter 10, the change of state follows a curved path, the curvature of which is a consequence of the heat transfer characteristics of the process, not of the construction of the psychrometric chart.

It can be seen from Fig. 3.4 that the moisture content of the air is reduced, as also is its enthalpy and dry-bulb temperature. The percentage saturation, of course, increases. It might be thought that the increase of humidity would be such that the 'off' state, represented by the point B, would lie on the saturation curve. This is not so for the very good reason that no air washer or cooler coil is a hundred per cent efficient. It is unusual to speak of the efficiency of a cooler coil. Instead, the alternative terms, *contact factor* and *by-pass factor* are used. They are complementary values and contact factor, sometimes denoted by β, is defined as

$$\beta = \frac{g_a - g_b}{g_a - g_c} = \frac{h_a - h_b}{h_a - h_c} \tag{3.1}$$

Similarly, by-pass factor is defined as

$$(1-\beta) = \frac{g_b - g_c}{g_a - g_c} = \frac{h_b - h_c}{h_a - h_c} \tag{3.2}$$

It is sufficient, for all practical purposes, to assume that both these expressions can be re-written in terms of dry-bulb temperature, namely that

$$\beta = \frac{t_a - t_b}{t_a - t_c} \tag{3.3}$$

and that

$$(1-\beta) = \frac{t_b - t_c}{t_a - t_c} \tag{3.4}$$

It is less true to assume that they can also be written, without sensible error, in terms of wet-bulb temperature, since the scale of wet-bulb values on the psychrometric chart is not at all linear. The assumption is, however, sometimes made for convenience, provided the values involved are not very far apart and that some inaccuracy can be tolerated in the answer.

Typical values of β are 0·80 to 0·90, for practical coil selection.

EXAMPLE 3.5. 1·5 m^3/s of moist air at a state of 28°C dry-bulb, 20·6°C wet-bulb and 1013·25 mbar flows across a cooler coil and leaves the coil at 12·5°C dry-bulb and 8·336 g/kg of dry air.

(*a*) Determine the apparatus dew point.
(*b*) Calculate the contact factor.
(*c*) Calculate the cooling load.

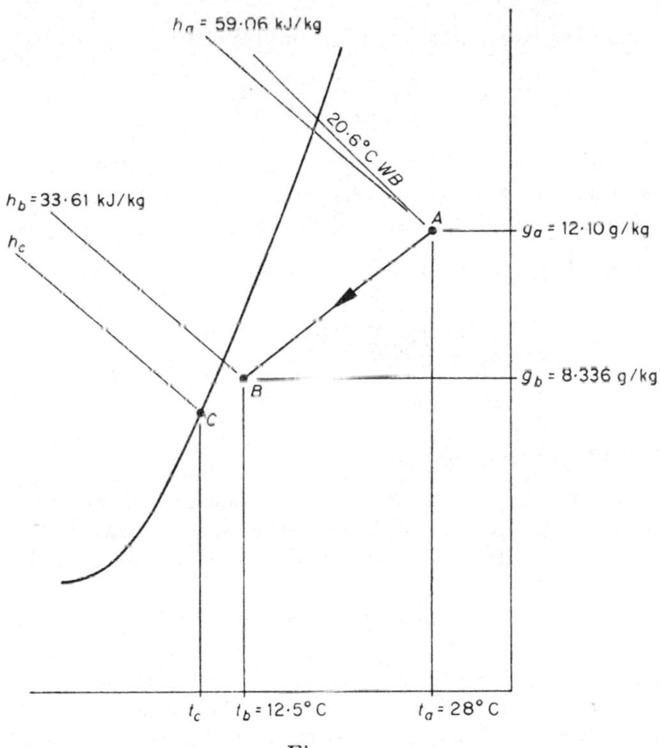

Fig. 3.5

Answer

Figure 3.5 shows the psychrometric changes involved and the values immediately known from the data in the question. From I.H.V.E. tables (or from a psychrometric chart) it is established that h_a = 59·06 kJ/kg and that h_b = 33·61 kJ/kg. The tables also give a value for the humid volume at the entry state to the coil, of 0·8693 m^3/kg.

(*a*) Mark state points A and B on a psychrometric chart. Join them by a straight line and extend the line to cut the saturation curve. Observation of the point of intersection shows that this is at a temperature of 10·25°C. It is not easy to decide this value with any exactness unless a large psychrometric chart is used. However, it is sufficiently accurate for present purposes (and for most practical purposes), to take the value read from an ordinary chart. Hence the apparatus dew point is 10·25°C.

(*b*) Either from the chart or from tables, it can be established that the enthalpy at the apparatus dew point is 29·94 kJ/kg. Using the definition of contact factor,

$$\beta = \frac{59{\cdot}06 - 33{\cdot}61}{59{\cdot}06 - 29{\cdot}94} = 0{\cdot}875$$

Less accurately, one might determine this from the temperatures involved:

$$\beta = \frac{28^\circ - 12{\cdot}5^\circ}{28^\circ - 10{\cdot}25^\circ} = 0{\cdot}874$$

Clearly, in view of the error in reading the apparatus dew point from the chart, the value of β obtained from the temperatures is quite accurate enough in this example and, in fact, in most other cases.

(*c*) Cooling load = mass flow × decrease of enthalpy

$$= \frac{1{\cdot}5}{0{\cdot}8693} \times (59{\cdot}06 - 33{\cdot}61)$$

$$= 43{\cdot}9 \text{ kW}$$

3.5 Humidification

This, as its name implies, means that the moisture content of the air is increased. This may be accomplished by either water or steam but the present section is devoted to the use of water only, section 3.7 being used for steam injection into moving airstreams.

There are three methods of using water as a humidifying agent: the passage of moist air through a spray chamber containing a very large number of small water droplets; its passage over a large wetted surface; or the direct injection of water drops of aerosol size into the room being conditioned. (A variant of this last technique is to inject aerosol-sized droplets into an airstream moving through a duct.) Which ever method is used the psychrometric considerations are similar.

It is customary to speak of the humidifying efficiency or the effectiveness of an air washer (although neither term is universally accepted) rather than a contact or by-pass factor. There are several definitions, some based on the extent to which the dry-bulb temperature of the entering moist airstream approaches its initial wet-bulb value, and others based on the change of state undergone by the air. In view of the fact that the psychrometric chart currently in use by the Institution of Heating and Ventilating Engineers is constructed with mass (moisture content) and energy (enthalpy) as oblique, linear co-ordinates, the most suitable definition to use with the chart is that couched in terms of these fundamentals. There is the further advantage that such a definition of effectiveness, E, is identical with the definition of contact factor, β, used for cooler coils.

Although humidifying efficiency is often expressed in terms of a process of adiabatic saturation, this is really a special case, and is so regarded here.

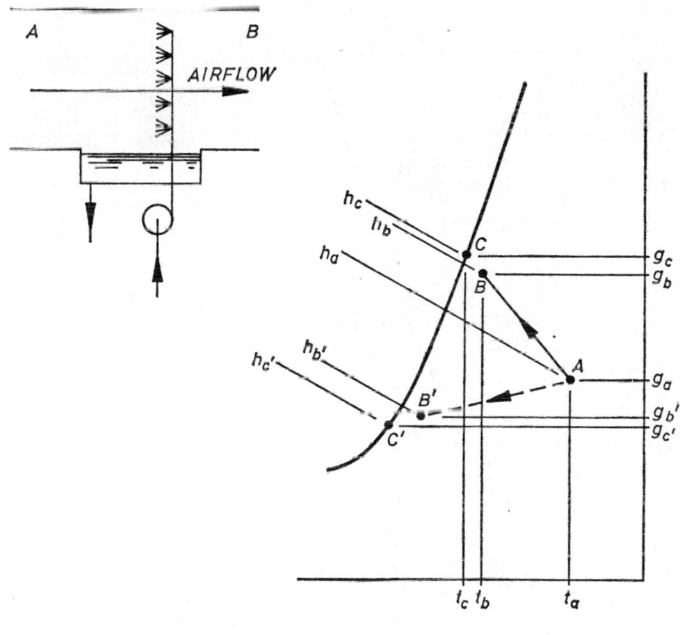

Fig. 3.6(*a*)

Figure 3.6(*a*) shows an illustration of the change of state experienced by an airstream as it passes through a spray chamber.

The effectiveness of the spray chamber is then defined by

$$E = \frac{h_b - h_a}{h_c - h_a} \tag{3.5}$$

and humidifying efficiency is defined by

$$\eta = 100\,E \tag{3.6}$$

It is evident that, because moisture content is the other linear co-ordinate of the psychrometric chart and the points A, B and C lie on the same straight line, C being obtained by the extension of the line joining AB to cut the saturation curve, then it is possible to put forward an alternative and equally valid definition of effectiveness, expressed in terms of moisture content:

$$E = \frac{g_b - g_a}{g_c - g_a} \tag{3.7}$$

and $$\eta = 100\,E, \text{ as before.}$$

Figure 3.6(*a*) shows that h_b is greater than h_a. This implies that there is a heat input to the spray water being circulated through the spray chamber. This could be easily accomplished by means of, say, a calorifier in the return

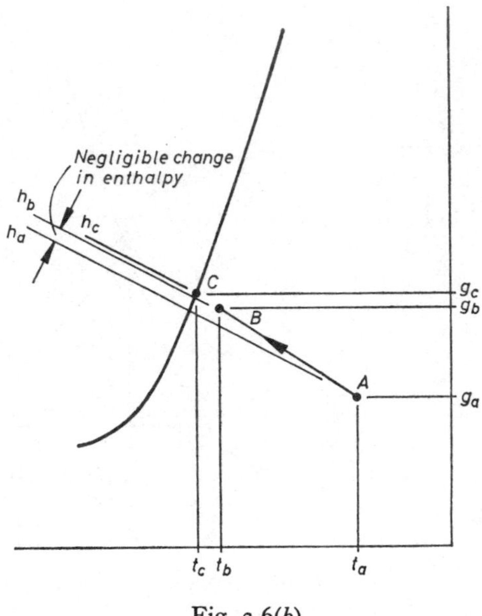

Fig. 3.6(*b*)

pipe from the washer tank. If the spray water had been chilled, instead of heated, then a change of state might have been as shown dotted on Fig. 3.6(*a*), from the point A to the point B'. Under these circumstances, the effectiveness would have been expressed by

$$E = \frac{h_a - h_{b'}}{h_a - h_{c'}}$$

or $$E = \frac{g_a - g_{b'}}{g_a - g_{c'}}$$

Consider the special case of adiabatic saturation: for this to occur it is necessary that

(*a*) the spray water is totally recirculated, no heat exchanger being present in the pipelines or in the washer tank;

(*b*) the spray chamber, tank and pipelines are perfectly lagged; and (*c*) the feed water supplied to the system to make good the losses due to evaporation, is at the temperature of adiabatic saturation (*see* § 2.18).

Under these conditions it may be assumed that the change of state follows a line of constant wet-bulb temperature (since the Lewis number for air-water vapour mixtures is unity), the system having been given sufficient time to settle down to steady-state operation. Because feed water is being supplied at the wet-bulb temperature (virtually) and this will not, as a general rule, equal the datum temperature for the enthalpy of water, 0°C, then there must be a change of enthalpy during the process. Strictly speaking then, it is incorrect to speak of the process as being an adiabatic one. The use of the term stems from the phrase 'temperature of adiabatic saturation' and so its use is condoned. One thing is fairly certain though: at the temperatures normally encountered the change of enthalpy during the process is negligible, and so effectiveness must be expressed in terms of change of moisture content. Figure 3.6(*b*) shows a case of adiabatic saturation, in which it can be seen that there is no significant alteration in enthalpy although there is a clear change in moisture content. It can also be seen that a fall in temperature accompanies the rise in moisture content. This temperature change provides an approximate definition of effectiveness or efficiency which is most useful, and, for the majority of practical applications, sufficiently accurate. It is

$$E \simeq \frac{t_a - t_b}{t_a - t_c} \tag{3.8}$$

The way in which the psychrometric chart is constructed precludes the possibility of this being an accurate expression; lines of constant dry-bulb temperature are not parallel and equally spaced, such properties being exclusive to enthalpy and moisture content.

EXAMPLE 3.6. 1·5 m^3/s of moist air at a state of 15°C dry-bulb, 10°C wet-bulb (sling) and 1013·25 mbar barometric pressure, enters the spray chamber of an air washer. The humidifying efficiency of the washer is 90 per cent, all the spray water is recirculated, the spray chamber and the tank are perfectly lagged, and a feed at 10°C from the Water Board main is supplied to make good the losses due to evaporation.

Calculate (*a*) the state of the air leaving the washer, (*b*) the rate of flow of make-up water from the mains.

Answer

(*a*) This is illustrated in Fig. 3.7.

Using the definitions of humidifying efficiency quoted in eqns. (3.6) and (3.7), and referring to tables of psychrometric data for the properties of moist air at states *A* and *C*, one can write

$$\frac{90}{100} = \frac{g_b - 5{\cdot}558}{7{\cdot}659 - 5{\cdot}558}$$

Hence, $g_b = 7{\cdot}449$ g/kg of dry air.

The state of the moist air leaving the air washer is 10°C wet-bulb (sling), 7·449 g/kg and 1013·25 mbar barometric pressure. The use of eqn. (3.8) shows that the approximate dry-bulb temperature at exit from the washer is 10·5°C.

(*b*) The amount of water supplied to the washer must equal the amount evaporated from it. From tables, the humid volume at state *A* is 0·8232 m^3/kg. Each kilogram of dry air passing through the spray chamber has its associated water vapour augmented by an amount equal to $g_b - g_a$, that is, by 1·891 g.

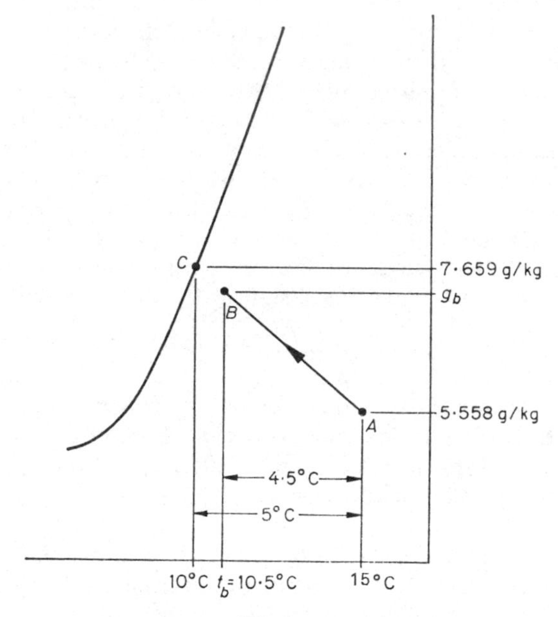

Fig. 3.7

Thus, the rate of make-up is

$$= \frac{1 \cdot 5 \times 0 \cdot 001891 \times 3600}{0 \cdot 8232}$$

= 12·41 kg of water per hour.

It is to be noted that this is not equal to the pump duty; the amount of water the pump must circulate depends on the humidifying efficiency and, to achieve a reasonable value for this, the spray nozzles used to atomise the recirculated water must break up the water into very small drops, so that there will exist a good opportunity for an intimate and effective contact between the airstream and the water. A big pressure drop across the nozzles results from the atomisation, if this is to be adequate. Spray water pumps used with this type of washer have to develop heads of the order of 2 bars, as a result.

3.6 Water injection

The simplest case to consider, and the one that provides the most insight into the change of state of the airstream subjected to humidification by the injection of water, is where all the injected water is evaporated. Figure 3.8 shows what happens when total evaporation occurs.

Air enters a spray chamber at state A and leaves it at state B, all the injected water being evaporated, none falling to the bottom of the chamber to run to waste or to be recirculated. The feed-water temperature is t_w. It is important to realise that since total evaporation has occurred, state B must lie nearer to the saturation curve, but just how much nearer will depend on the amount of water injected.

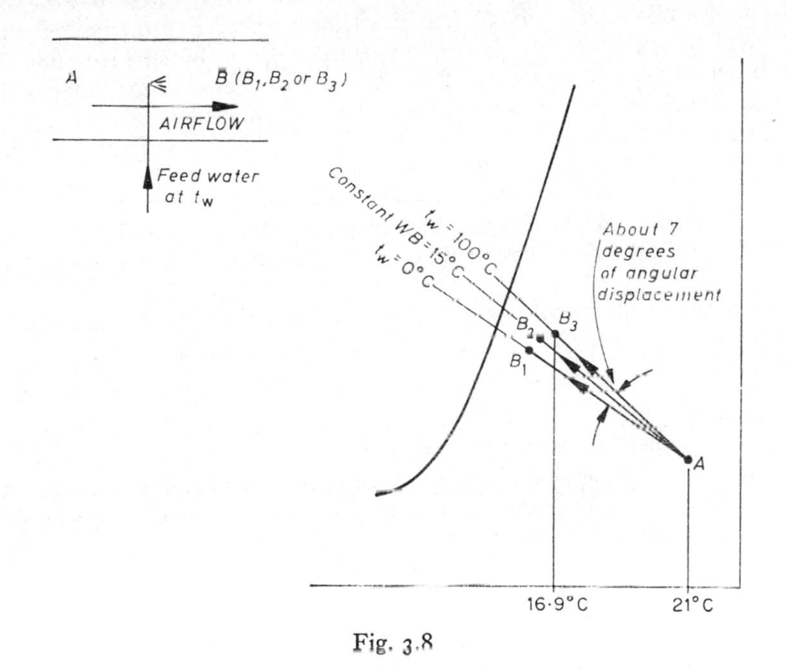

Fig. 3.8

Two equations, a heat balance and a mass balance, provide the answer required.

Striking a heat balance, we can write

$$h_a + h_w = h_b$$

and, in a similar way, the mass balance may be written as

$$m_a + m_w = m_b$$

Knowing the amount of feed water evaporated, the mass balance gives the necessary information about the moisture content of the airstream leaving the area of the water injection. Applying the mass balance to the water vapour only (since the associated kilogram of dry air may be ignored),

$$g_b = g_a + m_w$$

where m_w is the amount of feed water evaporated in kg/kg of dry air flowing through the spray chamber.

Applying the heat balance:

$$h_a + h_w = h_b$$

$$= (1{\cdot}007t - 0{\cdot}026) + g_b(2501 + 1{\cdot}84t)$$

One thing is immediately apparent: if feed water is injected into the airstream at a temperature of 0°C there will be no alteration in the enthalpy of the airstream, since 0°C is the temperature datum of zero enthalpy for the water associated with 1 kg of dry air. Under these circumstances the change of state between *A* and *B* will follow a line of constant enthalpy.

One further conclusion may be drawn: if the feed is at a temperature equal to the wet bulb of the airstream, the change of state will be along a line of constant wet-bulb temperature. This is a consequence of the fact that the Lewis number of air-water vapour mixtures at normally encountered temperatures and pressures, is virtually unity—as was discussed in sections 2.17 and 2.18.

To see what happens at other water temperatures, consider water at 100°C injected into a moving moist airstream and totally evaporated.

EXAMPLE 3.7. Moist air at a state of 21°C dry-bulb, 15°C wet-bulb (sling) and 1013·25 mbar barometric pressure enters a spray chamber. If, for each kilogram of dry air passing through the chamber, 0·002 kg of water at 100°C is injected and totally evaporated, calculate the moisture content, enthalpy and dry-bulb temperature of the moist air leaving the chamber.

Answer

From I.H.V.E. tables of psychrometric data,

$$h_a = 41{\cdot}88 \text{ kJ/kg dry air,}$$

$$g_a = 0{\cdot}008171 \text{ kg/kg dry air.}$$

Since the feed water has a temperature of 100°C, its enthalpy is 419·06 kJ/kg of water injected.

Use of the equation for mass balance yields the moisture content of the moist air leaving the spray chamber:

$$g_b = 0{\cdot}008171 + 0{\cdot}002$$

$$= 0{\cdot}010171 \text{ kg/kg dry air.}$$

Use of the energy balance equation gives the enthalpy of the air leaving the chamber and hence, also, its dry-bulb temperature:

$$h_b = 41{\cdot}88 + 0{\cdot}002 \times 419{\cdot}06$$
$$= 42{\cdot}718 \text{ kJ/kg dry air.}$$
$$= (1{\cdot}007t_b - 0{\cdot}026) + 0{\cdot}010171(2501 + 1{\cdot}84t_b)$$

thus,

$$42{\cdot}718 = 1{\cdot}007t_b - 0{\cdot}026 + 25{\cdot}44 + 0{\cdot}0187t_b$$
$$t_b = 16{\cdot}9°\text{C}$$

Reference back to Fig. 3.8 shows a summary of what happens with different feed-water temperatures: change of state from A to B_1 is along a line of constant enthalpy and is for a feed-water temperature of 0°C; change of state from A to B_2 occurs when the water is injected at the wet-bulb temperature of the entry air and takes place along a line of constant wet-bulb temperature; and change of state from A to B_3 (the subject of Example 3.6) is for water injected at 100°C. For all cases except that of the change A to B_1, an increase in enthalpy occurs which is a direct consequence of the enthalpy (and, hence, the temperature) of the injected water, provided this is all evaporated. In general, the condition line AB, will lie somewhere in between the limiting lines, $AB_1(t_w = 0°\text{C})$ and $AB_3(t_w = 100°\text{C})$. The important thing to notice is that the angular displacement between these two condition lines is only about 7 degrees and that one of the intermediate lines is a line of constant wet-bulb temperature. It follows that for all practical purposes the change of state for a process of so-called adiabatic saturation may be assumed to follow a line of wet-bulb temperature. It is worth noting that although the process line on the psychrometric chart still lies within the 7° sector mentioned, the warmer the water the faster the evaporation rate, since temperature influences vapour pressure (see eqns. (4.1) and (4.2)).

3.7 Steam injection

As in water injection, steam injection may be dealt with by a consideration of a mass and energy balance. If m_s kg of dry saturated steam are injected into a moving airstream of mass flow 1 kg/s of dry air, then we may write

$$g_b = g_a + m_s$$

and

$$h_b = h_a + h_s$$

If the initial state of the moist airstream and the condition of the steam is known, then the final state of the air may be determined, provided none of the steam is condensed. The change of state takes place almost along a line of constant dry-bulb temperature between limits defined by the smallest and largest enthalpies of the injected steam, provided the steam is in a dry, saturated condition. If the steam is superheated then, of course, the dry-bulb temperature of the airstream may increase by any amount, depending on the degree of superheat. The two limiting cases for dry saturated steam are easily considered. The lowest possible enthalpy is for dry saturated steam at 100°C. It is not possible to use steam at a lower temperature than this since the steam must be at a higher pressure than atmospheric if it is to issue from the nozzles. (Note, however, that steam could be generated at a lower temperature from a bath of warm water, which was not actually boiling.)

The other extreme is provided by the steam which has maximum enthalpy; the value of this is 2803 kJ/kg of steam and it exists at a pressure of about 30 bars and a temperature of about 234°C.

What angular displacement occurs between these two limits, and how are they related to a line of constant dry-bulb temperature? These questions are answered by means of two numerical examples.

EXAMPLE 3.8. Dry saturated steam at 100°C is injected at a rate of 0·01 kg/s into a moist airstream moving at a rate of 1 kg of dry air per second and initially at a state of 28°C dry-bulb, 11·9°C wet-bulb (sling) and 1013·25 mbar barometric pressure. Calculate the leaving state of the moist airstream.

Answer

From psychrometric tables, $h_a = 33 \cdot 11$ kJ/kg dry air,

$$g_a = 0 \cdot 001937 \text{ kg/kg}$$

From N.E.L. steam tables, $h_s = 2675 \cdot 8$ kJ/kg steam.

$$g_b = 0 \cdot 001937 + 0 \cdot 01$$

$$= 0 \cdot 011937 \text{ kg/kg dry air, by the mass balance.}$$

$$h_b = 33 \cdot 11 + 0 \cdot 01 \times 2675 \cdot 8$$

$$= 59 \cdot 87 \text{ kJ/kg dry air.}$$

Hence,

$$59 \cdot 87 = (1 \cdot 007 t_b - 0 \cdot 026) + 0 \cdot 011937(2501 + 1 \cdot 84 t_b)$$

$$t_b = 29 \cdot 2 °\text{C}.$$

EXAMPLE 3.9. Dry saturated steam with maximum enthalpy is injected at a rate of 0·01 kg/s into a moist airstream moving at a rate of 1 kg of dry air per second and initially at a state of 28°C dry bulb, 11·9°C wet bulb (sling) and 1013·25 mbar barometric pressure. Calculate the leaving state of the moist airstream.

Answer

The psychrometric properties of state A are as for the last example. The moisture content at state B is also as in Example 3.8.

From N.E.L. steam tables, $h_s = 2803$ kJ/kg of steam, at 30 bars and 234°C saturated.

Consequently,

$$h_b = 33{\cdot}11 + 0{\cdot}01 \times 2803$$
$$= 61{\cdot}14 \text{ kJ/kg}$$

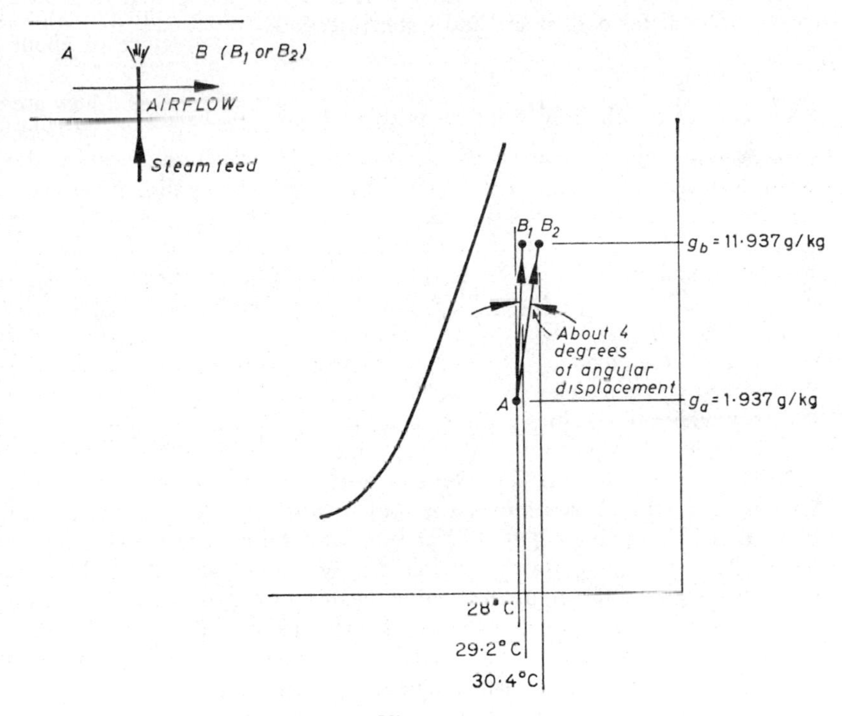

Fig. 3.9

Hence, as before:

$$61{\cdot}14 = (1{\cdot}007t_b - 0{\cdot}026) + 0{\cdot}011937(2501 + 1{\cdot}84t_b)$$
$$t_b = 30{\cdot}4°\text{C}.$$

Figure 3.9 illustrates what occurs. It can be seen that for the range of states considered, the change in dry-bulb value is not very great. In fact, the angular displacement between the two condition lines for the last two examples of steam injection, is only about 3 or 4 degrees. We can conclude that, although there is an increase in temperature, it is within the accuracy usually required in practical air conditioning to assume that the change of state following steam injection is up a line of constant dry-bulb temperature.

In the injection of superheated steam, every case should be treated on its merits, as the equation for the dry-bulb temperature resulting from a steam injection process shows:

$$t_b = \frac{h_b + 0{\cdot}026 - 2501g}{1{\cdot}007 + 1{\cdot}84g} \tag{3.9}$$

The final enthalpy is what counts, but it is worth noting that the above expression holds for both steam and water injection.

3.8 Cooling and dehumidification with re-heat

As was seen in section 3.4, when a cooler coil is used for dehumidification, the temperature of the moist air is reduced, but it is quite likely that under these circumstances this reduced temperature is too low. Although, as will be seen later (Chapter 6), we usually arrange that, under conditions of maximum loads, latent and sensible, the state of the air leaving the cooler coil is satisfactory in both these respects. This is not so for partial load operation. The reason for this is that latent and sensible loads are usually independent of each other. Consequently, it is necessary to arrange for the air that has been dehumidified and cooled by the cooler coil to be re-heated to a temperature consistent with the sensible cooling load; the smaller the sensible cooling load, the higher the temperature to which the air must be re-heated.

Figure 3.10(*a*) shows, in diagrammatic form, the sort of plant required. Moist air at a state *A* passes over the finned tubes of a cooler coil through which chilled water is flowing. The amount of dehumidification carried out is controlled by a dew-point thermostat, *C*1, positioned after the coil. This thermostat regulates the amount of chilled water flowing through the coil by means of the three-way mixing valve *R*1. Air leaves the coil at state *B*, with a moisture content suitable for the proper removal of the latent heat gains occurring in the room being conditioned. The moisture content has been reduced from g_a to g_b and the cooler coil has a mean surface temperature of t_c, Fig. 3.10(*b*) illustrating the psychrometric processes involved.

If the sensible gains then require a temperature of t_d, greater than t_b, the air is passed over the tubes of a heater battery, through which some heating medium such as low-pressure hot water may be flowing. The flow rate of this water is regulated by means of a two-port modulating valve, *R*2, controlled from a thermostat *C*2 positioned in the room actually being air-conditioned. The air is delivered to this room at a state *D*, with the correct temperature and moisture content.

The cooling load is proportional to the difference of enthalpy between h_a and h_b, and the load on the heater battery is proportional to h_d minus h_b. It follows from this that part of the cooling load is being wasted by the re-heat.

This is unavoidable in the simple system illustrated, and is a consequence of the need to dehumidify first, and heat afterwards. It should be observed, however, that in general it is undesirable for such a situation to exist during

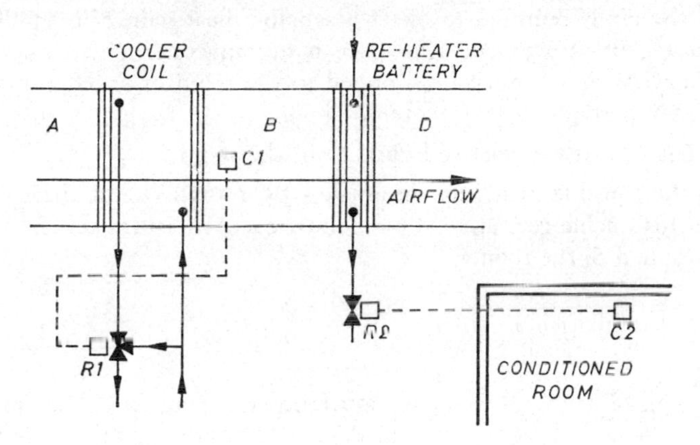

Fig. 3.10(*a*)

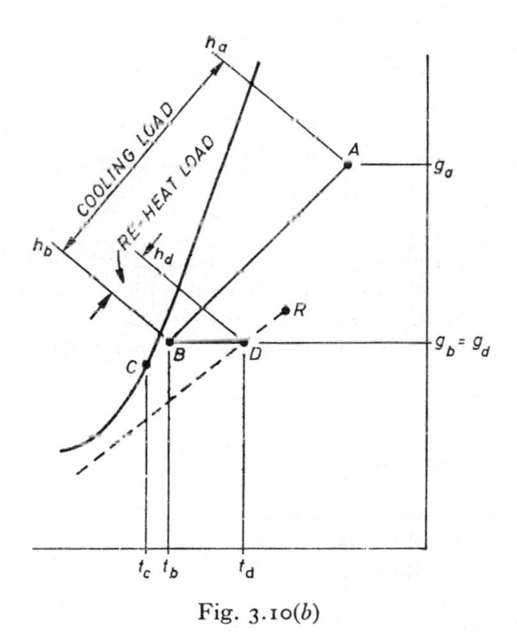

Fig. 3.10(*b*)

maximum load conditions. Re-heat is usually only permitted to waste cooling capacity under partial load conditions, that is, the design should be such that state *B* can adequately deal with both maximum sensible and maximum latent loads. These points are illustrated in the following example:

EXAMPLE 3.10. Moist air at 28°C dry-bulb, 20·6°C wet-bulb (sling) and 1013·25 mbar barometric pressure flows over a cooler coil and leaves it at a state of 10°C dry-bulb and 7·046 g/kg of dry air.

(*a*) If the air is required to offset a sensible heat gain of 2·35 kW and a latent heat gain of 0·31 kW in a space being air-conditioned, calculate the mass of dry air which must be supplied to the room in order to maintain a dry-bulb temperature of 21°C therein.

(*b*) What will be the relative humidity in the room?

(*c*) If the sensible heat gain diminishes by 1·175 kW but the latent heat gain remains unchanged, at what temperature and moisture content must the air be supplied to the room?

Answer

(*a*) If m_a kg/s of dry air are supplied at a temperature of 10°C, they must have a sensible cooling capacity equal to the sensible heat gain, if 21°C is to be maintained in the room.

Thus,

$$m_a \times c_a \times (21^\circ - 10^\circ) = 2{\cdot}35 \text{ kW}$$

then,

$$m_a = \frac{2{\cdot}35}{1{\cdot}012 \times 11^\circ} = 0{\cdot}211 \text{ kg/s}$$

(*b*) 0·211 kg/s of dry air with an associated moisture content of 7·046 g/kg of dry air must take up the moisture evaporated by the liberation of 0·31 kW of latent heat. Assuming a latent heat of evaporation of, say, 2454 kJ/kg of water (at about 21°), then the latent heat gain corresponds to the evaporation of

$$\frac{0{\cdot}31 \text{ kJ/s}}{2454 \text{ kJ/kg}} = 0{\cdot}0001262 \text{ kg/s of water.}$$

The moisture associated with the 0·211 kg/s of air delivered to the room each hour will increase by this amount. The moisture picked up by each kg of dry air supplied to the room will be

$$\frac{0{\cdot}1262}{0{\cdot}211} = 0{\cdot}599 \text{ g/kg}$$

Thus, the moisture content in the room will be equal to 7·046+0·599, that is, 7·645 g/kg. The relative humidity at this moisture content and 21°C dry bulb is found from tables of psychrometric data to be 49·3 per cent.

(*c*) If 0·211 kg/s of dry air is required to absorb only 1·175 kW of sensible heat then, if 21°C is still to be maintained, the air must be supplied at a higher temperature:

$$\text{temperature rise} \times 1{\cdot}012 \times 0{\cdot}211 = 1{\cdot}175 \text{ kW}$$

$$\text{temperature rise} = \frac{1{\cdot}175}{1{\cdot}012 \times 0{\cdot}211} = 5{\cdot}5\text{°C}$$

Thus, the temperature of the supply air is 15·5°C.

Since the latent heat gains are unaltered, the air supplied to the room must have the same ability to offset these gains as before; that is, the moisture content of the air supplied must still be 7·046 g/kg of dry air.

EXAMPLE 3.11. (*a*) For sensible and latent heat gains of 2·35 and 0·31 kW, respectively, calculate the load on the cooler coil in Example 3.10.

(*b*) Calculate the cooling load and the re-heater load for the case of 1·175 kW sensible heat gains and unchanged latent gains.

Answer

(*a*) The load on the cooler coil equals the product of the mass flow of moist air over the coil and the enthalpy drop suffered by the air. Thus, as an equation:

$$\text{cooling load} = m_a \times (h_a - h_b)$$

The notation adopted is the same as that used in Fig. 3.10. In terms of units, the equation can be written as

$$\text{kW} = \frac{\text{kg of dry air}}{\text{s}} \times \frac{\text{kJ}}{\text{kg of dry air}}$$

Using enthalpy values obtained from tables for the states A and B, the equation becomes

$$\begin{aligned}\text{cooling load} &= 0{\cdot}211 \times (59{\cdot}06 - 27{\cdot}81)\\ &= 0{\cdot}211 \times 31{\cdot}25\\ &= 6{\cdot}59 \text{ kW}\end{aligned}$$

(*b*) Since it is stipulated that the latent heat gains are unchanged, air must be supplied to the conditioned room at the same state of moisture content as before. That is, the cooler coil must exercise its full dehumidifying function, the state of the air leaving the coil being the same as in Example 3.11(*a*), above. To deal with the diminished sensible heat gains it is necessary to supply the air to the room at a temperature of 15·5°C, as was seen in Example 3.10(*c*). The air must, therefore, have its temperature raised by a re-heater battery from 10°C to 15·5°C. Since the moisture content is 7·046 g/kg at both these temperatures, we can find the corresponding enthalpies at states B and D directly from a psychrometric chart or by interpolation (in the case of state D) from psychrometric tables.

Hence, we can form an equation for the load on the re-heater battery:

$$\text{re-heater load} = 0{\cdot}211 \times (33{\cdot}41 - 27{\cdot}81)$$
$$= 0{\cdot}211 \times 5{\cdot}60$$
$$= 1{\cdot}18 \text{ kW}.$$

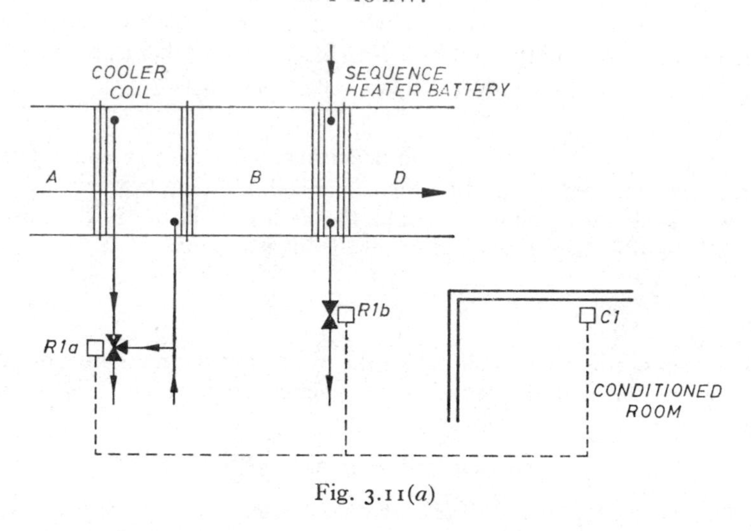

Fig. 3.11(a)

If the relative humidity in the room is unimportant there is no need to supply air to it at a controlled moisture content. This fact can be taken advantage of, and the running costs of the refrigeration plant which supplies chilled water to the cooler coil minimised. The method is to arrange for the room thermostat ($C1$ in Fig. 3.11(a)) to exercise its control in sequence over the cooler coil valve and the re-heater valve ($R1a$ and $R1b$). As the sensible heat gain in the conditioned space reduces to zero, the three-way mixing valve, $R1a$, gradually opens its by-pass port, reducing the flow of chilled water through the cooler coil and, hence, reducing its cooling capacity. If a sensible heat loss follows on the heels of the sensible heat gain, the need arises for the air to be supplied to the room at a temperature exceeding that maintained there, so that the loss may be offset. The plant shown in Fig. 3.11(a) deals with this by starting to open the throttling valve, $R1b$, on the heater battery.

The consequences of this are shown in Fig. 3.11(b), as far as a partial cooling load is concerned. Because its sensible cooling capacity has been diminished by reducing the flow of chilled water through its finned tubes, the cooler coil is able to produce an ‘ off ’ state B', at the higher temperature required by the lowered sensible heat gain, without having to waste any of its capacity in re-heat. There is a penalty to pay for this economy of control: because of the reduction in the flow rate of chilled water, the characteristics of the coil performance change somewhat and the mean coil surface temperature is higher ($t_{c'}$ is greater than t_c), there is less dehumidification and less sensible cooling, and, as a result, the moisture content of the air supplied to the room is greater. In consequence, for the same latent heat gain in the room a higher relative humidity will be maintained, the room state becoming R' instead of R.

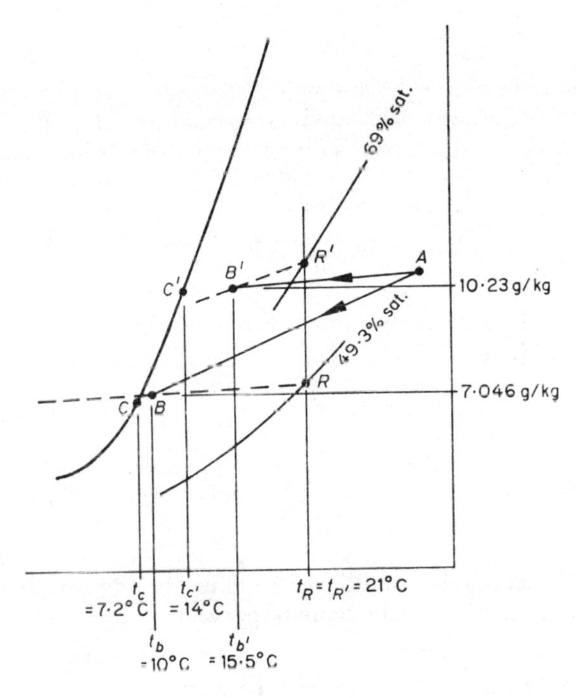

Fig. 3.11(*b*)

EXAMPLE 3.12. If the plant in Example 3.10 is arranged for sequence control over the cooler coil and heater battery, and if under the design conditions mentioned in that example the plant is able to maintain 21°C dry-bulb and 49·3 per cent relative humidity in the conditioned space, calculate the relative humidity that will be maintained there, under sequence control, if the sensible heat gains diminish to 1·175 kW, the latent gains remaining unaltered.

It is given that the mean coil surface temperature under the condition of partial load is 14°C.

Answer

As was mentioned in section 3.4, the mean coil surface temperature (identified by the state point C), lies on the straight line joining the cooler coil 'on' and 'off' states (points A and B) where it cuts the saturation curve. Figure 3.11(b) illustrates this, and also shows that the moisture content of state B' is greater than that of B. It is possible to calculate the moisture content of the supply air by assuming that the dry-bulb scale on the psychrometric chart is linear. By proportion, we can then assess a reasonably accurate value for the moisture content of the air leaving the cooler coil under the partial load condition stipulated in this example:

$$\frac{g_a - g_{b'}}{g_a - g_{c'}} = \frac{t_a - t_{b'}}{t_a - t_{c'}}$$

From tables, or from a psychrometric chart, $g_a = 12{\cdot}10$ g/kg and $g_c' = 10{\cdot}01$ g/kg. Also, it was calculated in Example 3.10(b) that the supply temperature must be 15·5°C dry-bulb for the partial sensible load condition. Hence we can write

$$\begin{aligned} g_{b'} &= g_a - (g_a - g_{c'}) \times \frac{(t_a - t_{b'})}{(t_a - t_{c'})} \\ &= 12{\cdot}10 - (12{\cdot}10 - 10{\cdot}01) \times \frac{(28 - 15{\cdot}5)}{(28 - 14)} \\ &= 12{\cdot}10 - 2{\cdot}09 \times \frac{12{\cdot}5}{14} \\ &= 12{\cdot}10 - 1{\cdot}87 \\ &= 10{\cdot}23 \text{ g/kg} \end{aligned}$$

Since the moisture pick-up has been previously calculated as 0·599 g/kg, the moisture content in the conditioned room will be 10·83 g/kg; at 21°C dry bulb the relative humidity will be about 69 per cent.

3.9 Pre-heat and humidification with re-heat

Air-conditioning plants which handle fresh air only are faced in winter with the task of increasing both the moisture content and the temperature of the air they supply to the conditioned space. Humidification is needed because the outside air in winter may have a very low moisture content, and if this air were to be introduced directly to the room there would be a correspondingly low moisture content there as well. The low moisture content may not be intrinsically objectionable, but when the air is heated to a higher temperature its relative humidity may become very low. For example, outside air in winter might be at −1°C, saturated (*see* Chapter 5). The moisture content at this state is only 3·484 g/kg of dry air. If this air is heated to 20°C dry-bulb, and if there is a moisture pick-up in the room of 0·6 g/kg of dry air, due to latent heat gains, then the relative humidity in the room will be as low as 27 per cent. This value may well be regarded as too low for proper comfort. The plant must also increase the temperature of the air, either to the value of the room temperature if there is background heating to offset fabric losses, or to a value in excess of this if it is intended that the air delivered shall deal with the fabric losses.

The methods by which the moisture content of air may be increased have been discussed earlier in this chapter. The ingenuity of the designer and the restrictions imposed on the design by the details of the application are the important factors influencing the method chosen. A popular and effective approach is to pre-heat the air, pass it through an air washer, where it undergoes adiabatic saturation, and then to re-heat it to the temperature at which it must be supplied to the room. Pre-heating and adiabatic saturation permit

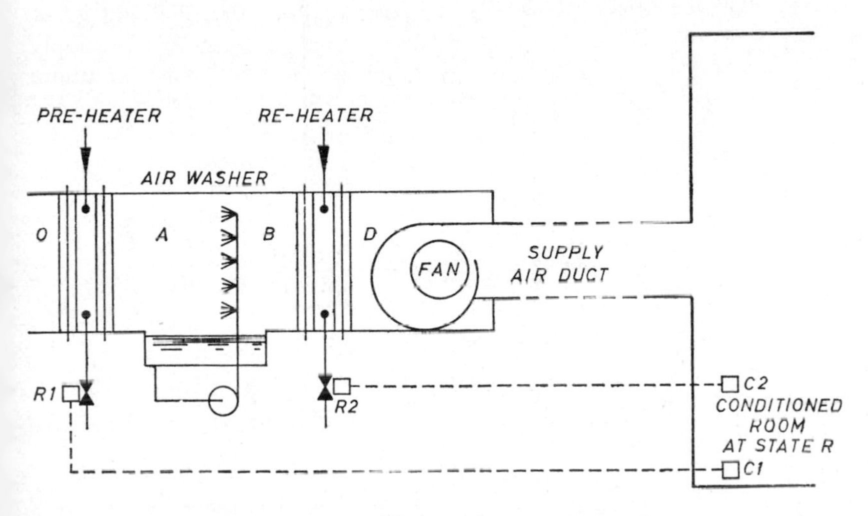

Fig. 3.12(*a*)

the relative humidity in the room to be controlled, and re-heating allows the temperature therein to be properly regulated, in winter.

Figure 3.12(*a*) shows, in diagrammatic form, a typical plant. Opening the modulating valve $R1$ in the return pipeline from the pre-heater increases the heating output of the battery and provides the necessary extra energy for the evaporation of more water in the washer, if an increase in the moisture content of the supply air is required. Similarly, opening the control valve $R2$, associated with the re-heater, allows air at a higher temperature to be delivered to the room being conditioned. $C1$ and $C2$ are room humidistat and room thermostat, respectively.

EXAMPLE 3.13. The plant shown in Fig. 3.12(*a*) operates as illustrated by the psychrometric changes in Fig. 3.12(*b*).

Air is pre-heated from −5·0°C dry-bulb and 86 per cent saturation to 23°C dry-bulb. It is then passed through an air washer having a humidifying efficiency of 85 per cent and using recirculated spray water. Calculate the following:

(*a*) The relative humidity of the air leaving the washer.

(*b*) The cold water make-up to the washer in litres/s, given that 2·5 m^3/s leaves the air washer.

(*c*) The duty of the pre-heated battery in kW.

(*d*) The temperature of the air supplied to the conditioned space if the sensible heat losses from it are 24 kW and 20°C dry-bulb is maintained there.

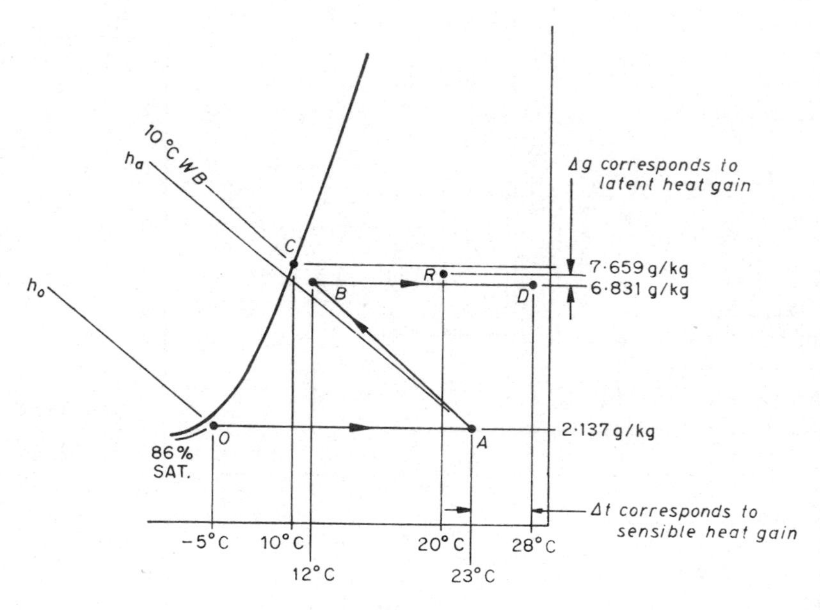

Fig. 3.12(*b*)

(*e*) The duty of the re-heater battery.

(*f*) The relative humidity maintained in the room if the latent heat gains therein are 5 kW.

Answer

(*a*) At state O the moisture content is found from tables to be 2·137 g/kg Consequently, at 23°C dry-bulb and 2·137 g/kg the wet-bulb (sling) value is found from tables to be 10°C. This state is represented by the point *A* in the diagram. Assuming that the process of adiabatic saturation occurs up a wet-bulb line, we can establish from tables that the moisture content at state *C*, the apparatus dew point, is 7·659 g/kg. If the further assumption is made that the dry-bulb scale is linear, we can evaluate the dry-bulb temperature at state *B*, leaving the washer, by proportion:

$$\begin{aligned} t_b &= 23 - 0.85(23 - 10) \\ &= 23 - 11.05 \\ &= 12°C \text{ dry-bulb.} \end{aligned}$$

We know that the dry-bulb temperature at the point *C* is 10°C because the air is saturated at *C* and it has a wet bulb temperature of 10°C.

The above answer for the dry-bulb at *B* is approximate (because the dry-bulb scale is not linear) but accurate enough for all practical purposes. On the other hand, we may calculate the moisture content at *B* with more exactness, since the definition of humidifying efficiency is in terms of moisture content changes and this scale is linear on the psychrometric chart.

$$\begin{aligned} g_b &= 7.659 - (1 - 0.85) \times (7.659 - 2.137) \\ &= 7.659 - 0.828 \\ &= 6.831 \text{ g/kg.} \end{aligned}$$

We can now refer to tables or to a chart and determine that the relative humidity at 12°C dry-bulb and 6.831 g/kg is 78 per cent. More accurately, we might perhaps determine the humidity at a state of 10°C wet-bulb (sling) and 6.831 g/kg. The yield in accuracy is of doubtful value and the method is rather tedious when using tables.

(*b*) The cold water make-up depends on the mass flow of dry air. It may be determined that the humid volume of the air leaving the washer at state *B* is 0.8162 m³/kg of dry air. The cold water fed from the mains to the washer serves to make good the losses due to evaporation within the washer. Since the evaporation rate is (6.183 − 2.137) g/kg, we may calculate the make-up rate:

$$\begin{aligned} \text{make-up rate} &= \frac{(6.183 - 2.137) \times 2.5}{1000 \times 0.8162} \\ &= 0.0124 \text{ kg/s or litres/s.} \end{aligned}$$

(*c*) The pre-heater must increase the enthalpy of the air passing over it from h_o to h_a. Reference to tables establishes that $h_o = 0.299$ kJ/kg and $h_a = 28.57$ kJ/kg. Then,

$$\begin{aligned} \text{pre-heater duty} &= \frac{2.5 \times (28.57 - 0.299)}{0.8162} \\ &= 3.063 \text{ kg/s} \times 28.27 \text{ kJ/s} \\ &= 86.59 \text{ kW} \end{aligned}$$

(*d*) The 3·063 kg/s of air delivered to the room duffuses throughout the space, and in the process its temperature falls from the value t_d, which is to be calculated, to the value being maintained, 20°C dry-bulb.

A heat balance equation can now be written—

sensible heat loss = weight of dry air plus associated moisture × specific heat of humid air × (supply temperature minus room temperature).

$$24 = 3{\cdot}063\ [1 \times 1{\cdot}012 + 0{\cdot}006831 \times 1{\cdot}89] \times (t_d - 20)$$

$$= 3{\cdot}063 \times 1{\cdot}025 \times (t_d - 20)$$

$$t_d = 20 + \frac{24}{3{\cdot}063 \times 1{\cdot}025}$$

$$= 28°\text{C}$$

(*e*) The re-heater battery thus has to heat the humid air from the state at which it leaves the washer to the state at which it enters the room, namely from 12°C dry-bulb and 6·831 g/kg to 28° dry-bulb and 6·831 g/kg.

$$\text{Re-heater duty} = 3{\cdot}063 \times 1{\cdot}025 \times (28 - 12)$$

$$= 50{\cdot}2\ \text{kW}$$

Alternatively, we can determine the enthalpies of the states on and off the heater, by interpolating from tables or reading directly from a psychrometric chart:

$$\text{re-heater duty} = 3{\cdot}063 \times (45{\cdot}6 - 29{\cdot}3)$$

$$= 49{\cdot}9\ \text{kW}$$

(*f*) A mass balance must be struck to determine the rise in moisture content in the room as a consequence of the evaporation corresponding to the liberation of the latent heat gains:

latent heat gain in kW = (kg of dry air per hour delivered to the room) × (the moisture pick-up in kg of water per kg of dry air) × (the latent heat of evaporation of water in kJ/kg of water).

$$5{\cdot}0 = 3{\cdot}063 \times (g_r - 0{\cdot}006831) \times 2454$$

$$g_r = 0{\cdot}006831 + \frac{5{\cdot}0}{3{\cdot}063 \times 2454}$$

$$= 0{\cdot}006831 + 0{\cdot}000666$$

$$= 0{\cdot}007497\ \text{kg/kg.}$$

From tables or from a chart it may be found that at a state of 20°C dry-bulb and 7·497 g/kg, the relative humidity is about 51 per cent, and h_r = 39·14 kJ/kg.

3.10 Mixing and adiabatic saturation with re-heat

Figure 3.13(*a*) shows a very common type of plant, in schematic form: air at state *R* is extracted from a conditioned room and partly recirculated, the remainder being discharged to atmosphere. The portion of the extracted air

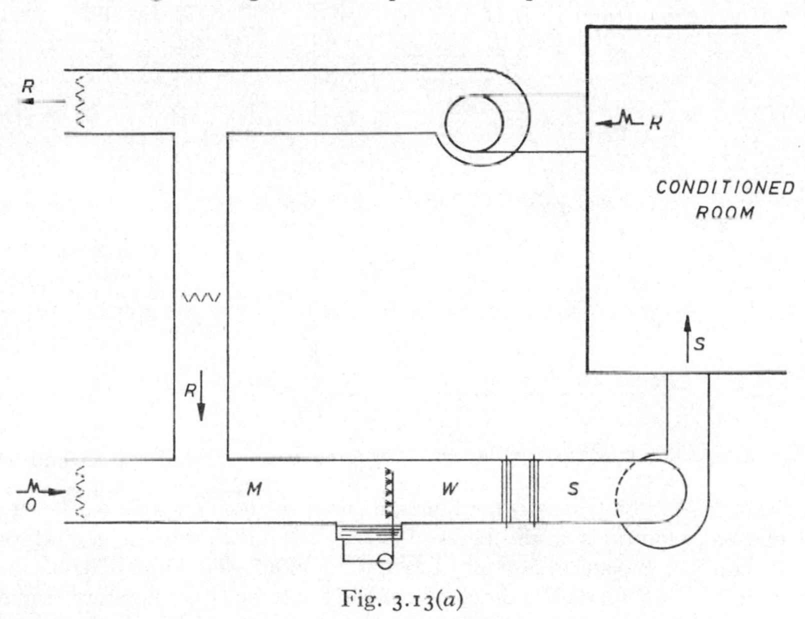

Fig. 3.13(*a*)

returned to the air-conditioning plant mixes with air at state O, drawn from outside, and forms a mixture state *M*. The air then passes through an air washer, only the spray water of which is recirculated, and adiabatic saturation occurs, the state of the air changing from *M* to *W* (*see* Fig. 3.13(*b*)) along a line of constant wet-bulb temperature (*see* §§ 3.5 and 3.6). An extension of the line *MW* cuts the saturation curve at a point *A*, the apparatus dew point. To deal with a particular latent heat gain in the conditioned room it is necessary to supply the air to the room at a moisture content g_s, it being arranged that the difference of moisture content $g_r - g_s$, in conjunction with the mass of air delivered to the room, will offset the latent gain. In other words, the air supplied must be dry enough to absorb the moisture liberated in the room.

It is evident that the moisture content of the air leaving the washer must have a value g_w, equal to the required value, g_s. This is amenable to calculation by making use of the definition of the effectiveness of an air washer, in terms of g_a, g_w and g_m (*see* § 3.5).

EXAMPLE 3.14. If the room mentioned in Example 3.13 is conditioned by means of a plant using a mixture of recirculated and fresh air, of the type illustrated in Fig. 3.13(*a*), calculate:

(*a*) the percentage of the air supplied to the room, by mass which is recirculated, and

(*b*) the humidifying efficiency of the air washer.

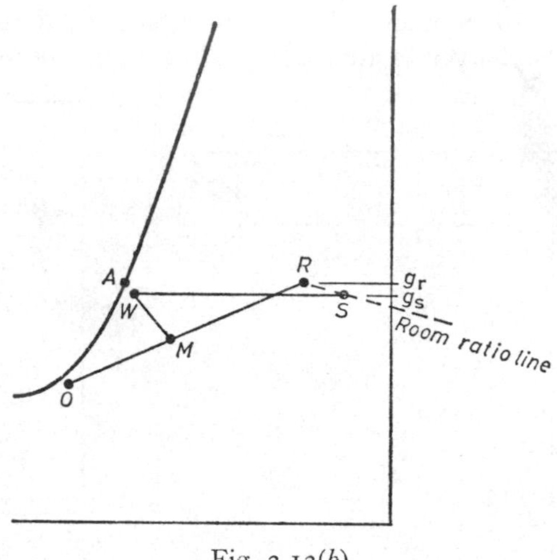

Fig. 3.13(*b*)

Answer

(*a*) Since the wet-bulb scale is not linear, it is not accurate enough to calculate the mixing proportions on this basis. Instead, one must make use of changes of enthalpy or moisture content. Bearing in mind that lines of constant enthalpy are not parallel to lines of constant wet-bulb temperature, some slight inaccuracy is still present if the assumption is made that the change of state accompanying a process of adiabatic saturation is along a line of constant enthalpy. However, this is unavoidable, and so such an assumption is made.

Referring to Fig. 3.13(*b*) it can be seen that

$$h_a \simeq h_w \simeq h_m.$$

From tables it is established that h_w (at 12°C dry-bulb and 6·831 g/kg) is 29·30 kJ/kg.

From the principles set out in section 3.2, governing the change of state associated with a mixing process, it is clear that the percentage of recirculated air, by mass,

$$= \frac{h_m - h_o}{h_r - h_o} \times 100$$

$$= \frac{29{\cdot}30 - 0{\cdot}299}{39{\cdot}14 - 0{\cdot}299} \times 100$$

$$= 75 \text{ per cent.}$$

Thus, 75 per cent of the air supplied to the room, if recirculated and mixed with 25 per cent of air from outside, will have an enthalpy of 29·30 kJ/kg and a wet bulb of 10°C (sling). If adiabatic saturation is then to produce a state of 12°C dry-bulb and 6·831 g/kg, the humidifying efficiency of the washer used can no longer be the value used for Example 3.13, namely, 85 per cent. An entirely different washer must be used for the above calculations to be valid and this must have an effectiveness which may be calculated as follows:

(*b*) Since efficiency is expressed in terms of moisture content, it is necessary to determine the value of g_m, the values of g_a and g_w being already known.

$$g_m = 0{\cdot}75\, g_r + 0{\cdot}25\, g_o$$

$$= 0{\cdot}75 \times 7{\cdot}497 + 0{\cdot}25 \times 2{\cdot}137$$

$$= 6{\cdot}159 \text{ g/kg}$$

$$\text{Humidifying efficiency} = \frac{g_w - g_m}{g_a - g_m} \times 100$$

$$= \frac{6{\cdot}831 - 6{\cdot}159}{7{\cdot}659 - 6{\cdot}159}$$

$$= 45 \text{ per cent.}$$

If the washer used in this example had an efficiency of 85 per cent, as in Example 3.13, then the calculations would not have been so easy. The line *AWM* would have had to have been at a lower wet-bulb value in order to fulfil two requirements:

(i) $$g_w = 6{\cdot}831 \text{ g/kg.}$$

(ii) $$\frac{g_w - g_m}{g_a - g_m} \times 100 = 85 \text{ per cent.}$$

For this to be the case, the dry-bulb temperature of *W* must obviously be less than 10°C. The easiest way to achieve a practical solution is by drawing a succession of lines representing processes of adiabatic saturation on a psychrometric chart and calculating several values of efficiency until one of acceptable accuracy is obtained.

BIBLIOGRAPHY

1. W. Goodman. *Air Conditioning Analysis with Psychrometric Charts and Tables.* The Macmillan Company, New York, 1943.

4

Comfort and Inside Design Conditions

4.1 The metabolic rate

Like any other machine, the human body works best at certain temperatures. Unlike many machines, however, which can tolerate a wide range of variations in their environmental temperatures, the human machine can tolerate only a comparatively narrow band of changes.

The body differs further from the insentient machine by having an appreciation of comfort. That is to say, although humans have been subjected to environmental air temperatures ranging from less than −50°C to more than 115°C and have survived, it is only within a small range of about 8°C (from, say, 18°C to 26°C) that most people can feel comfortable.

The term 'comfort' is not easily defined: it is ultimately a subjective quality, related to the ease with which an individual maintains a thermal balance between himself and his environment, and it is complicated by personal prejudice, sex and age, to such an extent that no formula for its expression can readily be devised. In the long run, comfort is only assessed by posing the question to the individual: 'Are you comfortable?'

Since the thermal balance, upon which a feeling of comfort depends, is related to the manner in which the body loses heat to its environment, a study of the modes whereby the body loses heat, and of the features in the environment that affect such losses, is relevant.

The food eaten by a human being is turned into energy by a process of oxidation effected by the air breathed into the lungs. The energy available to the body in this way is put to use in doing work and in keeping the individual in a state of good health. Since the body has an efficiency of the order of 20 per cent, it must reject to the environment a portion of its energy intake. The rate and the manner of rejection of this heat is under the control of a bodily system of automatic regulation, referred to later.

The rate at which the body produces heat is termed the *metabolic rate*, the value of which varies greatly between a lower limit of 60 W, the basal metabolic rate, equal to the heat production from a normal healthy person asleep, and a value of perhaps ten times this for a person carrying out sustained very hard work. Experimental measurements have established both the total output of heat from the body, for different rates of working, and their sensible and latent components.

The temperature of the body itself remains comparatively constant at a value of about 36·9°C for the tissues near the surface (mouth temperature) and about 37·2°C for the deep tissues (rectal temperature). Some departures from these values occur. It has been found that the body temperature when resting in bed, in the morning, is some 0·5°C less than the temperature in the afternoon. The temperature also varies a little with age and considerably with health. The excessive generation of heat within the body, due to illness, or the prevention of bodily heat loss, either artificially or because of a malfunctioning of the bodily thermostatic control system, results in a rise of the temperature of the body. Increases of a few degrees are serious and the rise of temperature acquires considerable gravity when 40·5°C is reached, at which value, for some apparently unknown reason, sweating stops and further rise of temperature is accelerated. A value of 43·5°C is usually fatal. There is some opinion that, at higher temperatures than normal, the metabolic rate is increased because of the stimulation which high temperatures afford to bodily chemical reactions.

At the other end of the scale, individuals subjected to exposure in extremely cold weather conditions, have had their body temperatures reduced to 27°C or below, and have recovered. A prolonged stay at such a temperature is usually fatal, although surgical techniques have been developed which take advantage of a temporary, artificially induced body temperature as low as 10°C for short periods of time. To combat a fall in temperature, the body arranges to generate more heat by muscular tension and, in a more extreme case, by shivering.

To sum up: the rate at which heat is liberated from the body depends on the rate of working; to maintain itself at a temperature consistent with health and comfort, the body must lose this heat to its environment.

4.2 Bodily mechanisms of heat transfer

There are four modes of heat transfer: evaporation, E, radiation, R, convection, C, and conduction. Of these, conduction can be neglected entirely; the amount of body in contact with a surface is too small and the period of contact usually too short for the mode to be worth considering. There is some small additional loss through the rejection of excreta from the body, but this also is usually ignored.

Evaporation losses from the body occur in three ways: by the exhalation of saturated water vapour from the lungs, by the continual (in suitable circumstances of environmental humidity) process known as insensible perspiration, and by the emergency mechanism termed sweating. Insensible perspiration is a consequence of body fluids oozing through the membrane of the skin under the influence of osmotic pressures and forming microscopic droplets of moisture on its surface. These droplets, because of their small size, evaporate very rapidly (in suitable environmental circumstances) and thus cannot be observed or felt. Hence the use of the adjective insensible.

An entirely different process is that of sweating. When subjected to a comfortable environment, an individual should not sweat. If there is a tendency for body temperature to rise, the automatic in-built regulating system attempts to increase the rate of heat loss from the body by evaporation if the losses by radiation and convection are proving inadequate. This increase in evaporation loss is achieved by bringing the sweat glands into operation and flooding strategic areas of the skin with sweat. In the extreme case, almost the entire surface of the body is covered by fluids from the sweat glands. Sweating is clearly inconsistent with a feeling of comfort.

The loss of heat by evaporation from a wetted surface is a function of the difference between the vapour pressure of the water (at its particular temperature) on the surface and the partial pressure of the water vapour in the ambient air. It also depends on the relative velocity of airflow over the surface and on the nature of the flow; that is, whether it is parallel, as over the surface of a lake, or transverse, as over the wick of a wet-bulb thermometer.

The following pair of empirical equations provide a means of obtaining a solution to numerical problems:

For parallel flow

$$\text{Evaporation rate} = (0{\cdot}0885 + 0{\cdot}0779\, v)(p_w - p_s)\ \text{W/m}^2 \qquad (4{\cdot}1)$$

For transverse flow

$$\text{Evaporation rate} = (0{\cdot}01873 + 0{\cdot}1614\, v)(p_w - p_s)\ \text{W/m}^2 \qquad (4{\cdot}2)$$

where v = relative velocity of airflow in m/s,

p_w = vapour pressure in N/m^2, corresponding to the temperature of the water,

p_s = partial pressure of the water vapour in the ambient air, in N/m^2.

The body will lose heat by radiation to its environment if the mean surface temperature of the body is higher than that of the surrounding surfaces. The mean value of body surface temperature is influenced by the type of clothing worn and by the extent to which it covers the surface of the individual. It has been found, by experimental measurement, that the surface temperature of the forehead and of the cheeks of the face, varies between 31·5°C and 33·5°C, that the back of the hand has a temperature between 25·5°C and 32·5°C and, also, that clothing surface temperatures vary from 24°C to 29·5°C. Generalisation on these values is risky since the type of clothing and the environmental conditions influence the surface temperatures, but the above figures are given as a guide only and refer to ambient temperatures of about 21°C. Nevertheless, attempts have been made to generalise, and opinion has it that mean body surface temperatures are about 24°C in the U.K. and about 27°C in the U.S., for normally clothed people. The differences in values may be ascribed to the different standards of clothing adopted in the two countries (indoor clothing in America in winter is lighter) and to the fact that a somewhat lower indoor temperature, about 21°C, is regarded as comfortable in this country, compared with that demanded in America, namely about 24°C. All the figures quoted refer to indoor conditions in winter. (In summer, the temperatures of the environment alter.) The thermostatic control mechanism built into the body is able to vary the skin temperature and so increase or decrease the rate at which the body is able to lose heat by radiation (and also by convection) to its environment. When the skin is subjected to a low ambient air temperature, the blood vessels in the surface tissues constrict and the supply of blood to the skin is reduced. This reduces the conductance of the skin and lowers the surface temperature. Conversely, a high ambient air temperature causes the surface blood vessels to dilate and the mean body surface temperature to rise.

Whether or not the body loses heat by radiation depends also—as was mentioned a little earlier on the mean temperature of the surfaces of the room in which the individual is placed. This average surface temperature is termed the *mean radiant temperature* and is defined as: the surface temperature of that sphere which, if it surrounded the point in question, would radiate to it the same amount of heat as the room surfaces which surround the point actually do. From this, it is clear that mean radiant temperature varies from place to place inside a given room. What is perhaps not quite so clear is that a failure to state a diameter for the sphere does not invalidate the definition. If a large sphere is chosen, its radiating surface is increased proportionally to the square of the increase in its diameter. At the same time, because of the inverse square law which governs radiation exchanges, the radiant energy received at the centre of the sphere diminishes in a like manner. The body will lose heat by convection provided that its mean surface temperature exceeds the ambient dry-bulb temperature, and heat exchange is increased by a rise in air velocity.

4·3 Environmental influences on comfort

From the foregoing, it can be inferred that the body maintains its thermal equilibrium with the environment by means of heat exchanges, involving—

(*a*) evaporation (about 25 per cent),
(*b*) radiation (about 45 per cent),
(*c*) convection (about 30 per cent).

It can also be deduced that the environmental properties which influence these modes of heat transfer are—

(i) dry-bulb temperature (affecting evaporation and convection),
(ii) relative humidity (affecting evaporation only),
(iii) air velocity (affecting evaporation and convection),
(iv) mean radiant temperature (affecting radiation).

Items (i) and (ii) are physical variables which are directly under the control of an air-conditioning system, and (iii) is a consequence of the method of air distribution adopted. It is a little unusual for item (iv) to be manipulated in an air-conditioning system. It is true that many systems have been designed and successfully installed which make use of a chilled ceiling, in conjunction with a supply of treated air, but this does not permit the mean radiant temperature to be directly controlled. The temperature of the chilled ceiling is commonly regulated from room dry-bulb temperature.

The way in which the individual's body maintains itself in comfortable equilibrium will be by its automatic use of one or more of the three modes of heat transfer. The percentage proportions quoted earlier are for seated, clothed subjects in an environment of about 18·5°C dry bulb and still air. The humidity would be in the vicinity of 50 per cent. It is possible to write an equation for the heat balance of the body:

$$M-W = E+R+C+S \tag{4.3}$$

where M = the metabolic rate,

W = the useful rate of working,

$M-W$ = the net surplus heat which must be liberated or stored; it is the quantity commonly quoted for the heat given off by individuals at different rates of working.

E = the heat lost by evaporation,

R = the heat lost or gained by radiation; if the mean radiant temperature exceeds the mean body surface temperature, then the body will undergo a radiant heat gain.

C = the heat lost or gained by convection; if the ambient dry bulb is greater than the mean body surface temperature, then a heat gain by convection will occur. It follows that, whereas increasing air velocity will increase convective losses if the ambient dry bulb is less than the body surface temperature, it will have the opposite effect if the ambient dry bulb is higher.

S = the rate at which heat is stored within the body. Under steady-state conditions, the body remains comfortable and healthy because S is zero. In the presence of an oppressively hot environment, the load imposed upon E, R and C may be so great that S is positive and the body temperature will rise, eventually resulting in heat-stroke.

Equation (4.3) shows that the body maintains itself in thermal equilibrium with its environment by the interplay of a number of component factors, E, R and C, related to features of the environment. It follows then, that it is not possible to specify a comfortable environment in terms of a single physical variable, such as dry-bulb temperature but, instead, account must be taken of all the physical variables affecting the three modes of heat transfer: dry-bulb temperature, relative humidity, air velocity and mean radiant temperature.

4.4 Synthetic comfort scales

Several attempts have been made, with considerable success, to correlate the factors in the environment influencing the thermal equilibrium of the body. The decision as to what combinations of factors produce a feeling of comfort has been based, in each case, on the statistical evidence accruing from experiment on large numbers of people. The four scales of comfort which have been established are:

(*a*) equivalent temperature,
(*b*) effective temperature,
(*c*) corrected effective temperature.
(*d*) dry or wet resultant temperature.

The first of these is a British concept arising from experimental work carried out using a sort of synthetic human being, termed a *eupatheoscope*. In essence, this was merely a blackened cylinder, 55 cm high and 20 cm in diameter. The cylinder was electrically heated internally and the power output of the heater was related to the manner in which the cylinder lost heat to its environment by radiation and convection. A thermostatic control was arranged to maintain a surface temperature on the cylinder of about 24°C in an environment of still air at 18·5°C. The cylinder's surface temperature thus corresponded to the mean surface temperature of a clothed human being in similar surroundings. From this, the scale of equivalent temperature (which takes into account mean radiant temperature, dry-bulb temperature and air velocity, but not relative humidity) was devised.

The scale has not proved very popular outside the U.K. and its place in modern air conditioning is taken by the American concept of effective temperature that has been in wide use for many years. In its original form, it specified the temperature of still, suturated air which gave a feeling of comfort equal to a particular combination of air movement, dry- and wet-bulb temperatures. As such, it overemphasised the effect of humidity at low temperatures and underestimated it at high values. A newer American concept of effective temperature refers to air at 50 per cent relative humidity and gives much better results, being associated with more commonly encountered environments. Independent work by Fanger has produced comfort charts which give good agreement with the newer effective temperature for lightly-clad, sedentary workers and an air velocity of 0·1 m/s. His results suggest that, regardless of age, sex or race, there is a unique, comfortable dry-bulb temperature in any given combination of environmental parameters, for a particular activity and mode of dress. Corrected effective temperature takes account of radiation by substituting the globe thermometer reading for that of the dry-bulb, thus taking notice of all four environmental factors. The dry resultant temperature is the temperature measured by a thermometer at the centre of a blackened sphere 100 mm in diameter (cf globe thermometer) and for an air velocity of 0·1 m/s it is $0{\cdot}5\ (t_{rm}+t)$. It is more widely used at present in Europe than are equivalent or globe temperatures.

It is clear from the above that, unfortunately, there is as yet no completely satisfactory index of comfort.

4.5 Instruments

In order to assess the degree of comfort in an air-conditioned space, the following instruments are necessary:
Mercury-in-glass thermometer,
Sling psychrometer,
Kata thermometer,
Stop-watch,
Globe thermometer.

Dry-bulb temperature is measured by the mercury-in-glass thermometer, *properly shielded from radiation*, wiped dry beforehand, and given sufficient time to settle down to give a steady-state reading. Some idea of the accuracy of the thermometer should be held and the scale of the instrument should be appropriate to the values likely to be read.

The sling psychrometer should have its reservoir filled with clean fresh water, preferably at about the same temperature as the wet-bulb value expected in the room. The wick on the bulb of the wet-bulb thermometer should be clean, and the stems of both thermometers should be wiped free of moisture. When the wick has become properly wet, the sling should be whirled vigorously for ten seconds or so, in the place where the readings are required. Whirling should be vigorous so that the air-flow over the two instruments is about 5·0 m/s or more. This minimises radiation errors, as was explained in section 2.17 and gives stable readings for both wet- and dry-bulb temperatures. As soon as whirling is stopped, the wet-bulb reading should be taken, as this will start to rise immediately. This procedure should be repeated several times until three sets of values, obtained in sequence, show close agreement.

The Kata thermometer was originally devised for use as an indicator of warmth and comfort. It is used today, however, as an anemometer for measuring low air velocities such as are encountered in assessing comfort conditions from, say, 0·05 up to 0·5 m/s, although it can be used for velocities up to 6 m/s. Several varieties are in use but the one currently most popular is an alcohol-in-glass thermometer, having a large (1·8 cm diameter ×4·0 cm long) silvered bulb (to minimise radiation errors) and having two marks on its stem, at 55°C and 52°C. The time the level of the alcohol in the stem takes to fall from the upper to the lower of these values is an indication of the rate at which the bulb of the thermometer is cooling. Since cooling is achieved almost entirely by convection, it is also an indication of the air velocity over the bulb. It is essential that the thermometer should be still and that it is free from moisture when measurements are being made.

The procedure for use is as follows:

1. Suspend an ordinary mercury-in-glass thermometer near the place where the Kata reading is to be taken.

2. Place the Kata thermometer in a thermos flask of hot water (of temperature substantially greater than 55°C) and keep it there until the column of alcohol has risen past the upper marking on the stem and is half way into the small reservoir at the top of the stem.
3. Remove the thermometer, carefully dry it and suspend it in the position where the measurement of air velocity is required. Make sure that the instrument is still.
4. Using a stop-watch, measure the time that elapses as the falling column of alcohol passes between the two markings on the stem.
5. Take note of the dry-bulb temperature.
6. Repeat the procedure several times until consistent readings of cooling time are obtained for three successive measurements, then take the average of these three readings.
7. Calculate the velocity from eqn. (4.4) or use a nomogram.

The equation used is:

$$V = \left[\frac{\dfrac{H}{(T-t)} - a}{b}\right]^2 \tag{4.4}$$

where t is the dry bulb temperature in °C, H is the 'cooling power' of the Kata thermometer and equals the Kata factor divided by the average cooling time in seconds. (The Kata factor is engraved on the stem of the instrument and is sometimes a three-figure number preceded by the letter F.)

T, a and b are constants, the values of which depend on the type of instrument and the range of velocity.

A Kata thermometer, with a cooling range from 55°C to 52°C is best used in air dry-bulb temperatures of less than 27°C. For use in higher air temperatures, an instrument having a cooling range from 66°C to 63°C is recommended, otherwise the response of the instrument is sluggish.

The Globe thermometer consists of an ordinary mercury-in-glass thermometer mounted with its bulb at the centre of a 150 mm diameter hollow copper sphere. The surface of the sphere is finished matt black. The Globe thermometer is influenced by dry-bulb temperature, air velocity and mean radiant temperature. In still air, it reads the mean radiant temperature exactly. In using the instrument, simultaneous readings of a Kata thermometer and a dry-bulb thermometer must be taken. Sufficient time—about 20 minutes—must also be allowed to pass for the Globe thermometer to settle down and give a steady-state reading. Either a nomogram or the following equation is used to determine the mean radiant temperature:

$$T_{rm}^{\ 4} \times 10^{-9} = T_g^{\ 4} \times 10^{-9} + 1{\cdot}442 \times (T_g - T_a)\sqrt{v} \tag{4.5}$$

where T_{rm} = the mean radiant temperature in K,
T_g = the reading of the Globe thermometer in K,
v = the air velocity in m/s,
T_a = the air dry-bulb temperature in K.

4.6 The choice of inside design conditions

It has been established by practical experiment that for a person to feel comfortable, the following conditions should be fulfilled.

1. The temperature of the air should have a lower value, rather than a higher one, if a choice is possible.

2. The average desirable air velocity in the room should be at a value which depends on the dry-bulb temperature and on whether winter heating or summer cooling is taking place. Table 4.1 provides a guide:

TABLE 4.1

	Winter heating			Summer cooling		
Air temperature in °C	19°	20°	21°	22°	23°	24°
Air velocity in the room in m/s	0·10–0·13	0·10–0·15	0·15–0·20	0·20–0·24	0·25–0·30	0·30–0·35

(Reproduced by kind permission of Haden Young Ltd.)

The part of the body subjected to the airflow influences the feeling of comfort; an airstream at a certain temperature and velocity might be acceptable on the face but it could well be objectionable on the back of the neck.

3. The relative humidity should preferably lie between 30 and 60 per cent and should certainly not exceed 70 per cent. There is some evidence to suggest that smells are more noticeable in humidities exceeding 60 per cent.

4. The mean radiant temperature should be higher rather than lower; cold room surfaces in a warm room cause complaints of stuffiness in winter time (but not in summer, when low mean radiant temperatures can be pleasant in a high dry-bulb temperature).

5. The temperature difference between the air at head level and at foot level should be as small as possible. The difference should preferably not exceed 1.5°C and should never be greater than 3°C.

6. Thermal radiation is best not directed on to the tops of heads.

7. The concentration of carbon dioxide in a conditioned space should not exceed 0·1 per cent. Beyond this figure there is no immediate ill effect but at 2 per cent the depth of respiration is 30 per cent greater than normal and the frequency of respiration is about 15 per cent more. Although the oxygen content can be reduced to 13 per cent without ill effect, nausea and other bad effects accumulate, resulting in death, if the carbon dioxide content rises to about 8 per cent or more.

The choice of inside design conditions, to be maintained for comfort purposes, by an air-conditioning system, depends on the physiological considerations already discussed and on economic considerations. Bearing his in mind, the designer bases his choice of inside conditions on the following:

(i) The outside design conditions.

(ii) The period of occupancy in the conditioned space.

(iii) The rate of working of the occupants in the space.

Using summer cooling as the important basis, an individual entering a conditioned room from outside senses, at once, colder conditions. Acclimatisation occurs very rapidly and, quite soon, comfort is experienced. This feeling persists for a while and will remain indefinitely, if the effective temperature in the room is consistent with comfort for the individual's rate of working. People living in a hot climate become acclimatised to it in a matter of weeks and, when entering a conditioned space, will accept, as comfortable, a higher effective temperature than people living in a more temperate climate. Consequently, whereas in the U.K., an inside dry-bulb of 22·5°C with 50 per cent relative humidity (giving an effective temperature of about 20°C) is comfortable for most people, in a tropical climate where outside design conditions are, say, 35°C dry-bulb and 26·5°C wet-bulb, an inside condition of 25·5°C dry-bulb with 50 per cent relative humidity (an effective temperature of 22°C) is acceptable.

It is usual, then, to choose as an inside design dry-bulb, a value of from 4°C to 11°C, less than the outside design dry-bulb, and to adopt a nominal relative humidity of about 50 per cent for comfort-conditioning in spaces where the occupants are engaged in sedentary work.

Human beings notice changes of temperature more readily, under normal conditions, than they do changes of humidity. Thus, changes of ± 1°C about 21°C may pass unnoticed, but changes of $+20$ per cent and -20 per cent or more, could well pass unnoticed at 50 per cent relative humidity with the same temperature. It follows that a much greater amount of deviation from the nominal value can be tolerated for humidity than for temperature in comfort applications. This fact has its impact in the choice of automatic controls and in the tolerances acceptable for them.

There used to be much talk of the 'shock' effect experienced when a person walked from a hot outside environment into a room conditioned at a relatively low temperature. In fact, unless the person's clothes and body are very wet from perspiration (in which case he suffers considerable evaporative cooling on entering a cooler environment) there is no damage done to the system for temperature drops of up to 10°C or even more, for people in good health. The human body is resilient, and acclimatisation is fairly rapid.

There is some economic sense, however, in not adopting too low a value of temperature when conditioning a space such as the foyer of a theatre or a shop where the occupancy (by the customer) is relatively short-term. For example, a foyer conditioned at 25°C when the outside is at 28°C gives an immediate impression (which fades with time) of being cooler than outside. It is cheaper to condition at 25°C than at 22·5°C and gives a feeling of immediate benefit. It has the further desirable (though not essential) effect of providing a transitional state between the exterior at 28°C and the auditorium at 22·5°C and so reducing the acclimatisation period needed upon entry to the latter. It is necessary to condition the auditorium to, say, 22·5°C because the occupation is sustained over several hours therein and 25°C would not prove comfortable over the whole of such a period.

There is a lower limit to the value maintained in a conditioned space, of course, and this is set by the winter. Thus, in outside conditions of −2°C saturated, in winter, the inside condition would be maintained at, say, 20°C. This value would be suitable until it was 20°C outside. For outside air temperature rising from 20°C to 28°C, say, the room temperature would be allowed to rise from 20°C to 22·5°C, say.

4.7 Environmental freshness

When breathing quietly, human beings inspire only about 0·16 to 0·20 litres of air per second. The amount of fresh air needed to sustain life is thus very small indeed, being of the order of 0·2 litres/s per person. It is highly desirable to keep the amount of fresh air handled by a conditioning plant to a minimum, for economic reasons. But this minimum is not dictated by the amount of air required for breathing purposes.

Two criteria show the need for larger air quantities. Sufficient fresh air should be supplied to dilute the carbon dioxide content in the room to below 0·1 per cent and to dilute the accumulation of body odours and other smells in the room to a socially acceptable level. It is generally considered that as little as 5 litres/s per person is required in spaces such as banking halls and churches, where there is no smoking. On the other hand, as much as 25 litres/s per person may be needed in areas such as board rooms where a high density of occupation may prevail and where heavy smoking may be likely. For offices, about 5 to 8 litres/s per person is desirable, depending on the usage and the degree of smoking.

EXERCISES

1. The human body adjusts itself, within limits, to maintain a relatively constant internal temperature of 37°C.

(*a*) How does the body attempt to compensate for a cool environment which tends to lower the internal temperature?

(*b*) How does the body attempt to compensate for a warm environment approaching body temperature or exceeding it?

(*c*) State, giving your reasons, whether an increased air motion as provided by a large rate of air change or the action of a ‘ punkah ’ fan is likely to be beneficial to comfort in a room at 29·5°C, 75 per cent relative humidity.

2. It is required to visit a site to obtain measurements of comfort conditions at a point 1·5 m above floor level in the centre of a particular room. The measurements required are dry-bulb and wet-bulb temperatures, globe temperature and mean air speed. State what simple instruments and ancillary equipment should be taken to site in order to obtain the required measurements. Briefly describe the instruments, with the aid of sketches and explain how to use each one, mentioning any precautions which should be taken to ensure accuracy of the values obtained.

3. (*a*) Write down an equation expressing the thermal balance between the human body and its environment.

(*b*) Under what conditions is the temperature of the deep tissues of the human body going to change? Discuss the physiological mechanisms which the body employs to adjust such imbalance. How can the air conditioning engineer, through an appropriate manipulation of the environment, assist the body in feeling comfortable?

4. List the factors in the environment which affect the body's feeling of comfort and describe how they influence the rate of heat loss from the body.

BIBLIOGRAPHY

1. T. Bedford. *Basic Principles of Ventilating and Heating.* H. K. Lewis, 1948.
2. T. N. Adlam. *Radiant Heating.* Industrial Press, 1947.
3. A.S.H.R.A.E. *Guide and Data Book, Fundamentals.*
4. F. A. Missenard. La température résultante d'un milieu, *Chal. Ind.*, July 1933.
5. P. O. Fanger. *Thermal Comfort*, Danish Technical Press, 1970.
6. J. D. Hardy. Thermal Comfort and Health *J.A.S.H.R.A.E.*, Feb. 1971, 43–51.

5

Climate and Outside Design Conditions

5.1 Climate

The variations in temperature, humidity and wind occurring throughout the world are due to several factors the integration of which, for a particular locality, provides the climate experienced. There is, first, a seasonal change in climatic conditions, varying with latitude and resulting from the fact that, because the earth's axis of rotation is tilted at about $23\frac{1}{2}°$ to its axis of revolution about the sun, the amount of solar energy received at a particular

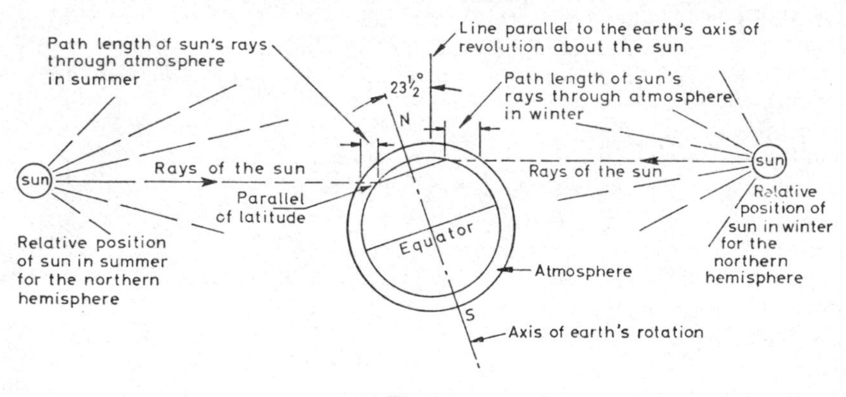

Fig. 5.1

place on the earth's surface alters throughout the year. The geography of the locality provides a second factor, influential in altering climate within the confines imposed by the seasonal variation.

Figure 5.1 illustrates the geometrical considerations which show that, at a particular latitude, the earth receives less solar radiation in winter than it does in summer. The importance of this in its effect on the seasons stems from the fact that, for all practical purposes, the sun is the sole supplier of energy to the earth.

The geography of a place determines how much solar energy is absorbed by the earth, how much is stored and how readily it is released to the atmosphere. The atmosphere is comparatively transparent to the flux of radiant solar energy (termed 'insolation') but the land masses which receive the energy are opaque to it and are fairly good absorbers of it, although this depends on the reflectivity of the surface. This means that the thermal energy from the sun warms up land surfaces on which it falls. Some of this energy travels inwards and is stored in the upper layers of the earth's crust; some is convected to the atmosphere and some is re-radiated back to space, but at a longer wavelength (about 10 micrometres), its mean surface temperature being very much less than that of the sun. Four-fifths of the earth's surface is water, not land, and water behaves in a different fashion as a receiver of insolation, being partially transparent to it; consequently, the energy is absorbed by the water in depth, with the result that its surface temperature does not reach such a high value during the daytime. On the other hand, at night, heat is lost from the land to the sky much more rapidly since less was stored in its shallow upper crust than was absorbed and stored in the deeper layers of water. The result is that land-surface temperatures tend to be lower at night than are water-surface temperatures. It is evident from this that places in the middle of large land masses will tend to have a more extreme annual variation of temperature than will islands in a large sea. Thus, the climate of places on the same latitude can vary enormously. To realise this we have but to compare the temperate seasons experienced in the British Isles with the extremes suffered in Central Asia and Northern Canada at about the same latitude. The exchanges of radiant energy cited above as responsible for the differences in maritime and continental climates are complicated somewhat by the amount of cloud. Cloud cover acts as an insulating barrier between the earth and its environment; not only does it reflect back to outer space a good deal of the solar energy incident upon it but it also stops the passage of the low-frequency infra-red radiation which the earth emits. Mountain ranges also play a part in altering the simple picture, presented above, of a radiation balance.

The effect of the unequal heating of land and sea is to produce air movement. This air movement results in adiabatic expansions and compressions taking place in the atmosphere, with consequent decreases and increases respectively in air temperature. These temperature changes, in turn, may result in cloud formation as values below the dew point are reached.

One overall aspect of the thermal radiation balance is prominent in affecting our weather and in producing permanent features of air movement such as the Trade Winds and the Doldrums. This is the fact that, for latitudes higher than about 40°, the earth loses more heat to space by radiation than it receives from the sun but, for latitudes less than 40°, the reverse is the case. The result is that the lower latitudes heat up and the higher ones cool down. This produces a thermal up-current from the equatorial regions and a corresponding down-current in the higher latitudes. While this is true for an ideal atmosphere, the fact that the earth rotates and that other complicating factors are present means that the true behaviour is quite involved and not yet fully understood.

5.2 Winds

It might be thought that wind flowed from a region of high pressure to one of low pressure, following the most direct path. This is not so. There are, in essence, three things which combine to produce the general pattern of wind flow over the globe. This pattern is complicated further by local effects such as the proximity of land and sea, the presence of mountains and so on. However, the three overall influences are:

(i) the unequal heating of land and sea;
(ii) the deviation of the wind due to forces arising from the rotation of the earth about its axis;
(iii) the conservation of angular momentum—a factor occurring because the linear velocity of air at low latitudes is less than at high latitudes.

The general picture of wind distribution is as follows. Over equatorial regions the weather is uniform; the torrid zone is an area of very light and variable winds with frequent calms, cloudy skies and violent thunderstorms. These light and variable winds are called the ' Doldrums '. Above and below the Doldrums, up to 30° north and south, are the Trade Winds, which blow with considerable steadiness, interrupted by occasional storms. Land and sea breezes (mentioned later) also affect their behaviour.

Above and below 30° of latitude, as far as the sub-polar regions, the Westerlies blow. They are the result of the three factors mentioned earlier, but their behaviour is very much influenced by the development of regions of low pressure, termed cyclones, producing storms of the pattern familiar in temperate zones. This cyclonic influence means that the weather is much less predictable in the temperate zones, unless the place in question is in a very large land mass—for example in Asia or in North America. In temperate areas consisting of a mixture of islands, broken coastline and sea, as in north-western Europe, cyclonic weather is the rule and long-term behaviour difficult to forecast. The whole matter is much complicated by the influence of warm and cold currents of water.

5.3 Local Winds

Exceptions exist to all the generalisations made in the preceding section. One important local effect occurs when, at coastlines, land and sea breezes result from the unequal heating. In the daytime, air rises over the hot land, and cool air comes in from the sea to displace it as the day advances. At night, the rapid cooling of the land chills the air in the vicinity of its surface which then moves out to sea to displace warmer air, which rises. This effect results in an averaging out of coastal air temperatures. The diurnal variation in dry-bulb temperature is less near the coast than it is further inland.

The seasonal variation of temperature (as well as the diurnal variation) is greatest over the middle of large land masses. This results in seasonal winds, called monsoons, blowing in from the sea to the land in summer and outwards from land to sea in winter. India, Asia and China have distinct monsoon effects. Tornadoes, the origin of which is obscure, are instances of the highest wind speeds encountered. They are vortices, a few hundred yards across, which move in a well-defined path. They can occur all over the world but, fortunately, are common only in certain areas. Their lives are short but violent. Wind speeds of 480 km/h or more are likely and pressures at their centre have reached values as low as about 800 millibars. (The lowest recorded barometric pressure in the United Kingdom is 925 mbar and the highest is 1055 mbar. In London the values are 950 mbar and 1049 mbar, respectively.)

At higher altitudes, the comparative absence of dust particles and the reduction in the amount of water vapour in the atmosphere mean that radiation outwards to space is unimpeded at night. The consequence is that the surface of the earth on high ground cools more rapidly at night than it does at lower altitudes. The air in contact with the colder surface becomes chilled (and hence denser) and slides down mountain sides to low-lying ground where it is apparent as a gentle wind, termed a katabatic or gravity wind. Examples are common and a typical case is the mistral, a katabatic wind into the Mediterranean from the high plateau of south and eastern France.

Other local winds result from the passage of air in front of an advancing depression, across a hot desert—the Sahara, for example, which gives a dry wind along the coast of north Africa. (The same wind changes its character in crossing the Mediterranean, becoming a warm moist wind to southern Europe.)

There is another local wind which results from a process of adiabatic expansion and cooling as air rises. Air striking the windward side of rising slopes expands as its ascends, cooling occurring meanwhile. A temperature below the dew point is eventually reached, clouds form and, in due course, descends. The same air then flows over the highest point of the slope and rain falls on the lee side, suffering adiabatic compression and an increase in temperature as it does so. It appears as a dry, warm wind on the lee side of the high ground. An example of this is the chinook wind, blowing down the eastern slopes of the Rocky mountains.

One last instance of a local wind of importance occurs in West Africa during winter in the northern hemisphere. The Sahara desert cools considerably at night in the winter, causing a progressive drop in air temperature as the season advances. This cool, dry air flows westwards to the African coast, displacing the warmer and lighter coastal atmosphere.

5.4 The formation of dew

At nightfall, the ground, losing heat by radiation, undergoes a continued fall in temperature and the air in contact with the ground also suffers a fall in emperature, heat transfer between the two taking place by convection. Eventually, the temperature of the ground drops below the dew point and condensation forms. Although the rate of heat loss from solid surfaces is roughly constant, depending on the fourth power of the absolute temperature of the surface, all solid objects do not fall in temperature at the same rate. A good deal of heat is stored in the upper layers of the earth, and heat flows outwards to the surface to make good radiation losses therefrom. Thus, good conductors of heat in good thermal contact with the ground will fall in temperature at a rate much the same as that of the main surface mass of the earth nearby. Bad conductors, or insulated objects, however, will not be able to draw heat from the earth to make good their losses by radiation and so their temperature will fall more rapidly and dew will tend to be deposited first on such objects. Examples of these two classes are rocks, which are good conductors and are in intimate contact with the earth and grasses which are poor conductors. Dew tends to form on grass before it does on rocks.

5.5 Mist and fog

For condensation to occur in the atmosphere the presence is required of small, solid particles termed condensation nuclei. Any small solid particle will not do; it is desirable that the particles should have some affinity for water. Hygroscopic materials such as salt and sulphur dioxide then, play some part in the formation of condensation. The present opinion appears to be that the products of combustion play an important part in the provision of condensation nuclei and that the size and number of these nuclei vary tremendously. Over industrial areas there may be several million per cubic centimetre of air whereas, over sea, the density may be as low as a few hundred per cubic centimetre.

These nuclei play an important part in the formation of rain as well as fog, but for fog to form the cooling of moist air must also take place. There are two common sorts of fog: advection fog, formed when a moist sea breeze blows inland over a cooler land surface, and radiation fog. Radiation fog forms when moist air is cooled by contact with ground which has chilled as the result of heat loss by radiation to an open sky. Cloud cover discourages such heat loss and, inhibiting the fall of surface temperature, makes fog formation less likely. Still air is also essential; any degree of wind usually dissipates fog fairly rapidly. Fog has a tendency to occur in the vicinity of industrial areas owing to the local atmosphere being rich in condensation nuclei. Under these circumstances, the absence of wind is helpful in keeping up the concentration of such nuclei. The dispersion of the nuclei is further impeded by the presence of a temperature inversion, that is, by a rise in air temperature with increase of height, instead of the reverse. This discourages warm air from rising and encourages the products of combustion and fog to persist, other factors being helpful.

5.6 Rain

Upward air currents undergo adiabatic expansion and cooling. The adiabatic lapse rate of air temperature which occurs thus is about 1°C for every 100 m of increase in altitude. Comparing this with the normal fall of temperature, 0·6°C in every 100 m, it can be seen that in due course the rising air, which was originally, perhaps, at a higher temperature than ambient air, is eventually at a lower temperature than its environment. At this point, upward motion ceases.

If, during the adiabatic cooling process, the air temperature in the rising current fell below its dew point, condensation would occur in the presence of adequate condensation nuclei. The liquid droplets so formed would tend to fall under the influence of the force of gravity but the rate of fall would be countered by the frictional resistance between the droplet and the rising current of air. Whether the cloud, formed as the result of the condensation, starts to rain or not depends on the resultant of the force of gravity downwards and the frictional resistance upwards. The rate at which rain falls depends on the size of the drops formed.

5·7 Diurnal temperature variation

The energy received from the sun is the source of heat to the atmosphere and so the balance of heat exchanges by radiation between the earth and its environment, which is reflected in changes of air temperature, must vary according to the position of the sun in the sky. That is to say, there will be a variation of air temperature against time.

The surface of the earth is at its coolest just before dawn, having had, in the absence of cloud cover, an opportunity of losing heat to the black sky during the whole of the night. Accordingly, it is usual to regard the lowest air temperature as occurring about one hour before sunrise. As soon as the sun rises, its radiation starts to warm the surface of the earth, and as the temperature of the ground rises heat is convected from the surface of the earth to the layers of air immediately above it. There is thus a progressive increase in air temperature as the sun continues to rise, and also for some little time after it has passed its zenith, because of the fact that some of the heat received by the ground from the sun and stored in its upper layers throughout the morning escapes upwards and is lost by convection in the early afternoon. It is, therefore, usual to find the highest air temperatures at about 2 or 3 pm (sun-time). In fact, between about 1 pm and 5 pm one would not expect very great changes of temperature.

On an average basis, it is not unreasonable to suppose that there is some sort of rough sinusoidal relationship between sun time and air dry-bulb temperature. The curve would not be wholly symmetrical since the time between lowest and highest temperature would not necessarily equal that between highest and lowest.

For the month of June, sunrise is at about 4 am and sunset at about 8 pm. The time of lowest temperature is at about 3 am and the time of highest temperature is at 3 pm. There is, thus, a lapse of 12 hours while the temperature is increasing. Since the night period for unrestricted cooling is only about 7 hours (8 pm to 3 am) the curve will be broader in the daytime than at night. The reverse would be the case in December.

For a symmetrical sine curve, the temperature six hours before or after the time of highest temperature would be exactly half-way between the lowest and highest diurnal values. For a time 5 hours before or after, it would be about $\frac{5}{8}$ths of the way between the lowest and highest values. For a broader daytime curve, such as one gets in June, it is reasonable to suppose that a higher value would prevail at this time. Consequently, the very rough conclusion can be drawn that, for the summer season, March to September, at 5 hours before or after the time of maximum daily temperature the value of the dry-bulb is about two-thirds of the diurnal range above the lowest daily value. Figure 5.2 illustrates this.

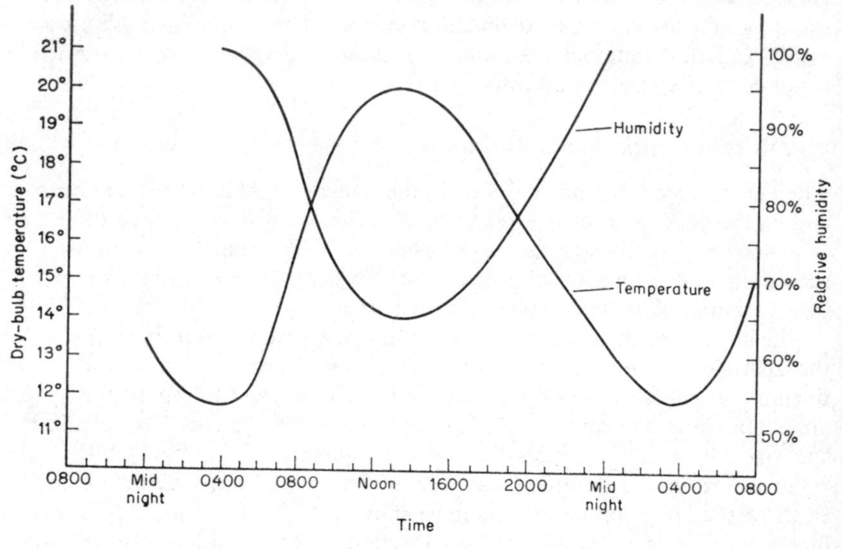

Fig. 5.2

Such a rough assumption is of value in choosing an outside air temperature in a month like June, when maximum solar heat gains through an eastwards facing window occur at 9 am.

5.8 Diurnal variation of humidity

If the humidity is known at 1 pm then the moisture content can be determined at this time since it is reasonable to suppose that at 1 pm the dry-bulb temperature is not far short of its maximum value. One o'clock is chosen as a time, rather than 2 pm or 3 pm (the times of maximum temperature) because meteorological observations of humidity are usually made then.

In the absence of weather changes it might be supposed that the moisture content of the air would stay constant throughout the day until the air reached its dew point. This is not strictly true. During hot weather vegetation releases a good deal of its moisture into the air, becoming dehydrated in the process. This causes an increase in the moisture content of the air, even though no change of weather has occurred. However, assuming a constant moisture content during most of the day, the relative humidity of the air will rise as the afternoon and evening pass and night falls, because of the drop in dry-bulb temperature. As cooling of the air continues the dew point is reached and the air being then saturated any further fall in temperature will cause dew or mist to form, reducing the moisture content of the atmosphere. With the dawn and sunrise, temperatures increase and the relative humidity of the air falls somewhat, dew evaporating as the morning proceeds. Eventually all the dew has disappeared and the moisture content is back at its assumed constant value of the day before. Figure 5.2 illustrates the change of relative humidity which accompanies the diurnal variation in temperature. The curves shown are idealised and are based on the following data and assumptions:

Place	Edgbaston, Birmingham
Month	August
Mean daily temperature (max.)	20°C (at about 2 pm)
Mean daily temperature (min.)	11·7°C (at about 4 am)
Relative humidity at 1 pm	65 per cent
Temperature at 9 am	$11{\cdot}7°C + \frac{2}{3}(20° - 11{\cdot}7°)$
	$= 11{\cdot}7{\cdot}C + 5{\cdot}5°$
	$= 17{\cdot}2°C$
Temperature at 7 pm	17·2°C (in a similar fashion).

A curve of temperature against time is then plotted, using the four values of temperature—time obtained above and assuming a roughly sinusoidal shape. From this curve, the temperature at 1 pm is read off as 19·9°C, which is virtually 20°C. Accordingly, at 1 pm with a temperature of 20°C and a relative humidity of 65 per cent the moisture content is found, from tables, to be 9·509 g/kg. Using this as a constant value, values of relative humidity can now be established for different times of the day, since the temperature graph gives temperatures at different times of the day.

Time	Moisture content	Temperature°C	RH (%)
8 am	9·509 g/kg	15·1°	86
10 am	,,	18·5°	72
12 noon	,,	19·75°	66
2 pm	,,	20°	65
4 pm	,,	19·5°	67

A curve of relative humidity against time can now be drawn. It should be observed, by reference to I.H.V.E. tables, that the dew-point temperature for a moisture content of 9·509 g/kg is 13·2°C. Dew will therefore form when this temperature is reached, and below this temperature the humidity will stay at 100 per cent until the temperature starts to rise again after 4 am.

5.9 Meteorological measurement

Stations for measuring various relevant atmospheric properties are set up all over the world by the meteorological authorities in different countries. The coverage of the surface of the earth is not complete and so there is a wealth of information for some areas and a scarcity of it for others.

Measurements are made of temperatures and humidity within louvred boxes, positioned in the open air. The louvres permit the ready circulation by natural means of outside air over the instruments but shield them from rain and sunshine. Since the air movement over any wet-bulb thermometer mounted in such a screened housing is by natural means, the velocity of airflow will be too low (less than 5 m/s) to minimise effectively interference from radiation (*see* § 2.17) and so, ' screened ' wet-bulb values will always be about ½ degree higher than 'sling' wet-bulb values. Whereas wet-bulb values on the psychrometric chart published by the I.H.V.E. are sling values data taken from meteorological tables are always based on screened values, and due allowance must be made for this when using such data.

In a meteorological screened housing, measurements are made daily of the highest and lowest values of dry-bulb temperature. These are averaged for each month of the year, and the two average daily values so obtained are tabulated for the particular month for each year. Over a period of several years a further average of the values can be taken, yielding a mean daily maximum temperature and a mean daily minimum temperature, for each month, for a particular place.

The extreme high and low values of temperature in each month are also noted, and averages of these are calculated over the number of years for which observations are made. This yields mean monthly maximum and minimum values of dry-bulb temperature. Other records of temperature are kept but these are largely irrelevant to a study of air conditioning.

As regards records of relative humidity, the picture is not so comprehensive. No records are kept of wet-bulb temperatures as a rule, but instead, measurements of relative humidity are made twice daily—one at about 9 am and another at about 1 pm. The readings are all taken at a height of about 1·5 m above ground level. Typical meteorological data for Kew are shown in Table 5.1.

5.10 The seasonal change of outside psychrometric state

Meteorological data can be used to provide a picture on a psychrometric chart of the way in which climate varies, as far as mean daily and mean monthly values of temperature are concerned. It is immediately evident from the previous section that values of temperature are available, but the manner in which the figures for relative humidity should be used requires some further comment.

The values of humidity quoted are the averages of daily observations. They should therefore be associated with mean daily temperatures at the same time. Since the mean daily maximum temperature occurs at between 1 pm and 4 pm it is reasonable to associate it with the daily average value of

TABLE 5.1

Temperature (°C)	Jan.	Feb.	Mar.	Apr.	May	June	July	Aug.	Sept.	Oct.	Nov.	Dec.
Mean daily maximum	7·2°	7·8°	9·4°	12·8°	17·2°	20°	21·7°	21·1°	18·3°	13·9°	9·4°	7·8°
Mean daily minimum	2·2°	2·2°	2·8°	4·4°	7·8°	10·6°	12·8°	12·2°	10°	7·2°	3·9°	2·8°
Mean monthly maximum	11·7°	12·2°	15·6°	19·4°	23·9°	26·7°	27·8°	27·2°	23·9°	18·3°	14·4°	12·2°
Mean monthly minimum	−5·6°	−4·4°	−3·3°	−1·1°	1·7°	5·6°	8·3°	7·2°	3·3°	0°	−2·2°	−3·9°
Relative humidity at 1 pm (%)	81	73	63	62	58	57	57	61	64	70	78	81

The above figures are approximate and are for observations made at Kew at a latitude of 51° 28′ N and a longitude of 0° 19′ W. The altitude of Kew is 5·5 m above sea level. The tabulated figures are for readings made over the period from about 1910 to about 1940. More up-to-date observations yield similar results.

humidity at about the same time. For these two values of dry-bulb temperature and humidity it is possible to get a value of moisture content in kg/kg from I.H.V.E. tables of psychrometric data. This permits values of mean daily maximum dry-bulb temperature, at their corresponding moisture contents, to be plotted on a psychrometric chart.

Using the information from Table 5.1, the following is tabulated:

Month	Mean daily max. (°C)	Humidity at 1 pm (%)	Moisture content (g/kg)	Mean monthly max. (°C)
January	7·2°	81	5·12	11·7°
February	7·8°	73	4·81	12·2°
March	9·4°	63	4·63	15·6°
April	12·8°	62	5·73	19·4°
May	17·2°	58	7·15	23·9°
June	20·0°	57	8·32	26·7°
July	21·7°	57	9·38	27·8°
August	21·1°	61	9·65	27·2°
September	18·3°	64	8·74	23·9°
October	13·9°	70	6·96	18·3°
November	9·4°	78	5·74	14·4°
December	7·8°	81	5·33	12·2°

We can now plot two sets of points on a psychrometric chart: set (*a*), referring to mean daily maximum temperatures, and set (*b*), which uses mean monthly maximum temperàtures, each with their corresponding moisture contents.

Figure 5.3 shows these two sets of points, plotted on an I.H.V.E. psychrometric chart.

Mean monthly and mean daily minimum values are usually on the 100 per cent curve, if these are plotted, since dew formation involves the attainment of a saturated condition.

An envelope can be drawn around the points on the assumption that cooling can occur either from radiation loss at night or by the adiabatic saturation resulting from rainfall. In the first case the cooling occurs along a line of constant moisture content until the dew point is reached, and then down the saturation curve. In the second case cooling will be up a wet-bulb line. It is possible that if the rain drops that fall from a cloud are at a temperature less than the wet-bulb some further cooling will occur. The resulting final state will then be at a lower wet-bulb than the original value. This does not invalidate a proposal to use a wet-bulb line through the August mean monthly maximum state as an average upper boundary to the envelope. The dotted line then encloses the area within which most of our weather lies. At the lower extreme, in winter, low temperatures are likely to be associated with saturated air, and so a winter design value of, say, −2°C saturated would be a typical lower extreme.

Such an envelope, although it gives a picture of the changes of state likely to occur on an average over 12 months of the year, is not the best guide to the

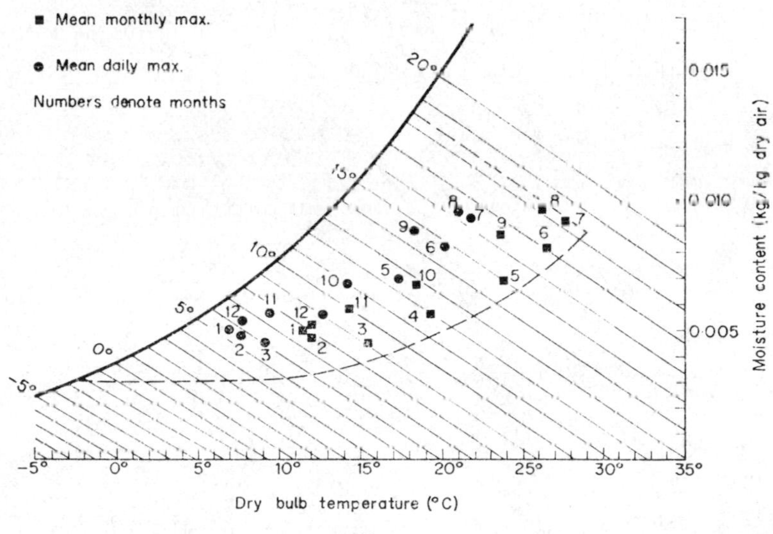

Fig. 5·3

establishment of outside design conditions. It will, however, provide a useful guide, in the absence of other information. In Fig. 5·3, for example, the summer design condition is 28°C dry-bulb and 18·5°C wet-bulb (sling) or 18°C wet-bulb (screen). This wet-bulb value is perhaps a little low, as the following section shows.

5·11 The choice of outside design conditions

The most logical way of doing this is in terms of the frequency of occurrence of certain values of dry-bulb and wet-bulb temperature. An air-conditioning system can then be specified to maintain certain inside conditions for a certain percentage of the year.

Figure 5.4 shows on a map of the United Kingdom isotherms of wet-bulb values which are not exceeded for more than 1 per cent of the summer months—June to September inclusive.

Figure 5.5 is a graph of the frequency of occurrence of wet-bulb and dry-bulb temperatures at Croydon for twelve months of the year. So far as high values are concerned, in the four summer months June to September inclusive, the relevant percentage should be three times 0·3 per cent, since four summer months represents only one-third of the year. On this basis, the outside design dry-bulb is about 27·2°C. One concludes that on a 1 per cent basis, for the summer months, an outside design condition of 28°C dry-bulb

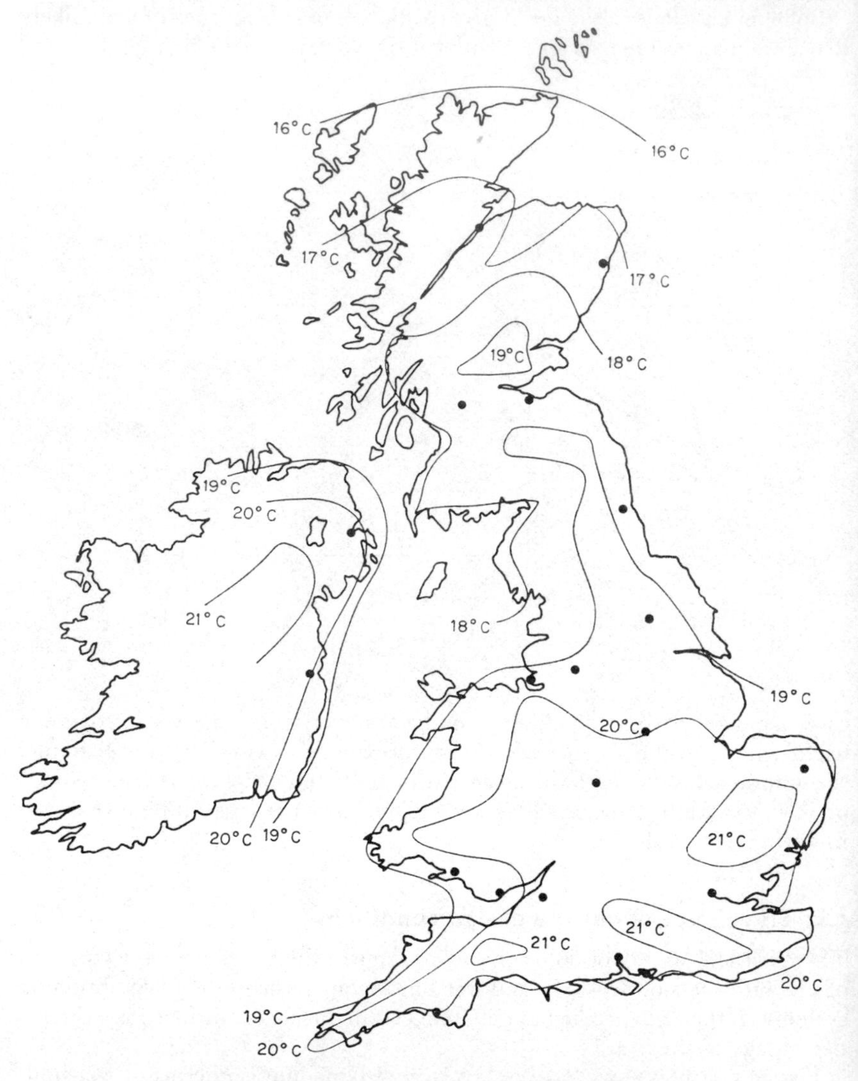

Fig. 5.4 High-design wet-bulb isotherms
The isotherms are exceeded for only 1 per cent of the months June to September inclusive

and 20·5°C wet-bulb (screen) or 20°C wet-bulb (sling) is satisfactory for the London area. On a similar basis, for the two-thirds of the year regarded as winter, an outside dry-bulb value of about −5°C should be chosen for design purposes, coupled with a relative humidity of 100 per cent. This is not a realistic way of selecting an outside winter design temperature. Whereas in summer the high values of temperature occur during the afternoon when the premises are occupied, low values in winter usually occur at night when offices are unoccupied or when people are in bed. A figure of −2°C is adequate for about two-thirds per cent of the year or about 2 per cent of the coldest winter months, November to February, inclusive.

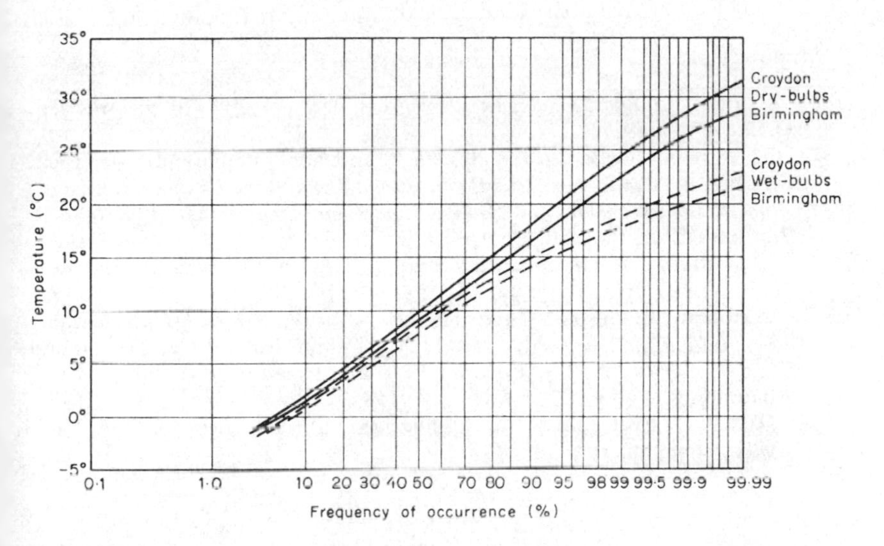

Fig. 5.5 Graph of wet- and dry-bulb frequency
The broken lines are wet-bulb temperatures uncorrected for altitude.
Based on hourly observations for the period 1946–55.

The advantage of this method of selecting an outside design state, particularly for summertime, is that any criterion of reliability can be chosen, making the method adaptable to both commercial comfort conditioning and to industrial conditioning. For example, in summer the $2\frac{1}{2}$ per cent criterion is about 2° less than the 1 per cent.

EXERCISES

1. (i) Briefly explain the causes of outside air diurnal variation of temperature, accounting for the times at which maximum and minimum values occur.

(ii) Briefly explain what differences there might be and why they occur, in the daily and annual temperature ranges between two places on the same latitude, one place being classified as a dry tropical region and the other a humid tropical region. Similarly, account for the differences between a temperate coastal region and a temperate inland region.

(iii) Tabulate factors affecting the choice of outside design condition for air conditioning a modern building.

2. Briefly describe a simple method of estimating summer outside air conditions for the design calculations of an air conditioning scheme. If a more accurate assessment of heat gains to a building is required, explain what further meteorological information is wanted.

BIBLIOGRAPHY

1. W. H. PICK. *A Short Course in Elementary Meteorology*, 5th edn, revised. H.M.S.O., 1938.
2. O. G. SUTTON. *Understanding Weather*, 2nd edn. Penguin Books, 1964.
3. A.S.H.R.A.E. *Guide and Data Book, Fundamentals and Equipment.*
4. *Evaluated Weather Data for Cooling Equipment Design.* Head Wrightson Processes Ltd., London, England.
5. H. C. JAMIESON. Meteorological Data and Design Temperatures. *J. Instn. Heat. Vent. Engrs.*, 1955, **22**, 465.
6. METEOROLOGICAL OFFICE: *Tables of Temperature, Relative Humidity and Precipitation for the World.* Part I, North America; Part 2, Central and South America; Part 3, Europe; Part 4, Africa; Part 5, Asia; Part 6, Australasia.
7. R. HARRISON. Air Conditioning at Home and Abroad. *J. Instn. Heat. Vent. Engrs.*, 1958, **26,** 177

6

The Choice of Supply Design Conditions

6.1 Sensible heat removal

Consider a room which is hermetically sealed and in which there is a source of heat operating continuously. In practice, the source of heat could be from electrical lighting and from people (*see* § 7.24) but, in this simple illustration, suppose that a group of electric fires is producing heat at a rate of Q watts. Suppose further, that the walls, ceiling and floor of the room are of area A, have an average overall thermal transmittance of U W/m² °C and that the environment outside the room is at a constant temperature of t_o. Assume that the temperature within the room is t_r.

The problem is to determine what the effect will be on the temperature within the room if its hermetic seal is broken and it is ventilated at a known rate. As a first step towards the solution of this problem, it is necessary to strike a heat balance for the room in its hermetically sealed state.

The heat gain to the room from the electric fires equals the heat lost from the room to its environment in the steady-state.

In mathematical form this equation becomes

$$Q = AU(t_r - t_o) \tag{6.1}$$

whence

$$t_r = t_o + \frac{Q}{AU} \tag{6.2}$$

Thus, t_r will always be greater than the outside air temperature by an amount Q/AU, after there has been sufficient passage of time for a steady-state condition to exist. It is clear that the amount by which t_r is greater than t_o will depend directly on Q, the heat gain within the room.

EXAMPLE 6.1. Calculate the steady-state temperature maintained within a room 3 m × 3 m × 3 m given that the average U-value for the room walls, floor and ceiling, is 1·1 W/m² °C, that the outside air temperature is 27°C, and that there is a sensible heat gain within the room, due to electric lights and people, of 2 kW.

Answer

Writing down the heat balance in the form of eqn. (6.2) we get

$$
\begin{aligned}
t_r &= 27^\circ + \frac{2000}{6 \times 3^2 \times 1 \cdot 1} \\
&= 27^\circ + 33 \cdot 7^\circ \\
&= 60 \cdot 7^\circ\text{C}
\end{aligned}
$$

This is clearly a highly undesirable level of temperature if people are expected to spend any length of time within the room. The obvious answer to the problem is the use of ventilation. This takes the solution a step further and means writing down a new heat balance equation. If the rate of air change is n per hour (signifying that the rate of passage of air through the room is equivalent to changing its cubical content n times in each hour) then this air will remove some of the heat given off by the electric fires and will assist in the maintenance of a lower value of t_r.

If the air enters the room at a temperature t_o and, in absorbing part of the heat given off by the electric fires within the room, its temperature rises to t_r, then the amount of heat it absorbs is given by the expression

$$
\frac{\rho_t c n V (t_r - t_o)}{3600}
$$

where ρ_t = the density of the air at temperature t_o in kg/m^3,

c = the specific heat of air in J/kg °C,

V = the volume of the room in m^3.

This expression can be combined with eqn. (6.1) to yield a new heat balance:

$$
Q = \frac{\rho_t c n V (t_r - t_o)}{3600} + AU(t_r - t_o) \tag{6.3}
$$

whence

$$
t_r = t_o + \frac{3600 Q}{\rho_t c n V + 3600 A U} \tag{6.4}
$$

It is again clear that t_r will always exceed t_o but that the excess will not be as large as with eqn. (6.2) for the unventilated room.

EXAMPLE **6.2.** For the room and conditions used in Example 6.1 calculate the steady-state value of t_r if the room is ventilated at a rate of 10 air changes per hour, given that the density of the outside air is 1·15 kg/m^3 at 27°C and that its specific heat is 1·034 kJ/kg °C.

Answer

From eqn. (6.4):

$$t_r = 27^\circ + \frac{3600 \times 2000}{1{\cdot}15 \times 1034 \times 10 \times 3^3 + 3600 \times 6 \times 3^2 \times 1{\cdot}1}$$

$$= 27^\circ + \frac{7\,200\,000}{321\,500 + 214\,000}$$

$$= 27^\circ + 13{\cdot}4^\circ$$

$$= 40{\cdot}4^\circ\text{C}$$

(Although the answers to Examples 6.1 and 6.2 are correct, they are also misleading. In practice, even for the conditions quoted, room temperatures would not rise to such high values because (*a*) the outside temperature falls at night, and (*b*) the thermal capacity of the room walls, floor and ceiling acts as a reservoir in which heat can be stored during the daytime.)

It is evident from eqn. (6.4) that, in the presence of a net heat gain, the temperature within the room will always exceed that of the outside air. Under certain conditions this may be satisfactory. In Example 6.2, a room temperature, of, say, 20°C will be maintained if the outside air temperature is low enough.

EXAMPLE 6.3. For the room and other conditions used in Example 6.2, calculate the value of the steady outside temperature which will result in an inside steady-state value of 20°C.

Answer

From eqn. (6.4):

$$20^\circ = t_o + 13{\cdot}4^\circ$$

$$\therefore t_o = 6{\cdot}6^\circ\text{C}$$

That is to say, the heat loss through the fabric of the room plus the cooling effect of 10 air changes of ventilating outside air at an entry temperature of 6·6°C exactly balance the internal heat gain of 2 kW from the electric fires.

For all temperatures in excess of 6·6°C in this example then, the inside temperature will exceed 20°C because of the presence of the internal heat gain.

It is clear from this that, if an inside temperature of 20°C must be maintained in the room throughout the year the temperature of the entering ventilating air must be artificially kept at a value which is less than 20°C. In general terms, if a room suffers a heat gain of Q watts then air must be supplied at a temperature t_s which is lower than the value of the room air temperature t_r. If the heat gain is known and if t_r and t_s are chosen, then the necessary amount of air which must be supplied to the room can be calculated.

The derivation of an expression for this quantity of air stems from the concept of specific heat capacity.

Heat removed in J = mass of air supplied × specific capacity heat of air × temperature rise of air

$$= mc(t_r - t_s) \tag{6.5}$$

The mass of air supplied can be converted to the volume supplied, if the density $\rho_{t,s}$ is known for the temperature of the air supplied.

Heat removed in J = (volume of air supplied × $\rho_{t,s}$) $c(t_r - t_s)$

$$= (V_{t,s}\,\rho_{t,s})\,c(t_r - t_s) \tag{6.6}$$

The expression $V_{t,s}\rho_{t,s}$ is another way of denoting the mass of the air. If this mass is fixed, then it can be expressed as any volume, provided the appropriate density of air is taken. Thus, we may say that

$$V_{t,s}\rho_{t,s} = m = V_t\rho_t \tag{6.7}$$

From Charles' law (*see* § 2·5) it is known that the density of air is inversely proportional to its absolute temperature. That is,

$$\rho_t = \rho_o \frac{T_o}{T_t} = \rho_o \frac{(273 + t_o)}{(273 + t\,)} \tag{6.8}$$

Equation (6.8) can be used with eqn. (6.7) to give an expression for the mass of the air (which is absorbing sensible heat) in terms of a generalised volume V_t and a corresponding generalised density, ρ_t.

$$m = V_t\rho_o \frac{(273 + t_o)}{(273 + t\,)} \tag{6.9}$$

$$\text{Heat removed in J} = mc(t_r - t_s)$$

$$= V_t\rho_o \frac{c(t_r - t_s)(273 + t_o)}{(273 + t)} \tag{6.10}$$

Introducing a time factor of one second for the heat supplied, and observing that the mass supplied in one second equals the volume (expressed at any temperature t) per second multiplied by the density at a temperature t, we can then re-write eqn. (6.10).

Taking a value of 1·026 kJ/kg °C for the specific heat capacity of humid air and 1·191 kg/m^3 for its density at 20°C dry-bulb and 50 per cent saturation, we can say

$$\text{heat gain in kW} = \frac{\text{flow rate (m}^3\text{/s)} \times 1{\cdot}19 \times 1{\cdot}026 \times (273+20) \times (t_r - t_s)}{(273+t)}$$

$$= \frac{\text{flow rate } (m^3/s) \times 358 \times (t_r - t_s)}{(273+t)} \qquad (6.11)$$

This is conveniently re-expressed as

$$\text{flow rate (m}^3\text{/s)} = \frac{\text{sensible heat gain (kW)}}{(t_r - t_s)} \times \frac{(273+t)}{358} \qquad (6.12)$$

It is important to note that if a net heat loss is to be offset, the ventilating air must supply heat and so its temperature must be greater than the value maintained in the room.

EXAMPLE 6.4. A room measuring 3 m × 3 m × 3 m suffers sensible heat gains of 2 kW and is to be maintained at 22°C dry-bulb by a supply of cooled air. If the supply air temperature is 13°C dry-bulb, calculate

(*a*) the mass of air which must be supplied in kg/s
(*b*) the volume which must be supplied in m^3/s,
(*c*) the volume which must be extracted, assuming that no natural infiltration or exfiltration occurs.

Take the specific heat of air as 1·012 kJ/kg °C.

Answer

(*a*) The heat gain is to be offset by a mass of m kg/s of air being supplied at 13°C and rising in temperature to 22°C. A basic relationship can immediately be written down:

(Heat removed by air in kJ/s) = (mass of air supplied in kg/s) × (specific heat of air) × (temperature rise of air)

Inserting the known values, this becomes

$$2 = m \times 1{\cdot}012 \times (22^\circ - 13^\circ)$$

whence

$$m = \frac{2}{1{\cdot}012 \times 9}$$

$$= 0{\cdot}2195 \text{ kg/s.}$$

(*b*) The rate of mass flow of air supplied can be expressed as a volumetric flow rate at any temperature desired, by making use of Charles' law. This has already been taken account of in the derivation of eqn. (6.12). Making use of this equation, a ready answer can be found:

Air quantity supplied (at temperature 13°C)

$$= \frac{2}{(22^\circ - 13^\circ)} \times \frac{(273+13)}{358}$$

$$= 0{\cdot}177 \text{ m}^3/\text{s at } 13^\circ\text{C}.$$

It is important to observe that, for a given mass flow rate, a corresponding volumetric flow rate has meaning only if its temperature is also quoted. This principle is not always adhered to since it is often quite clear from the context what the temperature is. It may also be splitting hairs to mention temperature when only a small variation in temperature is likely.

(*c*) That a variation of calculable magnitude exists is shown by evaluating the amount of air leaving the room. Since no natural exfiltration or infiltration occurs, the mass of air extracted mechanically must equal that supplied. Hence,

$$\text{rate of extraction} = \frac{2}{22^\circ - 13^\circ} \times \frac{(273+22)}{358}$$

$$= 0{\cdot}183 \text{ m}^3/\text{s at } 22^\circ\text{C}.$$

This is just the same as using Charles' law:

$$\text{rate of extraction} = \text{flow rate in m}^3/\text{s at } 13^\circ\text{C} \times \frac{(273+22)}{(273+13)}$$

$$= 0{\cdot}177 \times \frac{295}{286} = 0{\cdot}183 \text{ m}^3/\text{s at } 22^\circ\text{C}.$$

Equation (6.12) can be used then to express the volumetric flow rate at any desired temperature for a given constant mass flow rate. The equation is sometimes simplified. A value of 13°C is chosen for the expression of the volumetric flow rate. We then have

$$\text{flow rate (m}^3/\text{s)} = \frac{\text{sensible heat gain (kW)}}{(t_r - t_s)} \times \frac{(273+13)}{358}$$

$$= \frac{\text{sensible heat gain (kW)}}{(t_r - t_s)} \times 0{\cdot}8 \qquad (6.13)$$

This version suffers from the slight disadvantage that a Charles' law correction for temperature cannot be made and so the answers it gives are correct only for air at 13°C. The inaccuracy that may result from this is usually unimportant in most comfort conditioning applications.

6.2 The specific heat capacity of humid air

The air supplied to a conditioned room in order to remove sensible heat gains occurring therein, is a mixture of dry air and superheated steam (*see* Chapter 2). It follows that these two gases, being always at the same temperature because of the intimacy of their mixture, will rise together in temperature as both offset the sensible heat gain. They will, however, offset differing amounts of sensible heat because, first, their masses are different, and secondly, so are their specific heats.

Consider 1 kg of dry air with an associated moisture content of g kg of superheated steam, supplied at temperature t_s in order to maintain temperature t_r in a room in the presence of sensible heat gains of Q kW. A heat balance equation can be written thus:

$$Q = 1 \times 1{\cdot}012 \times (t_r - t_s) + g \times 1{\cdot}890 \times (t_r - t_s)$$

where 1·012 and 1·890 are the specific heats at constant pressure of dry air and steam respectively. Rearranging the equation:

$$Q = (1{\cdot}012 + 1{\cdot}89\ g)(t_r - t_s)$$

The expression (1·012+1·89 g) is sometimes called the specific heat of humid air.

Taking into account the small extra sensible cooling or heating capacity of the superheated steam present in the supply air (or its moisture content) provides a slightly more accurate answer to certain types of problem. Such extra accuracy may not be warranted in most practical cases but it is worthy of consideration as an exercise in fundamental principles.

EXAMPLE 6.5. Calculate accurately the weight of dry air that must be supplied to the room mentioned in Example 6.4, given that its associated moisture content is 7·500 g/kg of dry air and that the specific heat at constant pressure of superheated steam is 1·890 kJ/kg°C.

Answer

As before, the fundamental equation is:

$$\text{heat removed by air in kJ/s} = \text{mass in kg/s} \times \text{specific heat capacity} \times \text{temperature change}$$

$$2 = m \times (1 \times 1{\cdot}012 + 0{\cdot}0075 \times 1{\cdot}890) \times (22^\circ - 13^\circ)$$

where m is the amount of dry air supplied in kg/s.

$$m = \frac{2}{(1{\cdot}012 + 0{\cdot}014)(22^\circ - 13^\circ)}$$

Hence

$$m = \frac{2}{1{\cdot}026 \times 9}$$

$$= 0{\cdot}217 \text{ kg dry air per second.}$$

This should be compared with 0·2195 kg/s, the answer to Example 5.4(*a*). Note that, for the moisture content quoted, the specific heat of the humid air is 1·026 kJ/kg °C.

6.3 Latent heat removal

If the air in a room is not at saturation, then water vapour may be liberated in the room and cause the moisture content of the air in the room to rise. Such a liberation of steam is effected by any process of evaporation as, for

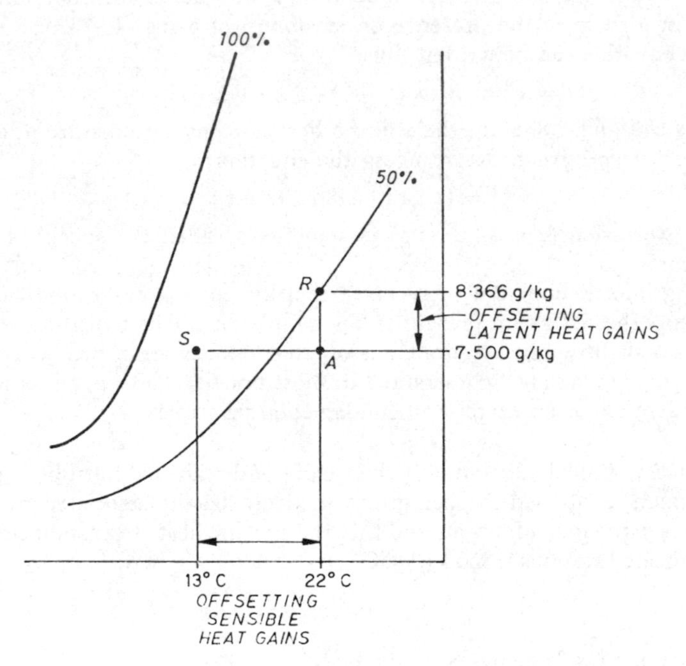

Fig. 6.1

example, the case of insensible perspiration and sweating on the part of the people present. Since it is necessary to provide heat to effect a process of evaporation, it is customary to speak of the addition of moisture to a room as kW of latent heat rather than as kg/s of water evaporated.

The heat gains occurring in a room can be considered in two parts: sensible gains and latent gains. The mixture of dry air and associated water vapour supplied to a room has therefore a dual role; it is cool enough initially to suffer a temperature rise up to the room dry-bulb temperature in offsetting the sensible gains, and its initial moisture content is low enough to permit a rise to the value of the room moisture content as latent heat gains are offset. Figure 6.1 illustrates this.

If m kg/s of dry air with an associated moisture content of g kg per kg of dry air is supplied at state S to a conditioned space wherein there are only sensible heat gains, then its temperature will rise from 13°C to 22°C as it diffuses through the room, offsetting the sensible gains. The resultant room condition would be typified by state point A. If there were then a latent gain in the room caused by some evaporation taking place (say from peoples' skin) without any further increase in sensible heat gain, then the state in the room would change from A to R, more or less up a line of constant dry-bulb temperature (*see* § 3.7). Exactly how far the moisture content of R would be above $g_a (= g_s)$ would depend on the latent heat gain. If the mass of dry air supplied and its associated moisture content is known, then it is possible to calculate the rise in room moisture content corresponding to given latent heat gains:

kW of latent heat gain = (kg of water evaporated/s) × (latent heat of evaporation in kJ/kg of water).

= (kg of dry air supplied/s) × (the increase in its specific humidity in kg of water/kg of dry air) × (latent heat of evaporation in kJ/kg of water).

In symbols, this is written

$$\text{latent heat gain} = m_a \times (g_r - g_s) \times h_{fg}$$

where m_a = mass flow rate in kg dry air/s,

g_r = moisture content maintained in the room in kg/kg

g_s = moisture content of the supply air in kg/kg,

h_{fg} = latent heat of evaporation in kJ/kg.

As in the derivation of eqn. (6.12), the mass flow rate may be expressed volumetrically, with a Charles' law correction for temperature incorporated:

$$m_a = (\text{flow rate in m}^3/\text{s}) \times \rho_o \times \left(\frac{273 + t_o}{273 + t}\right)$$

Taking a value of 1·191 kg/m³ for ρ_o and 2454 kJ/kg for h_{fg} at 20°C, we get

$$\text{latent heat gain in kW} = \text{flow rate (m}^3/\text{s)} \times 1{\cdot}191 \times \frac{293}{(273+t)} \times (g_r - g_s) \times 2454$$

$$= \text{flow rate (m}^3/\text{s)} \times 0{\cdot}856 \times 10^6$$

For convenience the moisture content can be expressed in g/kg and the equation becomes

$$\text{flow rate (m}^3/\text{s)} = \frac{\text{latent heat gain (kW)}}{(g_r - g_s)} \times \frac{(273+t)}{856} \qquad (6.14)$$

It is emphasised that the air quantity calculated in m^3/s at temperature t is the same air quantity calculated by means of eqn. (6.12). We calculate the necessary air quantity by means *either* of eqn. (6.12), *or* eqn. (6.14), *or* by first principles. It is usual to use sensible gains to establish the required supply air quantity, by means of eqn. (6.12) first of all and then to establish by means of eqn. (6.14) the rise in moisture content which will result from supplying this air, with its associated moisture content (g_s) to the room. Alternatively, we may use eqn. (6.14) to establish what the moisture content of the supply air must be in order to maintain a certain moisture content in the room in the presence of known latent heat gains. The temperature in the room would then be the secondary consideration.

EXAMPLE 6.6. A room measures 20 m × 10 m × 3 m high and is to be maintained at a state of 20°C dry-bulb and 50 per cent saturation. The sensible and latent heat gains to the room are 7·3 kW and 1·4 kW, respectively.

(*a*) Calculate from first principles, the mass and volume of dry air that must be supplied at 16°C to the room, each second. Also calculate its moisture content. Take the specific heats of dry air and superheated steam as 1·012 and 1·890 kJ/kg °C, respectively, the density of air as 1·208 kg/m^3 at 16°C and the latent heat of evaporation as 2454 kJ/kg of water.

(*b*) Making use of eqns. (6.12) and (6.14), calculate the supply air quantity in m^3/s and its moisture content in g/kg.

Answer

From I.H.V.E. tables, the moisture content in the room is found to be 7·376 g/kg of dry air.

$$(a)\ \text{Mass flow of dry air} = \frac{7{\cdot}3}{(20^\circ - 16^\circ) \times (1 \times 1{\cdot}012 + 0{\cdot}007376 \times 1{\cdot}890)}$$

$$= \frac{7{\cdot}3}{4 \times 1{\cdot}026}$$

$$= 1{\cdot}779\ \text{kg/s}$$

$$\text{Volumetric flow rate} = \frac{1{\cdot}779}{1{\cdot}208} = 1{\cdot}473\ \text{m}^3\text{/s at } 16^\circ\text{C}$$

$$\text{Latent heat gain} = 1{\cdot}779\ \text{kg/s} \times \text{moisture pick-up in kg/kg} \times 2454$$

$$1{\cdot}4 = 1{\cdot}779 \times (0{\cdot}007376 - g_s) \times 2454$$

$$g_s = 0{\cdot}007376 - 0{\cdot}000321$$

$$= 0{\cdot}007055\ \text{kg/kg.}$$

(*b*) From eqn. (6.12)

$$\text{Supply air quantity} = \frac{7\cdot3}{(20^\circ-16^\circ)} \times \frac{(273+16)}{358}$$

$$= 1\cdot473 \text{ m}^3\text{/s at } 16^\circ\text{C}$$

From eqn. (6.14), and making use of the air quantity found above,

$$g_s = g_r - \frac{1\cdot4}{1\cdot473} \times \frac{(273+16)}{856}$$

$$= 7\cdot376 - 0\cdot321$$

$$= 7\cdot055 \text{ g/kg dry air.}$$

6.4 The slope of the room ratio line

The room ratio line is the straight line, drawn on a psychrometric chart, joining the points representing the state maintained in the room and the initial condition of the air supplied to the room.

The slope of this line is an indication of the ratio of the latent and sensible heat exchanges taking place in the room, and the determination of its value plays a vital part in the selection of economical supply states.

Any supply state which lies on the room ratio line differs from the room state by a number of degrees of dry-bulb temperature and by a number of grammes of moisture content. The values of this pair of differences are directly proportional to the mass of air supplied to the room for offsetting given sensible and latent gains or losses. Thus, in order to maintain a particular psychrometric state in a room, the state of the air supplied must always lie on the room ratio line. If it does not, then either the wrong temperature or the wrong humidity will be maintained in the room.

EXAMPLE 6.7. For the room mentioned in Example 6.6, calculate the condition which would be maintained therein if the supply air were not at a state of 16°C dry-bulb and 7·055 g/kg but at (*a*) 16°C dry-bulb and 7·986 g/kg and (*b*) 17°C dry-bulb and 7·055 g/kg.

The heat gains are the same as in Example 6.6.

Answer

(*a*) Making use of eqn. (6.14),

$$g_r = g_s + \frac{1\cdot4}{1\cdot473} \times \frac{(273+16)}{856}$$

$$= 7\cdot986 + 0\cdot321$$

$$= 8\cdot307 \text{ g/kg.}$$

∴ the room state is 20°C dry-bulb and 57 per cent relative humidity, by reference to I.H.V.E. tables.

Note that the supply of 1·473 m³/s at 16°C continues to maintain 20°C because (by calculation in Example 6.6 (*b*)) this offsets the unchanged sensible heat gains of 7·3 kW.

$$t_r = t_s + \frac{7{\cdot}3}{1{\cdot}473} \times \frac{(273+16)}{358}$$

$$= 17°C + 4°C$$

$$= 21°C$$

Therefore the state maintained in the room will be 21°C dry-bulb and 7·376 g/kg, or 21°C and 48 per cent relative humidity, by reference to I.H.V.E. tables.

Note that the moisture content maintained in the room remains at 7·376 g/kg because, by calculation in Example 6.6 (*b*), the supply of 1·473 m³/s at a moisture content of 7·055 g/kg offsets unchanged latent gains and suffers a rise of 0·321 g/kg.

Note also that, in the use made of eqn. (6.12) above, it is legitimate to use (273+16) not (273+17), because 1·473 m³/s is expressed at 16°C, and we are considering the corresponding mass flow rate. In other words, this example assumes that some unmentioned device, such as a re-heater battery, is used to elevate the supply air temperature by 1°C, but that the mass flow rate is unchanged.

The calculation of the slope of the room ratio line is clearly a matter of some importance since it appears that one has a choice of a variety of supply air states. It seems that if any state can be chosen and the corresponding supply air quantity calculated, then the correct conditions will be maintained in the room. This is not so: Economic pressures restrict the choice of supply air state to a value fairly close to the saturation curve. It must, of course, still lie on the room ratio line.

The calculation of the slope of the line can be done in one of two ways:

(*a*) by calculating it from the ratio of the sensible to total heat gains to the room,

(*b*) by making use of the ratio of the latent to sensible gains in the room.

Method (*a*) merely consists of making use of the protractor at the top left-hand corner of the chart.

EXAMPLE 6.8. Calculate the slope of the room ratio line if the sensible gains are 7·3 kW and the latent gains are 1·4 kW.

Answer

$$\text{Total gain} = 7{\cdot}3 + 1{\cdot}4 = 8{\cdot}7 \text{ kW}$$

$$\text{Ratio: } \frac{\text{sensible gain}}{\text{total gain}} = \frac{7{\cdot}3}{8{\cdot}7} = 0{\cdot}84$$

This value can be marked on the outer scale of the protractor on the chart and with the aid of a parallel rule or a pair of set squares the same slope can be transferred to any position on the chart. Figure 6.2 illustrates this. The line O O′ in the protractor is parallel to lines drawn through R_1 and R_2. The important point to appreciate is that the room ratio line can be drawn *anywhere* on the chart. Its slope depends only on the heat gains occurring in the room and *not on the particular room state.*

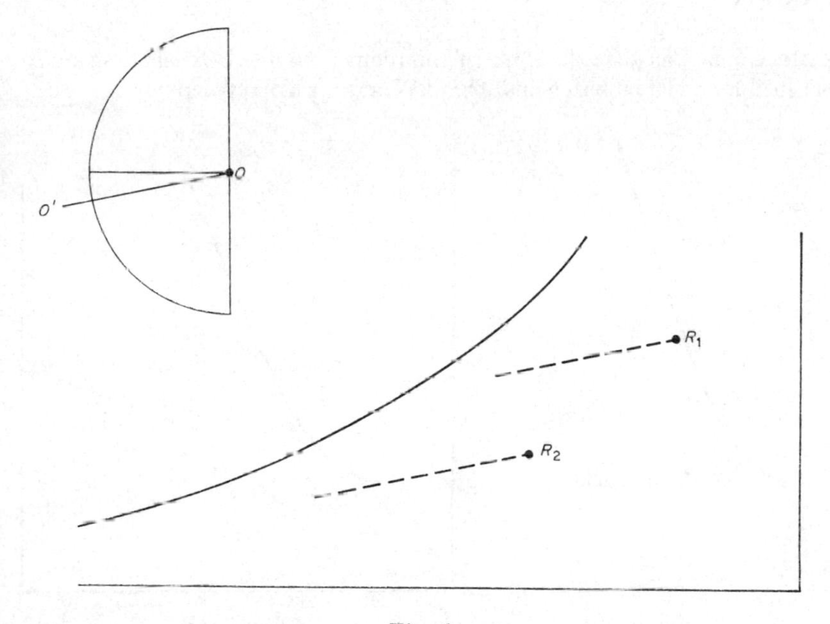

Fig. 6.2

The slope of the line can also be expressed in terms of the ratio of latent gain to sensible gain. This, in itself, is correct but a slightly inaccurate version of this is to calculate the ratio in terms of the moisture picked up by the air supplied to the room and the rise in dry-bulb temperature which it suffers.

Making use of eqns. (6.14) and (6.12), we can write,

$$\frac{\text{latent gain in kW}}{\text{sensible gain in kW}} = \frac{\text{flow rate (m}^3\text{/s)} \times (g_r - g_s) \times 856 \times (273 + t)}{\text{flow rate (m}^3\text{/s)} \times (t_r - t_s) \times 358 \times (273 + t)}$$

Hence
$$\left(\frac{g_r - g_s}{t_r - t_s}\right) = \frac{\text{latent gain in kW}}{\text{sensible gain in kW}} \times \frac{358}{856}$$

or
$$\frac{\Delta g}{\Delta t} = \frac{\text{latent gain}}{\text{sensible gain}} \times \frac{358}{856} \qquad (6.15)$$

This is slightly inaccurate because it expresses sensible changes in terms of dry-bulb temperature changes. Since the scale of dry-bulb temperature is not linear on the psychrometric chart, the linear displacement corresponding to a given change in dry-bulb temperature is not constant all over the chart. It follows that the angular displacement of the slope will vary slightly from place to place on the chart. This inaccuracy is small over the range of values of psychrometric state normally encountered and so the use of eqn. (6.15) is tolerated.

EXAMPLE 6.9. Calculate the slope of the room ratio line by evaluating $\Delta g/\Delta t$ for sensible and latent heat gains of 7·3 kW and 1·4 kW respectively.

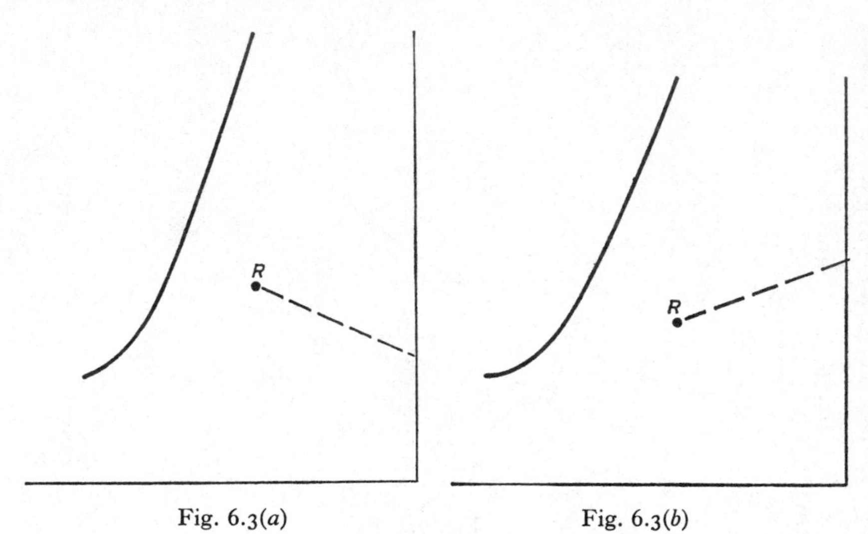

Fig. 6.3(*a*) Fig. 6.3(*b*)

Answer

$$\frac{\Delta g}{\Delta t} = \frac{1{\cdot}4}{7{\cdot}3} \times \frac{358}{856} \text{ by eqn. (6.15)}$$
$$= 0{\cdot}0802 \text{ g/kg °C.}$$

The room ratio line does not necessarily have to slope downwards from right to left as in Fig. 6.2. A slope of this kind indicates the presence of sensible and latent heat gain. Gains of both kinds do not always occur. In winter, for example, a room might have a sensible heat loss coupled with a latent heat gain. Under these circumstances, the line might slope downwards from left to right as in Fig. 6.3 (*a*). It is also possible, if a large amount of outside air having a moisture content lower than that in the room infiltrates, for the room to have a latent heat loss coupled with a sensible heat loss (or even with a sensible heat gain). The room ratio line for both latent and sensible losses would slope upwards from left to right, as in Fig. 6.3 (*b*).

EXAMPLE 6.10. If the room mentioned in Example 6.6 suffers a sensible heat loss of 3 kW and a latent heat gain of 1·2 kW in winter, calculate the necessary supply state of the air delivered to the room. It is assumed that the supply fan handles the same amount of air in winter as in summer and that the battery required to heat the air in winter is positioned on the suction side of the fan.

Answer

Supply air quantity = 1·473 m^3/s.

This volumetric flow rate is the same in winter even though its temperature will be at a much higher value than the 16°C used for its expression in summer because the fan handles a constant amount of air (*see* § 15.21).

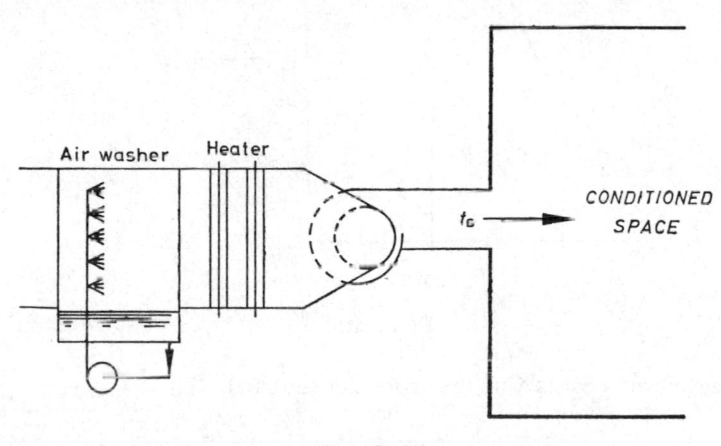

Fig. 6.4(*a*)

A diagram of the plant is shown in Fig. 6.4(*a*) and of the psychrometry in Fig. 6.4(*b*).

Since a heat loss is to be offset, the air must be heated by the heater battery to temperature t_s and the value of t_s must exceed t_r the room temperature.

From eqn. (6.12)

$$1{\cdot}473 = \frac{3}{(t_s - 20)} \times \frac{(273 + t_s)}{358}$$

There are two ways of solving this linear equation:

1.
$$1{\cdot}473 t_s - 1{\cdot}475 \times 20 = \frac{3}{358} \times (273 + t_s)$$

$$t_s = 21{\cdot}67°C$$

2. Guess a value of t, say 25°C, and use in the expression $(273+t)$. Then, from eqn. (6.12)

$$t_s = 20° + \frac{3}{1{\cdot}473} \times \frac{(273+25)}{358}$$
$$= 20° + 1{\cdot}70°$$
$$= 21{\cdot}70°\text{C}.$$

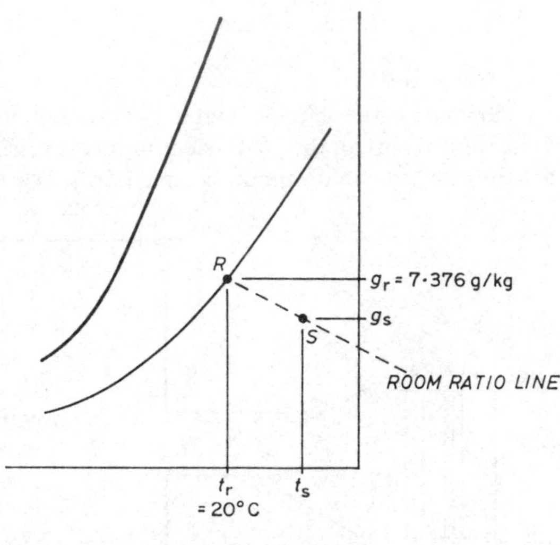

Fig. 6.4(*b*)

Try again with $t = 22$°C in the expression $(273+t)$. Then

$$t_s = 20° + \frac{3}{1{\cdot}473} \times \frac{(273+22)}{358}$$
$$= 21{\cdot}68°\text{C}.$$

From eqn. (6·14)

$$g_s = 7{\cdot}376 - \frac{1{\cdot}2}{1{\cdot}473} \times \frac{(273+21{\cdot}67)}{856}$$
$$= 7{\cdot}376 - 0{\cdot}280$$
$$= 7{\cdot}096 \text{ g/kg dry air.}$$

This establishes the supply state, S, as 21·67°C dry-bulb and 7·096 g/kg, moisture content.

S must lie on the room ratio line. This is self-evident since we have just calculated

$$\Delta t = 1{\cdot}67°\text{C and } \Delta g = 0{\cdot}280 \text{ g/kg}$$

Thus

$$\frac{\Delta g}{\Delta t} = \frac{0{\cdot}280}{1{\cdot}67}$$
$$= 0{\cdot}168 \text{ g/kg °C.}$$

Strictly speaking, if the room ratio line is regarded as having a positive slope when both sensible and latent heat gains are occurring, its slope should be regarded as negative for this case. That is,

$$\frac{\Delta g}{\Delta t} = \frac{+0{\cdot}280}{-1{\cdot}670} = -0{\cdot}168 \text{ g/kg °C}.$$

On the other hand, if both a latent heat loss and a sensible heat loss occurred, the slope would be positive again:

$$\frac{\Delta g}{\Delta t} = \frac{-0{\cdot}280}{-1{\cdot}670} = +0{\cdot}168 \text{ g/kg °C}.$$

It is best not to adhere to a sign convention but to express the slope always as positive and to use one's understanding of fundamentals to decide on the way in which the line slopes.

6.5 Heat gain due to fan power

The flow of air along a duct results in the airstream suffering a loss of energy, and for the flow to be maintained a fan must make good the energy loss (*see* § 15.3). The energy dissipated through the ducting system is apparent as a change in the total pressure of the airstream and the energy input by the fan is indicated by the fan total pressure.

Ultimately, all energy losses appear as heat (although, on the way to this, some are evident as noise, in duct systems). So an energy balance equation can be formed involving the energy supplied by the fan and the energy lost in the airstream. That is to say, the loss of pressure suffered by the airstream as it flows through the ducting system and past the items of plant (which offer a resistance to airflow) constitutes an adiabatic expansion which must be offset by an adiabatic compression at the fan.

So, all the power supplied by the fan is regarded as being converted to heat and causing an increase in the temperature of the air handled, Δt, as it flows through the fan.

A heat balance equation can be written, accepting an expression for air power derived later in section 15·3.

$$\text{Air power} = \text{fan total pressure (N/m}^2) \times \text{volumetric flow rate (m}^3\text{/s)}$$

The rate of heat gain corresponding to this is the volumetric flow rate $\times$ $\rho \times c \times \Delta t$, where ρ and c are the density and specific heat of air, respectively.

Hence
$$\Delta t = \frac{\text{fan total pressure (N/m}^2)}{\rho c}$$

The air quantities have cancelled, indicating that the rise in air temperature is independent of the amount of air handled, and using $\rho = 1{\cdot}2$ kg/m^3 and $c = 1026$ J/kg °C we get

$$\Delta t = \frac{\text{fan total pressure (N/m}^2\text{)}}{1231} \text{ in °C.} \tag{6.16}$$

Thus, the air suffers a temperature rise of 0·000813°C for each N/m^2 of fan total pressure.

The energy the fan receives is in excess of what it delivers to the airstream, since frictional and other losses occur as the fan impeller rotates the airstream. The power input to the fan shaft is termed the *fan power* (*see* section 15.3) and the ratio of the air power to the fan power is termed the *total fan efficiency* and is denoted by η. Not all the losses occur within the fan casing. Some take place in the bearings external to the fan, for example. For this reason, full allowance for efficiency should not be made. It is suggested that a compromise be adopted.

Taking the fan efficiency as 70 per cent (a modest allowance), instead of 30 per cent losses being absorbed by the airstream, suppose that 15 per cent are transmitted to the air handled. Equation (6.16) becomes

$$\Delta t = \frac{\text{fan total pressure (N/m}^2\text{)}}{1231 \times 0{\cdot}85}$$

$$\Delta t = \frac{\text{fan total pressure (N/m}^2\text{)}}{1045} \tag{6.17}$$

Thus almost one thousandth of a degree rise in temperature for each N/m^2 of fan total pressure results from the energy input at the fan. Put another way: one tenth of a degree rise occurs for each millibar of fan total pressure.

6.6 Wasteful re-heat

If an arbitary choice of supply air temperature is made, it is almost certain that the resultant design will not be an economical one, either in capital or in running cost. The reason for this is that wasteful re-heat may have to be offset by the refrigeration plant. The point is best illustrated by means of an example.

EXAMPLE 6.11. The sensible and latent heat gains to a room are 10 kW and 1 kW, respectively. Assuming that the plant illustrated in Fig. 6.5(*a*) is used and that the cooler coil has a contact factor of 0·85, calculate the cooling load and analyse its make-up if the outside condition is 28°C dry-bulb, 20·9°C wet-bulb (sling) and the condition maintained in the room is 22°C dry-bulb and 50 per cent saturation. The supply air temperature is arbitrarily fixed at 16°C. Assume that the temperature rise due to fan power is 1°C.

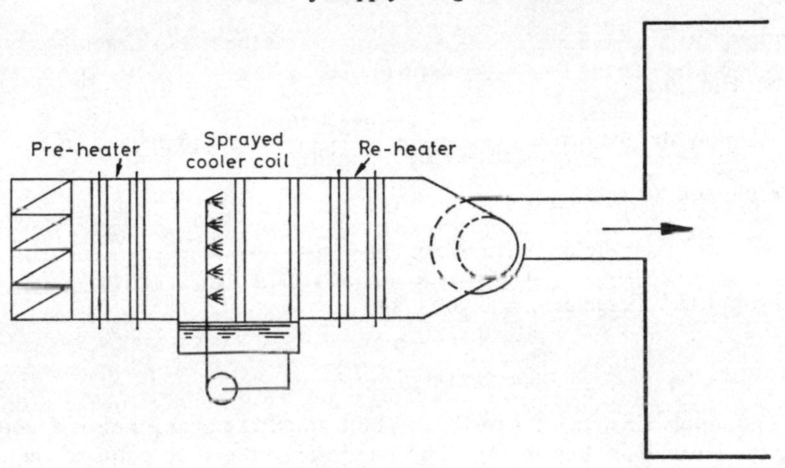

Fig. 6.5(*a*)

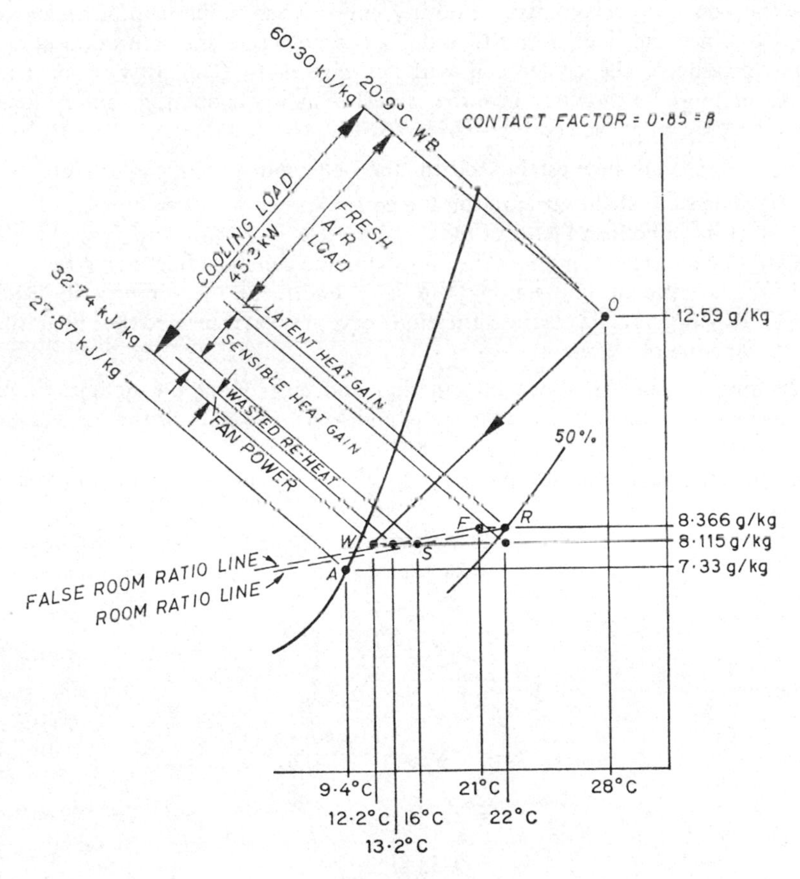

Fig. 6.5(*b*)

Answer

From eqn. (6.12):

$$\text{supply air quantity} = \frac{10}{(22^\circ - 16^\circ)} \times \frac{(273+16)}{358} = 1{\cdot}346\ \text{m}^3/\text{s at } 16^\circ\text{C}$$

From eqn. (6.14):

$$\text{supply air moisture content} = g_r - \frac{1}{1{\cdot}346} \times \frac{(273+16)}{856}$$

From I.H.V.E. tables $g_r = 8{\cdot}366$ g/kg

$$\therefore g_s = 8{\cdot}366 - 0{\cdot}251$$
$$= 8{\cdot}115\ \text{g/kg}.$$

The supply air state is thus 16°C dry-bulb and 8·115 g/kg. This is denoted by the letter *S* in Fig. 6.5(*b*). The line joining the state point of the air entering a cooler coil to the state point of the air leaving the coil must always cut the 100 per cent relative humidity curve when suitably produced (*see* § 3.4). Since the line joining *O* and *S* does not cross the saturation curve when produced, the cooler coil will not give state *S* directly. The coil must therefore be chosen to give the correct moisture content, g_s, and re-heat must be used to get the correct temperature, t_s.

Four things are now established for the behaviour of the cooler coil:

(i) The state of the air entering the coil is known.
(ii) It has a contact factor of 0·85.
(iii) The moisture content of the air leaving the coil must be 8·115 g/kg.
(iv) The straight line joining the state points of the air entering and leaving the coil cuts the saturation curve at what is termed the apparatus dew point.

Making use of the above information and employing the definitions of contact factor expressed by eqns. (3.1) and (3.3), the state of the air leaving the cooler coil *W* can be worked out.

Denote the apparatus dew point by *A*. Then, on a moisture content basis—

$$0{\cdot}85 = \frac{g_o - g_w}{g_o - g_a}$$
$$= \frac{12{\cdot}59 - 8{\cdot}115}{12{\cdot}59 - g_a}$$

Whence—

$$g_a = 12{\cdot}59 - \frac{(12{\cdot}59 - 8{\cdot}115)}{0{\cdot}85}$$
$$= 12{\cdot}59 - 5{\cdot}26$$
$$= 7{\cdot}33\ \text{g/kg}.$$

Thus, A is located at 100 per cent relative humidity and 7·33 g/kg.

W can now be marked in on the line joining O to A at a moisture content of 8·115 g/kg.

The temperature of A, from tables (or from a chart) is 9·4°C and its enthalpy is 27·87 kJ/kg.

Using the definition of contact factor once more, but in terms of differences of enthalpy:

$$0{\cdot}85 = \frac{h_o - h_w}{h_o - h_a}$$

$$= \frac{60{\cdot}30 - h_w}{60{\cdot}30 - 27{\cdot}87}$$

h_o is found from tables or a chart.

Then
$$h_w = 60{\cdot}30 - 0{\cdot}85(60{\cdot}30 - 27{\cdot}87)$$
$$= 60{\cdot}30 - 0{\cdot}85 \times 32{\cdot}43$$
$$= 60{\cdot}30 - 27{\cdot}56$$
$$= 32{\cdot}74 \text{ kJ/kg.}$$

Making use of the approximate form of the definition of contact factor one can establish the value of the temperature at W.

$$0{\cdot}85 = \frac{28^\circ - t_w}{28^\circ - 9{\cdot}4^\circ}$$
$$t_w = 28^\circ - 0{\cdot}85(28^\circ - 9{\cdot}4^\circ)$$
$$= 28^\circ - 0{\cdot}85 \times 18{\cdot}6^\circ$$
$$= 28^\circ - 15{\cdot}8^\circ$$
$$= 12{\cdot}2^\circ\text{C.}$$

The enthalpy at O and at W is known and the amount of air flowing over the coil is also known. The cooling load can therefore be calculated—taking the humid volume of the air supplied at state S to be 0·8294 m³/kg, from tables:

$$\text{cooling load} = \frac{1{\cdot}346}{0{\cdot}8294} \times (60{\cdot}30 - 32{\cdot}74)$$

$$= \frac{1{\cdot}346 \times 27{\cdot}56}{0{\cdot}8294}$$

$$= 44{\cdot}73 \text{ kW.}$$

It is instructive to analyse the load:

$$\text{fresh-air load} = \frac{1{\cdot}346 \times (60{\cdot}30 - 43{\cdot}39)}{0{\cdot}8294} = 27{\cdot}44\ \text{kW}$$

$$\text{sensible heat gain} = \frac{1{\cdot}346 \times 1{\cdot}026 \times (22^\circ - 16^\circ)}{0{\cdot}8294} = 10{\cdot}00\ \text{kW}$$

$$\text{latent heat gain} = \frac{1{\cdot}346 \times 2454 \times (8{\cdot}366 - 8{\cdot}115)}{0{\cdot}8294 \times 1000} = 1{\cdot}00\ \text{kW}$$

$$\text{fan power} = \frac{1{\cdot}346 \times 1{\cdot}026 \times (13{\cdot}2^\circ - 12{\cdot}2^\circ)}{0{\cdot}8294} = 1{\cdot}66\ \text{kW}$$

$$\text{re-heat wasted} = \frac{1{\cdot}346 \times 1{\cdot}026 \times (16^\circ - 13{\cdot}2^\circ)}{0{\cdot}8294} = \underline{4{\cdot}66\ \text{kW}}$$

$$\text{Total} = 44{\cdot}76\ \text{kW}$$

(This is to be compared with 44·73 kW, computed by using the difference of enthalpy across the cooler coil.)

The analysis is illustrated in Fig. 6.5(*b*). It can be seen that about 4·66 kW plays no useful thermodynamic part. The fact that re-heat was needed arose from the arbitrary decision to make the supply air temperature 16°C.

A more sensible approach would be to dispense with re-heat entirely and to use the lowest practical temperature for the supply air consistent with the contact factor of the cooler coil. The temperature difference between t_r and t_s would then be increased and less supply air would be needed. The consequence of this would be to reduce the cooling load because, although the sensible and latent heat gains are unaffected, the fresh-air load and the cooling dissipated in offsetting fan power are reduced, since less air is handled. This is clarified in the next section.

6.7 The choice of a suitable supply state

The supply air temperature chosen should be as low as possible, provided air distribution difficulties do not arise in the conditioned room. In practice, this means that the temperature of the air supplied to the room will be about 8°C to 11°C less than the temperature maintained therein. In making the selection, due allowance must be given to the temperature rise from fan power and duct heat gains and it must be borne in mind that a cooler coil never has a contact factor of 1·0. A practical value of this is 0·8 to 0·9. The larger the value of the contact factor, the deeper and more expensive the cooler coil becomes.

Consider Fig. 6.6. If the temperature rise due to fan power etc. were nothing, then the state of the air leaving the cooler coil W' would lie on the room ratio line, if re-heat were to be entirely avoided. Air would enter the coil at state O and leave it at state W'. The apparatus dew point would be A' and the contact factor would be

$$\beta = \frac{h_o - h_{w'}}{h_o - h_{a'}}$$

State W' could be profitably used as state S; the sensible and latent gains could be offset by $(t_r - t_{w'})$ and $(g_r - g_{w'})$, respectively.

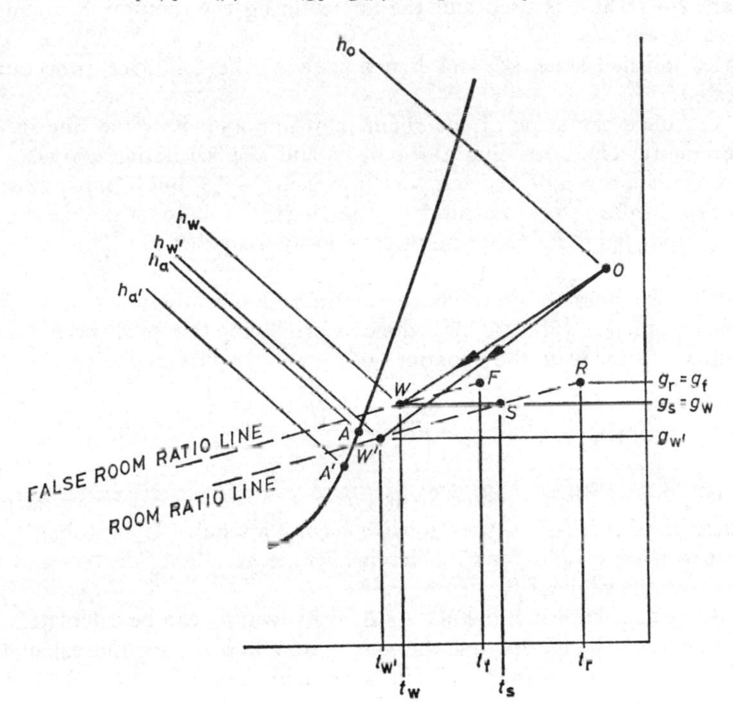

Fig. 6.6

The value $(t_r - t_{w'})$ would be used to calculate the supply air quantity by means of eqn. (6.12) (or from first principles) and $g_{w'}$ would be determined either by reference to the position of W' on the room ratio line or by calculation, knowing a value for the slope of the room ratio line, namely,

$$\Delta g = \frac{\Delta g}{\Delta t} \cdot \Delta t$$

or

$$(g_r - g_{w'}) = \frac{\Delta g}{\Delta t}(t_r - t_{w'})$$

whence

$$g_{w'} = g_r - \frac{\Delta g}{\Delta t}(t_r - t_{w'})$$

If an allowance is made for the temperature rise due to fan power and duct heat gains, the picture becomes a little more complicated.

The state of the air leaving the coil must be of the correct moisture content but must be a few degrees cooler than the required supply temperature. That is $(t_s - t_w)$ must correspond to fan power etc.

The change of state suffered by the air in its passage through the air-handling plant is from O to W across the cooler coil and from W to S through the fan. No re-heat is used and the air, entering the room at S, maintains state R therein.

To established state W, and hence state S, the following procedure is suggested:

1. Calculate the slope of the room ratio line and draw the line in on a psychrometric chart, passing through R and the saturation curve.

2. Calculate the expected temperature rise $(t_s - t_w)$ due to fan power etc. and draw a false room line ratio on the chart, parallel to the real one and starting from point F. State point F is located so that $g_f = g_r$ and $(t_r - t_f) = (t_s - t_w)$.

3. Choose a point W on the false room ratio line so that the contact factor will have a practical value. This is done by producing OW to A and evaluating the contact factor from the equation

$$\beta \simeq \frac{t_o - t_w}{t_o - t_a}$$

If the value of β is between 0·80 and 0·90 then the coil will be a reasonably practical one. If it is not, then make a second attempt. It is probably convenient to select a value for t_w which is a whole number of degrees and then to check on the value of β.

4. Since t_w is known and since $\Delta g/\Delta t$ is known, g_w can be calculated.

The state of the air leaving the coil is now known and the calculations for air quantity and cooling load can be done in the usual way.

EXAMPLE 6.12. Using the relevant information in Example 6.11 choose a suitable supply air temperature which avoids the use of re-heat. Calculate the supply air quantity, the cooling load and the contact factor of the cooler coil.

Answer

As for Example 6.11, the room ratio line is fixed by the sensible and latent heat gains and is, therefore, as shown in Fig. 6.5(*b*).

Draw in a false room ratio line, through F, parallel to RS, so that $t_r - t_f = 1°C$ (as allowed in Example 6.11 for fan power). Figure 6.7 illustrates this.

Choose a point W, on the false room ratio line so that when OW is joined and produced to A, the ratio $t_o - t_w / t_o - t_a$ is of the order of 0·85.

In fact, choose $t_w = 12°C$, then by joining OW and producing to A, it is found that $t_a = 9°C$.

Then
$$\beta \simeq \frac{28° - 12°}{28° - 9°}$$
$$= \frac{16°}{19°}$$
$$= 0{\cdot}84$$

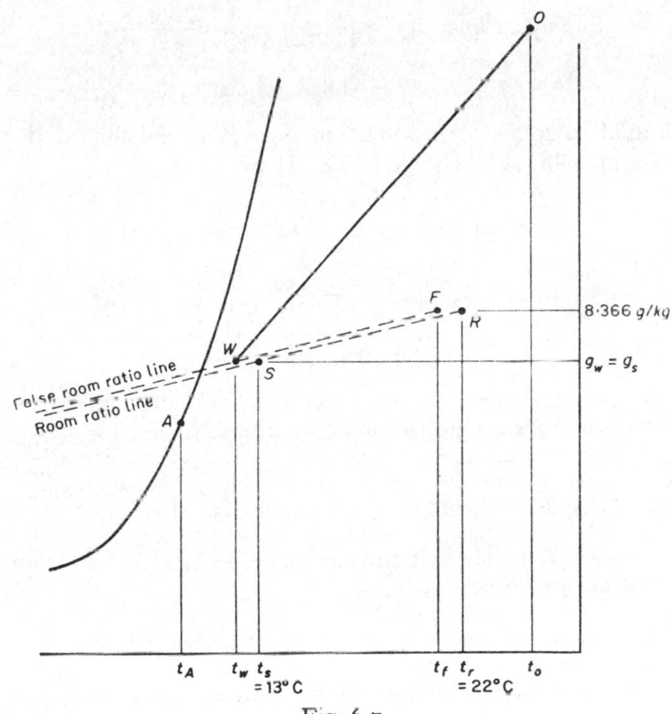

Fig. 6.7

This is satisfactory so the choice of $t_w = 12°C$ is accepted.

Since $t_w = 12°C$, and the temperature rise due to fan power is 1°C, then the supply air temperature is $12° + 1° = 13°C$.

$$\frac{\Delta g}{\Delta t} = \frac{1}{10} \times \frac{358}{856} = 0{\cdot}0418 \text{ g/kg °C}$$

$$\Delta g = \Delta t \times \frac{\Delta g}{\Delta t}$$

$$(g_r - g_s) = (t_r - t_s)\frac{\Delta g}{\Delta t}$$

$$g_s = g_r - (22° - 13°) \times 0{\cdot}0418$$
$$= 8{\cdot}366 - 9 \times 0{\cdot}0418$$
$$= 8{\cdot}366 - 0{\cdot}376$$
$$= 7{\cdot}990 \text{ g/kg.}$$

This must also equal g_w.

Hence, the state of the air leaving the cooler coil W is 12°C dry-bulb and 7·990 g/kg.

$$\text{Supply air quantity} = \frac{10}{(22° - 13°)} \times \frac{(273+13)}{358}$$
$$= 0{\cdot}898 \text{ m}^3\text{/s at } 13°\text{C.}$$

The humid volume at S is 0·8206 m^3/kg. The enthalpy at W is 32·23 kJ/kg. The enthalpy at O is 60·30 kJ/kg. Hence,

$$\text{cooling load} = \frac{0{\cdot}898}{0{\cdot}8206} \times (60{\cdot}30 - 32{\cdot}23)$$
$$= 30{\cdot}72 \text{ kW.}$$

EXERCISES

1. A room is to be maintained at a condition of 20°C dry-bulb and 7·376 g/kg moisture content by air supplied at 15°C dry-bulb when the heat gains are 7 kW sensible and 1·4 kW latent. Calculate the weight of air to be supplied to the room and the moisture content at which it should be supplied. Take the latent heat of evaporation at room condition as 2454 kJ/kg.

(*Answers:* 1·365 kg/s dry air, assuming a value of 1·026 kJ/kg °C for the specific heat of humid air; 6·958 g/kg.

2. Briefly discuss the factors to be considered when selecting
(i) a design room temperature,
(ii) the temperature differential betwen supply and room air.

3. An air-conditioning plant comprising filter, cooler coil, fan and distributing ductwork uses only fresh air for the purpose of maintaining comfort conditions in summer. Using the information listed below, choose a suitable supply air temperature, calculate the cooler coil load and determine its contact factor.
Sensible heat gains to conditioned space: 11·75 kW.
Latent heat gains to conditioned space: 2·35 kW.
Outside design state: 28°C dry-bulb, 19·9°C wet-bulb (sling).
Inside design state: 21°C dry-bulb, 50 per cent saturation.
Temperature rise due to fan power and duct heat gains: 1°C. A psychrometric chart is provided for your use.

(*Answers:* Suitable supply air temperature is about 11°C; corresponding cooler coil load is 33·1 kW; corresponding contact factor is 0.88.)

7

Heat Gains from Solar and Other Sources

7.1 The composition of heat gains

A broad distinction is conveniently drawn between heat gains arising from sources outside the conditioned space and those originating within it. These two categories may be sub-divided as follows, and so present an analysis of the overall sensible heat gains:

(a) *External Gains*

(i) Direct and scattered solar radiation through windows.
(ii) Heat transmitted through the glass and also through the non-glass fabric of the room enclosure, by virtue of the air-to-air temperature difference. Such gains are often termed ' transmission gains '.
(iii) Solar radiation eventually causing a heat gain to the room through the non-glass fabric of the walls.
(iv) Sensible heat gain arising from the infiltration of warm air from outside.

(b) *Internal Gains*

(i) Electric lighting.
(ii) Occupancy.
(iii) Power dissipation.
(iv) Process work.

There can be latent heat gains to the conditioned space, in addition to the sensible gains. These may arise from the following sources:

(i) Infiltration of outside air.
(ii) Occupancy.
(iii) Process Work.

Included in the term ' process work ' are such diverse activities as cooking by gas or oil, the evaporation of any water and lighting by gas or, just conceivably, by oil.

Many of the items listed are easily determined, but some of them—notably solar heat gain through walls—pose difficulties, as is indicated later in section 7.18. The solution to such problems is governed by the degree of accuracy acceptable and by the magnitude or importance of the item in question. A precise answer to the problems is often impossible, but it is clear that the air-conditioning engineer must be made aware of all possible sources of heat gain if the system he is to design may have a capacity equal to the load likely to be imposed on it for his assumed set of design conditions. Such awareness puts him in a position to make an intelligent guess when a precise solution is impossible or exceedingly difficult.

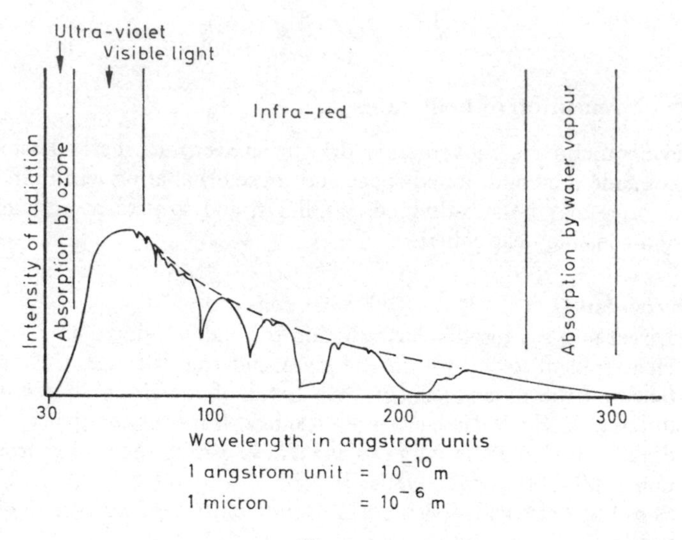

Fig. 7.1

7.2 The physics of solar radiation

For all practical purposes the sun is the only source of energy available for the continuation of life on earth. It radiates energy as a black body having a surface temperature of 6000°C over a spectrum stretching from wavelengths of 0·29 to 4·75 micrometres. As Fig. 7.1 shows, most of the energy is concentrated in the visible portion of sunlight and in the infra-red region. The broken boundary curve in the figure indicates the distribution of energy in the radiation reaching the fringe of the atmosphere. The full line shows a number of troughs, each of which represents an absorption of energy as the sunlight passes through the atmosphere. The total intensity of the radiation at the upper limit of the atmosphere is about 1·362 kW/m^2, incident upon a surface normal to the path of the rays. This figure is a mean, since the earth is slightly closer (about 3 per cent) to the sun in January than it is in July, the orbit of the earth being elliptical. Only about 1·025 kW/m^2 reaches the surface of the earth when the sun is vertically overhead in a cloudless sky. Of this figure, about 0·945 kW/m^2 is from radiation received directly from the sun and the remainder from indirect solar radiation received from the vault of the sky.

Outside the earth's atmosphere, direct solar radiation is composed of about 5 per cent ultra-violet, 52 per cent visible light and 43 per cent infra-red. At the surface of the earth, its approximate composition is 1 per cent ultra-violet, 39 per cent visible light and 60 per cent infra-red. It is clear from this, and from Fig. 7.1, that a good deal of selective absorption is taking place as the sunlight traverses the atmosphere.

7.3 Sky radiation

This is otherwise known as 'diffused' radiation or 'scattered' radiation. Its presence constitutes a heat gain to the earth, in addition to that from direct radiation, and it arises from the translucent nature of the atmosphere.

The atmospheric losses in direct radiation which produce the sky radiation stem from four principal phenomena:

(i) A scattering of the direct radiation in all directions which occurs when the radiant energy encounters the actual molecules of the ideal gases (nitrogen and oxygen) in the atmosphere. This effect is more pronounced for the shorter wavelengths, and accounts for the blue appearance of the sky.
(ii) Scattering resulting from the presence of molecules of water vapour.
(iii) Selective absorption by the ideal gases and by water vapour. Asymmetrical gaseous molecules such as ozone, water vapour, carbon-dioxide etc. have a greater ability to absorb (and hence to emit) radiation than do gaseous molecules of symmetrical structure such as nitrogen and oxygen.
(iv) Scattering caused by dust particles.

The losses incurred are largely due to the permanent gases and to water vapour. There is also some diffusion of the direct sunlight caused when direct (and indirect) radiation encounters cloud. Not all the solar energy removed from direct radiation by the processes of scattering and absorption reaches the surface of the earth. Some is scattered back to space, and the absorbed energy which is re-radiated at a longer wavelength (mostly from water vapour) is also partly lost to outer space. Some direct energy is additionally lost by reflection from the upper surfaces of clouds. However, a good deal of the energy which direct radiation loses does eventually reach the surface of the earth in a scattered form.

The amount of sky radiation varies with the time of day, the weather, the cloud cover and the portion of the sky from which it is received, the amount from the sky in the vicinity or the sun exceeding that from elsewhere, for example. Sky radiation cannot, however, be assigned a specific direction and, hence, it casts no directional shadow outdoors.

The A.S.H.R.A.E. Guide gives equations for assessing the total scattered radiation received by a surface from the sky and the ground. The amount received depends on seasonal variations of moisture constant and earth-sun distance, the angular relation with the surroundings and the relevant surface reflectances.

The I.H.V.E., on the other hand, quote values for clear skies only. The values in Table 7.1 express the intensity of sky radiation received by horizontal and vertical surfaces in W/m^2 for various months of the year and various altitudes of the sun. The sun is higher in the sky at midday, for example, and there is consequently a good deal more direct radiation available for scattering. Thus, values of sky radiation depend on the altitude of the sun above the horizon as well as (although to a lesser extent) on orientation. In spite of this, the intensity is so low that, as has been remarked, sky radiation casts no directional shadow.

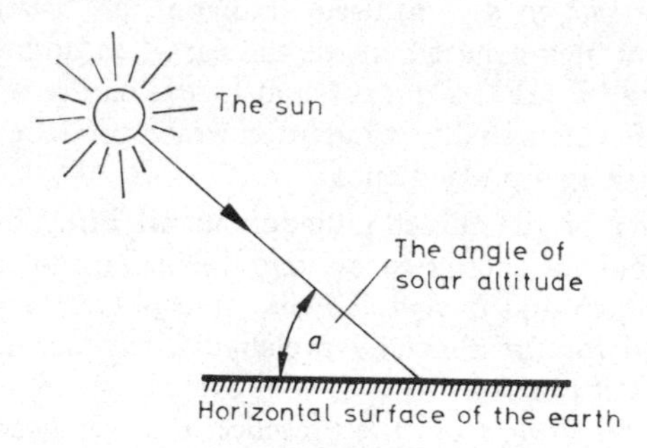

Fig. 7.2(*a*)

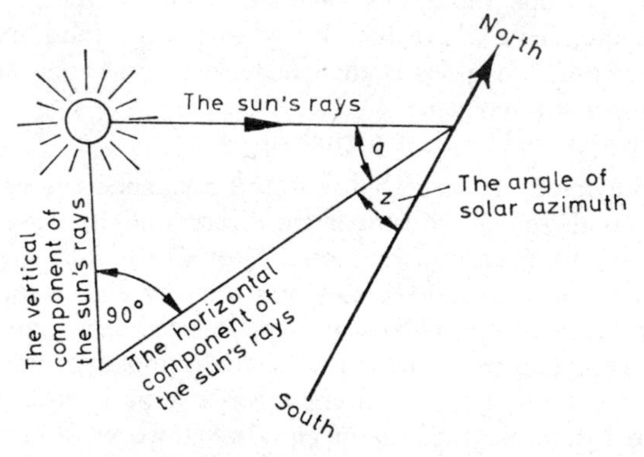

Fig. 7.2(*b*)

7.4 Definitions

There are a few basic terms in common use to describe the attributes of direct solar radiation. These are defined below, without lengthy explanation. The full implications of their meaning will become clear later in the text.

Altitude of the sun (*a*). This is the angle a direct ray from the sun makes with the horizontal at a particular place on the surface of the earth. It is illustrated in Fig. 7.2(*a*). For a given date and time, the sun's altitude is different at different places over the world.

Azimuth of the sun (z). This is the angle the horizontal component of a direct ray from the sun makes with the true south in the northern hemisphere. It is illustrated in Fig. 7.2(*b*). Solar azimuth is in degrees of angular displacement east and west of south. (Some authorities choose to express this in continuous angular displacement through 360° from north.)

Solar Wall Azimuth (n). This is the angle the horizontal component of the sun's ray makes with a direction normal to a particular wall. It is illustrated

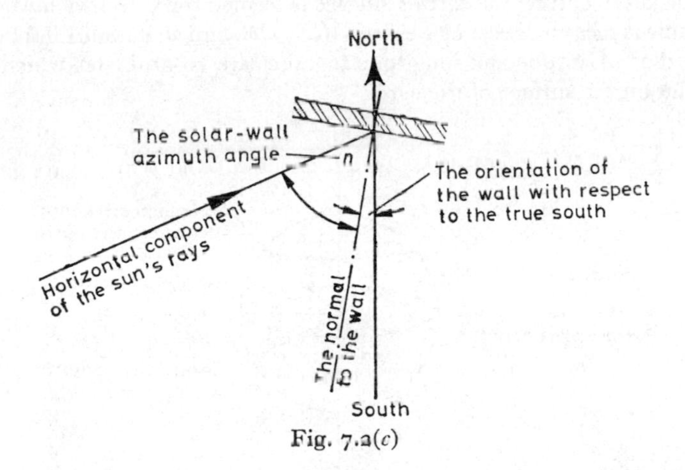

Fig. 7.2(*c*)

in Fig. 7.2(*c*), where it can be seen that this angle, referring as it does to direct radiation, can have a value between 0° and 90° only. When the wall is in shadow ($n > 90°$) the value of n has no meaning.

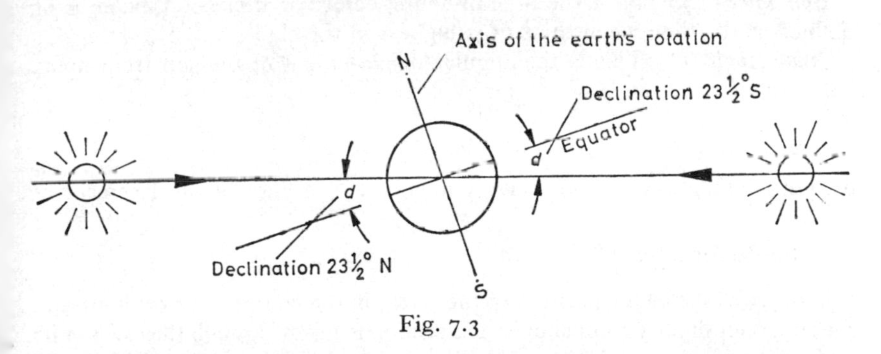

Fig. 7.3

Declination (d). This is the angular displacement of the sun from the plane of the earth's equator. Because the axis of the earth is tilted at an angle of about $23\frac{1}{2}°$ to the axis of the plane in which it orbits the sun, the value of the declination will vary throughout the year between $+23\frac{1}{2}°$ and $-23\frac{1}{2}°$. Declination is expressed in degrees north or south (of the equator). The geometry of declination is illustrated in Fig. 7.3.

Latitude (L). The latitude of a place on the surface of the earth is its angular displacement above or below the plane of the equator, measured from the centre of the earth. It is illustrated in Fig. 7.4. The latitude at London is about 51° North of the equator.

Longitude. This is the angle which the semi-plane through the poles, and a particular place on the surface, makes with a similar semi-plane through Greenwich. The semi-plane through Greenwich is an arbitrary zero, and the line it makes in cutting the earth's surface is termed the Greenwich Meridian. Longitude is measured east or west of Greenwich and so its value lies between 0° and 180°. Latitude and longitude together are co-ordinates which locate any point on the surface of the earth.

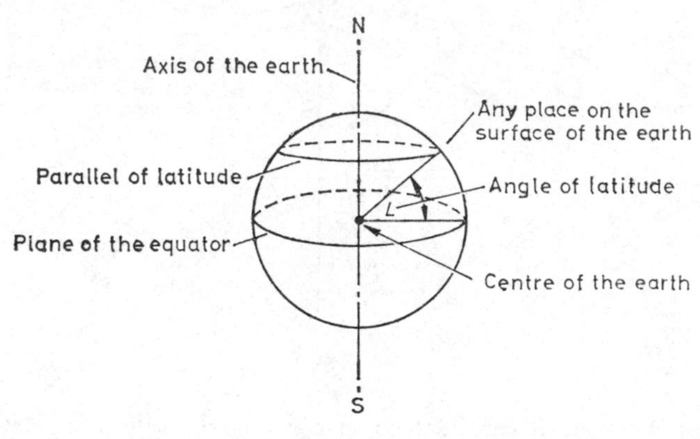

Fig. 7.4

Sun Time (T). This is the time in hours, before or after noon, noon being defined as the time when the sun is highest in the sky.

Hour Angle (h). This is the angular displacement of the sun from noon.

Thus,
$$h = \frac{360°}{24} \times T$$

or, put another way, 1 hour corresponds to 15° of angular displacement.

7.5 The declination of the sun

Figure 7.5(a) shows the position of the earth, in the course of its orbit around the sun, seen slightly from above. If a section is taken through the earth at its two positions at mid-winter and mid-summer, the concept of declination is seen more clearly. Figure 7.5(b) is such a section, illustrating that the sun is vertically overhead at noon on a latitude of $23\frac{1}{2}°$ S in mid-winter and also on $23\frac{1}{2}°$ N at noon in mid-summer. These two values of latitude correspond to the sun's declination at those dates and are a consequence of the earth's pole being tilted at an angle of $23\frac{1}{2}°$ to the axis of the plane in which it orbits the sun. Figure 7.5(b) also illustrates this.

The angular inclination of the pole does not vary substantially from this value owing to the gyroscopic effect of the earth's rotating mass. Because of this, the sun can be vertically overhead only between the latitudes $23\frac{1}{2}°$ N and $23\frac{1}{2}°$ S. These two parallels of latitude are called the tropics of Cancer

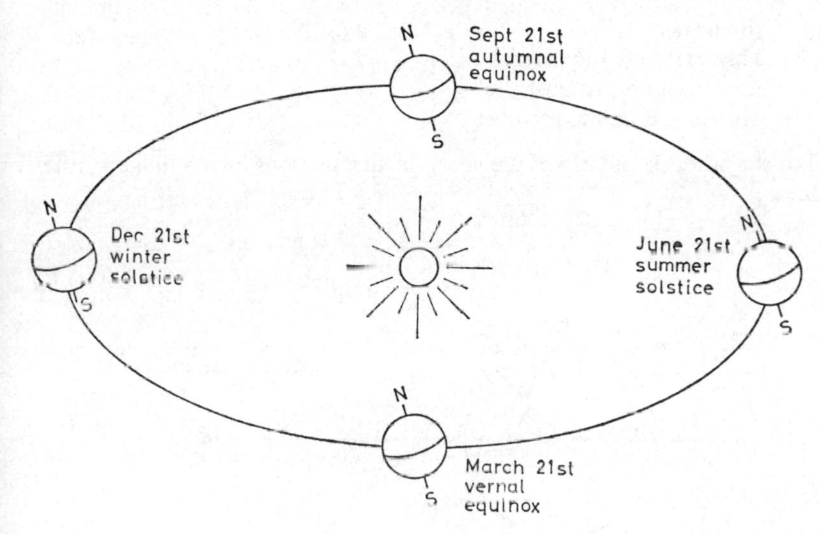

Fig. 7.5(*a*)

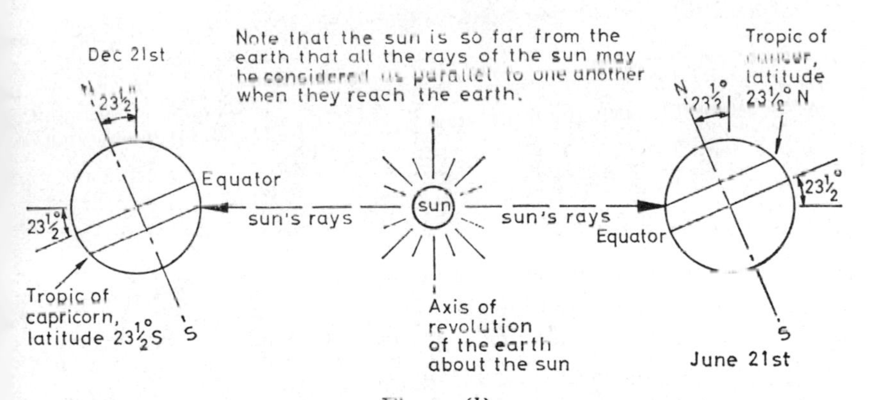

Fig. 7.5(*b*)

and Capricorn, respectively. They define the bounds of the Torrid Zone, often referred to as 'the Tropics'. It follows that, at dates other than the winter and summer solstices, the sun will be vertically overhead at noon in the tropics, at some latitude between $23\frac{1}{2}°$ N and S. In fact, the sun is vertically overhead at noon on the equator itself on two dates in the year—the autumnal equinox and the vernal (spring) equinox. On these occasions the declination is zero.

The dates on which the solstices and the equinoxes occur are not always the same because the earth's year is not exactly 365 days. Thus, the presence of a leap year introduces a slight variation in the dates quoted.

Approximate values of declination are:

June 21st	$23\frac{1}{2}°$ N
May 21st and July 21st	$20\frac{1}{4}°$ N
April 21st and August 21st	$11\frac{1}{2}°$ N
March 21st and September 21st	0°

For the other six months of the year, the declinations have similar southerly values.

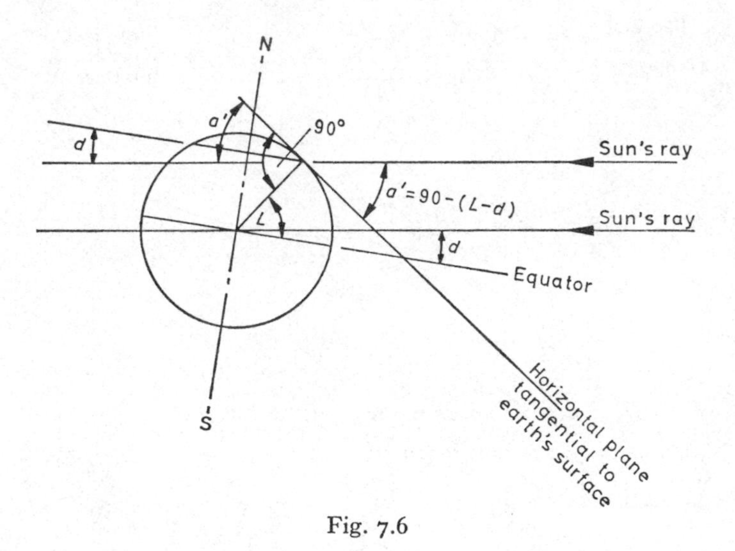

Fig. 7.6

7.6 The altitude of the sun

It is worth considering two cases:

(*a*) the special case for the altitude at noon;

(*b*) the general case for the altitude at any time of the day.

(*a*) The altitude at noon (a')

This is easily illustrated in geometrical terms, as in Fig. 7.6, in which it can be seen that a very simple expression is derived from a consideration of the latitude of the place in question and the declination for the particular date:

$$a' = 90-(L-d) \tag{7.1}$$

It should be borne in mind when considering Fig. 7.6 that all parallels of latitude are parallel to the plane of the equator, and that, because the sun is so far away from the earth, all its rays are parallel to one another when they strike the earth.

EXAMPLE 7.1. Calculate the maximum and minimum altitudes of the sun at noon, in London, given that the latitude of London is 51° N.

Answer

The greatest altitude occurs at noon on 21st June and the least at noon on 21st December. Hence, from eqn. (7.1), and knowing that the declinations are $23\frac{1}{2}°$ N and S respectively in June and December, we can write

$$a'_{(max)} = 90° - (51° - 23\tfrac{1}{2}°) = 62\tfrac{1}{2}°$$

$$a'_{(min)} = 90° - (51° - (-23\tfrac{1}{2}°)) = 15\tfrac{1}{2}°$$

(*b*) The general case of the sun's altitude (*a*)
Although a geometrical illustration is possible, it is in three dimensions and consequently, being difficult to illustrate, is not shown. The result, obtained from such geometrical considerations, is that

$$\sin a = \sin d \times \sin L + \cos d \times \cos L \times \cos h \tag{7.2}$$

This equation degenerates, to give the special result obtained from eqn. (7.1), if a value of $h = 0$ is taken for noon.

EXAMPLE 7.2. Calculate the altitude of the sun in London on 21st June (*a*) at 1 pm and (*b*) at 5 pm (British Summer Time), making use of eqn. (7.2) in both cases.

Answer

(*a*) At noon in London on 21st June

$$d = 23\tfrac{1}{2}° \text{ N (i.e.} +23\tfrac{1}{2}°)$$

$$L = 51° \text{ N (i.e.} +51°)$$

$$T = 0 \text{ (sun time is British Summer Time minus 1 h)}$$

$$\therefore h = 0$$

Then

$$\sin a = \sin 23\tfrac{1}{2}° \times \sin 51° + \cos 23\tfrac{1}{2}° \times \cos 51° \times \cos 0$$

$$= 0{\cdot}399 \times 0{\cdot}777 + 0{\cdot}917 \times 0{\cdot}629 \times 1{\cdot}0$$

$$= 0{\cdot}310 + 0{\cdot}577$$

$$= 0{\cdot}887$$

$$\therefore a = 62\tfrac{1}{2}°.$$

(*b*) At 5 pm $T = 4$ pm, hence $h = 60°$

Then

$$\sin a = \sin 23\tfrac{1}{2}° \times \sin 51° + \cos 23\tfrac{1}{2}° \times \cos 51° \times \cos 60°$$
$$= 0{\cdot}399 \times 0{\cdot}777 + 0{\cdot}917 \times 0{\cdot}629 \times 0{\cdot}5$$
$$= 0{\cdot}310 + 0{\cdot}288$$
$$= 0{\cdot}598$$
$$\therefore \quad a = 36\tfrac{3}{4}°.$$

Note that a more exact value for the latitude of London is 51·5° N and using this would give slightly different answers.

7.7 The Azimuth of the sun

There is no special case that can easily be illustrated geometrically except, perhaps, the trivial one of the sun at noon, when the aximuth is zero by definition. The general case is capable of geometrical treatment but, three dimensions being involved, it is preferred here to give the result:

$$\tan z = \frac{\sin h}{\sin L \cos h - \cos L \tan d} \tag{7.3}$$

EXAMPLE 7.3. Calculate the azimuth of the sun in London on 21st June at 5 pm British Summer Time, making use of eqn. (7.3).

Answer

As before, in Example 7.2.

$$L = +51°, d = +23\tfrac{1}{2}°, h = 60°$$

Then, from eqn. (7.3)

$$\tan z = \frac{0{\cdot}866}{0{\cdot}777 \times 0{\cdot}5 - 0{\cdot}629 \times 0{\cdot}435}$$
$$= 7{\cdot}53$$
$$\therefore \quad z = 82\tfrac{1}{2}° \text{ W of S.}$$

That is, the sun is almost due West.

7.8 The intensity of direct radiation on a surface

If the intensity of direct solar radiation incident on a surface normal to the rays of the sun is I W/m^2, then the component of this intensity in any direction can be easily calculated. Two simple cases are used to illustrate this, followed by a more general case.

Case 1 The component of direct radiation normal to a horizontal surface (I_h), *as illustrated in* Fig. 7.7(*a*).

If the angle of altitude of the sun is a, then simple trigonometry shows that the component in question is $I \sin a$.

$$I_h = I \sin a \tag{7.4}$$

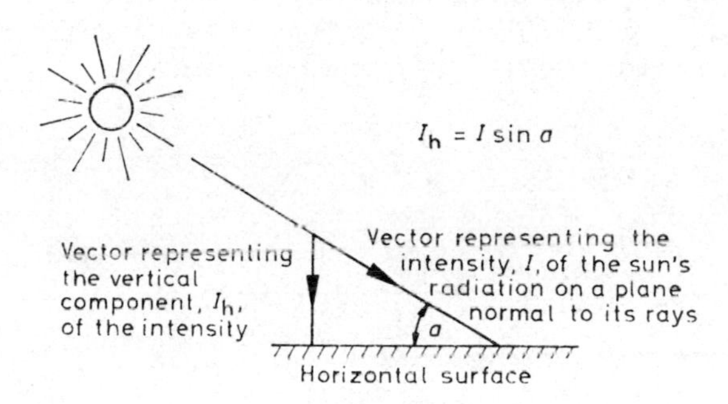

Fig. 7.7(*a*)

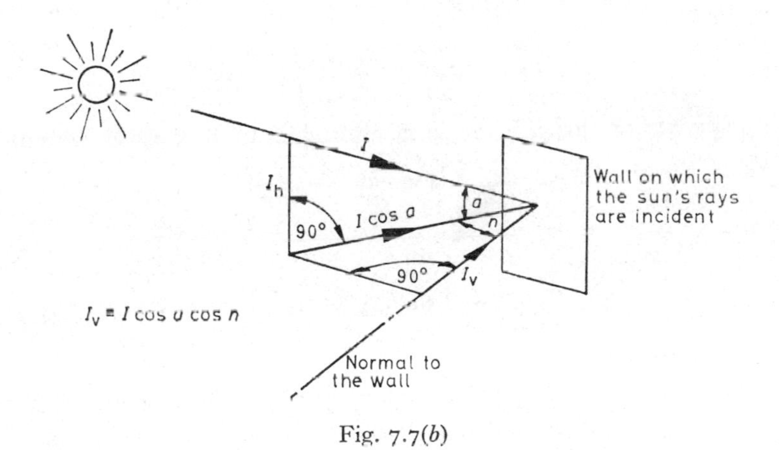

Fig. 7.7(*b*)

Case 2 The component of direct radiation normal to a vertical surface (I_v), *as illustrated in* Fig. 7.7(*b*).

The situation is a little more complicated since the horizontal component of the sun's rays, $I \cos a$, has first to be obtained and then in its turn, has to be further resolved in a direction at right angles to a vertical surface which has a particular orientation: that is, the solar wall azimuthal angle n must be worked out. It then follows that the resolution of $I \cos a$ normally to the wall is $I \cos a \cos n$.

$$I_v = I \cos a \cos n \tag{7.5}$$

EXAMPLE 7.4. If the altitude and azimuth of the sun are $62\frac{1}{2}°$ and $82\frac{1}{2}°$ W of S respectively, calculate the intensity of direct radiation normal to (*a*) a horizontal surface and (*b*) a vertical surface facing south-west.

Answer

Denote the intensity of direct radiation normal to the rays of the sun by the symbol I, having units of W/m^2, and denote its components normal to a horizontal and a vertical surface by I_h and I_v, respectively.

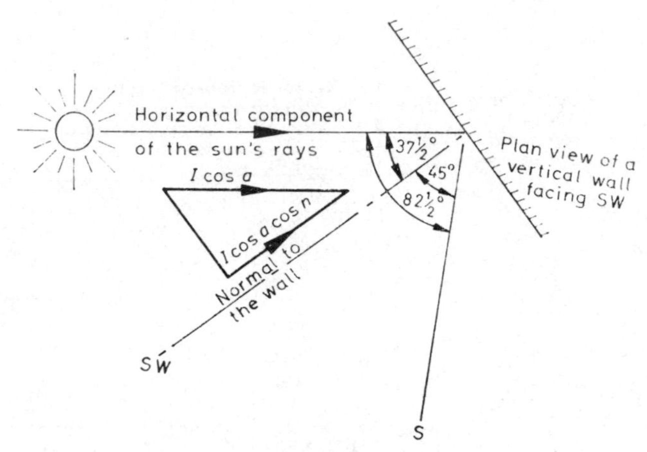

Fig. 7.8

(*a*)

$$I_h = I \sin a \tag{7.4}$$

$$= I \sin 62\tfrac{1}{2}°$$

$$= 0.887\, I$$

(*b*) Draw the diagram shown in Fig. 7.8. From this it is clear that the value of the solar wall azimuthal angle is $37\frac{1}{2}°$.

$$I_v = I \cos a \cos n \tag{7.5}$$

$$= I \cos 62\tfrac{1}{2}° \cos 37\tfrac{1}{2}°$$

$$= I \times 0.462 \times 0.793$$

$$= 0.366\, I$$

Case 3 The component of direct radiation normal to a tilted surface (I_δ).

Figure 7.9 illustrates the case and shows, in section, a surface that is tilted at an angle δ to the horizontal. The surface has an orientation which gives it a solar wall azimuthal angle of n.

If the surface were vertical, then the resolution of I, normal to its surface, would be $I \cos a \cos n$. The sun's rays also have a resolution normal to a horizontal surface of $I \sin a$. It follows that the rays normally incident on a tilted surface instead of a vertical one have a vector sum of $I \sin a$ and $I \cos a \cos n$, in an appropriate direction. The components of $I \sin a$ and $I \cos a \cos n$ in this appropriate direction are $I \sin a \cos \delta$ and $I \cos a \cos n \sin \delta$, respectively, both normal to the tilted surface. Their sum is the total value of the intensity of radiation normally incident on the tilted surface—the answer required.

Hence

$$I_\delta = I \sin a \cos \delta \pm I \cos a \cos n \sin \delta \qquad (7.6)$$

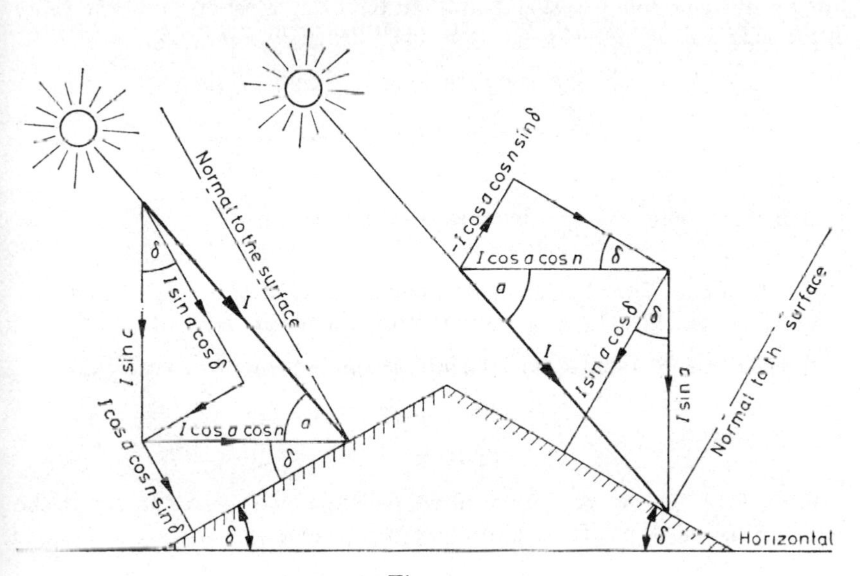

Fig. 7.9

Equation (7.6) degenerates to eqn. (7.4) for a horizontal surface where δ is zero, and to eqn. (7.5) for a vertical surface where δ is 90°. The significance of the alternative signs, is that if the tilted surface faces the sun, a positive sign should be used and *vice versa* if the surface is tilted away from the sun. Naturally enough, if the surface is tilted so much that it is in shadow, I is zero, since the angle of incidence of the radiation will be 90°.

Equation (7.6) can be written in a more general but less informative manner:

$$I_\delta = I \cos i \qquad (7.7)$$

where i is the angle of incidence of the ray on the surface. This is true of all surfaces, but it must be borne in mind that the angle of incidence is between the incident ray and the *normal* to the surface. It is *not* the glancing angle.

All that eqn. (7.7) does is to state in succinct form what eqn. (7.6) does more explicitly. This means that cos i is the same as sin a cos $\delta \pm$ cos a cos n sin δ.

EXAMPLE 7.5. Calculate the component of direct solar radiation which is normally incident upon a surface tilted at 30° to the horizontal and which faces south-west, given that the solar altitude and azimuth are $62\frac{1}{2}°$, and $82\frac{1}{2}°$ west of south, respectively. The tilted surface is facing the sun.

Answer

For a wall facing south-west and an azimuth of $82\frac{1}{2}°$ west of south, the value of the solar wall azimuthal angle, n, is $37\frac{1}{2}°$; hence, by eqn. (7.6),

$$\begin{aligned} I_\delta &= I \sin 62\tfrac{1}{2}° \cos 30° + I \cos 62\tfrac{1}{2}° \cos 37\tfrac{1}{2}° \sin 30° \\ &= I(0{\cdot}887 \times 0{\cdot}866 + 0{\cdot}462 \times 0{\cdot}793 \times 0{\cdot}5) \\ &= 0{\cdot}95\, I. \end{aligned}$$

Doing the same example by means of the data in the I.H.V.E. Guide (*see* Table 7.1) yields the following:

$$\left.\begin{aligned} I_h &= 794\ \text{W/m}^2 \\ I_v &= 327\ \text{W/m}^2 \end{aligned}\right\} \text{by interpolation for a solar altitude of } 62\tfrac{1}{2}° \text{ and a solar wall azimuthal angle of } 37\tfrac{1}{2}°.$$

Intensity on the tilted surface facing the sun is then

$$\begin{aligned} I_\delta &= 0{\cdot}866 \times 794 + 0{\cdot}500 \times 327 \\ &= 852\ \text{W/m}^2 \end{aligned}$$

From Table 7.1 the intensity of direct radiation on a surface normal to the sun's rays is 898 W/m^2, for an altitude of $62\frac{1}{2}°$. Hence,

$$I_\delta = \frac{852}{898} I = 0{\cdot}95\, I.$$

This is the same as the result obtained by first principles.

Similarly, for the case of the surface tilted away from the sun:

$$\begin{aligned} I &= I(0{\cdot}887 \times 0{\cdot}866 - 0{\cdot}462 \times 0{\cdot}793 \times 0{\cdot}5) \\ &= 0{\cdot}584\, I. \end{aligned}$$

7.9 The numerical value of direct radiation

In order to evaluate the amount of solar radiation normally incident upon a surface, it is first necessary to know the value of the intensity, I, which is normally incident on a surface held at right angles to the path of the rays. The values of I can be established only by referring to experimental results for different places over the surface of the earth. These seem to suggest that I

TABLE 7.1 *Intensity of direct solar radiation with a clear sky*

Inclination and orientation of surface	Sun altitude (degrees)											
	5°	10°	15°	20°	25°	30°	35°	45°	50°	60°	70°	80°
	Intensity of direct solar radiation with a clear sky for place 0–300 m above sea level (W/m²)											
1. Normal to sun	210	388	524	620	688	740	782	824	860	893	912	920
2. Horizontal roof	18	67	136	212	290	370	450	523	660	773	857	907
3. Vertical wall: orientation from sun in degrees (wall solar azimuth angle) 0°	210	382	506	584	624	642	640	624	553	447	312	160
10°	207	376	498	575	615	632	630	615	545	440	307	158
20°	197	360	475	550	586	603	602	586	520	420	293	150
30°	182	330	438	506	540	556	555	540	480	387	270	140
40°	160	293	388	447	478	492	490	478	424	342	240	123
45°	148	270	358	413	440	454	453	440	390	316	220	113
50°	135	246	325	375	400	413	412	400	355	287	200	103
55°	120	220	290	335	358	368	368	358	317	256	180	92
60°	105	190	253	292	312	210	320	312	277	224	156	80
65°	90	160	214	247	264	270	270	264	234	190	132	68
70°	72	130	173	200	213	220	220	213	190	153	107	55
75°	54	100	130	150	160	166	165	160	143	116	80	40
80°	36	66	88	100	108	110	110	103	96	78	54	28

(Reproduced by kind permission from the I.H.V.E. Guide)

is independent of the place and that it depends only on the altitude of the sun. There is sense in this, since the amount of direct radiation reaching the surface of the earth will clearly depend on how much is absorbed in transit

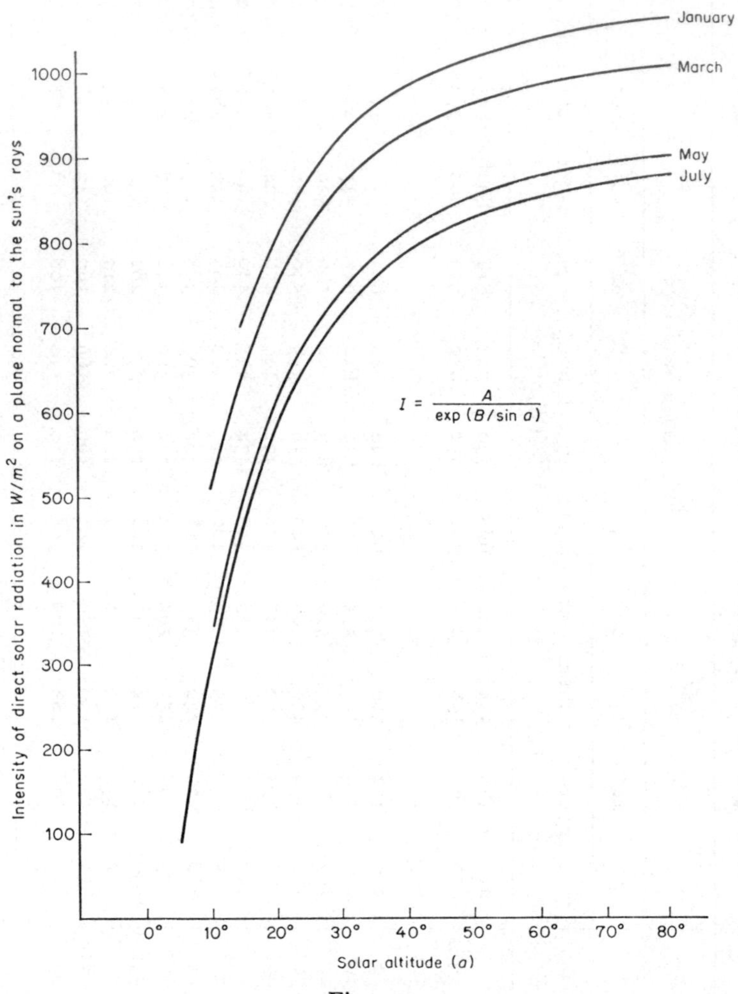

Fig. 7.10

through the atmosphere, and the atmospheric path length is greater when the sun is lower in the sky; that is, when its altitude is less.

There are four sources of information to which reference may be made: *Solar Radiation at Kew Observatory*, by J. M. Stagg, *Proposed Standard Solar Radiation Curves for Engineering Use*, by P. Moon, *Direct Solar Radiation Available on Clear Days*, by J. L. Threlkeld and R. C. Jordan, and the A.S.H.R.A.E. Guide.

Table 7.1 shows data from the second source and Table 7.2 gives correction factors which take account of the increase in intensity with height.

TABLE 7.2 *Percentage increase in direct solar radiation at varying heights above sea level*

Height above sea level	Solar altitude									
	10°	20°	25°	30°	35°	40°	50°	60°	70°	80°
1000 m	32	22	17	16	14	13	12	11	10	10
1500 m	51	32	26	23	21	19	16	15	14	14
2000 m	66	40	34	29	27	24	21	19	18	18
3000 m	89	51	43	37	34	34	31	27	22	22

Note that sky radiation *decreases* by approximately 30 per cent at 1000 m and by about 60 per cent at 1500 m above sea level.

Figure 7.10 shows curves based on the two American sources of information, using the equation:

$$I = A/\exp(B/\sin a) \text{ kW/m}^2 \qquad (7.8)$$

A and B have values (given in Table 7.6) which depend upon seasonal variations in the earth-sun distance, atmospheric moisture content and dust pollution. The information provided by eqn. (7.8) is more accurate than that originally published by Moon, and gives values for average, cloudless days; in very clear atmospheres values could be 15% higher.

7.10 External shading

Any object interposed between the sun's rays and a receiving surface casts a shadow on the latter. If the object is completely opaque to the solar radiation, then no direct radiation is received by the shaded area; but radiation from the sky will be received by the whole area, shaded and unshaded. If the object interposed is not completely opaque, such as a translucent plastic sheet, then some direct radiation is transmitted, but not as much as if the shading object were absent. A knowledge of the effect of shading is vital to calculating solar heat gains through glass and, to this end, a knowledge of the geometry involved when a shadow is cast, is useful.

7.11 The geometry of shadows

This is best illustrated by considering the dimensions of the shadow cast on a vertical window by a reveal and an overhanging lintel.

Figure 7.11(*a*) shows in perspective a window which is recessed from the wall surface by an amount R. The front elevation of the window, seen in Fig. 7.11(*b*), shows the pattern of the shadow cast by the sunlight on the glass. The position of the point P' on the shadow, cast by the sun's rays

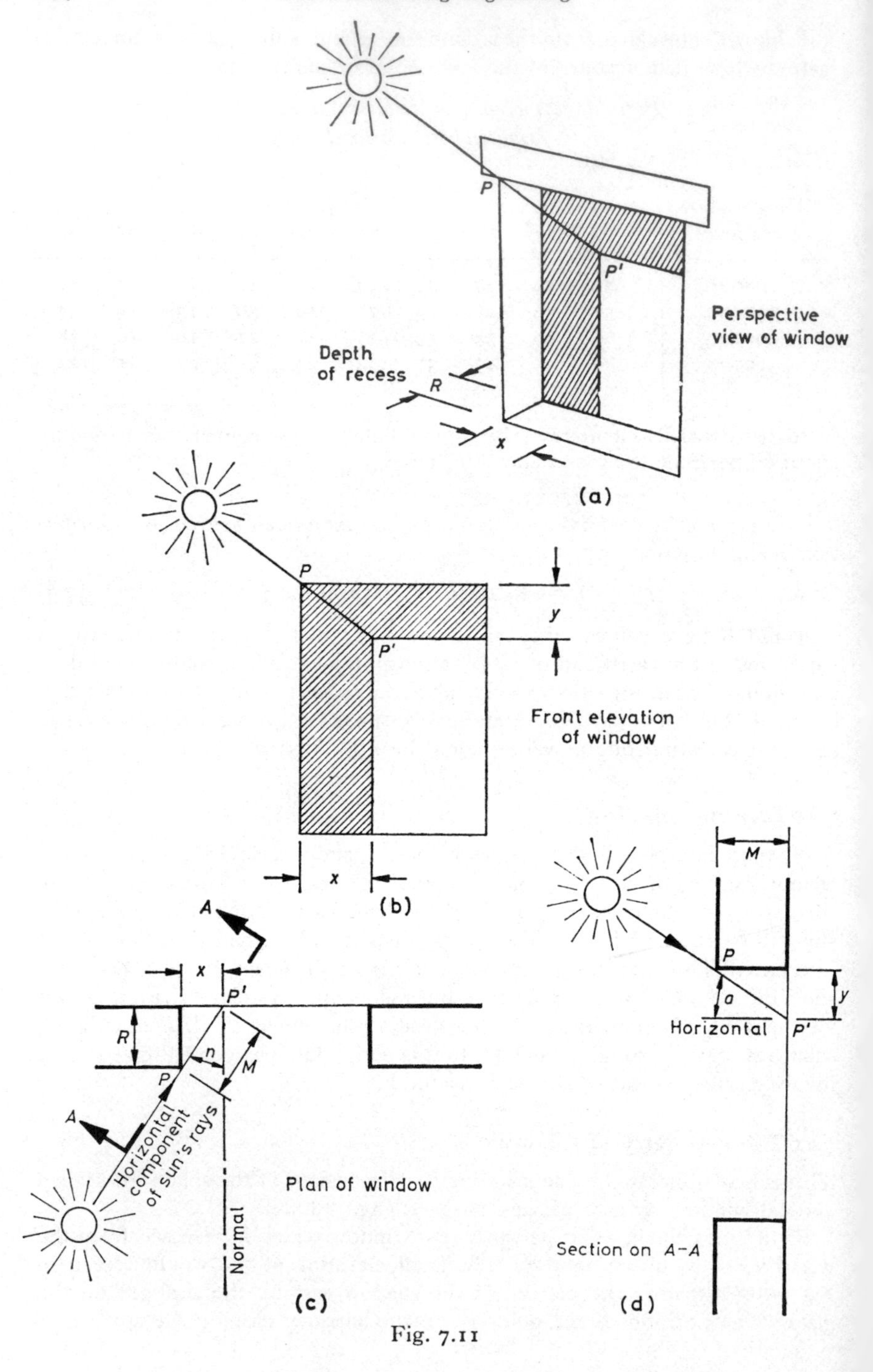
P
P'
Perspective view of window
Depth of recess
R
x
(a)
P
P'
y
Front elevation of window
x
(b)
A
x
P'
R
n
M
P
A
Horizontal component of sun's rays
Plan of window
Normal
(c)
M
P
a
y
Horizontal
P'
Section on A–A
(d)

Fig. 7.11

passing the corner point P in the recess, is the information required. In other words, if the co-ordinates x and y of the point P' on the glass can be worked out, the dimensions of the shaded area are known.

Figure 7.11(*c*) shows a plan of the window. In this, the solar wall azimuth angle is n and co-ordinate x is immediately obtainable from the relationship:

$$x = R \tan n \qquad (7.9)$$

To see the true value of the angle of altitude of the sun, we must not take a simple sectional elevation of the window but a section in the plane of the sun's rays; that is, along $A-A$, Fig. 7.11(*d*).

If the hypotenuse of the triangle formed by R and x in Fig. 7.11(*c*) is denoted by M, then the value of M is clearly $R \sec n$. Reference to Fig. 7.11(*d*) shows that the value of the co-ordinate y can be obtained from

$$\begin{aligned} y &= M \tan a \\ &= R \sec n \tan a \qquad (7.10) \end{aligned}$$

The I.H.V.E. Guide publishes data in tabular form on the shading cast on a recessed window.

EXAMPLE 7.6. Calculate, by means of eqns. (7.9) and (7.10), the area of the shaded portion of a window which is recessed by 50 mm from the surface of the wall in which it is fitted, given the following information:

Altitude of the sun	43° 30′
Azimuth of the sun	66° west of south
Orientation of the window	facing south-west
Dimensions of the window	2·75 m high × 3 m wide

Answer

The angle n is $66° - 45° = 21°$.

From eqn. (7.9)

$$\begin{aligned} x &= 50 \tan 21 \\ &= 50 \times 0{\cdot}384 \\ &= 19 \text{ mm.} \end{aligned}$$

From eqn. (7.10)

$$\begin{aligned} y &= 50 \tan 43° 30' \sec 21° \\ &= 50 \times 0{\cdot}949 \times 1{\cdot}071 \\ &= 51 \text{ mm.} \end{aligned}$$

It is to be noted that the depth of the recess is the only dimension of the window which governs values of x and y. The actual size of the window is only relevant when the area of the shadow is required. It follows that eqns. (7.9) and (7.10) are valid for locating the position of any point, regardless of whether it is the corner of a recessed window or not. Thus, as an instance, the position of the end of the shadow cast by a flagpole on a parade ground could be calculated.

Continuing with the answer to Example 7.6,

$$\begin{aligned}\text{sunlit area} &= (2750-y)\times(3000-x)\\ &= 2699\times 2981\\ &= 8{\cdot}05\ \text{m}^2\\ \text{total area} &= 2{\cdot}75\times 3{\cdot}00\\ &= 8{\cdot}25\ \text{m}^2\\ \text{shaded area} &= 8{\cdot}25-8{\cdot}05\\ &= 0{\cdot}2\ \text{m}^2\end{aligned}$$

This does not seem to be very significant, but it must be borne in mind that it is the depth of the recess which governs the amount of shading.

EXAMPLE 7.7. Using, where necessary, the information given in Example 7.6, calculate the depth of recess which would result in the window being entirely shaded.

Answer

By eqns. (7.9) and (7.10) it has been previously calculated that

$$x = 19\ \text{mm and}$$

$$y = 51\ \text{mm.}$$

Since both x and y are directly proportional to the depth of recess, y is the significant variable, being greater than x, for the given values of altitude, azimuth and wall orientation.

For the window to be in shadow, y would have to equal 2·75 m, the height of the glass. Hence,

$$2{\cdot}75 = R\tan 43^\circ\,30'\times\sec 21^\circ$$

$$R = \frac{2{\cdot}75}{0{\cdot}949\times 1{\cdot}071}$$

$$= 2{\cdot}71\ \text{m.}$$

For a different time of the day or for a different orientation, the window might quite well be extensively in shadow, even from a 50 mm recess. Conversely, even a very deep recess may still permit direct sunlight to fall on a window of a certain orientation at certain times of the day and in certain months of the year.

7.12 The transmission of solar radiation through glass

Equation (7.7) merely gave the value of the direct solar radiation which was normally incident upon a surface:

$$I_\delta = I \cos i \tag{7.7}$$

This equation does not indicate how much of the quantity I_δ actually enters the surface, to cause heat gain or, in the case of glass, how much is transmitted through the glass.

Of the energy which is incident upon the glass, some is reflected and lost, some is transmitted through the glass, and some is absorbed by the glass as the energy passes through it. This small amount of absorbed energy raises the temperature of the glass, and the glass eventually transmits this heat, by convection, partly to the room and partly to the exterior. It follows that some knowledge of the extent to which glass reflects, transmits and absorbs solar radiation is necessary.

Table 7.3 gives figures for transmissibility and absorption but, in round figures, for angles of incidence between 60° and 0°, ordinary single window glass transmits about 85 per cent of the energy incident upon it. About 6 per cent is absorbed and the remaining 9 per cent is reflected. As the angle of incidence increases beyond 60°, the transmitted radiation falls off to zero, the reflected amount increasing. The absorption figure remains fairly constant at about 6 per cent for angles of incidence up to 80°.

For double glazing, the picture is more complicated but, in approximate terms, only about 90 per cent of what passes through single glazing is transmitted. Thus, the transmitted percentage for double glazing is about 76 per cent of the incident energy.

TABLE 7.3 *Transmissivity and absorptivity for direct solar radiation through glass.*

(The absorptivity for indirect radiation is taken as 0·06, regardless of the angle of incidence. The transmissivity for indirect radiation is taken as 0·79, regardless of the angle of incidence.)

	Angle of incidence							
	0°	20°	40°	50°	60°	70°	80°	90°
Transmissivity	0·87	0·87	0·86	0·84	0·79	0·67	0·42	0
Absorptivity	0·05	0·05	0·06	0·06	0·06	0·06	0·06	0

(Reproduced by kind permission from the A.S.H.R.A.E. Guide)

EXAMPLE 7.8. For the window and information given in Example 7.6, calculate the instantaneous gain from solar radiation to a room, given that the transmission coefficient for direct radiation τ is 0·85, that the intensity of direct solar radiation on a surface normal to the rays is 832 W/m^2, that the intensity of sky radiation is 43 W/m^2, normal to the window, and that the transmission coefficient τ' for sky radiation is 0·80.

Answer

Total transmitted solar radiation is given by

$$Q = \text{(transmitted direct radiation} \times \text{sunlit area of glass)} + \text{(transmitted sky radiation} \times \text{total area of glass)}$$

$$Q = \tau I_\delta A_{sun} + \tau' \times 43\, A_{total}$$

By eqn. (7.5), for a vertical window

$$I_\delta = I_v = I \cos a \cos n$$
$$= I \cos 43^\circ\, 30' \cos 21^\circ$$
$$= 832 \times 0{\cdot}725 \times 0{\cdot}934$$
$$= 832 \times 0{\cdot}676.$$

Thus, transmitted total radiation is

$$Q = 0{\cdot}85 \times 832 \times 0{\cdot}676 \times A_{sun} + 0{\cdot}8 \times 43 \times A_{total}$$

From Example 7.6 it is known that

$$A_{sun} = 8{\cdot}05 \text{ m}^2$$
$$A_{total} = 8{\cdot}25 \text{ m}^2.$$

Hence, the transmitted total radiation is

$$Q = 0{\cdot}85 \times 832 \times 0{\cdot}676 \times 8{\cdot}05 + 34{\cdot}4 \times 8{\cdot}25$$
$$= 3850 + 283$$
$$= 4133 \text{ W}.$$

As is seen later, not all of this energy constitutes an immediate load on the air-conditioning system.

To summarise the procedure for assessing the instantaneous solar heat transmission through glass:

1. For the given date, time of day and latitude, establish the altitude, a, and azimuth, z, of the sun either by means of eqns. (7.2) and (7.3) or by reference to tables.

2. Determine the intensity of direct radiation, I, on a surface normal to the sun's rays for the particular altitude, either by reference to tables or by means of eqn. (7.8).

3. For the orientation of the window in question, calculate the solar wall azimuth angle, n.

4. Using eqn. (7.6) directly, or the simplified versions, eqns. (7.4) or (7.5), calculate the component of the direct radiation, I_δ, which is normally incident on the surface. If the surface is vertical, I_δ is I_v, and if it is horizontal, I_δ is I_h.

5. For the particular angle of incidence on the glass, determine the transmission factor τ. If i is between 0° and 60°, take τ as 0·85, in the absence of other information.

6. Establish the dimensions of any shadows cast by external shading, hence, knowing the dimensions of the window, calculate the sunlit and total areas.

7. Knowing the altitude of the sun and the month of the year, refer to tables and find the value of the sky radiation normally incident on the glass.

8. Calculate the direct transmitted radiation by multiplying τ, I_δ, and the sunlit area, and add to this the product of τ', the sky radiation and the total area. This is the total instantaneous solar radiation transmitted through the window.

7.13 The heat absorbed by glass

The amount of the solar energy absorbed by the glass during the passing of the direct rays of the sun through it depends largely on the absorption characteristics of the particular type of glass.

Ordinary glass does not have a very large coefficient of absorption, but certain specially made glasses absorb a good deal of heat. The heat absorbed causes an increase in the temperature of the glass, and heat then flows by conduction through the glass to both its surfaces. At the indoor and outdoor surfaces the heat is convected and radiated away at a rate dependent on the value of the inside and outside surface film coefficients of heat transfer, h_{si} and h_{so}.

If values are assumed for the temperature in the room, t_r, and for the temperature outside, t_o, a heat balance equation can be drawn up and a value calculated for the mean temperature of the glass. It is assumed in doing this that, because the glass is so thin, the surface temperatures are virtually the same as the mean, i.e.,

$$\text{heat gain to glass} = \text{heat loss from glass.}$$

Refer to Fig. 7.12 and take the mean glass temperature as t_g and the coefficient of absorption as α.

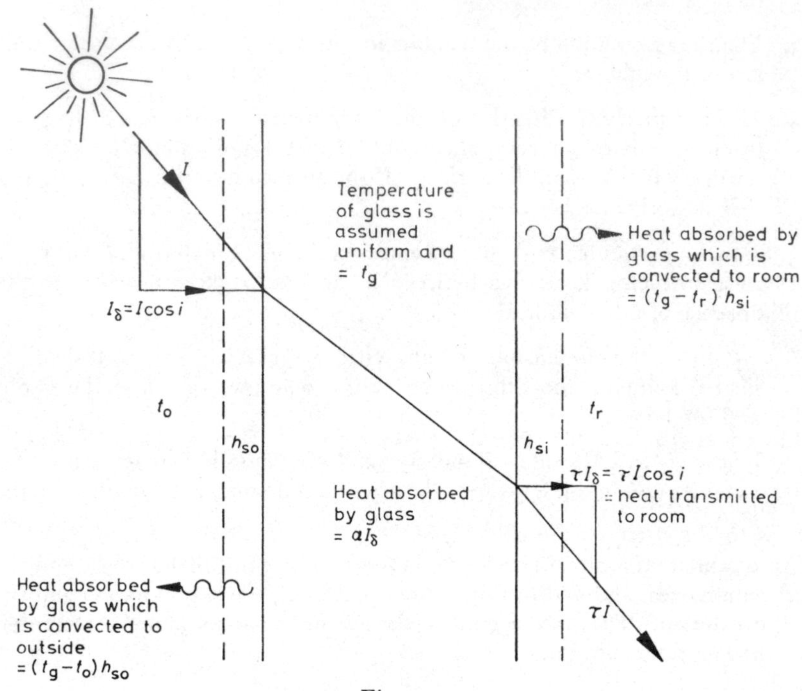

Fig. 7.12

$$\alpha I_\delta = (t_g - t_o)h_{so} + (t_g - t_r)h_{si}$$

Take

$$h_{so} = 22{\cdot}7 \text{ W/m}^2\ {}^\circ\text{C}$$

$$h_{si} = 7{\cdot}9 \text{ W/m}^2\ {}^\circ\text{C}$$

$$\alpha I_\delta = 22{\cdot}7t_g - 22{\cdot}7t_o + 7{\cdot}9t_g - 7{\cdot}9t_r$$

$$= 30{\cdot}6t_g - (22{\cdot}7t_o + 7{\cdot}9t_r)$$

$$t_g = \frac{\alpha I_\delta + 22{\cdot}7t_o + 7{\cdot}9t_r}{30{\cdot}6} \qquad (7.11)$$

EXAMPLE 7.9. Calculate the temperature of a single sheet of ordinary glass, entirely in sunlight, given that $I_\delta = 631$ W/m², $\alpha = 0{\cdot}06$, $t_o = 28$°C, $t_r = 22$°C.

Answer

From eqn. (7.11)

$$t_g = \frac{0{\cdot}06 \times 631 + 22{\cdot}7 \times 28 + 7{\cdot}9 \times 22}{30{\cdot}6}$$

$$= \frac{37{\cdot}8 + 636 + 174}{30{\cdot}6}$$

$$= \frac{847{\cdot}8}{30{\cdot}6}$$
$$= 27{\cdot}7°\text{C}.$$

Note that if no sunlight were absorbed, the glass temperature would be

$$t_g = \frac{636+174}{30{\cdot}6}$$
$$= \frac{810}{30{\cdot}6}$$
$$= 26{\cdot}4°\text{C}.$$

If there were no absorption effect, the heat transmitted to the interior would be

$$7{\cdot}9\times(26{\cdot}4-22) = 34{\cdot}8 \text{ W/m}^2$$

When there is an absorption effect, the total amount absorbed will be 6 per cent of 631 W/m^2 or 37·8 W/m^2 and the amount of heat transmitted to the room will be

$$7{\cdot}9\times(27{\cdot}7-22) = 45 \text{ W/m}^2$$

Hence, 10·2 W/m^2 of the total absorbed quantity (or about 27 per cent), passes to the room in the case of single glazing for the figures considered. With double glazing, it can be shown in a similar way that about 75 per cent of the absorbed heat passes to the room.

It can be seen that the heat gain due to the quantity of solar radiation absorbed by ordinary glass, is very small. Heat-absorbing glasses are quite a different story, as is seen in the following table.

TABLE 7.4 *Thermal transmission coefficients for glass and blinds.*

Glass or Shading Element	Absorption coefficient	Reflection coefficient	Transmission coefficient
4 mm clear glass	0·08	0·08	0·84
6 mm plate glass	0·14	0·08	0·80
6 mm heat-absorbing glass	0·40	0·06.	0·54
6 mm silver laminate glass	0·45	0·41	0·14
Light-coloured Venetian blinds with their slats at 45°	0·37	0·51	0.12

The figures are for angles of incidence of less than about 45° and so are relevant for most cases. When used to obtain the values in Table 7.5 outside and inside heat transfer coefficients of 15 and 10 W/m^2 °C, respectively, were adopted.

7.14 Internal shading and double glazing

The effect of internal shading devices in reducing solar heat gain is considerable but not so great as the effect of external shades. In tropical and sub-tropical climates, external shades are used extensively but in temperate climates the weather is often so adverse in winter that the maintenance of external shades is expensive. For this reason, internal shading is much adopted and is essential for sunlit windows if air conditioning is to be satisfactory. Table 7.5 indicates the effect of shading windows from solar radiation.

TABLE 7.5 *Shading factors for windows*

	Solar thermal radiation					Shading factors		
	light trans. %	refl. %	absd. %	trans. %	total trans. %	radn.	conv.	total
Single unshaded glass								
Ordinary 4 mm glass	87	8	8	84	87	0·96	0·04	1·00
Plate glass	85	8	12	80	84	0·92	0·05	0·97
Heat-absorbing (bronze)	50	10	34	56	67	0·64	0·13	0·77
Heat-reflecting (bronze)	12	45	47	8	24	0·09	0·18	0·27
Heat-reflecting (gold)	15	48	43	9	23	0·10	0·17	0·27
Single glass + internal Venetian blinds								
Ordinary glass (4 mm)	—	45	44	11	47	0·12	0·41	0·53
Plate glass	—	42	48	10	47	0·11	0·42	0·53
Heat-absorbing (bronze)	—	27	66	7	44	0·08	0·43	0·51
Heat-reflecting (gold)	—	48	—	—	22	0·01	0·25	0·26
Double unshaded glass								
Ordinary glass (4 mm)	76	13	16	71	76	0·77	0·08	0·85
Plate glass	72	13	23	64	73	0·74	0·10	0·84
Heat-absorbing (bronze)*	42	13	42	45	55	0·51	0·12	0·63
Heat-reflecting (bronze)*	11	46	48	6	17	0·08	0·10	0·18
Heat-reflecting (gold)*	13	49	44	7	16	0·09	0·09	0·18
Double glass + internal Venetian blinds								
Ordinary glass (4 mm)	—	38	—	—	48	0·12	0·35	0·55
Plate glass	—	38	54	8	46	0·10	0·43	0·53
Heat-absorbing (bronze)*	—	23	71	6	37	0·07	0·36	0·43
Double glass + Venetian blinds between the sheets								
Ordinary glass (4 mm)	—	42	50	5	25	0·08	0·17	0·29
Plate glass	—	42	50	8	23	0·09	0·17	0·26
Heat-absorbing (bronze)*	—	27	67	6	22	0·07	0·18	0·25

*Inner leaf ordinary or plate glass.
Ordinary glass is 4 mm but all other glass is 6 mm.

Notes for Table 7·5

1. Except where stated otherwise, all sheets of glass are 6 mm thick.
2. The figures quoted for heat-absorbing and heat-reflecting glasses are for current proprietary brands, easily obtainable.
3. Venetian blinds are assumed to be light coloured with their slats at 45° and incident solar radiation at an angle of incidence of 30° to the glass. Their reflectivity is taken as 0·51, transmissivity as 0·12 and absorptivity as 0·37.
4. The external wind speed is assumed as about 2 m/s.
5. Of the heat absorbed by the glass, two-thirds is assumed to be lost to the outside and one-third to the room.
6. The total transmitted solar thermal radiation is the actual gain to the room and includes some of the heat absorbed by the glass (and blinds).
7. The total shading factor merely compares the total transmitted solar thermal radiation with that of ordinary, clear, 4 mm glass.
8. The convection shading factor is not influenced by storage effects for the building and therefore corresponds to an instantaneous load on the air conditioning system. The radiation shading factor is susceptible to building storage effects and so does not correspond to an instantaneous load. Hence, the total shading factor corresponds to a mixed load, part instantaneous and part delayed.
9. It is usually not worth-while using Venetian blinds with heat-reflecting glass, although the large amount of heat absorbed by such glass can give it a surface temperature exceeding 40°C. Similarly, from the opposite point of view, it is not worth using heat-absorbing glass if internal Venetian blinds are to be used anyway. People cannot feel comfortable in direct strong sunlight in a conditioned room; so internal blinds or strongly reflective glass is essential for comfort.

Internal shades are very often of the Venetian blind type and for these to give most benefit, they should be of a white or aluminium colour and should have polished surfaces. They should be adjusted so that they reflect the rays of the sun back to the outside and so that no direct rays pass between the slats into the room. Under such conditions, only about 53 per cent of the direct and sky radiation normally incident upon a window is transmitted to the room. The blinds themselves heat up in due course and provide a source of convective and radiative heat gain to the room. Such blinds are most effective when fitted *between* sheets of double glazing.

As Table 7.5 shows, the presence of two sheets of glass reduces the net solar transmitted heat by about 15 per cent only. It usually follows from this that the saving in capital and running costs of the air-conditioning system are not enough to pay for the extra cost of double glazing.

The main favourable feature of double glazing is the reduction in noise transmission, but this is effective only if the air gap between the sheets exceeds about 100 mm.

The obvious advantage that a higher relative humidity can be maintained in the room in winter (because the inside surface temperature of double glazing is higher than that of single glazing, thus permitting a higher room dew point) is often spurious: poorly fitting double glazing permits the air in the room to enter the space between the sheets, and objectionable condensation then takes place on the inside of the glass sheets, where it cannot readily be wiped away. On the other hand, a small amount of ventilation through the outer sheet and frame is sometimes deliberately arranged and is most successful in preventing condensation.

7.15 Numerical values of scattered radiation

The total amount of solar radiation (I_t) normally incident on a surface is given by

$$I_t = I \cos i + I_s + I_r \tag{7.12}$$

where I_s is sky radiation and I_r is radiation reflected from surrounding surfaces. The A.S.H.R.A.E. Guide evaluates I_s and I_r according to

$$I_s + I_r = CIF_s + \rho I(C + \sin a)F_g \tag{7.13}$$

The value of the dimensionless constant, C, varies through the year and is given in Table 7.6. The angle factor for the ground with relation to the particular surface, F_g, is complementary to F_s, the factor between the surface and the sky and is given by

$$F_g = 0.5\,(1 - \cos \delta) \tag{7.14}$$

Note that the sum of all the angle factors between a surface and its surroundings is unity. Hence, $F_s = 1 - F_g$ for any surface seeing only the ground and the sky.

The reflectivity, ρ, of the ground depends on the type of surface and varies from 0·1 for dark asphalt to 0·7 for white stone or concrete. Average grass and earth surroundings correspond to a value of 0·2 for ρ.

EXAMPLE 7.10 Calculate the total scattered radiation normally incident on a vertical surface in July for a solar altitude of 40°, a height above sea level of 2000 m and a ground reflectivity of 0·2.

Answer

From Table 7.1 or eqn. (7.8), $I = 814\ \text{W/m}^2$ or $786\ \text{W/m}^2$ at sea level and from Table 7.2 the increase in these values is 24 per cent at a height of 2000 m. Also, from Table 7.6, $C = 0.136$ and by eqn. (7.14), $F_g = 0.5(1 - \cos 90°) = 0.5$.

Hence, by eqn. (7.13)

$$\begin{aligned} I_s + I_r &= 1.24\{0.136 \times 814 \times (1 - 0.5) + 0.2 \times 814 \times (0.136 + \sin 40) \times 0.5\} \\ &= 1.24\,(55.4 + 63.4) \\ &= 147\ \text{W/m}^2. \end{aligned}$$

TABLE 7.6 *Constants for determining the values of direct and scattered radiation at sea level, to be used in eqns.* (7.8) and (7.13).

Month	Jan.	Feb.	Mar.	Apr.	May.	Jun.	Jul.	Aug.	Sep.	Oct.	Nov.	Dec.	Units
C	0·058	0·060	0·071	0·097	0·121	0·134	0·136	0·122	0·092	0·073	0·063	0·057	—
A	1·230	1·213	1·186	1·135	1·104	1·088	1·085	1·107	1·152	1·192	1·220	1·233	kW/m^2
B	0·142	0·144	0·156	0·180	0·196	0·205	0·207	0·201	0·177	0·160	0·149	0·142	—

TABLE 7.7 *The intensity of scattered radiation from a clear sky in* W/m².

Month	Surface	Radiation	Solar Altitude 5°	10°	15°	20°	25°	30°	35°	40°	50°	60°	70°	80°
Jan.	Horizontal	Sky	14	32	41	47	51	54	56	57	59	61	62	62
	Vertical	Sky	7	16	21	24	26	27	28	29	30	30	31	31
	Vertical	Ground	3	13	22	32	42	52	61	69	84	96	106	109
Feb.	Horizontal	Sky	14	32	42	48	52	55	57	58	60	62	62	63
	Vertical	Sky	7	16	21	24	26	27	28	29	30	31	31	32
	Vertical	Ground	3	12	22	32	42	51	60	68	83	95	104	110
Mar.	Horizontal	Sky	14	38	46	53	58	62	64	66	69	70	72	72
	Vertical	Sky	7	19	23	27	29	31	32	33	34	35	36	36
	Vertical	Ground	3	13	21	31	40	50	58	66	81	93	102	107
Apr.	Horizontal	Sky	14	39	55	65	72	77	80	83	87	90	91	92
	Vertical	Sky	7	20	27	32	36	38	40	42	44	45	45	46
	Vertical	Ground	3	11	20	30	39	47	56	64	77	89	97	102
May	Horizontal	Sky	14	43	61	75	84	91	95	98	104	107	109	110
	Vertical	Sky	7	21	31	38	42	46	47	49	52	53	54	55
	Vertical	Ground	2	10	19	29	38	46	55	62	75	87	95	100

	Horizontal	Sky	14	45	66	80	90	97	102	106	112	115	118	119
June	Vertical	Sky	7	22	33	40	45	48	51	53	56	57	59	59
	Vertical	Ground	2	10	19	29	37	46	54	62	75	86	94	99
	Horizontal	Sky	14	45	66	81	90	98	103	107	113	116	118	120
July	Vertical	Sky	7	22	33	40	45	49	52	53	56	58	59	60
	Vertical	Ground	2	10	19	23	37	46	54	61	75	86	94	99
	Horizontal	Sky	14	42	62	75	84	90	95	99	104	107	109	110
Aug.	Vertical	Sky	7	21	31	33	42	45	48	50	52	53	55	55
	Vertical	Ground	2	10	19	29	38	46	54	62	76	86	95	100
	Horizontal	Sky	14	38	53	63	70	74	78	80	84	86	88	89
Sep.	Vertical	Sky	7	19	27	32	35	37	39	40	42	43	44	44
	Vertical	Ground	3	11	20	30	39	48	56	64	78	90	98	104
	Horizontal	Sky	14	35	47	55	60	63	66	68	71	72	74	75
Oct.	Vertical	Sky	7	17	24	27	30	32	33	34	35	36	37	38
	Vertical	Ground	3	12	21	31	40	50	58	67	81	93	102	107
	Horizontal	Sky	14	33	43	50	54	57	59	61	64	65	65	66
Nov.	Vertical	Sky	7	16	22	25	27	29	30	31	32	32	33	33
	Vertical	Ground	3	12	22	32	42	51	60	69	84	96	104	110
	Horizontal	Sky	14	31	41	45	50	53	55	57	58	60	60	61
Dec.	Vertical	Sky	7	16	20	23	25	26	27	28	29	30	30	30
	Vertical	Ground	4	13	23	33	42	52	61	69	85	97	106	112

Note that Table 7.7 quotes values of 53 and 61 W/m^2 at sea level for sky and ground radiation. These are less than the values calculated above because they are based on $I = 786$ W/m^2, the American-derived figure, used when compiling the table.

If we take a solar altitude of 40° as occurring at 0836 hours in July at latitude 50°N, then the I.H.V.E. Guide quotes a figure of 43 W/m^2, normally incident on a vertical surface, for diffuse radiation. By implication this is sky radiation only and is to be compared with the figure of 55·4 W/m^2 from the American source.

7.16 Minor factors influencing solar gains

There are five relatively minor factors that should be taken into account when calculating the instantaneous heat transmission resulting from solar radiation:

1. Atmospheric haze.
2. The type of window frame.
3. The height of the place above sea level.
4. Variation of the dew point.
5. The hemisphere.

1. *Atmospheric Haze*

This is most noticeable in industrial areas and, in this context, is regarded as resulting from industrial contaminants carried aloft by thermal up-currents. It is usually more pronounced in the afternoon, owing to the build up of the ground surface temperatures which produce the up-currents as the day progresses. Haze can reduce the value of I_δ, the direct radiation normally incident on a surface, so much that a diminution of 15 per cent in the total radiation received (scattered plus direct) may occur. A conservative factor of 0·95 may be applied for cities such as London.

2. *The Type of Window Frame*

Tables of the actual heat gain occurring through windows are often published (*see* Table 7.8). Generally, tables of this sort are for wooden-framed windows. If, as in many modern buildings, the framework is steel, then, because steel is a much better thermal conductor than wood, the heat gains are increased by about 17 per cent.

A more convenient way of taking account of this effect is to apply the tabulated figure for the heat gain to the glass area, in wooden-framed windows, but to the area of the opening in the wall, in steel-framed windows.

3. *The Height of the Place above Sea Level*

There is some difference of opinion in published data as to the influence of increased height above sea level on the intensity of direct solar radiation. Two indisputable facts are that the intensity is about 1362 W/m^2 on a surface normal to the sun's rays, at the limits of the atmosphere, but only about

950 W/m^2, as a maximum, at the surface of the earth. Thus, the intensity is reduced by about 30 per cent when the path-length of the sun's rays is at a minimum.

This minimum path-length occurs when the sun is at its zenith or, put another way, when the ' air-mass ' is unity. The bulk of the earth's atmosphere is below 3000 m and hence the major part of the reduction in intensity occurs below this level. The reduction in intensity, being clearly bound up with the air mass, must also be dependent on the altitude of the sun, since this affects the path-length of the sun's rays. (It is approximately true to assume that the path-length is proportional to the cosecant of the angle of altitude of the sun for values of this from 90° to 30°.)

TABLE 7.8 *Solar air conditioning loads through windows in the U.K. only, for floor slabs of* 500 kg/m^2 *surface density, expressed in* W/m^2.

The values quoted are for single plate glass and, where correction is necessary, should be multiplied by those in Table 7.9. The area to be used is the opening in the wall, in the case of metal-framed windows, and the area of the glass for wooden-framed windows. A haze factor of 0·9 has been allowed. It is assumed that shades are not provided on windows facing north or on horizontal glazing and that on all other exposures the blinds will be raised when the windows are not in direct sunlight. Scattered radiation is included, the storage effect of the building has been taken into account, but air-to-air transmission gain is excluded.

Windows with internal Venetian blinds of a light colour

		Sun time												
	Exposure	6	7	8	9	10	11	N	1	2	3	4	5	6
Date	N	3	9	22	22	22	25	25	25	25	28	28	28	28
	NE	117	145	139	113	73	60	47	47	41	38	38	32	25
	E	126	183	205	199	164	110	76	69	63	57	50	44	38
June 21	SE	6	73	123	158	173	170	148	113	76	63	54	47	38
	S	6	6	41	69	95	113	126	129	123	107	82	47	41
	SW	16	19	19	19	25	57	98	132	161	170	154	120	63
	W	22	25	25	26	28	28	28	57	113	170	208	214	189
	NW	16	19	22	22	25	25	25	25	38	82	126	161	157
	horizontal	28	63	104	155	206	259	310	354	391	420	436	443	434
	N	3	9	10	19	22	22	22	22	22	22	25	25	25
	NE	107	136	130	104	69	54	44	44	38	35	35	28	25
July 23	E	127	183	205	199	164	111	76	69	63	57	50	44	38
and	SE	9	79	130	171	186	180	158	123	82	66	60	50	41
May 21	S	9	9	44	79	107	130	142	145	142	120	91	54	44
	SW	19	22	22	22	28	66	111	155	183	196	177	139	85
	W	22	25	25	28	28	28	28	57	114	171	208	215	189
	NW	16	19	19	19	22	22	22	22	35	76	117	148	145
	horizontal	9	32	70	114	164	219	269	314	350	380	391	391	376

TABLE 7.8—*continued*

	Windows with internal Venetian blinds of a light colour													
		Sun time												
	Exposure	6	7	8	9	10	11	N	1	2	3	4	5	6
	N	0	6	16	16	16	16	16	19	19	19	19	19	19
	NE	88	111	104	85	54	44	38	35	32	28	28	25	19
Aug. 24	E	123	177	199	193	158	107	73	66	60	54	47	44	38
and	SE	9	85	142	186	205	199	174	133	88	73	63	54	44
April 20	S	9	9	60	101	139	167	186	189	183	158	120	69	60
	SW	22	25	25	25	32	73	120	167	202	212	196	152	79
	W	19	25	25	28	28	28	28	54	111	164	202	208	183
	NW	13	16	16	16	19	19	19	19	28	63	95	117	117
	horizontal	0	9	32	70	114	161	209	253	285	306	316	310	285
	N	0	6	9	13	13	13	13	13	13	13	13	13	13
	NE	54	66	63	50	35	28	22	22	19	19	16	16	13
Sept. 22	E	107	155	174	167	139	95	63	60	54	47	44	38	32
and	SE	9	88	148	193	212	205	180	139	91	76	66	57	47
March 22	S	13	13	66	117	158	193	215	218	212	180	139	79	66
	SW	22	25	25	25	32	76	127	174	208	221	202	158	82
	W	19	22	22	25	25	25	25	47	98	145	177	183	161
	NW	6	9	9	9	13	13	13	13	19	38	57	73	73
	horizontal	0	0	9	32	63	104	145	180	209	225	225	209	190
	N	0	3	6	6	6	6	6	9	9	9	9	9	9
	NE	28	35	32	25	16	13	13	9	9	9	9	6	6
Oct. 23	E	82	117	133	127	107	73	47	44	41	38	32	28	25
and	SE	9	85	142	186	205	199	174	133	88	73	63	54	44
Feb. 20	S	13	13	73	123	167	205	228	231	224	189	145	85	73
	SW	22	25	25	25	32	73	123	167	202	212	196	152	79
	W	16	16	16	19	19	19	19	38	73	111	136	139	123
	NW	3	3	6	6	6	6	6	6	9	19	28	38	35
	horizontal	0	0	6	19	41	66	92	114	133	143	143	133	120
	N	0	3	6	6	6	6	6	6	6	6	6	6	6
	NE	9	9	9	9	6	3	3	3	3	3	3	3	3
Nov. 21	E	50	73	82	79	63	44	28	28	25	22	19	19	16
and	SE	6	69	117	152	164	161	139	107	73	60	50	44	38
Jan. 21	S	13	13	66	114	155	186	208	212	205	174	133	76	66
	SW	16	19	19	19	25	60	98	136	161	174	158	123	63
	W	9	9	9	13	13	13	13	22	44	66	82	85	76
	NW	0	0	0	0	3	3	3	3	3	6	9	13	13
	horizontal	0	0	0	3	13	28	44	57	66	70	66	60	54

TABLE 7.8—*continued*

Windows with internal Venetian blinds of a light colour

	Exposure	Sun time												
		6	7	8	9	10	11	N	1	2	3	4	5	6
	N	0	3	3	3	3	3	3	3	6	6	6	6	6
	NE	6	9	6	6	3	3	3	3	3	3	3	3	0
	E	38	54	60	57	47	32	22	19	19	16	16	13	9
Dec. 22	SE	6	63	104	136	152	145	130	98	66	54	47	41	35
	S	9	9	60	104	142	171	193	193	189	161	123	69	60
	SW	16	19	19	19	22	54	88	123	148	158	145	114	60
	W	6	6	6	9	9	9	9	16	32	50	60	63	54
	NW	0	0	0	0	0	0	0	0	3	3	6	9	9
	horizontal	0	0	0	0	3	13	25	35	41	41	38	35	32

(Reproduced by kind permission of Haden Young Ltd)

Experimental observations show that for altitude variations from 0 to 3000 m above sea level, the maximum intensity of direct radiation normal to the sun's rays alters from about 950 W/m^2 to about 1170 W/m^2. That is to say, an average increase of about $2\frac{1}{2}$ or 3 per cent is occurring for each 300 m of increase of height above sea level. It should be noted that as the intensity of direct solar radiation increases with height the intensity of sky radiation falls off. Table 7.2 takes this into account.

4. *Variation of the Dew Point*

Although the dew point of the air falls with increasing height above sea level (*see* § 5.6), the effect of this lapse rate is already taken account of in the altitude correction mentioned above. That a variation in dew point has any effect at all stems from the fact that a change in dew point means a change in moisture content and this, in turn, means a change in the absorption capacity of the air water vapour mixture which constitutes the lower reaches of our atmosphere.

However, dew point varies over the surface of the earth between places of the same height above sea level. Some variation in intensity is, therefore, to be expected on this count. Some authorities suggest that an increase in the intensity of direct radiation of 7 per cent for each 5°C reduction in dew point occurs.

5. *Hemisphere*

As has been mentioned in section 7.1, the sun is 3 per cent closer to the earth in January than it is in summer. This results in an increase of about 7 per cent in the value of the intensity of radiation reaching the upper part of the earth's atmosphere in January. Hence, calculations carried out for summer in the southern hemisphere should take account of this increase.

TABLE 7.9 *Correction factors for Table 7.8.*

Type of glazing and shading	Correction factor
Single glass + internal Venetian blinds	
Ordinary or plate glass	1·00
Heat-absorbing (bronze)	0·96
Heat-reflecting (gold)	0·49
Double glass + internal Venetian blinds	
Ordinary	1·08
Plate	1·00
Heat-absorbing (bronze)*	0·81
Double glass + Venetian blinds between the sheets	
Ordinary	0·55
Plate	0·49
Heat-absorbing (bronze)*	0·47

*Inner leaf ordinary or plate glass.
The above factors are obtained from the ratios of the shading factors to ordinary plate glass from Table 7.5.

EXAMPLE 7.11. Calculate the instantaneous transmission of solar radiation through a window recessed 300 mm from the outer surface of the wall in which it is set. Use the following data:

Single ordinary glazing, facing south-west.

Latitude	40° N.
Outside air temperature (t_o)	32°C
Room air temperature (t_r)	24°C.
Sun time	1 pm on 23rd July.
Altitude of location	600 m above sea level.

Moderate industrial haze.

The window framework is steel and the size of the opening in the wall is 3 m × 3 m. The surrounding ground is covered with grass.

$$U \text{ glass} = 5{\cdot}86 \text{ W/m}^2 \text{ °C}$$

$$h_{so} \text{ glass} = 22{\cdot}7 \text{ W/m}^2 \text{ °C}$$

$$h_{si} \text{ glass} = 7{\cdot}9 \text{ W/m}^2 \text{ °C}$$

Answer

It is convenient to divide the calculations into a number of steps, in order to illustrate methods.

(*a*) *Calculation of solar altitudes*

Declination (d) = $20\frac{1}{4}°$ N (*see* § 7.4)

Hour angle (h) = 15° (*see* § 7.3)

$$\sin a = \sin 20\tfrac{1}{4}° \sin 40° + \cos 20\tfrac{1}{4}° \cos 40° \cos 15°$$

(*see* eqn. (7.2))

$$= 0{\cdot}3461 \times 0{\cdot}6428 + 0{\cdot}9382 \times 0{\cdot}7660 \times 0{\cdot}9659$$

$$= 0{\cdot}2225 + 0{\cdot}6940$$

$$= 0{\cdot}9165$$

Thus $a = 66°\ 25'$.

The I.H.V.E. table of altitude quotes 66°.

(*b*) *Calculation of solar azimuth*

$$\tan z = \frac{\sin 15°}{\sin 40° \cos 15° - \cos 40° \tan 20\tfrac{1}{4}°}$$

(*see* eqn. (7.3))

$$= \frac{0{\cdot}2588}{0{\cdot}6428 \times 0{\cdot}9659 - 0{\cdot}7660 \times 0{\cdot}3689}$$

$$= \frac{0{\cdot}2588}{0{\cdot}6200 - 0{\cdot}2825}$$

$$= \frac{0{\cdot}2588}{0{\cdot}3375}$$

$$= 0{\cdot}7680.$$

Thus $z = 37°\ 31'$, west of south.

The I.H.V.E. Guide quotes 217° measured clockwise from north.

(*c*) *Determination of solar-wall azimuth*

Figure 7.13 shows that the solar-wall azimuth angle, n, is $45° - 37°\ 31' = 7°\ 29'$.

(d) Calculation of sunlit area

By eqn. (7.9)

$$x = 300 \tan 7°29'$$
$$= 39{\cdot}4 \text{ mm}$$

By eqn. (7.10)

$$y = 300 \sec 7° 29' \tan 66° 25'$$
$$= 300 \times 1{\cdot}009 \times 2{\cdot}2907$$
$$= 694 \text{ mm.}$$

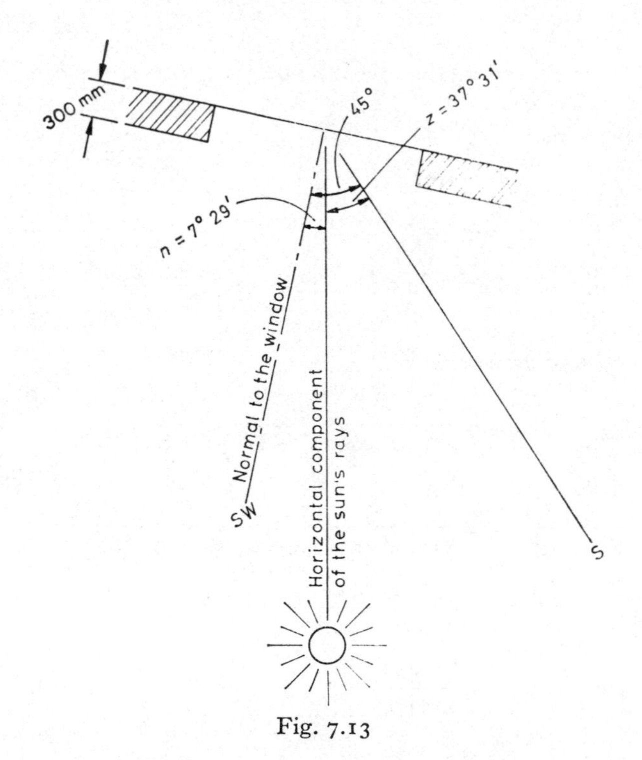

Fig. 7.13

The table in the I.H.V.E. Guide yields values of 40·5 and about 704 mm for x and y respectively.

$$\text{Sunlit area} = (3{\cdot}0 - 0{\cdot}0394)(3{\cdot}0 - 0{\cdot}694)$$
$$= 2{\cdot}9606 \times 2{\cdot}306$$
$$= 6{\cdot}84 \text{ m}^2$$
$$\text{Total area} = 9 \text{ m}^2.$$

(*e*) *Determination of the intensity of direct radiation*

By eqn. (7.8)

$$I = A\ \exp(-B/\sin 66°25')$$
$$= 1{\cdot}085\ \exp(-0{\cdot}207/0{\cdot}9165)$$
$$= 1{\cdot}085 \times 0{\cdot}7977$$
$$= 0{\cdot}865\ \text{kW/m}^2 \text{ on a surface normal to the sun's rays at sea level.}$$

Table 7.1 gives a value of 0·905 kW/m^2, by interpolation.

(*f*) *Determination of the intensity of sky radiation*

By interpolation, Table 7.7 (for the intensity of scattered solar radiation from a clear sky on a vertical surface) gives a value of 59 W/m^2 at sea level. The footnote to Table 7.2 quotes a decrease of about 20 per cent, by extrapolation, for a height of 600 m above sea level where the value of sky radiation therefore becomes 47 W/m^2.

There should be a correction applied to this figure to take account of the additional scattering resulting from industrial haze. However, data are scanty, the effect is small, and so it is ignored here.

(*g*) *Determination of the effect of haze and height above sea level upon direct radiation*

The same table quotes an increase of 6 per cent for direct radiation and, by inference, for ground radiation also. A haze correction is 0·9 and so the total correction factor is $1{\cdot}06 \times 0{\cdot}9 = 0{\cdot}954$.

(*h*) *Calculation of the intensity of ground radiation*

From Table 7.7, the intensity of radiation scattered from grass at sea level and for a solar altitude of 66° 25′ is 91 W/m^2, by interpolation. This intensity is normally incident on the window.

At 600 m the figure would be $1{\cdot}06 \times 91 = 96$ W/m^2.

(*i*) *Calculation of the intensity of direction radiation normal to the window*

By eqn. (7.5)

$$I_v = 865 \cos 66°\ 25' \cos 7°\ 29'$$
$$= 865 \times 0{\cdot}4000 \times 0{\cdot}9914$$
$$= 343\ \text{W/m}^2\text{, uncorrected for height and haze.}$$

Table 7.1 gives 352 W/m^2, by interpolation.

(*j*) *Direct radiation transmitted through the window*

By eqns. (7.5) and (7.7)

$$\cos i = \cos a \cos n$$
$$= \cos 66°\ 25' \cos 7°\ 29'$$
$$= 0{\cdot}3966$$
$$\therefore i = 66°\ 38'$$

The transmissivity for this angle of incidence is 0·71 according to Table 7.3. Thus, the transmission is:

$$= 0{\cdot}71 \times 343$$

$$= 244\ \mathrm{W/m^2},\ \text{uncorrected for haze and height.}$$

(*k*) *Scattered radiation transmitted*

Taking a value of 0·79 for the transmissivity of scattered radiation, as recommended in Table 7.3, then, at sea level, the scattered radiation transmitted would be

$$0{\cdot}79\,(59+91) = 118\ \mathrm{W/m^2}$$

At an altitude of 600 m, ignoring the effect of haze, the transmission would be:

$$0{\cdot}79\,(47+96) = 113\ \mathrm{W/m^2}.$$

(*l*) *Direct plus scattered radiation transmitted*

Through glass in sunlight, the radiation transmitted is due to direct and scattered sources and it is

$$=(244+118) = 362\ \mathrm{W/m^2}.$$

This value, even when corrected for altitude and haze, is of little use if only part of the window is in sunlight. Scattered radiation is transmitted through the entire window, but direct radiation is transmitted through the sunlit portion only.

Hence, transmitted direct radiation, corrected for haze and height

$$= 0{\cdot}954 \times 244 \times \text{sunlit area}$$

$$= 0{\cdot}954 \times 244 \times 6{\cdot}84$$

$$= 1592\ \mathrm{W}.$$

Transmitted scattered radiation, corrected for height but not for haze

$$= 113 \times \text{total area}$$

$$= 113 \times 9$$

$$= 1017\ \mathrm{W}.$$

Hence, the transmitted radiation is 2609 W. As is seen later, this does not constitute an immediate load on the air-conditioning system.

(*m*) *Transmission due to absorbed solar radiation*

Assume a value of 0·06 for the absorption coefficient of single glazing with an angle of incidence equal to 66° 38′ in the case of both scattered and direct radiation (*see* Table 7.3).

By eqn. (7.11)

$$t_g = \frac{0{\cdot}06 \times (0{\cdot}954 \times 343 + 47 + 96) + 22{\cdot}7 \times 32 + 7{\cdot}9 \times 24}{30{\cdot}6}$$

$$= \frac{0{\cdot}06 \times 470 + 726 + 190}{30{\cdot}6}$$

$$= \frac{944}{30{\cdot}6}$$

$$= 30{\cdot}9^\circ\text{C}.$$

The convected and radiated heat gain to the interior due to air-to-air transmission and also due to the solar heat absorbed in the glass is

$$6{\cdot}84 \times 7{\cdot}9\,(30{\cdot}9 - 24) + (9 - 6{\cdot}84) \times 5{\cdot}86 \times (32 - 24)$$

$$= 372 + 101$$

$$= 473 \text{ W}.$$

It is worth noting that the heat gain due to the solar energy absorbed by the glass is quite small. In the figure of 473 W due to the transmission by virtue of air-to-air temperature difference, plus absorbed solar energy, the former is

$$9 \text{ m}^2 \times 5{\cdot}86 \times (32^\circ - 24^\circ) = 422 \text{ W}.$$

So the gain to absorption is only 51 W, i.e.,

$$\frac{51 \times 100}{2609} = 1{\cdot}87 \text{ per cent of the gain due to solar radiation,}$$

$$\text{or } \frac{51 \times 100}{(2609 + 473)} = 1{\cdot}7 \text{ per cent of the total gain.}$$

The gain due to absorption is almost negligible, for ordinary glass.

7.17 Heat gain through walls

The heat gain through a wall is the sum of the relatively steady-state flow (often simply termed ‘ transmission ’) that occurs because the inside air temperature is less than that outside, and the unsteady-state gain resulting from the varying intensity of solar radiation on the outer surface of the wall. The phenomenon of unsteady-state heat flow through a wall is complicated by the fact that a wall has a thermal capacity, and so a certain amount of the heat passing through it is stored, being released to the interior (or exterior) at some later time.

Two environmental factors are to be considered when assessing the amount of heat entering the outer surface of a wall:

(i) the diurnal variation of air temperature, and
(ii) the sinusoidal-type variation of solar intensity.

Figure 7.14 presents a simplified picture. At (*a*), under steady-state conditions, the graph of temperature through the wall is a straight line, the slope of which depends on the difference between the temperatures of the inner and outer surfaces and on the thickness of the wall. The calculation of heat gain under such circumstances is exactly the same as for the more familiar case of steady-state loss:

$$Q = U(t_o - t_r) \tag{7.15}$$

where Q is the rate of heat flow in W/m^2,

U is the thermal transmittance coefficient in W/m^2 °C, and

t_r and t_o are the air temperatures in the room and outside, respectively. A heat gain occurs if t_o exceeds t_r.

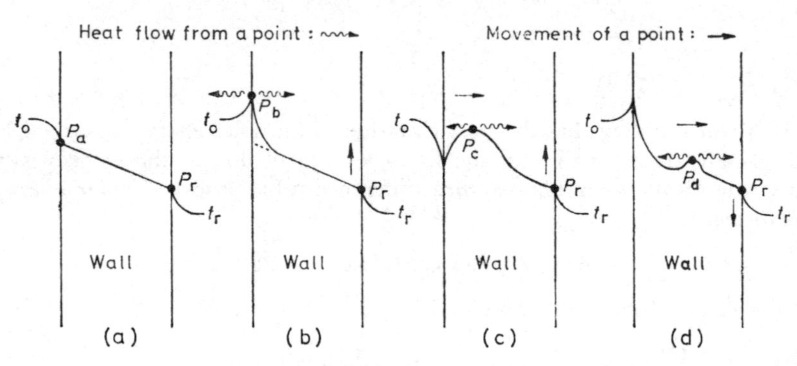

Fig. 7.14

Figure 7.14(*b*) shows the effect of raising the outer surface temperature. The temperature at the point P_b is greater than that at the point P_a. Such an increase could be caused by the outer wall surface receiving solar radiation. Heat flows away from P_b in both directions because its temperature is higher than both the air and the material of the wall in its vicinity. If the intensity of solar radiation then diminishes, the situation in Fig. 7.14(*c*) arises.

Since heat flowed away from P_b, the value of the temperature at P_c is now less than it was at P_b. When the surface temperature rises again, the situation is as shown at (*d*). It can be seen that the crest of the wave, represented by points P_a, P_b, P_c, P_d etc., is travelling to the right and that its magnitude is reducing. Further crests will follow the original one because a wave is being propagated through the wall as a result of an oscillation in the value of the outside surface temperature. Eventually the wave will reach the inner surface of the wall and will produce similar fluctuations in surface temperature. The inner surface temperature will have a succession of values corresponding to the point P_r, as it rises and falls. Thick walls with a large thermal capacity will damp the temperature wave considerably, whereas thin walls of small capacity will have little damping effect, and fluctuations in outside surface temperature will be apparent, almost immediately, as similar changes in inner surface temperature.

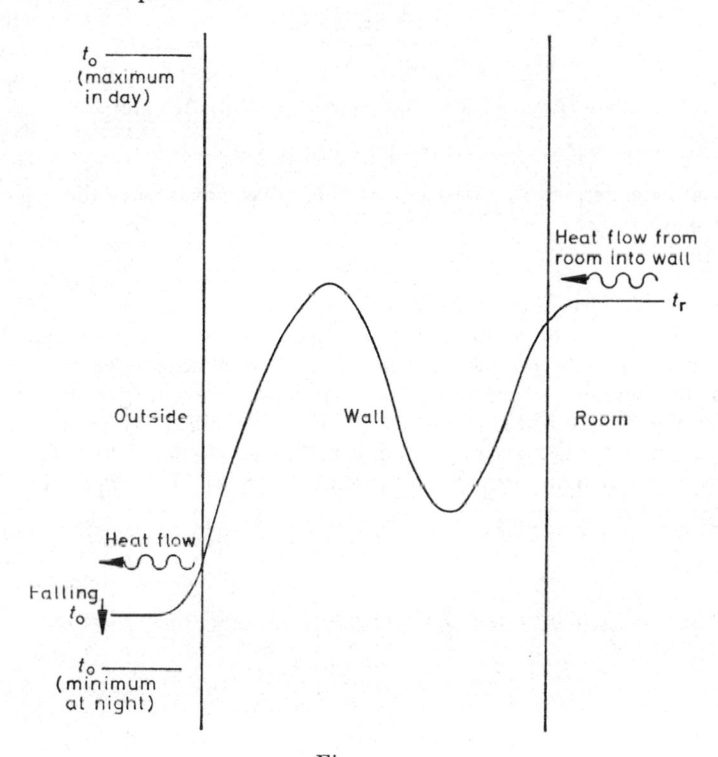

Fig. 7.15

It is possible for the inner surface temperatures to be less than room air temperatures, at certain times of the day, for walls of sufficiently heavy construction. A wide diurnal range of temperature can give this result. The outside surface temperature falls at night, by radiation to the black vault of the sky, and the effect of this is felt as a low inside surface temperature at some time later. At such a time, the air-conditioning load will be reduced because of the heat lost into the wall from the room. Figure 7.15 presents a simplified picture of this.

7.18 Sol-air temperature

In the first instance, heat flow through a wall depends on the rate at which heat enters its outer surface. The concept of ' Sol-air temperature ' has been in use for some time as an aid to the determination of the initial rate of entry of heat. It is defined as the value of the outside air temperature which would, in the absence of all radiation exchanges, give the same rate of heat flow into the outer surface of the wall as the actual combination of temperature differences and radiation exchanges really does.

The sol-air temperature, t_e, is a value, greater than t_o, which can be inserted in the equation

$$Q' = h_{so}(t_e - t_{so}) \tag{7.16}$$

where Q' = the rate of entry of heat, in W/m²,

h_{so} = the outside surface conductance, in W/m² °C, and

t_{so} = the outside surface temperature in °C.

Q' can be expressed in another way which does not involve the use of the sol-air temperature:

$$Q' = \alpha I_\delta + \alpha' I_s + h_{so}(t_o - t_{so}) + R \tag{7.17}$$

In this basic heat entry equation, α and α' are the absorption coefficients (usually about the same value) for direct, I_δ, and scattered, I_s, radiation which is normally incident on the wall surface. R is a remainder term which covers the complicated relatively low temperature exchanges by radiation between the wall and nearby surfaces. The value of R is difficult to assess; in all probability it is quite small and, if neglected, results in little error.

Equations (7.16) and (7.17) can be combined to yield an expression for t_e in useful form:

$$t_e = t_o + \frac{\alpha I_\delta + \alpha' I_s + R}{h_{so}} \tag{7.18}$$

If α' is made equal to α, and R is ignored, this expression becomes

$$t_e = t_o + \frac{\alpha(I_\delta + I_s)}{h_{so}} \tag{7.19}$$

The I.H.V.E. Guide quotes an equation virtually the same as this.

EXAMPLE 7.12. Calculate the sol-air temperature for the following set of conditions:

Vertical wall facing south-west

Sun time: 1 pm

$$\left.\begin{aligned} I_\delta = I_v &= 343\ \text{W/m}^2 \\ I_s &= 118\ \text{W/m}^2 \end{aligned}\right\}\text{(\textit{see} Example 7.11)}$$

$$\alpha = \alpha' = 0{\cdot}9$$

$$h_{so} = 22{\cdot}7\ \text{W/m}^2\ {}^\circ\text{C}$$

$$R = 0$$

$$t_o = 32^\circ\text{C}$$

Answer

Making use of eqn. (7.19)

$$t_e = 32^\circ + \frac{0{\cdot}9(343+118)}{22{\cdot}7}$$

$$= 32^\circ + \frac{415}{22{\cdot}7}$$

$$= 50{\cdot}3^\circ\text{C}.$$

TABLE 7.10 *Sol-air temperatures for 51·7 °N latitude.*

Based on a typical hot day in July, 90 per cent solar radiation absorbed by the outside surface of the roof or wall and a rate of heat transfer between the outside surface and the environment of 22·7 W/m² °C.

		Sol-air temperature in °C				
Sun time	Air temperature	Horizontal roof	North wall	East wall	South wall	West wall
1	19·1	19·9	19·9	19·9	19·9	19·9
2	19·1	19·1	19·1	19·1	19·1	19·1
3	18·3	18·3	18·3	18·3	18·3	18·3
4	17·6	17·6	17·6	17·6	17·6	17·6
5	17·2	22·2	23·3	32·8	18·0	18·0
6	17·6	28·2	26·1	42·5	18·9	18·9
7	18·3	35·1	23·2	47·2	20·1	20·1
8	19·4	43·1	21·8	48·6	25·9	21.8
9	20·7	49·7	23·8	46·8	32·7	23·7
10	21·9	56·0	25·3	42·4	39·0	25·3
11	23·3	60·5	26·9	31·3	43·5	26·9
12	24·7	62·2	28·3	28·3	46·2	28·3
13	25·8	63·0	29·3	29·3	46·1	39·4
14	26·8	60·9	30·1	30·1	43·9	47·3
15	27·5	56·5	30·5	30·5	39·5	53·6
16	27·8	51·4	30·1	30·1	34·3	57·0
17	27·5	44·3	32·4	29·3	29·3	56·4
18	26·7	37·3	35·1	27·9	27·9	51·5
19	25·6	30·6	32·8	26·3	26·4	42·8
20	24·4	24·4	24·4	24·4	24·4	24·4
21	23·3	23·3	23·3	23·3	23·3	23·3
22	22·2	22·2	22·2	22·2	22·2	22·2
23	21·4	21·4	21·4	21·4	21·4	21·4
24	20·6	20·6	20·6	20·6	20·6	20·6
24-h mean	22·9	37·0	25·2	29·8	28·3	29·9

Table 7.10 gives values of sol-air temperatures for average walls at different times and orientations at 51·7° N. From such a table the sol-air temperature for the above example is 42·8°C. So a difference of 10 degrees more southerly latitude and about 5 degrees more air temperature produces an increase of about 7½ degrees in the value of t_e, in this particular comparison.

7.19 Calculation of heat gain through a wall

The use of analytical methods to determine the transfer of heat through a wall into a room, in the presence of an unsteady rate of heat entry into the outer wall surface, poses very considerable difficulties. The determination of the sol-air temperature is of no value unless proper account is taken of the thermal capacity of the wall. No exact solution is possible but the I.H.V.E. Guide presents sufficient information to enable the heat gain to a room conditioned at a constant temperature to be calculated with more than sufficient accuracy for most practical purposes.

Most building materials have specific heats of about 0·84 kJ/kg °C, and so their thermal capacities depend largely on their density and thickness. The Institution's Guide gives a set of three curves (*see* Fig. 7.16(*a*)) for building materials of different densities which, in conjunction with a knowledge of the wall thickness, permit a ready determination of the time lag, φ, in hours (referred to in section 7.17). The diminution of the heat gain is also readily determined from another figure in the I.H.V.E. Guide (*see* Fig. 7.16(*b*)). This gives a decrement factor, f, against wall thickness.

On an average basis, the mean flow of heat, Q_m, through a wall into a room conditioned at a constant temperature, t_r, is given by

$$Q_m = AU(t_{em} - t_r) \tag{7.20}$$

t_{em} is the mean value of the sol-air temperature during a period of 24 hours.

If thermal capacitive effects were ignored, the instantaneous heat gain to the room at any time, θ, would be given by

$$Q_\theta = AU(t_e - t_r) \tag{7.21}$$

If capacitive effects are considered, then the actual heat flow into the room, at some later time $(\theta + \varphi)$, where φ is the time lag, is given by

$$Q_{\theta+\varphi} = AU(t_{em} - t_r) + AU(t_e - t_{em})f \tag{7.22}$$

In this equation, t_e is the sol-air temperature at time θ and f is the decrement factor.

This equation shows that $Q_{\theta+\varphi}$ can be greater or less than Q_m, depending, on whether t_e is greater or less than t_{em}. It is clear that sometimes (at night for instance) t_e must be less than its mean value t_{em}.

If a wall is very thick indeed—say greater than 600 mm—then the decrement factor will be very small and the effect of the second term in eqn. (7.22) is negligible. Equation (7.20) then gives a good approximation to the answer. If the wall is thin and has a small thermal capacity then the heat gain will vary considerably over 24 hours.

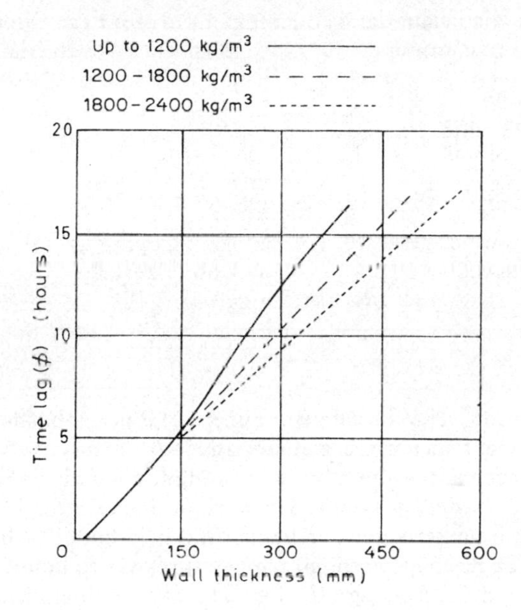

Fig. 7.16(*a*)

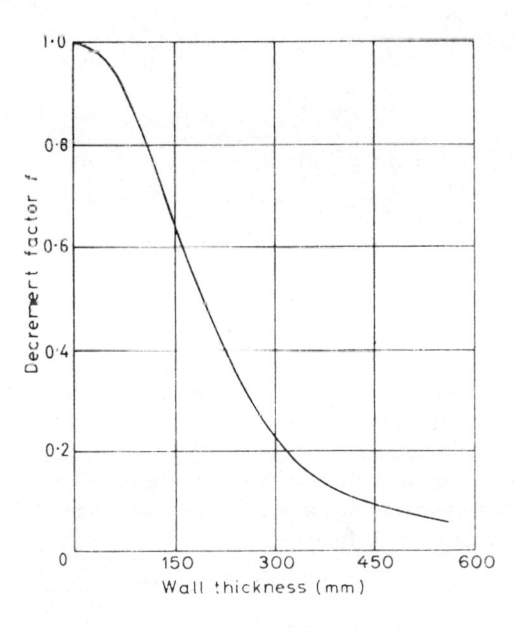

Fig. 7.16(*b*)

(*Reproduced by kind permission from the I.H.V.E. Guide*)

EXAMPLE 7.13. Calculate the heat gain at 3 pm sun time through a wall to a room conditioned continuously at 22°C, given the following data.

Orientation of wall	east
Wall thickness	150 mm
Density of wall	1200 kg/m^3
Latitude	51·7 °N
Declination	23½ °N
Outside air temperature	27·5°C at 3 pm
Colour of wall	dark outer surface
U-value	2·27 W/m^2 °C.

Answer

From the relevant I.H.V.E. data (*see* Fig. 7.16) it is established that the time lag and decrement factor are 5 hours and 0·65 respectively. This means that the heat received by the room at 3 pm first entered the outer surface of the wall 5 hours earlier, at 10 am. Hence the value of the sol-air temperature at 10 am must be determined. At this time it is found to be 42·4°C from Table 7.10. The mean value sol-air temperature over 24 hours for an eastward facing wall is 29·8°C. Making use of eqn. (7.22) we obtain the heat flow into the room at 3 pm.

$$
\begin{aligned}
Q_{(10\text{ am}+5\text{ h})} &= 2{\cdot}27(29{\cdot}8-22)+2{\cdot}27(42{\cdot}4-29{\cdot}8)\times 0{\cdot}65 \\
&= 2{\cdot}27\times 7{\cdot}8+2{\cdot}27\times 12{\cdot}6\times 0{\cdot}65 \\
&= 17{\cdot}7+18{\cdot}6 \\
&= 36{\cdot}3\text{ W/m}^2.
\end{aligned}
$$

EXAMPLE 7.14. For the same wall, calculate the heat gain to the room at 9 am sun time.

Answer

The decrement factor remains as 0·65 and the time lag stays at 5 hours. The heat gain at 9 am from the wall to the room first entered the outer surface of the wall at 4 am, five hours earlier. At 4 am the sol-air temperature is 17·6°C, and this is the same as the outside air temperature. The mean value of the sol-air temperature remains at 29·8°C, for an east wall. Equation (7.22) gives the heat gain at 9 am as follows:

$$
\begin{aligned}
Q_{(4\text{ am}+5\text{ h})} &= 2{\cdot}27(29{\cdot}8-22)+2{\cdot}27(17{\cdot}6-29{\cdot}8)\times 0{\cdot}65 \\
&= 17{\cdot}7-18{\cdot}0 \\
&= -0{\cdot}3\text{ W/m}^2.
\end{aligned}
$$

Thus, a heat loss is taking place from the room into the wall, at 9 am!

The A.S.H.R.A.E. Guide uses a method which differs from that in the I.H.V.E. Guide. Although both methods are the same in principle, that used by the I.H.V.E., it is claimed, is the more accurate, taking into account, as it does, certain second-order terms. The American work derives from work done by Mackey and Wright initially, and by Stewart later, who expressed the information in terms of 'equivalent temperature difference'. This, denoted by Δt, is the value of $(t_o - t_r)$ which, at any particular time of the day, allows the calculation of the heat gain to the room by means of the equation

$$Q = AU\Delta t \tag{7.23}$$

The original work by Mackey and Wright consisted of an analysis which yielded an answer in the form of an infinite series:

$$t_{si} = t_r + \frac{0{\cdot}107(t_{em} - t_e)}{0{\cdot}152 + L/k} + \sum_{n=1} \lambda_n t_n \cos(15n\theta - a_n - \varphi_n) \tag{7.24}$$

where t_{si} = the inside surface temperature of the wall at any time θ,

L = the wall thickness in m,

k = the thermal conductivity of the wall in W/m °C,

λ_n = the decrement factor for each harmonic (i.e. λ_1 is the fundamental decrement factor),

t_n = the harmonic temperature coefficient,

a_n = the harmonic phase angle, expressed in degrees, (i.e. a_1 for the first harmonic, etc.),

φ_n = the harmonic lag angle in degrees.

This is clearly cumbersome to use. So an approximate form of eqn. (7.24) was developed:

$$t_{si} = t_{sm} + \lambda_1(t_{e\theta} - t_{em}) \tag{7.25}$$

Here, t_{sm} is the mean inside surface temperature at a time $(\theta + \varphi_n/15)$ and $t_{e\theta}$ is the sol-air temperature at a time θ, i.e. $\varphi_n/15$ hours earlier than the time $(\theta + \varphi_n/15)$.

The simplification is obvious and, to make its use easier, graphs were published expressing values for the fundamental decrement factor, λ_1, and for the fundamental angle of time lag, φ_1. (15 degrees of angular displacement corresponds to one hour of time.) This information on λ_1 and φ_1 was couched in terms of $k\rho c$ and k/L, these being the relevant properties of the wall. The answers were not intended to be other than approximate.

Stewart took the results a step further by adopting an equivalent decrement factor, λ_e, as suggested by Mackey and Wright and given by the mean of the first and second harmonic decrement factors. The second-term effect thus introduced yielded greater accuracy, and data on this basis forms the subject of the tables of equivalent temperature difference published by the A.S.H.R.A.E.

The curves for time lag and decrement factor published by the I.H.V.E. are based on work by Danter. They give results which differ from those

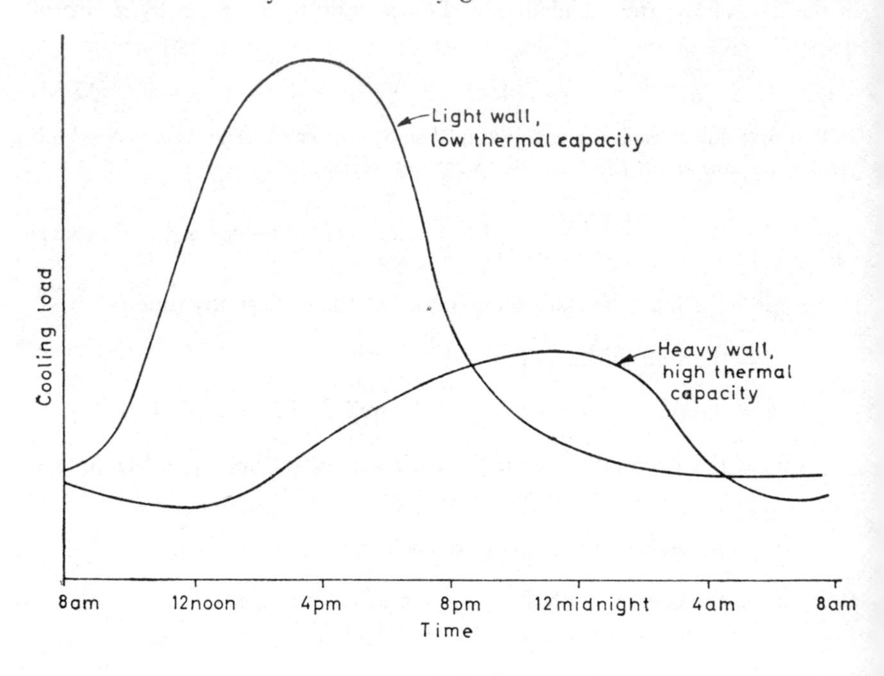

Fig. 7.17

obtained by the other methods mentioned because they take account of the fact that a curve of cooling load against time (*see* Fig. 7.17) has a different shape for each different wall structure. The American data are based on load curves assumed to have a sinusoidal shape. It is claimed that greater accuracy results from this refinement. However, a comparative example shows some discrepancy between the results achieved by the different methods. Which yields the most accurate answer it is difficult to decide. It is fortunate, therefore, that heat gains through walls usually constitute only a small percentage of the gross sensible heat gain. The same cannot always be said for heat gains through roofs exposed to sunlight, and it is in this direction that there is a great need for more accurate information. For very light-weight roofs it is sufficiently accurate to assume no time lag and a decrement factor of 1·0, adopting eqn. (7.21).

EXAMPLE 7.15. Making use of the information given below, calculate the heat gain through the wall of a room by the methods of Mackey and Wright, and of the I.H.V.E.

Material	150 mm of brickwork (therefore $L =$ 0·150 m).
Density	1760 kg/m^3 $= \rho$.
Specific heat	0·796 kJ/kg °C $= c$.
Conductivity	1·154 W/m °C $= k$.
Outside surface conductance	22·7 W/m^2 °C $= h_{so}$.
Inside surface conductance	9·4 W/m^2 °C $= h_{si}$.
Steady room air temperature	22°C $= t_r$.
Orientation	East.
Time of heat gain to the interior	3 pm sun time.

Take variations of sol-air and dry-bulb temperatures according to the information provided in Table 7.10 and make use of the curves published in reference 3 in the bibliography for the determination of time lag and decrement factor, for use with the method of Mackey and Wright.

Answer

Preliminary calculations:

$$k\rho c = 1{\cdot}154 \times 1760 \times 0{\cdot}796 = 1617\ (\text{kJ/m}^{2\circ}\text{C})^2/\text{s}$$

$$k/L = 1{\cdot}154/0{\cdot}150 = 7{\cdot}7\ \text{W/m}^{2\circ}\text{C}$$

$$t_{em} = 29{\cdot}8^\circ\text{C, from Table 7.10.}$$

$$\text{Fundamental time lag} = \varphi_1/15 = 3{\cdot}25\ \text{h (Mackey and Wright)}$$

$$\text{Fundamental decrement factor} = \lambda_1 = 0{\cdot}3\ \text{(Mackey and Wright)}$$

$$\text{Time lag} = 5\ \text{h, from Fig. 7.16}(a).$$

$$\text{Decrement factor} = f = 0{\cdot}64\text{, from Fig. 7.16}(b).$$

$$\text{U-value} = 3{\cdot}55\ \text{W/m}^2\ {}^\circ\text{C}.$$

(*a*) By the Method of Mackey and Wright

The mean daily inside surface temperature for steady flow is determined according to the formula:

$$t_{sm} = t_r + \frac{0{\cdot}107(t_{em} - t_r)}{0{\cdot}152 + L/k} \tag{7.26}$$

(This is a modification of eqn. (7.24).)

Inserting the appropriate values,

$$t_{sm} = 22 + \frac{0{\cdot}107(29{\cdot}8 - 22)}{0{\cdot}152 + 0{\cdot}132}$$

$$= 24{\cdot}9^\circ\text{C}.$$

If $t_{e\theta}$ is the sol-air temperature at time θ, the inside surface temperature t_{si} at a time $(\theta+\varphi_1/15)$ can be determined by eqn. (7.25):

$$t_{si} = 24{\cdot}9+0{\cdot}3(t_{e\theta}-29{\cdot}8)$$
$$= 16{\cdot}0+0{\cdot}3t_{e\theta}$$

The heat flow, Q/A, at time $(\theta+\varphi_1/15)$ can be determined by the equation

$$\frac{Q}{A}_{(\theta+\varphi_1/15)} = 9{\cdot}4(t_{si}-t_r) \tag{7.27}$$
$$= 9{\cdot}4(16{\cdot}0+0{\cdot}3t_{e\theta}-22)$$
$$= 2{\cdot}82t_{e\theta}-56{\cdot}4 \text{ W/m}^2 \tag{7.28}$$

This is the heat gain to the interior at a specified time, according to the approximate method of Mackey and Wright.

(*b*) Using the I.H.V.E. *Guide*

$$\frac{Q}{A}_{(\theta+\varphi_1/15)} = U\{(t_{em}-t_r)+f(t_{e\theta}-t_{em})\}$$

This is according to eqn. (7.22).
Inserting the appropriate values,

$$\frac{Q}{A}_{(\theta+\varphi_1/15)} = 3{\cdot}55\{(29{\cdot}8-22)+0{\cdot}65(t_{e\theta}-29{\cdot}8)\}$$
$$= 2{\cdot}31t_{e\theta}-41{\cdot}1 \text{ W/m}^2 \tag{7.29}$$

This equation is directly comparable with eqn. (7.28), and the use of these two formulas answers the question, yielding the curves shown in Fig. 7.18 for the methods of Mackey and Wright and the I.H.V.E.

The figure shows some general correspondence between the two methods; the method of Mackey and Wright gives a larger answer, occurring about two hours earlier than the peak value obtained by the I.H.V.E. technique. A direct comparison with Stewart's method (A.S.H.R.A.E.) is not possible because this does not cover directly a wall which is 150 mm in thickness. However, curves are shown for walls of 100 and 200 mm in thickness. The time of the peak value for a 100 mm wall is the same as that obtained by Mackey and Wright's method, for a 150 mm wall.

7.20 Air-conditioning load due to solar gain through glass

The solar radiation which passes through a sheet of window glazing does not constitute an immediate load on the air-conditioning system. This is because
(*a*) air is transparent to radiation of this kind, and
(*b*) a change of load on the air-conditioning system is indicated by an alteration to the air temperature within the room.

For the temperature of the air in the room to rise, solar radiation entering through the window must first warm up the solid surfaces of the furniture, floor slab and walls, within the room. These surfaces are then in a position to liberate some of the heat to the air by convection. Not all the heat will be liberated immediately, because some of the energy is stored within the depth of the solid materials. The situation is analogous to that considered in section

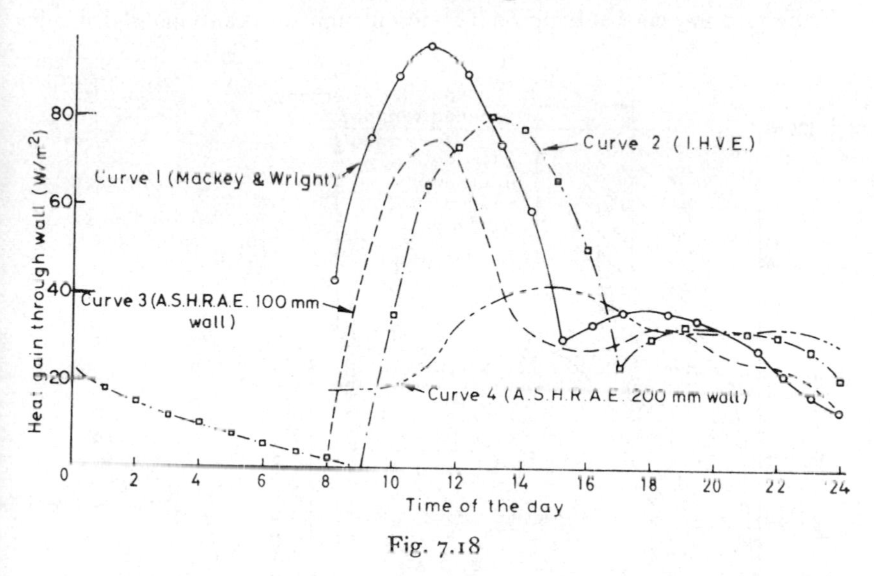

Fig. 7.18

Data for curves 1 and 2

East wall 50 °N latitude	150 mm thickness
$t_o = 27{\cdot}5^\circ$ at 3 pm	1760 kg/m^3
$t_r = 22^\circ$	$U = 3{\cdot}55$ W/m^2 °C

7.18 for heat gain through walls. There is, thus, a decrement factor to be applied to the value of the instantaneous solar transmission through glass, and there is also a time lag to be considered.

Figure 7.19 illustrates that, in the long run, all the energy received is returned to the room, but, because of the diminution of the peak values, the maximum load on the air-conditioning system is reduced.

Modern buildings have most of their mass concentrated in the floor slab, which will, therefore, have a big effect on the values of the decrement factor and the time lag. Since the specific heat of most structural materials is about 0·8 kJ/kg °C, the precise composition of the slab does not matter very much. Although most of the solar radiation entering through a window does strike the floor slab and get absorbed, the presence of furniture and floor coverings, particularly carpeting, reduces the influence of the slab. Wooden furniture has a smaller mass, hence any radiation received by it and absorbed will be subjected to only a small time lag and will be convected back to the room quite soon. The insulating effect of carpets means that the floor behaves as if it were thinner, resulting in a larger decrement factor. There is, thus, a tendency for a furnished, carpeted room to impose a larger load on the air-conditioning system, and to do so sooner than will an empty room.

Another factor of some importance is the time for which the plant operates. Figure 7.19 shows what happens if an installation runs continuously. Under

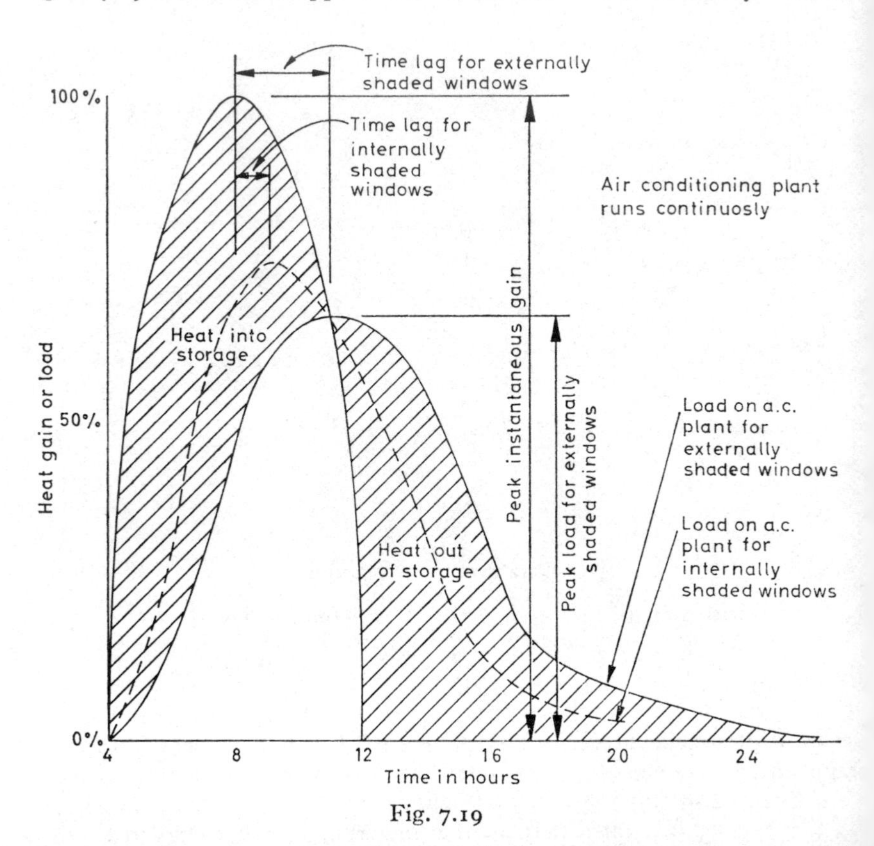

Fig. 7.19

these circumstances there is no, so-called, 'pull-down' load. If the plant operates for only, say, 12 hours each day, then the heat stored in the fabric of the building is released to the inside air during the night and, on start up next morning, the initial load may be greater than expected. This surplus is termed the pull-down load. Figure 7.20 illustrates the possible effect of such a surplus load. The importance of pull-down load is open to question: outside dry-bulb temperatures fall at night and, in the presence of clear skies, the building is then likely to lose a good deal of the stored heat by radiation. There is an initial load when the sun rises, but the major increase in load is unlikely to occur, in an office block for example, until people enter at 9 am and lights are switched on. This may swamp the effect of pull-down load and render its presence less obvious.

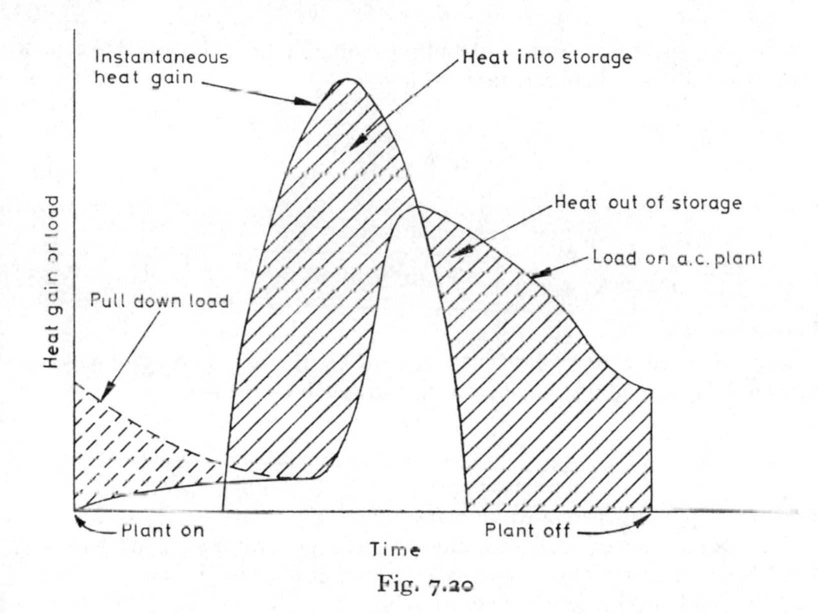

Fig. 7.20

An analysis by Bull yields an equation which may be used to estimate the air conditioning load if it is assumed that: the instantaneous solar gain (q_i) varies sinusoidally with time (t), the room air temperature is constant, the structural materials of the room have similar thermal properties and temperature gradients in them are negligible. His argument is as follows.

The instantaneous gain through the window equals the heat convected from the room surfaces to the air in the room, plus the energy stored in the structural materials irradiated by the sun. Then, interpreting the irradiated surface as the floor:

$$q_i A_w \, dt = h A_f \theta \, dt + A_f x \rho c \, d\theta \qquad (7.30)$$

where A_w and A_f are the window and floor areas, respectively, h is the coefficient of heat transfer from the floor to the room air, θ is the temperature difference between the surface and the air, x is the thickness of the floor slab in which heat is stored (sometimes taken as about the upper 100 mm), ρ is its density, c its specific heat capacity, dt an infinitesimally small element of time and $d\theta$ one of temperature difference.

If the concept of a time constant ($1/K$), equal to the thermal capacity of the relevant part of the floor divided by its rate of heat emission to the room air, is introduced, then eqn. (7.30) can be re-written as

$$\frac{d\theta}{dt} + K\theta = \frac{A_w}{A_f} \cdot \frac{K}{h} \cdot q_i \qquad (7.31)$$

For the instantaneous gain we can write

$$q_i = q_{max} \sin(\omega t - \varphi) \tag{7.32}$$

in which q_{max} is the maximum instantaneous gain, ω is the angular velocity of the sun and φ is a phase constant. Then

$$\frac{d\theta}{dt} + K\theta = \frac{A_w}{A_f} \cdot \frac{K}{h} \cdot q_{max} \sin(\omega t - \varphi)$$

for which the solution is

$$\theta = \frac{A_w}{h A_f} \cdot \frac{K}{(K^2 + \omega^2)} \cdot [K \sin a - \omega \cos a + \omega\, e^{-Ka/\omega}] \cdot q_{max} \tag{7.33}$$

where $a = \omega t - \varphi$.

We can see from eqn. (7.30) that the load on the air conditioning system (q_{ac}) is $h\,\theta\, A_f/A_w$ per unit area of window in sunshine. Then

$$q_{ac} = \frac{K}{(K^2 + \omega^2)} \cdot [K \sin a - \omega \cos a + \omega\, e^{-Ka/\omega}] \cdot q_{max} \tag{7.34}$$

for $0 \leqq a \leqq \pi$, per unit area of window in sunshine.

When the sun moves off the glazing $q_i = 0$ (in general at $a \geqq \pi$ or $t \geqq (\varphi + \pi)/\omega$) and Newtonian cooling is assumed to occur, eqn. (7.31) becoming $d\theta/dt = -K\theta$ for which, in this case, the solution is

$$\theta = \theta_o\, e^{-K(a-\pi)/\omega}$$

but for $a \geqq \pi$, from eqn. (7.33)

$$\theta_o = \frac{A_w}{h A_f} \cdot \frac{\omega K}{(K^2 + \omega^2)} \cdot [1 + e^{-K\pi/\omega}] \cdot q_{max}$$

and so

$$\theta = \frac{A_w \omega K}{h A_f (K^2 + \omega^2)} \cdot (1 + e^{-K\pi/\omega}) \cdot e^{-K(a-\pi)/\omega} \cdot q_{max}$$

Hence, by eqn. (7.30)

$$q_{ac} = \frac{\omega K}{(K^2 + \omega^2)} \cdot (1 + e^{-K\pi/\omega}) \cdot e^{-K(a-\pi)/\omega} \cdot q_{max} \tag{7.35}$$

for $a \geqq \pi$ per unit area of window.

EXAMPLE 7.16 Calculate the air conditioning load at 0800 in June resulting from solar heat gain through an east-facing window, single glazed with ordinary glass and unshaded. Assume that only the upper 75 mm of floor slab is significant, that it has a density of 2150 kg/m^3 and that its specific heat capacity is 0·85 kJ/kg °C. Take the surface heat transfer coefficient as 9·5 W/m^2 °C. Sunrise is at 0400 and sunset at 2000.

Answer

From Table 7.4 the transmission factor for single ordinary glass is 0.84 and from Table 7.11 the maximum total solar radiation normally incident on east-facing glass in June is 695 W/m². Also

$$\frac{1}{K} = \frac{0{\cdot}075 \times 2150 \times 850}{9{\cdot}5 \times 3600} = 4 \text{ hours.}$$

The period from 0400 to 2000 is 16 hours and equals $2\pi/\omega$. Thus $\omega = \pi/8$ hours^{-1}.

At 0400: $\sin(\omega t - \varphi) = 0 \therefore \varphi = \omega t - \pi/2$

More generally, $a = \omega t - \varphi = \omega t - \pi/2 = \pi(t-4)/8$

At 0800: $a = \pi(8-4)/8 = \pi/2$ and by eqn. (7.34)

$$q_{ac} = \frac{1}{4\left\{\frac{1}{16}+\frac{\pi^2}{64}\right\}} \cdot \left[\frac{1}{4}\sin\frac{\pi}{2} - \frac{\pi}{8}\cos\frac{\pi}{2} + \frac{\pi}{8}\exp\left(-\frac{1}{4}\cdot\frac{\pi}{2}\Big/\frac{\pi}{8}\right)\right] \cdot q_{max}$$

$$= \frac{16}{13{\cdot}87}\left[\frac{1}{4} + \frac{\pi}{8}e^{-1}\right] \cdot q_{max}$$

$$= 0{\cdot}455\, q_{max}$$

$$= 0{\cdot}455 \times 0{\cdot}84 \times 695 = 266 \text{ W/m}^2.$$

It is to be noted that Table 7.12 quotes a figure of 0·54 for the storage factor for bare glass at 0800 in June, with 12 hour operation and a slab density of 500 kg/m². The above example, of course, assumes 24 hour operation.

When the window has internal blinds these absorb part of the radiation and convect and re-radiate it back to the room. The remaining part is considered as direct transmission and so is susceptible to storage effects. Because the mass of the blinds is small and air is not entirely transparant to the long wave-length emission from the relatively low temperature blinds, the load imposed by convection and re-radiation is virtually instantaneous. The same argument holds for heat-absorbing glass.

For cases where the windows have internal Venetian blinds fitted the air conditioning cooling load may be calculated directly by means of Tables 7.8 and 7.9.

EXAMPLE 7.17. Calculate the load arising from the solar heat gain through a double-glazed window, shaded by internal Venetian blinds, facing south-west, at 3 pm sun time, in June, at latitude 51·7 °N, by means of Tables 7.8 and 7.9.

Answer

Reference to Table 7.8 shows that the load is 170 W/m² for single-glazed windows shaded internally by Venetian blinds of a light colour. Reference to Table 7.9 gives a factor of 1·08 to be applied to the value of 170 W/m² when the window is double glazed with ordinary glass. The load on the air-conditioning system is, therefore, 1·08 × 170 = 184 W/m².

Note that if the blinds had been fitted between the sheets of glass, the factor would have been 0·55, and the load would then have been 0·55 × 170 = 94 W/m². Compare the simplicity of this with Example 7.11.

For windows not fitted with internal Venetian blinds, where the direct use of Tables 7.8 and 7.9 is not appropriate, the air conditioning load may be determined by taking the maximum total solar intensity normal to a surface (Table 7.11) and multiplying this by factors for haze, dew-point, altitude, hemisphere, storage (Table 7.12) and shading (Table 7.5).

TABLE 7.11 *Maximum total solar intensities normal to surfaces for latitude* 51·7° N in W/m².

Orientation	Dec	Jan/ Nov	Feb/ Oct	Mar/ Sep	Aug / Apr	Jul/ May	Jun
North	45	55	75	100	120	135	135
NE/NW	45	55	110	240	405	505	530
E/W	205	255	410	545	660	695	695
SE/SW	425	490	625	695	685	645	625
Horizontal	175	235	385	560	725	820	850

Note that the above figures should be increased by 7 per cent for every 5° by which the dew-point is less than 15° for an application other than in London.

EXAMPLE 7.18 Calculate the air conditioning load arising from solar gain through a window fitted with unshaded, single, heat-reflecting (bronze) glass, facing SW, at 3 pm sun time in June in London, for a floor slab density of 500 kg/m².

Answer

The maximum total solar intensity normal to a SW surface in June is 625 W/m². Assume a haze factor of 0·95 for London. The shading co-efficient (Table 7.5) is 0·27 and the storage factor (Table 7.12) is 0·62. Then

$$q_{ac} = 625 \times 0{\cdot}95 \times 0{\cdot}27 \times 0{\cdot}62$$

$$= 99 \text{ W/m}^2 \text{ of glass surface.}$$

7.21 Heat transfer to ducts

A heat balance equation establishes the change of temperature suffered by a ducted airstream under the influence of a heat gain or loss:

$$Q = PLU\left\{\left(\frac{\mp t_1 \pm t_2}{2}\right) \pm t_r\right\} \tag{7.36}$$

where Q = heat transfer through the duct wall in W,

P = external duct perimeter in m,

L = duct length in m,

U = overall thermal transmittance in W/m² °C,

t_1 = initial air temperature in the duct in °C,

t_2 = final air temperature in the duct in °C,

t_r = ambient air temperature in °C.

Also,

$$Q = \text{flow rate } (\text{m}^3/\text{s}) \times (\mp t_1 \pm t_2) \times \frac{358}{(273+t)}$$

The factor in eqn. (7.36), which is of considerable importance, is the U-value. This is defined in the usual way by

$$\frac{1}{U} = r_{si} + \frac{l}{k} + r_{so} \tag{7.37}$$

In this equation the thermal resistance of the metal is ignored and the symbols used have the following meanings:

r_{si} = thermal resistance of the air film inside the duct, in m² °C/W,

r_{so} = thermal resistance of the air film outside the duct, in m² °C/W,

l = thickness of the insulation on the duct, in metres,

k = thermal conductivity of the insulation in W/m °C.

The value of k is usually easily determined but it is customary to take a value between 0·03 and 0·07. We have here selected a value of 0·045 as typical. Small alterations in the value of k are not significant within the range mentioned, but changing the thickness is, of course, very influential in altering the heat gain. Values of r_{so} are difficult to establish with any certainty; the proximity of the duct to a ceiling or wall has an inhibiting effect on heat transfer and tends to increase the value of r_{so}. A value of 0·1 is suggested.

The internal surface resistance, on the other hand, may be calculated with moderate accuracy if the mean velocity of airflow in the duct is known.

TABLE 7.12 *Storage load factors, solar heat gain through glass. 12 hour operation, constant space temperature.*

Exposure (north lat.)	Mass per unit area of floor kg/m²	Sun time 6	7	8	9	10	11	N	1	2	3	4	5
North and shade	500	0·98	0·98	0·98	0·98	0·98	0·98	0·98	0·98	0·98	0·98	0·98	0·98
	150	1·00	1·00	1·00	1·00	1·00	1·00	1·00	1·00	1·00	1·00	1·00	1·00
NE	500	0·59	0·68	0·64	0·52	0·35	0·29	0·24	0·23	0·20	0·19	0·17	0·15
	150	0·62	0·80	0·75	0·60	0·37	0·25	0·19	0·17	0·15	0·13	0·12	0·11
E	500	0·52	0·67	0·73	0·70	0·58	0·40	0·29	0·26	0·24	0·21	0·19	0·16
	150	0·53	0·74	0·82	0·81	0·65	0·43	0·25	0·19	0·16	0·14	0·11	0·09
SE	500	0·18	0·40	0·57	0·70	0·75	0·72	0·63	0·49	0·34	0·28	0·25	0·21
	150	0·09	0·35	0·61	0·78	0·86	0·82	0·69	0·50	0·30	0·20	0·17	0·13
S	500	0·26	0·22	0·38	0·51	0·64	0·73	0·79	0·79	0·77	0·65	0·51	0·31
	150	0·21	0·29	0·48	0·67	0·79	0·88	0·89	0·83	0·56	0·50	0·24	0·16
SW	500	0·33	0·28	0·25	0·23	0·23	0·35	0·50	0·64	0·74	0·77	0·70	0·55
	150	0·29	0·21	0·18	0·15	0·14	0·27	0·50	0·69	0·82	0·87	0·79	0·60
W	500	0·67	0·33	0·28	0·26	0·24	0·22	0·20	0·28	0·44	0·61	0·72	0·73
	150	0·77	0·34	0·25	0·20	0·17	0·14	0·13	0·22	0·44	0·67	0·82	0·85
NW	500	0·71	0·31	0·27	0·24	0·22	0·21	0·19	0·18	0·23	0·40	0·58	0·70
	150	0·82	0·35	0·25	0·20	0·18	0·15	0·14	0·13	0·19	0·41	0·64	0·80

Internal shade

Bare glass or external shade

North and shade	500	0·81	0·84	0·86	0·89	0·91	0·93	0·93	0·94	0·94	0·95	0·95	0·95
	150	1·00	1·00	1·00	1·00	1·00	1·00	1·00	1·00	1·00	1·00	1·00	1·00
NE	500	0·35	0·45	0·50	0·49	0·45	0·42	0·34	0·30	0·27	0·26	0·23	0·20
	150	0·40	0·52	0·69	0·64	0·48	0·34	0·27	0·22	0·18	0·16	0·14	0·12
E	500	0·34	0·44	0·54	0·58	0·57	0·51	0·44	0·39	0·34	0·31	0·28	0·24
	150	0·36	0·56	0·71	0·76	0·70	0·54	0·39	0·28	0·23	0·18	0·15	0·12
SE	500	0·29	0·33	0·41	0·51	0·58	0·61	0·61	0·56	0·49	0·44	0·37	0·33
	150	0·14	0·27	0·47	0·64	0·75	0·79	0·73	0·61	0·45	0·32	0·23	0·18
S	500	0·44	0·37	0·39	0·43	0·50	0·57	0·64	0·68	0·70	0·68	0·63	0·53
	150	0·28	0·19	0·25	0·38	0·54	0·68	0·78	0·84	0·82	0·76	0·61	0·42
SW	500	0·53	0·44	0·37	0·35	0·31	0·33	0·39	0·46	0·55	0·62	0·64	0·60
	150	0·48	0·32	0·25	0·20	0·17	0·19	0·39	0·56	0·70	0·80	0·79	0·69
W	500	0·60	0·52	0·44	0·39	0·34	0·31	0·29	0·28	0·33	0·43	0·51	0·57
	150	0·77	0·56	0·38	0·28	0·22	0·18	0·16	0·19	0·33	0·52	0·69	0·77
NW	500	0·54	0·49	0·41	0·35	0·31	0·28	0·25	0·23	0·24	0·30	0·39	0·48
	150	0·75	0·53	0·36	0·28	0·24	0·19	0·17	0·15	0·17	0·30	0·50	0·66

Theoretical considerations suggest that the value of r_{si} is a function of the Reynolds number, and experimental evidence suggests that

$$r_{si} = 0{\cdot}286 \frac{D^{0\cdot25}}{v^{0\cdot8}}, \text{ for circular ducts} \tag{7.38}$$

$$r_{si} = 0{\cdot}286 \frac{(2AB/(A+B))^{0\cdot25}}{v^{0\cdot8}}, \text{ for rectangular ducts} \tag{7.39}$$

where D = internal duct diameter in m,

v = mean duct velocity in m/s,

A, B = internal duct dimensions in m.

The change of temperature suffered by the air as it flows through the duct is of prime importance and the A.S.H.R.A.E. give an expression for this.

$$t_2 = \frac{t_1(y-1)+2t_r}{(y+1)}, \text{ for circular ducts} \tag{7.40}$$

and

$$y = \frac{503\,\rho DV}{\text{UL}} \tag{7.41}$$

where ρ is the density of the air in kg/m^3, and V is the air velocity in m/s.

The value of r_{si} is not sensitive to changes of air velocity for the range of duct sizes and velocities in common use; for example, r_{si} is 0·0284 m^2 °C/W for 8 m/s in a 75 mm diameter duct and 0·0243 m^2 °C/W for 20 m/s and 750 mm. Hence the U-value of a lagged duct is almost independent of the air velocity. From eqn. (7.37) the U-values are 1·47, 0·81 and 0·56 W/m^2 °C for lagging with thicknesses of 25, 50 and 75 mm, respectively, assuming r_{si} is 0·026 m^2 °C/W and k is 0·045 W/m °C. Then by eqn. (7.41) y has values of 400 DV, 726 DV and 1046 DV, respectively, for the three thicknesses mentioned, if we take ρ = 1·165 kg/m^3 (as typical for air at 30°C) and L = 1 metre.

Equation (7.40) can then be re-written to give the temperature drop, Δt, along one metre of duct:

$$\Delta t = t_1 - t_2 = \frac{2(t_1 - t_r)}{(y+1)} \tag{7.42}$$

Then, for most practical purposes:

$$\Delta t = \frac{(t_1 - t_r)}{200\,DV} \quad \text{for 25 mm lagging} \tag{7.43}$$

$$\Delta t = \frac{(t_1 - t_r)}{363\,DV} \quad \text{for 50 mm lagging} \tag{7.44}$$

$$\Delta t = \frac{(t_1 - t_r)}{523\,DV} \quad \text{for 75 mm lagging} \tag{7.45}$$

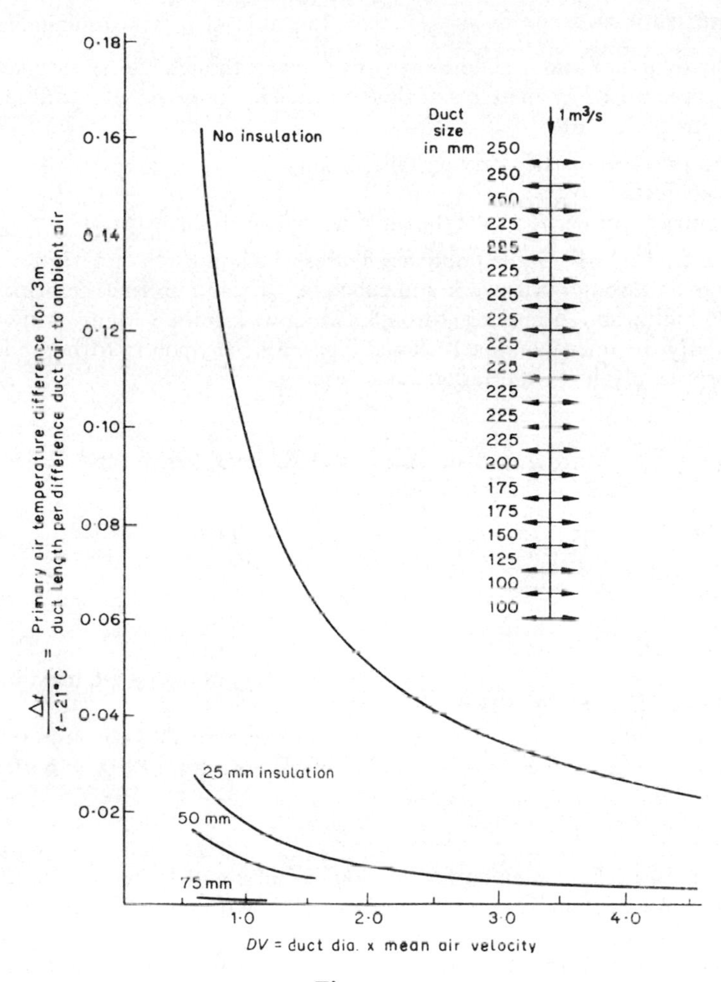

Fig. 7.21

These equations show that it is important to assess properly the ambient temperature, t_r, and that the temperature drop along the duct is inversely proportional to DV, being independent of the method of duct sizing adopted.

The equations are easy to use and they indicate that the rate of temperature drop is considerable once the value of DV falls below about $1{\cdot}5$ m^2/s. It is usually impractical to attempt to keep DV above this value by sizing or by increasing the airflow; it follows that the last few lengths of ductwork will suffer a considerable temperature drop—particularly a difficulty with the last four or five induction units in a non-changeover, perimeter induction system. The only sure way of virtually stopping the drop is to use 75 mm of lagging.

7.22 Infiltration

Outside air filters into a conditioned space, even though the space may be slightly pressurised by an excess of air supplied over air extracted. Infiltration is principally due to:

(*a*) wind pressure, particularly on tall buildings,
(*b*) stack effect,
(*c*) the entry of the occupants of the building, who introduce dirt.

The A.S.H.R.A.E. Guide publishes figures which indicate that wind causes airflow even through brickwork and concrete. It also publishes information on the infiltration occurring through window frames. Many windows, particularly the openable metal-framed types, are very poorly fitting indeed, and surprisingly high infiltration rates result.

EXAMPLE 7.19. Making use of A.S.H.R.A.E. data (*see* below), calculate the air change rate for the following room:

Height	2·9 m
Width	3·6 m
Depth	5·6 m
Cube	58·4 m^3
Wind velocity	9·0 m/s
Window size	1·5 m high × 2·6 m wide
Wall area	3·3 m^2
Wall material	200 mm of plastered brick.
Perimeter of unweatherstripped window crack	$2 \times 2{\cdot}6 + 3 \times 1{\cdot}5 = 9{\cdot}7$ m

Answer

Assuming an average wooden frame, double-hung and locked, the infiltration rate is 8×10^{-4} m^3/s for each metre of window frame crack, according to the A.S.H.R.A.E. Guide, for 9 m/s.

This represents a rate of infiltration through the window frame of $8 \times 10^{-4} \times 9{\cdot}7$ or $77{\cdot}6 \times 10^{-4}$ m^3/s, for a wind velocity of 9 m/s.

From the same source, the infiltration rate through a 200 mm plastered brick wall is $8{\cdot}0 \times 10^{-6} \times 3{\cdot}3$ or $26{\cdot}4 \times 10^{-6}$ m^3/s, for a wind velocity of 9 m/s. This is a trivial amount, but it is seen that half an air change per hour can result from wind effect alone.

If it is argued that the plenum effect will offset this, it must be remembered that wind velocities may be much higher than 9 m/s near the tops of buildings, even in summertime.

An Imperial version of the metric formula

$$q = 0{\cdot}172\, A\sqrt{\{h(t_1 - t_2)\}} \qquad (7{\cdot}46)$$

is published in the A.S.H.R.A.E. Guide and is easily derivable from first principles, by considering the difference of weight of two columns of air having a common height h metres, but different temperatures, t_1 inside the building and t_2 outside the building. A is the area in square metres available for the flow of air into, or out of the building, between the two columns, and q is the air quantity flowing, in m^3/s.

Assuming an effectiveness of 0·827,

$$\frac{q}{A} = 0{\cdot}172\sqrt{\{h(t_1 - t_2)\}}$$

is the velocity of airflow, and the corresponding pressure difference inducing the airflow is

$$\Delta p = 0{\cdot}043\, h(t_1 - t_2) \qquad (7.47)$$

EXAMPLE 7.20. A building is 65 m high and is conditioned uniformly throughout its interior at a temperature of 22°C when the mean outside air temperature is 28°C. Calculate the pressure likely to cause exfiltration from the entrance hall, assuming that the full height of 65 m is available for stack effect.

Answer

Making use of eqn. (7.47), available pressure difference, inside to outside,

$$\begin{aligned}\Delta p &= 0{\cdot}043 \times 65 \times (28 - 22)\\ &= 16{\cdot}75\ \text{N/m}^2.\end{aligned}$$

If it is assumed that air can leave the building at ground floor level, air must enter the building at the 65 m level, in a similar quantity, to make good the loss.

EXAMPLE 7.21. Calculate the air change rate of infiltration through the window mentioned in Example 7.19 for the conditions given in Example 7.20. Assume that frictional resistance to air flow at the window crack is given by $Q = 0{\cdot}125\,(\Delta p)^{0{\cdot}63}$ litres/s metre.

Answer

$$\begin{aligned}Q &= 0{\cdot}125 \times (16{\cdot}75)^{0{\cdot}63} \times 9{\cdot}7\\ &= 0{\cdot}737 \times 9{\cdot}7\\ &= 7{\cdot}15\ \text{litres/s.}\end{aligned}$$

$$\begin{aligned}\text{Air change rate} &= \frac{7{\cdot}15 \times 3600}{1000 \times 58{\cdot}4}\\ &= 0{\cdot}44\ \text{per hour.}\end{aligned}$$

The amount of dirt and air people carry with them, trapped in their clothing, constitutes a further small infiltration load which it is impossible to assess without experimental evidence.

The conclusion to be drawn is that air change rates due to infiltration may be surprisingly large, if all the factors are adverse. In the average case, it is customary to take a figure of half an air change per hour of infiltration as being a typical load on the room air-conditioning plant, for modern buildings, in summer design weather. In tall air-conditioned buildings it is not uncommon to try to prevent some or all of the stack effect by automatically monitoring the difference between the inside and outside pressures, near ground level, and regulating the capacity of the extract fan accordingly.

7.23 Electric lighting

Luminous intensity is defined, by international agreement, in terms of the brightness of molten platinum at a temperature of 1755°C, and the unit adopted for its expression is the candela. A point source of light delivers a flow of luminous energy which is expressed in lumens; the quotient of this luminous flux and the solid angle of the infinitesimal cone, in any direction, is the intensity of illumination expressed in candela. The density of luminous flux is the amount of luminous energy uniformly received by an area of one square metre. Thus, illumination, or density of luminous flux, is expressed in lumens per square metre, otherwise termed lux.

Electric lighting is usually chosen to produce a certain standard of illumination and, in doing so, electrical energy is liberated. Most of the energy appears immediately as heat, but even the small proportion initially dissipated as light eventually becomes heat after multiple reflections and reactions with the surfaces inside the room.

The standard of illumination produced depends not only on the electrical power of the source but also on the method of light production, the area of the surfaces within the room, their colour and their reflective properties. The consequence is that no straightforward relation exists between electrical power and standard of illumination. For example, fluorescent tube light fittings are more efficient than are tungsten filament lamps. This means that for a given room and furnishings, more electrical power, and hence more heat dissipation, is involved in maintaining a given standard of illumination if tungsten lamps are used. Table 7.13 gives some approximate guidance on power required for various intensities of illumination.

The table is based on a room of dimensions 9 m × 6 m, in plan, with light fittings mounted 3 m above floor level, for tungsten lamps and fluorescent fittings, and of dimensions 15 m × 9m × 4 m high, for mercury lamps. Light-coloured decorations and a reasonably clean room are assumed in each case. Generally speaking, larger rooms require fewer W/m^2 than smaller rooms do, for the same illumination.

The efficiency of fluorescent lamps deteriorates with age. Whereas initially a 40 W tube might produce 2000 lux with the liberation of 48 watts of power, the standard of illumination might fall to 1600 lux with the liberations of 48 watts at each tube after 7500 hours of life.

A further comment on fluorescent fittings is that the electrical power absorbed at the fitting is greater than that necessary to produce the light at the

TABLE 7.13

Illumination in lux	Wattage per m² of floor area, including any power required for control gear						
	Filament lamps		Colour-corrected mercury	80 W white fluorescent			
	Open industrial reflector (300 W)	General diffusing fitting (200 W)	Open industrial reflector (250 W)	Enamel plastic trough	Enclosed diffusing fitting	Louvred ceiling panel
150	19–28	28–36	8–14	6	8	8–11
200	28–36	36–50	11–15	8	11	11
300	38–55	50–69	19–28	11–14	11–16	14–19
500	66–88	—	33–34	16–22	22–28	22–33
1000	—	—	60–82	33–44	36–55	44–66

(*Reproduced by kind permission of "The Steam and Heating Engineer"*)

tube. A fitting which has 80 watts printed on the tube will need 100 watts of power supplied to it; the surplus 20 watts is liberated directly from the control gear of the fitting as heat into the room.

As a rough guide it can be assumed that a typical modern lighting standard of 500 lux in an office involves a power supply of about 33 W/m^2 of floor area. Thus, an office measuring 3 m × 6 m will require 600 watts at its fluorescent light fittings to produce a standard of 500 lux. Six tubes, each of 80 watts, will be needed.

The heat liberated when electric lights are switched on is not felt immediately as a load by the air-conditioning system since the heat transfer is largely by radiation. As with solar radiation, time must pass before a convective heat gain from solid surfaces causes the air temperature to rise. With an air-conditioning system running for 12 hours a day and a floor slab of 500 kg/m^2, the decrement (storage) factor, applicable to the heat gain from lights, is about 0·6 immediately and 0·9 one hour after the lights have been turned on, for projecting light fittings. If the light fittings are recessed, the corresponding factors are about 0·5 and 0·85. The decrement becomes 0·9 after 4 hours.

Where advantage can be taken is if the lights are recessed and the ceiling space above is used as part of the air extract system. Under these circumstances the storage factors are 0·7 initially, 0·8 after one hour and 0·9 after 8 hours. If the light fitting itself is used as an air outlet, for the extraction of air from the room, then the effect is a long-term one and a permanent allowance can be made for the heat liberated at the light and carried away by the extracted air, provided adequate information is available from the manufacturers of the light fitting. If all the extracted air is discharged to atmosphere then full value is credited for the reduction on the air-conditioning load. However, if as is more likely, a good deal of the extracted air is re-circulated, the full effect is not felt and due allowance must be made by increasing the temperature of the mixture air (outside air plus the recirculated air). Where the ceiling void is used as an extracted air path but ventilated light fittings are not used, 30 per cent of the heat may be removed by the extracted air. If extract ventilation light fittings, unducted, are used, 40 per cent may be removed.

Finally, lighting standards have risen enormously since the Second World War. Whereas 10 W/m^2 was once acceptable, most designers assume 25 or 35 W/m^2 today, if more precise information is not forthcoming from the architects. Intensities of 60 or even 120 W/m^2 are not uncommon.

7.24 Occupants

As was mentioned in section 4.1, human beings give off heat at a metabolic rate which depends on their rate of working. The sensible and latent proportiont of the heat liberated for any given activity depend on the value of the ambient dry-bulb temperature: the lower the dry-bulb temperature the larger the proportion of sensible heat dissipated.

Typical values of the sensible and latent liberations of heat are given in Table 7.14.

The figures for eating in a restaurant include the heat given off by the food.

Deciding on the density of occupation is usually a problem for the air-conditioning designer. A normal density for an office block is 10 m^2 per person, as an average over the whole conditioned floor area. The density of occupation may be as low as 20 m^2 per person in executive offices or as high as 6 m^2 per person in open office areas.

Some premises may have much higher densities than this; for restaurants, 2 m^2 per person is reasonable, but for department stores, at certain times of the year, densities may reach values of 1·5 to 1·0 m^2 per person, even after allowance has been made for the space occupied by goods. In concert halls, cinemas and theatres, the seating arrangement provides the necessary information but in dance halls and night clubs estimates are open to conjecture. Occupation may be very dense indeed. A figure of 0·5 m^2 per person is suggested tentatively.

TABLE 7.14

Activity	Metabolic rate W	Heat liberated in W, Room dry-bulb temperature (°C) 20° S	20° L	22° S	22° L	24° S	24° L	26° S	26° L
Seated at rest	115	90	25	80	35	75	40	65	50
Office work	140	100	40	90	50	80	60	70	70
Standing	150	105	45	95	55	82	68	72	78
Eating in a restaurant	160	110	50	100	60	85	75	75	85
Light work in a factory	235	130	105	115	120	100	135	80	155
Dancing	265	140	125	125	140	105	160	90	175

7.25 Power dissipation from motors

Deciding on the heat liberated from motors hinges on the following:

(i) The frequency with which the motors will be used, if there is more than one in the conditioned space, so that the maximum simultaneous liberation may be assessed.

(ii) The efficiency of the motor.

(iii) Whether the motor and its driven machine are in the conditioned space.

All the power drawn from the electricity mains is ultimately dissipated as heat. If both the motor and the machine are in the conditioned space then the total amount of power drawn from the mains appears as a heat gain to the room. If only the driven machine is in the room, the motor being outside, then the product of the efficiency (motor plus drive) and the power drawn from the mains is the heat liberated to the space conditioned. Similarly, when only the driving motor is within the room, 100 minus the efficiency is the factor to be used.

EXERCISES

1. (*a*) Why do the instantaneous heat gains occurring when solar thermal radiation passes through glass not constitute an immediate increase on the load of the air-conditioning plant? Explain what sort of effect on the load such instantaneous gains are likely to have in the long run.

(*b*) A single glass window in a wall facing 30° west of south is 2·4 m wide and 1·5 m high. If it is fitted flush with the outside surface of the wall, calculate the instantaneous heat gain due to direct solar thermal radiation, using the following data:

Intensity of direct radiation on a plane normal to the sun's rays:	790 W/m^2
Altitude of the sun:	60°
Azimuth of the sun:	70° west of south
Transmissivity of glass:	0·8

(*c*) If the decrement factor is 0·8, what will be the actual load on the air-conditioning system due to the instantaneous gain calculated in (*b*)?

Answers: (*b*) 870 W, (*c*) 696 W.

2. A window 2·4 m long × 1·5 m high is recessed 300 mm from the outer surface of a wall facing 10° west of south. Using the following data, determine the temperature of the glass in sun and shade and hence the instantaneous heat gain through the window.

Altitude of sun	60°
Azimuth of sun	20° east of south
Intensity of sun's rays	790 W/m^2
Sky radiation normal to glass	110 W/m^2
Transmissivity of glass	0·6
Reflectivity of glass	0·1
Outside surface coefficient	23 W/m^2 °C
Inside surface coefficient	10 W/m^2 °C
Outside air temperature	32°C
Inside air temperature	24°C

Answers: 37·9°C, 29·6°C, 983 W.

3. An air conditioned room measures 3 m wide, 3 m high and 6 m deep. One of the two 3 m by 3 m walls faces west and contains a single glazed window of size 1·5 m by 1·5 m. The window is shaded internally by Venetian blinds and is mounted flush with the external wall. There are no heat gains through the floor, ceiling, or walls other than that facing west and there is no infiltration. Compute the sensible and latent heat gains which constitute a load on the air conditioning system at 4 pm in June, given the following information.

Outside state	28°C dry-bulb, 19·5°C wet-bulb (sling)
Inside state	22°C dry-bulb, 50% saturation
Electric lighting	33 W per m^2 of floor area
Number of occupants	4
Heat liberated by occupants	90 W sensible, 50 W latent
Solar heat gain through window with Venetian blinds fully closed	258 W/m^2

U-value of wall	1·7 W/m² °C
U-value of glass	5·7 W/m² °C
Time lag for wall	5 hours ($=\varphi$)
Decrement factor for wall	0·62 ($=f$)

Diurnal variations of air temperature and sol-air temperature are as follows:

Sun time	9 am	10 am	11 am	Noon	1 pm	2 pm	3 pm	4 pm
Air temperature (°C)	20·6°	22·0°	23·3°	24·7°	25·8°	26·8°	27·5°	28·0°
Sol-air temperature (°C)	23·7°	25·3°	26·8°	28·3°	39·4°	47·3°	53·6°	57·0°

The mean sol-air temperature over 24 hours is 29·9°C ($=t_{em}$).

The heat gain through a wall, $q_{(\theta+\varphi)}$, at any time $(\theta+\varphi)$, is given by the equation:

$$q_{(\theta+\varphi)} = UA(t_{em} - t_i) + UA(t_e - t_{em})f$$

where t_e is the sol-air temperature.

t_i is the inside air temperature.

θ is the time in hours.

Answers: 1662 W, 200 W.

BIBLIOGRAPHY

1. A.S.H.R.A.E. *Guide and Data Books, Fundamentals and Equipment.*
2. *Air Conditioning System Design Manual, Part 1: Load Estimating.* Carlyle Air Conditioning and Refrigeration Ltd., London, S.W.1.
3. C. O. Mackey and L. T. Wright, Jr. Periodic heat flow—homogeneous walls or roofs, *Trans. A.S.H.V.E.*, 1944, **50,** 293–312.
4. J. P. Stewart. Solar heat gain through walls and roofs for cooling load calculations, *Trans. A.S.H.V.E.*, 1948, **54,** 361–388.
5. E. Danter. Periodic heat flow characteristics of simple walls and roofs, *J.I.H.V.E.*, 1960, **28,** 136–146.
6. Fluorescent luminaire calorimetry: principles and procedures, *J.A.S.H.R.A.E.*, Feb. 1965, 31–38.
7. I. S. Groundwater. *Solar Radiation in Air Conditioning.* Crosby Lockwood, 1957.
8. Olgyay and Olgyay. *Solar Control and Shading Devices.* Princeton University Press, 1957.
9. N. Robinson. *Solar Radiation.* Elsevier Publishing Company, Amsterdam, 1966.
10. D. C. Pritchard. Lighting in industry. *Steam and Heating Engineer*, Nov. 1964.
11. *Windows and Environment.* © 1969 Pilkington Brothers Ltd.
12. L. C. Bull. Solar radiation and air conditioning. *Heat. Vent. Engr*, April 1961.
13. A. W. Boeke and F. Zimmermann. A manual method for the calculation of the maximum air conditioning load of large buildings taking account of thermal storage, *Heiz-Lüft-Haustech.*, 1969, **20,** 400–406.

8

Cooling Load

8.1 Cooling load and heat gains

As was made clear in Chapter 6, air must be cooled and dehumidified to values of dry-bulb temperature and moisture content low enough to maintain a design state in a conditioned space. It should be noted that the load imposed by this duty on the cooler coil or air washer is not a simple summation of the sensible and latent heat gains. In addition to dealing with these loads, the fresh air handled must be reduced from the outside state to the room state; heat gains to the air ducting must be offset and the temperature rise in the air-stream resulting from fan power (*see* Chapters 6 and 15) must be countered. Figure 8.1 shows how the load may be analysed into these components.

The total load on the cooler coil is equal to

$$\begin{pmatrix}\text{mass flow of air in}\\ \text{kg dry air/s}\end{pmatrix} \times \begin{pmatrix}\text{change of enthalpy across}\\ \text{the coil in kJ/kg of dry air}\end{pmatrix}$$

$$= m(h_o - h_w)$$

$$\text{Fresh air load} = m(h_o - h_r)$$

$$\text{Latent gain} = m(h_r - h_c)$$

$$\text{Sensible gain} = m(h_c - h_s)$$

$$\text{Duct gain} = m(h_s - h_b)$$

$$\text{Fan power} = m(h_b - h_w)$$

EXAMPLE 8.1. An air-conditioning plant comprising outside air intake, re-circulated air connexion, mixing chamber, sprayed cooler coil, supply fan and ducting, handles a total of 1·26 kg/s of dry air (with its associated moisture). If the cooler coil receives a mixture of 20 per cent by mass of fresh air at a state of 28°C d.b., 19·5°C w.b. (sling), and 80 per cent of recirculated air at 22°C d.b., 50 per cent saturation, and reduces this to 10°C d.b., 7·352 g/kg, calculate the total cooling load and analyse this into its basic components. Fan power produces a temperature rise of 1°C and heat gains to the supply duct cause a further 2°C rise.

Answer

From I.H.V.E. Tables and by calculation where necessary,

$$\text{Outside enthalpy, } h_o = 55{\cdot}36 \text{ kJ/kg}$$

$$\text{Room enthalpy, } h_r = 43{\cdot}39 \text{ kJ/kg}$$

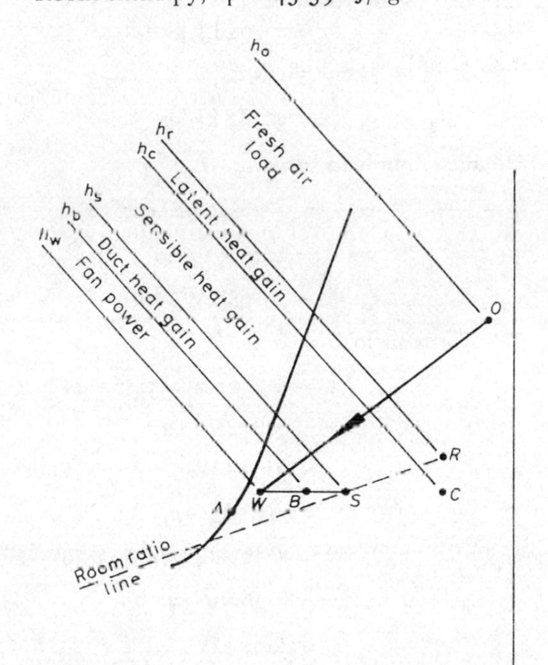

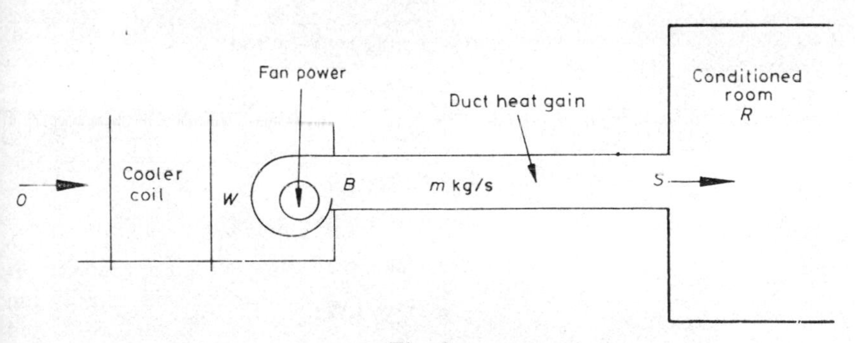

Fig. 8.1

Enthalpy of air entering the cooler coil, h_m

$$= 0{\cdot}2 \times 55{\cdot}36 + 0{\cdot}8 \times 43{\cdot}39$$

$$= 45{\cdot}78 \text{ kJ/kg}$$

Enthalpy of air leaving the cooler coil, h_w

$$= 28{\cdot}58 \text{ kJ/kg}$$

Enthalpy of air leaving the fan, $h_b = 29{\cdot}60$ kJ/kg

Enthalpy of air supplied to the room, h_s

$$= 31{\cdot}64 \text{ kJ/kg}$$

Enthalpy of air at 22°C and 7·352 g/kg, h_c

$$= 40{\cdot}82 \text{ kJ/kg}$$

$$\begin{aligned}
\text{total cooling load} &= m(h_m - h_w) \\
&= 1{\cdot}26(45{\cdot}78 - 28{\cdot}58) \\
&= 1{\cdot}26 \times 17{\cdot}20 \\
&= 21{\cdot}68 \text{ kW} \\
\text{fresh air load} &= 0{\cdot}2m(h_o - h_r) \\
&= 0{\cdot}2 \times 1{\cdot}26(55{\cdot}36 - 43{\cdot}39) \\
&= 0{\cdot}252 \times 11{\cdot}97 \\
&= 3{\cdot}01 \text{ kW} \\
\text{or} \qquad &= m \times (h_m - h_r) \\
&= 1{\cdot}26 \times (45{\cdot}78 - 43{\cdot}39) \\
&= 1{\cdot}26 \times 2{\cdot}39 \\
&= 3{\cdot}01 \text{ kW} \\
\text{latent load} &= m(h_r - h_c) \\
&= 1{\cdot}26(43{\cdot}39 - 40{\cdot}82) \\
&= 1{\cdot}26 \times 2{\cdot}57 \\
&= 3{\cdot}24 \\
\text{sensible load} &= m(h_c - h_s) \\
&= 1{\cdot}26(40{\cdot}82 - 31{\cdot}64) \\
&= 1{\cdot}26 \times 9{\cdot}18 \\
&= 11{\cdot}57 \text{ kW} \\
\text{duct gain} &= m(h_s - h_b) \\
&= 1{\cdot}26(31{\cdot}64 - 29{\cdot}60) \\
&= 1{\cdot}26 \times 2{\cdot}04 \\
&= 2{\cdot}57 \text{ kW}
\end{aligned}$$

Alternatively the calculation might be:

$$\text{duct gain} = 1{\cdot}26 \times 1{\cdot}026 \times 2^\circ\text{C}$$
$$= 2{\cdot}59 \text{ kW}$$

$$\text{fan power} = m(h_h - h_w)$$
$$= 1{\cdot}26 \times (29{\cdot}60 - 28{\cdot}58)$$
$$= 1{\cdot}26 \times 1{\cdot}02$$
$$= 1{\cdot}29 \text{ kW}.$$

Alternatively, $\text{fan power} = 1{\cdot}26 \times 1{\cdot}026 \times 1^\circ\text{C}$
$$= 1{\cdot}29 \text{ kW}.$$

Summarising:

	kW	Per cent
Fresh air load	3·01	13·9
Latent heat gain	3·24	14·9
Sensible heat gain	11·57	53·4
Duct heat gain	2·57	11·9
Fan power	1·29	5·9
Total cooling load	21·68	100·0

This example illustrates that a check may be made on the cooling load to verify the performance of the cooler coil. It also follows that a good estimate of the likely load can be made, before any psychrometry is done at all, simply by adding sensible and latent gains, calculating the fresh air load on a basis of the allowance per square metre of floor area, or per occupant and, after making an educated guess at the supply air quantity and fan total pressure, assessing the contribution likely from fan power and duct gains.

It should be clear from this example that heat gains are not the same as cooling load.

The foregoing was for a simple case where only one room was presumed to be air conditioned. If two or more rooms are air conditioned they are quite likely to have maximum sensible heat gains at different times of the day. For example, if one room has an eastern exposure and another a western orientation, their peak individual sensible gains will probably be at about 9 am and 3 pm, respectively. It is important to appreciate that the cooling load does not necessarily involve the sum of these two maximum gains. It all depends on the type of system to be used and the way in which it is to be controlled. With some systems it is necessary to calculate the heat gains at several different times for each different room and to determine at what time the maximum combination occurs.

One further important point to consider is that the fresh air load is an important part of the cooling load (13·9 per cent in the last example); with some installations, which handle 100 per cent fresh air, it is of overriding importance in its effect on the cooling load. Accordingly, one often finds that maximum cooling load occurs when the outside enthalpy, and hence the outside wet-bulb, is at a maximum. This usually occurs at from 2 to 5 pm in July or August in the United Kingdom.

8.2 Partial load

Most air-conditioning systems operate at partial load for most of their lives. It is common (and good) practice to design systems, which are of more than about 100 kW of refrigeration in size, so that maximum advantage is taken of the natural cooling affect of the outside air. The method of doing this for systems that use recirculated air and fresh air mixed is described in section 3.10, but if 100 per cent outside air is considered as a simple case, it is obvious that the load will diminish to zero and, in fact, become a heating load as summer passes and winter approaches.

EXAMPLE 8.2. A system comprising fresh air intake, filter, pre-heater, sprayed cooler coil, re-heater and fan, handles 1·26 kg/s of dry air and its associated moisture. The outside design state is 28°C dry-bulb, 19·5°C wet-bulb (sling) in summer and 0°C saturated in winter. The room state is 22°C dry-bulb with 50 per cent saturation and air at a constant state of 10°C dry-bulb, 9·7°C wet-bulb is obtained from the cooler coil. Calculate (*a*) the summer design cooling load, (*b*) the cooling load when the outside state is 13°C dry-bulb with 10°C wet-bulb (sling), and (*c*) the pre-heater load in winter design conditions.

Answer

The psychrometric operations are illustrated in Fig. 8.2.

(*a*) From I.H.V.E. tables, the enthalpy of the fresh air, h_o, is 55·36 kJ/kg and the enthalpy of the air leaving the cooler coil, h_w, is 28·58 kJ/kg.

$$\begin{aligned}\text{Then summer design cooling load} &= m(h_o - h_w)\\ &= 1{\cdot}26 \times (55{\cdot}36 - 28{\cdot}58)\\ &= 33{\cdot}7\ \text{kW}.\end{aligned}$$

(*b*) From I.H.V.E. tables, h_o is 29·16 kJ/kg for the outside condition given.

$$\begin{aligned}\text{Partial cooling load} &= 1{\cdot}26(29{\cdot}16 - 28{\cdot}58)\\ &= 0{\cdot}73\ \text{kW}.\end{aligned}$$

(*c*) From I.H.V.E. tables, h_o is 9·475 kJ/kg at the design winter condition. Air at this state must be pre-heated to a wet-bulb temperature of 9·7°C (sling) if adiabatic saturation is assumed as the air passes over the sprayed cooler coil. Making the further assumption that the humidifying efficiency is adequate, air will leave the sprayed coil at 10°C dry-bulb, 7·352 g/kg and 9·7°C wet-bulb. From a psychrometric chart it may be established that air will leave the pre-heater with a state of 3·789 g/kg (the same as the outside air moisture content), 9·7°C wet-bulb and an enthalpy, h_d, of 28·25 kJ/kg. Its dry-bulb temperature is 18·7°C.

$$\text{Pre-heater load} = 1{\cdot}26 \times (28{\cdot}25 - 9{\cdot}475)$$
$$= 23{\cdot}65 \text{ kW.}$$

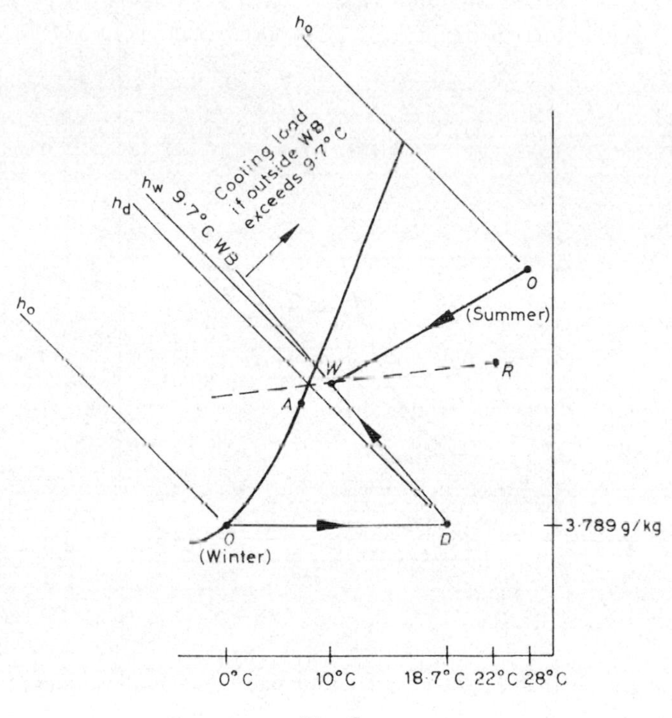

Fig. 8.2

If, on the other hand, fixed proportions of fresh and recirculated air are used throughout the year, the case is rather different and the cooling load may never disappear entirely.

EXAMPLE 8.3. For the system used in Example 8.1, calculate the cooling load when the outside condition is 0°C saturated.

Answer

Refer to Fig. 8.3. The line from M to W represents the summer design operation and for this the cooling load is 21·68 kW of refrigeration, as found in Example 8.1. The line from M' to W represents the winter operation of the sprayed cooler coil. At any outside state between O' and O, the mixture state will have a higher enthalpy than $h_{m'}$ but a lower one than h_m. The mixture state is fixed by the proportions of 20 per cent fresh air mixed with 80 per cent of recirculated air. Thus, in winter design weather the mixture enthalpy is

$$h_{m'} = 0{\cdot}2 \times 9{\cdot}475 + 0{\cdot}8 \times 43{\cdot}39$$
$$= 36{\cdot}60 \text{ kJ/kg.}$$

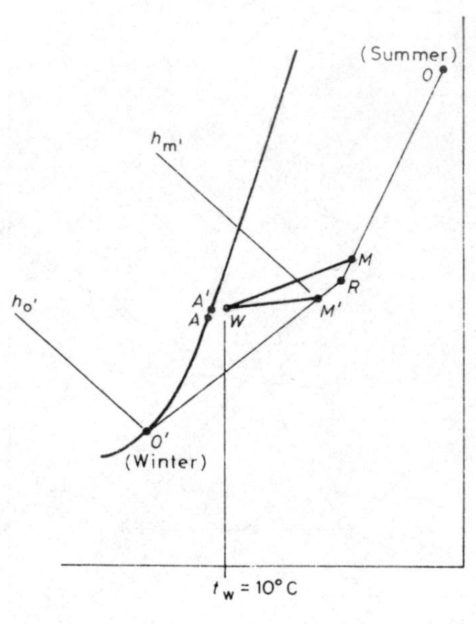

Fig. 8.3

This is greater than h_w, hence a cooling load exists at all seasons:

$$\text{winter minimum cooling load} = 1{\cdot}26(36{\cdot}60 - 28{\cdot}58)$$
$$= 10{\cdot}1 \text{ kW.}$$

This represents nearly 47 per cent of the summer design cooling load.

8.3 Cooling load offset by re-heat

From the last two examples, it is clear that a partial load condition exists as a result of a reduction in the enthalpy of the air entering a cooler coil, consequent upon seasonal and diurnal weather changes. It is evident that some steps must be taken to control the output of the coil to give a constant leaving air state, in the face of varying entering air states.

There is another reason why a partial load condition may exist: if sensible or latent gains reduce within the conditioned space then air is no longer required to be supplied to the space at the same values of dry-bulb temperature and moisture content. Supply at a higher dry-bulb temperature and a higher moisture content is needed for load reductions like this.

EXAMPLE 8.4. If the system mentioned in Example 8.1 is required to deal with a sensible gain of 5 kW under partial load conditions, calculate the necessary dry-bulb temperature of the supply air.

Answer

The weight of dry air handled is 1·26 kg/s. The state of the air handled by the fan is 10°C dry-bulb and 7·352 g/kg, for which the humid volume is 0·8112 m^3/kg. Hence, the volume handled by the fan at this state is 1·022 m^3/s. This is the volume necessary to deal with the summer design load. The volume handled by a fan running at constant speed is constant for a given duct system (*see* § 15.21). This volume is thus fixed for all seasonal variations and for all load variations.

Making use of eqn. (6.12),

$$t_s = 22 - \frac{5}{1{\cdot}022} \times \frac{(273+18)}{358}$$
$$= 22^\circ\text{C} - 4^\circ\text{C}$$
$$= 18^\circ\text{C dry-bulb.}$$

The value of 18°C in the brackets is an educated guess (*see* section 6.4). This is to be compared with the supply temperature required during summer design conditions:

$$t_s = 22 - \frac{11{\cdot}57}{1{\cdot}022} \times \frac{(273+13)}{358}$$
$$= 22^\circ\text{C} - 9^\circ\text{C}$$
$$= 13^\circ\text{C}.$$

It is apparent that if a supply state of 18°C dry-bulb and 7·352 g/kg (the latent gain being unaltered) is required, the performance of the sprayed cooler coil is too great.

EXAMPLE 8.5. Calculate the net cooling load for the partial load consideration of Example 8.4.

Answer

From the I.H.V.E. psychrometric chart, the enthalpy of the required supply condition is 36·7 kJ/kg, at state point S'. The net cooling load is, therefore

$$= 1{\cdot}26 \times (45{\cdot}78 - 36{\cdot}7)$$
$$= 1{\cdot}26 \times 9{\cdot}08$$
$$= 11{\cdot}4 \text{ kW.}$$

Add to this 3·86 kW for fan power etc., and the load is 15·26 kW or about 70 per cent of the summer design load.

Reference to Fig. 8.4(*a*) shows that the cooler coil will never be able to give an off condition at the required supply state. The condition line joining states M and W' may not cut the saturation curve.

There are many ways in which the required supply state can be achieved for a partial load condition, and some of these are considered in the following sections. The most obvious way is to allow the cooler coil to change the state

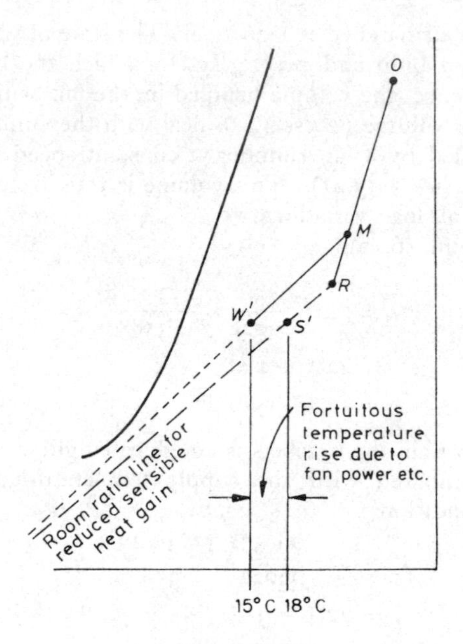

Fig. 8.4(*a*)

of the air from M to W and to impose a false sensible load on the system by means of a re-heater. This heater battery is positioned after the cooler coil or fan, as convenient (*see* Fig. 8.4(*b*)), and it heats the air from state S, at which the enthalpy is 31·6 kJ/kg (from the psychrometric chart), to S'.

$$\text{Re-heat load} = 1{\cdot}26 \times (36{\cdot}7 - 31{\cdot}6)$$
$$= 6{\cdot}43 \text{ kW.}$$

This figure should equal the reduction in sensible gain (11·57 − 5·0 = 6·57 kW). That it does not, exactly, is due to slight inaccuracies in reading enthalpy values from the psychrometric chart. In other words, the re-heat has given the cooler coil a false sensible load, equal to the drop in the room sensible load.

This method of operation is quite common. It has the virtue that air at a constant dew point (moisture content) can be supplied to a number of rooms, each of which can have its own re-heater and may, therefore, be under individual temperature control. For one of the rooms, humidity can also be controlled by changing the capacity of the cooler coil so that the moisture content of state W, off the coil, varies. This is dealt with in Chapters 10 and 13.

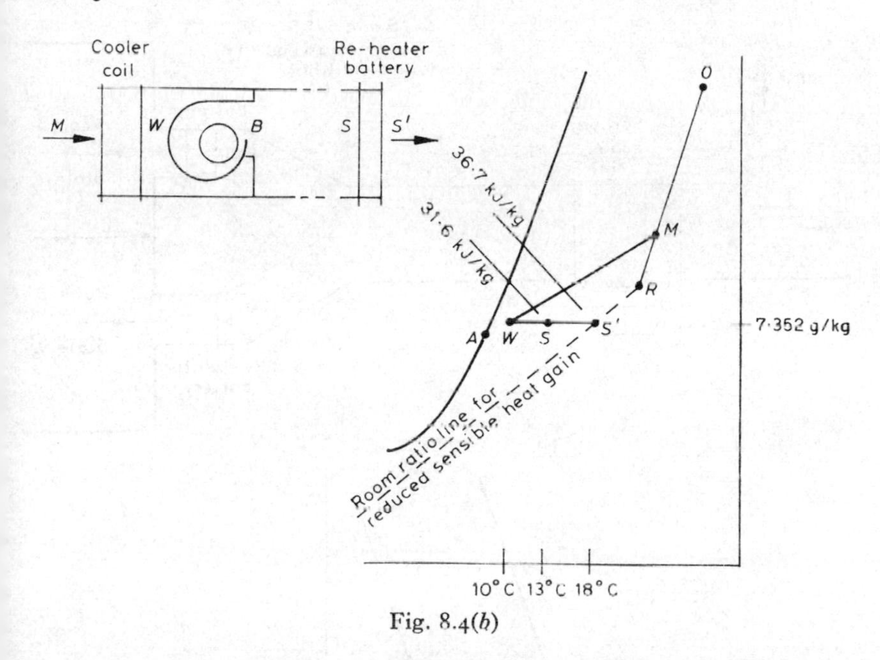

Fig. 8.4(b)

EXAMPLE 8.6 (*see* Fig. 8.5). If the plant described in Example 8.1 is used to serve two rooms A and B (A having loads as in Example 8.1 but B having a maximum sensible gain of 5 kW and zero latent gain, under summer design conditions), calculate the supply air quantities and states to each room and the total refrigeration load. Assume that 22°C dry-bulb and 50 per cent saturation is maintained in the first room and that 22°C dry-bulb is also maintained in the second, but the humidity therein is uncontrolled.

What will be the humidity in the second room under summer design conditions?

Answer

As in Example 8.1, for room A, 1·26 kg/s of air at 13°C and 7·352 g/kg will offset 11·57 kW of sensible heat gain and 3·24 kW of latent heat gain.

For room B,

$$\text{supply air quantity} = \frac{5}{(22-13)} \times \frac{(273+13)}{358}$$

$$= 0{\cdot}447 \text{ m}^3/\text{s at } 13°\text{C}.$$

The humid volume at the supply state is 0·8198 m³/kg dry air and so the mass flow rate is

$$\frac{0{\cdot}447}{0{\cdot}8198} = 0{\cdot}545 \text{ kg/s}$$

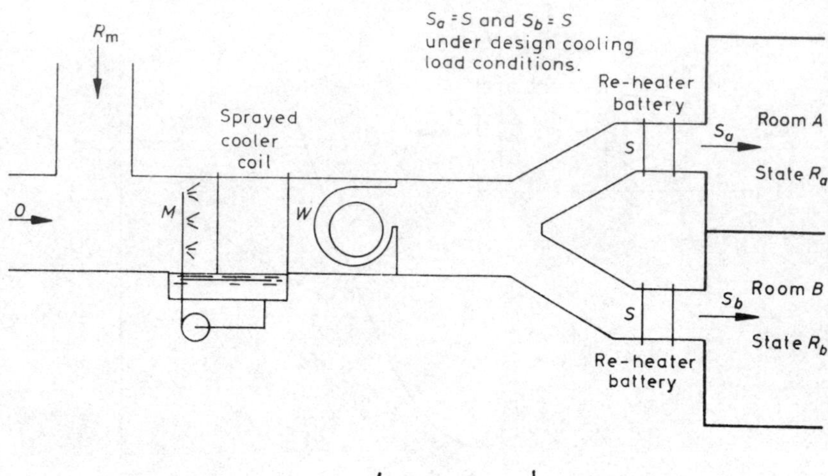

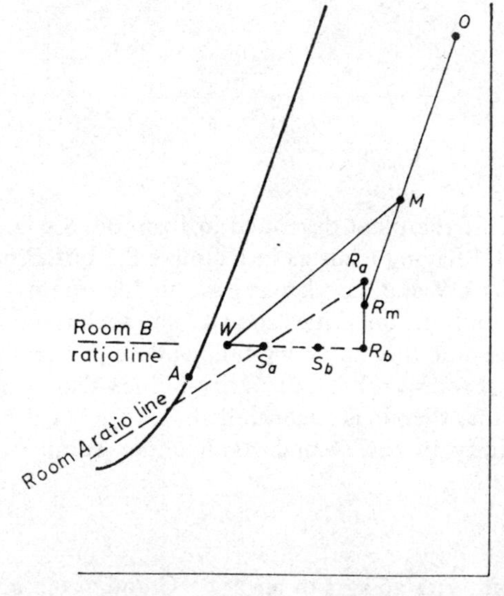

Fig. 8.5

Each room is supplied with air at the same state, but different flow rates because of the different sensible gains.

The humidity in the second room will correspond to a state of 22°C and 7·352 g/kg and will be 44 per cent, approximately.

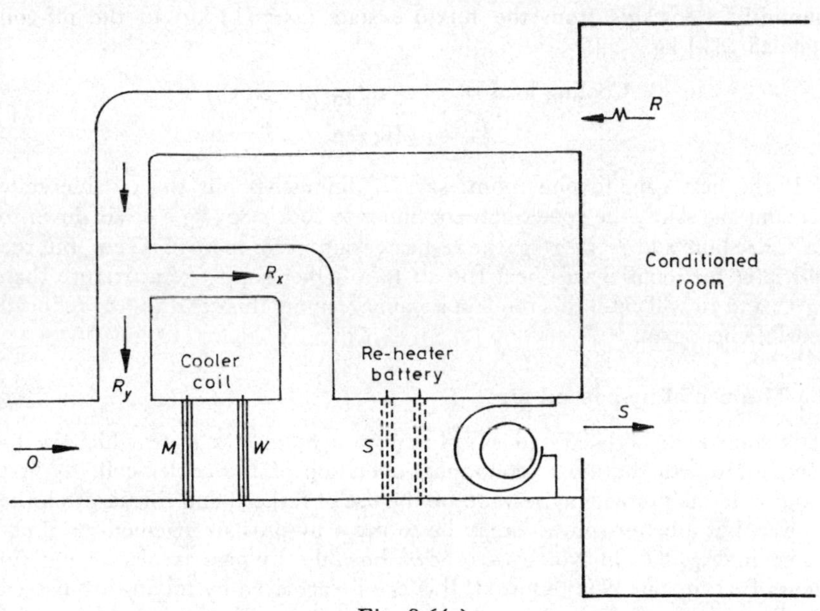

Fig. 8.6(*a*)

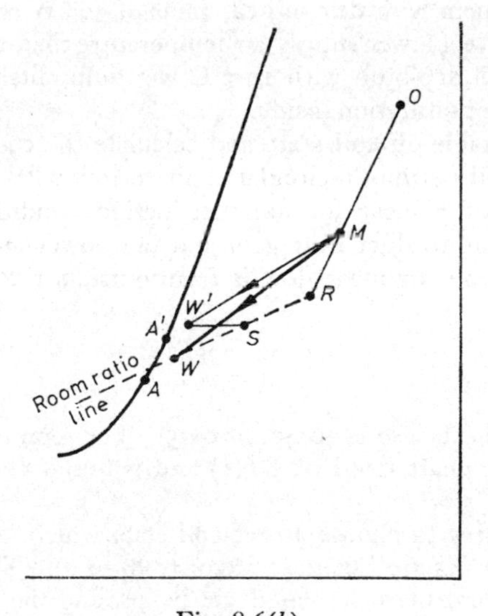

Fig. 8.6(*b*)

The refrigeration load is the result of the cooler coil having to cool and dehumidify 1·805 kg/s from the mixture state (45·78 kJ/kg) to the off-coil state 28·58 kJ/kg.

$$\text{Cooling load} = 1{\cdot}805 \times (45{\cdot}78 - 28{\cdot}58)$$
$$= 31 \text{ kW}.$$

If the heat gain to one room, say B, diminishes but the outside state remains the same, the cooler coil continues to cool 1·805 kg/s of air down to 10°C dry bulb and 7·352 g/kg, the re-heater output for room A is zero, but the re-heater for room B will heat the air to a higher supply temperature than 13°C, and air will enter this room at a state S_b, under this condition of reduced sensible heat gain.

8.4 The use of by-passed air

It is sometimes necessary to supply air to a room at a state which is incompatible with the most economical operation of the cooler coil. A first solution to the problem appears to be the use of re-heat, and this is often the answer, but a better solution may be to use a by-pass arrangement as illustrated in Fig. 8.6, in which (*a*) shows how the by-pass is placed and (*b*) shows that, if state S is required, this can be achieved by mixing by-passed air, R_x, with air off the cooler coil at state W. This avoids the use of wasteful re-heat from W' to S.

EXAMPLE 8.7. A room with design heat gains of 30 kW sensible and 10 kW latent, must not have a lower supply air temperature than 15°C. The design conditions are 28°C dry-bulb with 19·5°C wet-bulb outside and 22°C dry-bulb with 50 per cent saturation inside.

Determine a suitable off-coil state and calculate the cooling load if a by-pass around the coil permits recirculated air to mix with air off the coil, in avoiding the use of re-heat for summer design conditions. Ignore any temperature rise due to duct heat gain and fan power and assume that 15 per cent of the supply air mass flow is from outside, the balance being recirculated air.

Answer

The sensible-total heat ratio is 30/40, or 0·75. The room ratio line is drawn on a psychrometric chart (*see* Fig. 8.6(*b*) and is found to cut the saturation curve at 7·5°C.

It is now necessary to choose an off-coil state which intersects the room ratio line and has a practical contact factor (0·8 – 0·9). This is not quite so easy since the amount of recirculated air by-passing the coil is not known and, hence, the amount of recirculated air mixing with fresh air is not known either.

That is, the mixture state M, on the coil, is not known. Since some latitude is possible in the value of contact factor, a sensible guess at the state of M willnot have an undesirable effect on state W (off the coil) if M is later changed.

Suppose the mixture state M is mid-way between O and R, and has a value of 25°C dry-bulb. A line joining this assumed mixture state, M to W at 9°C on the room ratio line, cuts the saturation curve at about 7·2°C when produced. This line, MWA, which represents an assumed coil performance, gives an approximate contact factor of $(25° - 9°)/(25° - 7{\cdot}2°)$ or 0·9.

This is a practical value, so an off-coil state of 9°C is accepted.

Only air that goes through the cooler coil has the ability to offset the latent and sensible heat gains, so the quantity of air and the moisture content of W can now be assessed:

$$\text{Air flowing over coil} = \frac{30}{(22°-9°)} \times \frac{(273+15°)}{358} = 1{\cdot}856 \text{ m}^3/\text{s at } 15°\text{C}$$

(Note it is stipulated that the air to the room must be supplied at not less than 15°C and that this is therefore the temperature at which the fan is handling the air.)

Moisture content of air leaving the cooler coil

$$= 8{\cdot}366 - \frac{10}{1{\cdot}856} \times \frac{(273+15)}{856}$$

$$= 8{\cdot}366 - 1{\cdot}814 = 6{\cdot}552 \text{ g/kg.}$$

Thus the off-coil state, W, is 9°C dry-bulb and 6·552 g/kg.

$$\text{Total supply air quantity} = \frac{30}{22-15} \times \frac{(273+15)}{358}$$

$$= 3{\cdot}45 \text{ m}^3/\text{s at } 15°\text{C.}$$

$$\text{Humid volume at state } S = 0{\cdot}826 \text{ m}^3/\text{kg from the chart}$$

$$\therefore \text{ mass flow through coil} = 1{\cdot}856/0{\cdot}826 = 2{\cdot}247 \text{ kg/s}$$

$$\text{Enthalpy at } W\text{, from tables} = 25{\cdot}55 \text{ kJ/kg.}$$

$$\text{Mass flow of supply air} = \frac{3{\cdot}45}{0{\cdot}826} = 4{\cdot}18 \text{ kg/s.}$$

$$\text{Mass flow of fresh air} = 0{\cdot}15 \times 4{\cdot}18$$

$$= 0{\cdot}627 \text{ kg/s.}$$

Actual mixture proportions are 0·627 at state *O*, with 2·247−0·627, or 1·620 kg/s at state *R*.

$$\text{Hence, enthalpy at } M = \frac{0{\cdot}627}{2{\cdot}247}\times 55{\cdot}36+\frac{1{\cdot}620}{2{\cdot}247}\times 43{\cdot}39$$

$$= 15{\cdot}45+31{\cdot}25$$

$$= 46{\cdot}70 \text{ kJ/kg.}$$

$$\text{Approximately, dry-bulb at } M = \frac{0{\cdot}627}{2{\cdot}247}\times 28+\frac{1{\cdot}620}{2{\cdot}247}\times 22$$

$$= 7{\cdot}82^\circ+15{\cdot}85^\circ$$

$$= 23{\cdot}67^\circ\text{C.}$$

This information establishes the state of the air, *M*, entering the coil, and *M* is plotted on the psychrometric chart.

The cooling load is thus

$$2{\cdot}247(46{\cdot}70-25{\cdot}55) = 47{\cdot}5 \text{ kW.}$$

The supply state to the room is obtained by mixing 2·247 kg/s of air at *W* with 1·933 kg/s (= 4·18 − 2·247) of air at *R*.

Supply temperature is approximately

$$\frac{2{\cdot}247}{4{\cdot}18}\times 9+\frac{1{\cdot}933}{4{\cdot}18}\times 22$$

$$= 4{\cdot}84^\circ+10{\cdot}16^\circ = 15^\circ\text{C (as was stipulated).}$$

Supply moisture content is

$$\frac{2{\cdot}247}{4{\cdot}18}\times 6{\cdot}552+\frac{1{\cdot}933}{4{\cdot}18}\times 8{\cdot}366$$

$$= 3{\cdot}52+3{\cdot}87$$

$$= 7{\cdot}39 \text{ g/kg.}$$

One last point: although a re-heater is not necessary for operation under summer design conditions, it may be necessary to include one for winter operation. In Fig. 8.6(*b*) a re-heater is shown in broken outline for this reason.

8.5 Face and by-pass dampers

This is a method adopted to vary the output of the coil under changing load conditions. Figure 8.7(*a*) illustrates the way the dampers are arranged. Air at state *O* by-passes the face of the cooler coil to an extent governed by the position of the dampers.

When the face dampers are fully open, for design load conditions, air leaves the coil at W and enters the room at S, maintaining state R therein. Under a condition of reduced sensible load, a thermostat instructs the damper motor to open partly the by-pass dampers and, because the two are linked, to close, partly, the face dampers. Air at O mixes with air at W (assuming the coil continues to condition air to W, even though the air flow is reduced)

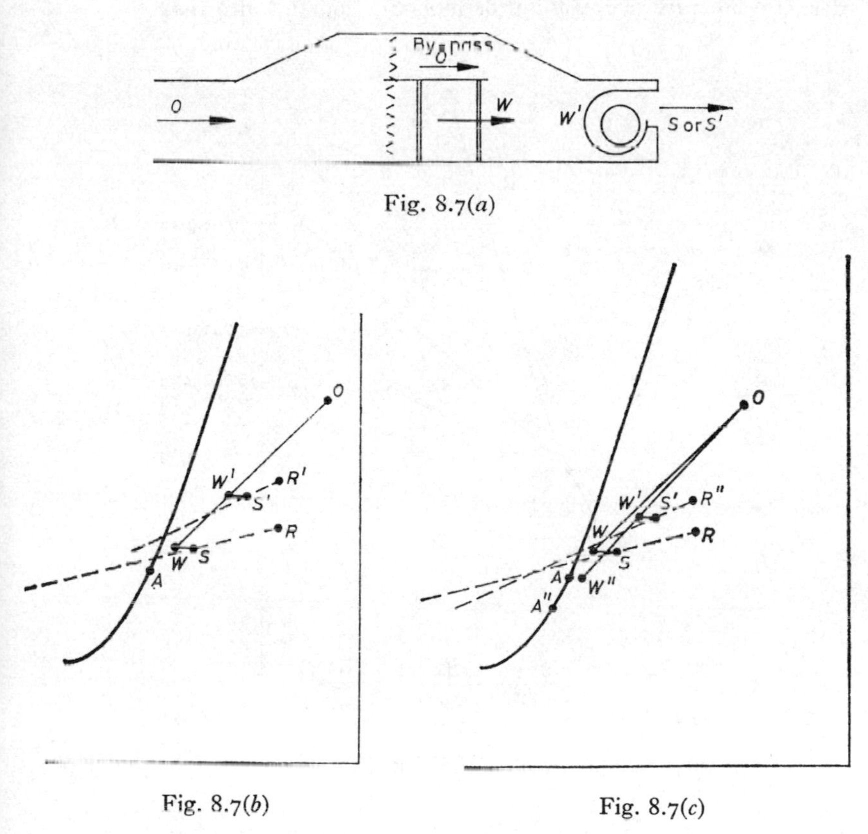

Fig. 8.7(a)

Fig. 8.7(b)

Fig. 8.7(c)

Because A'' is lower than A, the moisture content of S' is less in Fig. 8.7(c) than in Fig. 8.7(b) and so the relative humidity of R'' is less than that of R' in Fig. 8.7(b).

and a mixture state W', is produced. Fan power etc., changes this to state S' for supply to the room.

It can be seen from Fig. 8.7(b) that mixing O and W causes the supply state to have a higher moisture content. Inevitably, the room relative humidity rises even though the correct dry-bulb temperature is maintained and state R' prevails in the room instead of state R.

The true behaviour is slightly different since, when the airflow diminishes the contact factor improves (*see* section 10.4) and, because the load on the coil has fallen, the mean coil surface temperature also drops. A becomes A''.

Figure 8.7(*c*) illustrates the more correct behaviour. If the system had used a mixture of recirculated and fresh air, then air at such a mixture state would have by-passed the coil. This is a more complicated situation and is more difficult to illustrate because as the room temperature (or moisture content) rises, the mixture temperature (or moisture content) also rises.

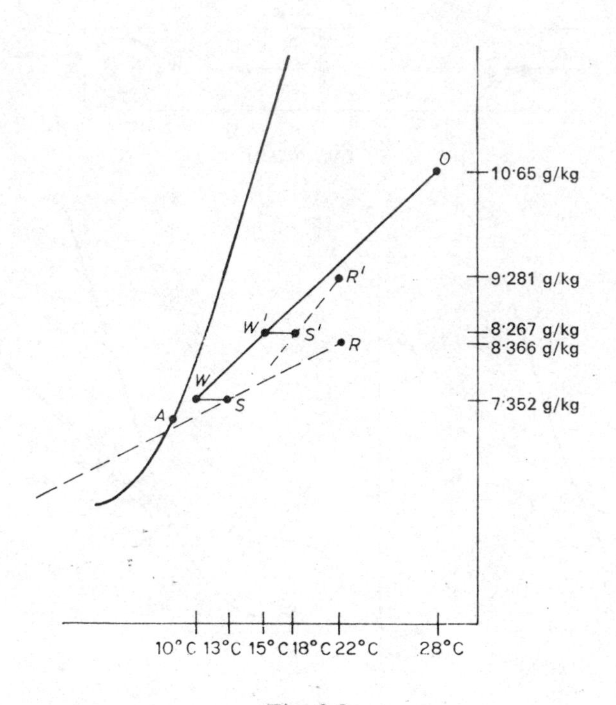

Fig. 8.8

EXAMPLE 8.8. If a room suffers sensible gains of 11·57 kW and latent gains of 3·24 kW when 22°C dry-bulb with 50 per cent saturation is maintained therein during outside design conditions of 28°C dry-bulb with 19·5°C wet-bulb (sling), calculate the inside conditions if the sensible gain diminishes to 5 kW, the latent gain remaining at 3·24 kW and the outside condition staying at the design value.

The plant comprises air intake, pre-heater, cooler coil with face and by-pass dampers (used for thermostatic control of room temperature), supply fan handling 1·26 kg/s of fresh air, and the usual distribution ducting. Assume that the performance of the cooler coil is unaltered by variations in air flow across it and that the temperature rise due to fan power etc. is 3°C.

Answer

From earlier calculations, the supply air state under design conditions is 13°C dry-bulb with 7·352 g/kg. The supply temperature for dealing with a sensible gain of 5 kW is 18°C dry-bulb, and so the air temperature needed after the face and by-pass dampers must be 15°C at partial load.

The psychrometric changes involved are illustrated in Fig. 8.8.

Assuming that the dry-bulb scale is linear, a good approximate answer is as follows:

$$g_{w'} = g_w + \left(\frac{15^\circ - 10^\circ}{28^\circ - 10^\circ}\right) \times (g_o - g_w)$$

$$= 7{\cdot}352 + \frac{5}{18} \times (10{\cdot}65 - 7{\cdot}352)$$

$$= 7{\cdot}352 + 0{\cdot}915$$

$$= 8{\cdot}267 \text{ g/kg.}$$

$$\therefore g_{r'} = g_{w'} + \text{moisture pick-up for the design latent load}$$

$$= 8{\cdot}267 + (8{\cdot}366 - 7{\cdot}352)$$

$$= 9{\cdot}281 \text{ g/kg.}$$

The room relative humidity is about 56 per cent.

8.6 Cooling in sequence with heating

When it is necessary to control only temperature in the conditioned space, re-heat can be avoided by arranging that the heater battery is used only when the cooler coil is off. Figure 8.9(*a*) illustrates the control operation: *Ra* and *Rb* are three-port motorised mixing valves which are controlled from a room thermostat *C*, in sequence. As the room temperature falls, *C* progressively reduces the capacity of the cooler coil, the by-pass port of *Ra* opening and, upon further fall in room temperature, progressively increases the capacity of the heater, the by-pass port of *Rb* closing. Thus, the two work in sequence, and the heater never being on when the cooler is also on, none of the cooling capacity is ever wasted by cancellation with re-heat.

The effect of a reduction in sensible heat gain, while the outside state remains constant, is shown in Fig. 8.9(*b*). If a supply temperature $t_{s'}$, is required instead of t_s, then, presuming the latent gains to be unchanged, the relative humidity in the room will rise, state *R′* being maintained.

If objectionable increases in humidity are to be prevented, a high limit humidistat is necessary. This overrides the action of *C*, partly closing the by-pass port of *Ra* for the purpose of doing some latent cooling. The temperature fall which would result is prevented by *C* partly closing the by-pass port of *Rb*. Thus, under these circumstances, the heater is behaving as a re-heater rather than as a sequence heater.

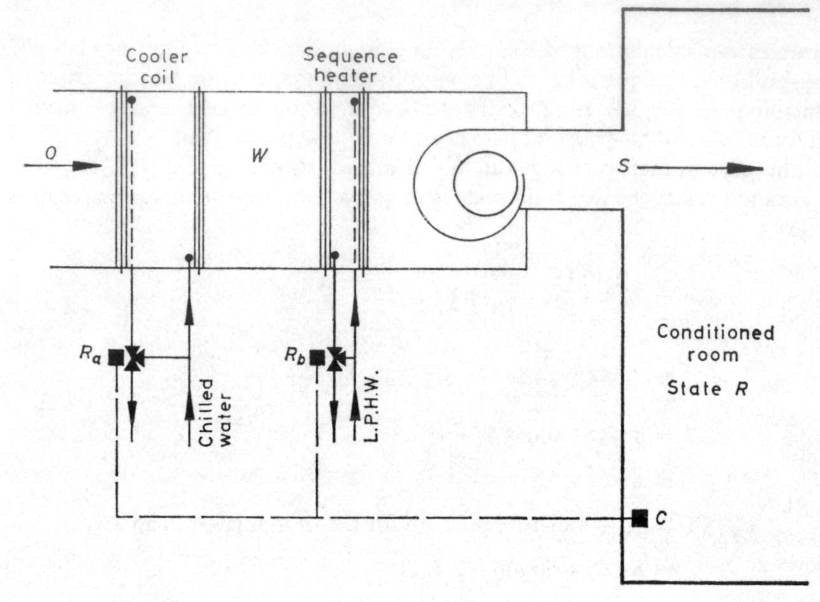

Fig. 8.9(*a*)

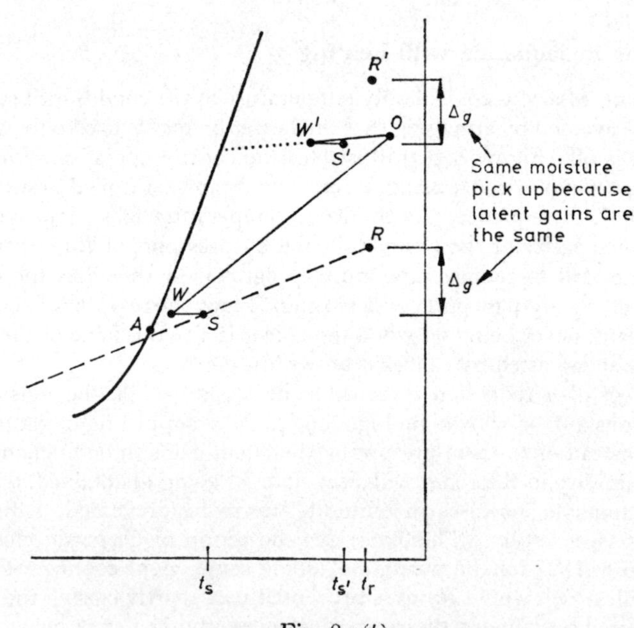

Fig. 8.9(*b*)

8.7 Hot deck–cold deck systems

Certain types of packaged air handling plant offer the facility of providing multiple zone control of temperature without the use of multiple re-heaters. In some respects there is a similarity here with double-duct systems.

The type of plant is illustrated in Fig. 8.10(*a*). Air is filtered and drawn through a fan which blows the air over a pair of coils, one heating and one cooling, arranged in parallel. The downstream part of the plant is split by a horizontal dividing plate into two chambers, an upper ' hot deck ' and a lower ' cold deck '. The outlets from these decks are dampered, and a number of ducts may accept air from the outlet. Any particular duct outlet can then have air at a controlled temperature by acccepting the appropriate proportions of air from the two decks, through the motorised dampers. The particular damper group for one zone would be thermostatically controlled.

Figure 8.10(*b*) illustrates the psychrometry, the temperature rise due to fan power etc., being ignored. Air comes off the hot deck at state H and off the cold deck at state C. The two airstreams fed to zone 1 are dampered automatically and mix to form state S_1. This is the correct state to give condition R_1 in zone 1. Similarly H and C mix to give state S_2 for supply to zone 2, maintaining state R_2 therein. It is seen that differences in relative humidity are possible between zones.

As with sequence heating, there is no direct cancellation of cooling, hence account can be taken of cooling load diversity in assessing refrigeration load.

8.8 Double-duct cooling load

Whether load diversity may be taken advantage of in estimating the required refrigeration capacity depends on which one of two systems is used (*see also* section 16.13).

The most common form of system (Fig. 8.11(*a*)) has the cooler coil in the cold duct and the heater battery in the hot duct on the discharge side of the fan. The advantage is that no air is first cooled and then re-heated. If a higher supply air temperature than for summer design conditions is required (say t_s rather than t_c), some hot air is mixed with cold air. No wasteful re-heat is involved. Ultimately, all the air would be from the hot duct and none from the cold duct. The disadvantage of the system is that the moisture content of the supply air varies. With a supply state S, for the same moisture pick-up in the room, a condition R' rather than R is maintained, as shown in Fig. 8.11(*b*).

The alternative system has a cooler coil on the suction side of the fan, as Figs. 8.11(*c*) and (*d*) show. The advantage of such a system is that the supply moisture content is steady and the relative humidity in the conditioned room does not alter greatly. Its disadvantage is that all the air in the hot duct has been re-heated, and diversity of heat gain cannot be exploited to give a reduced refrigeration plant size.

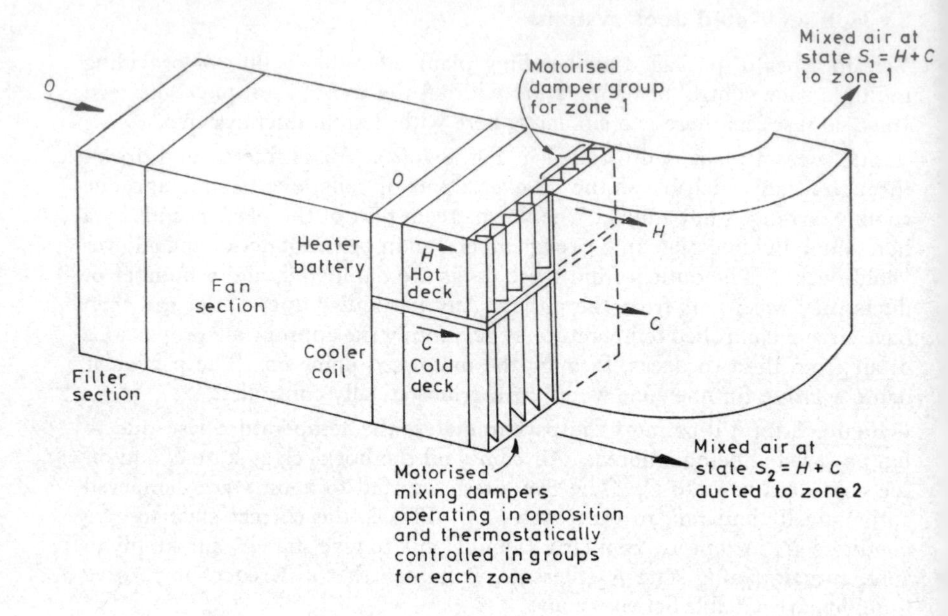

Fig. 8.10(*a*)

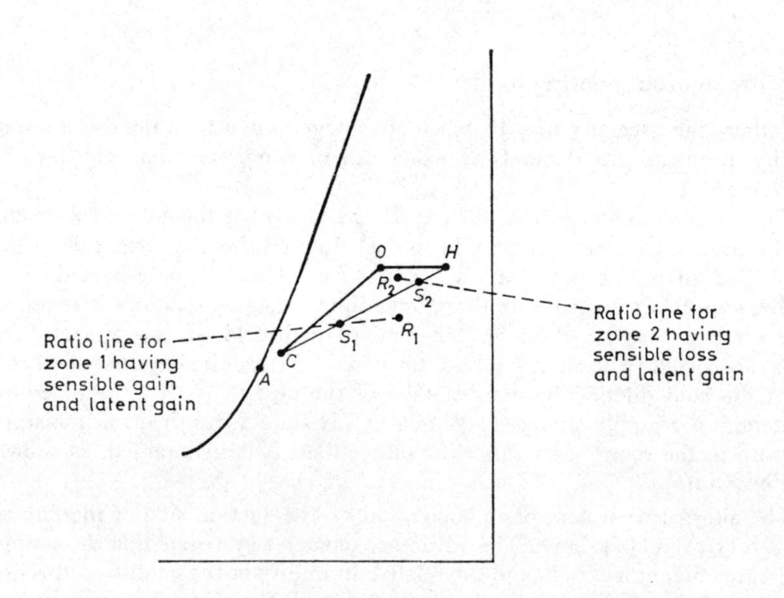

Fig. 8.10(*b*)

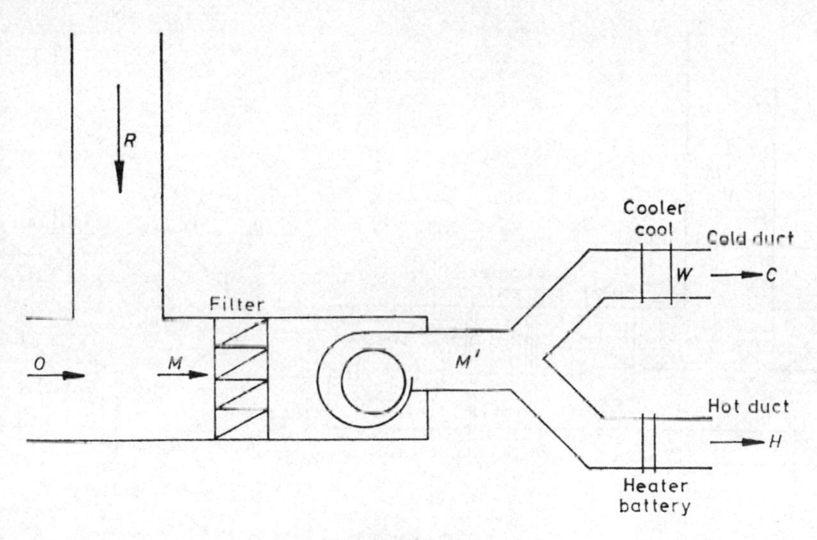

Fig. 8.11(*a*)

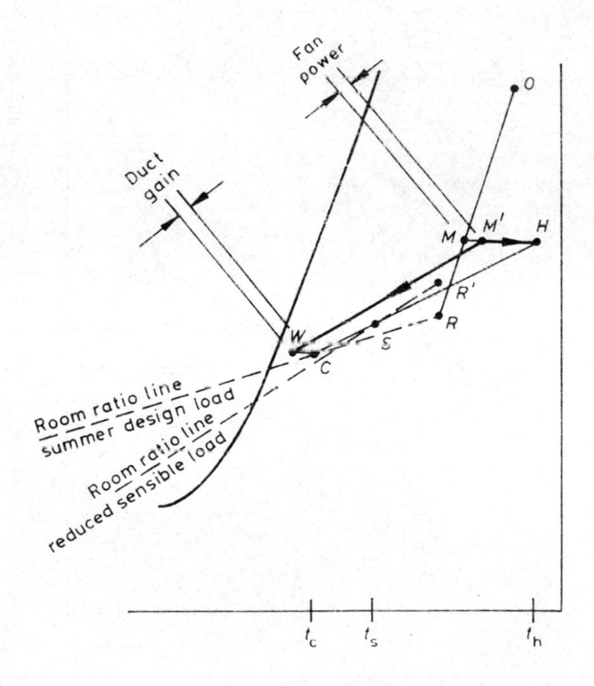

Fig. 8 11(*b*)

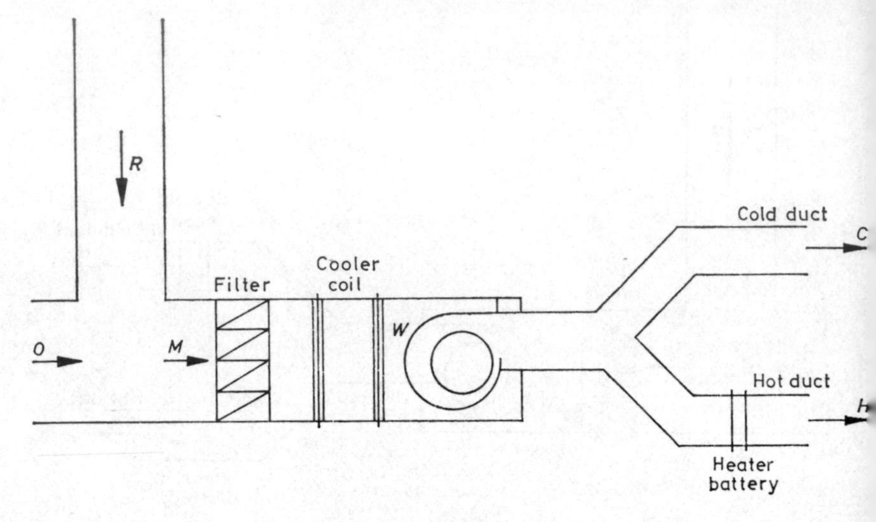

Fig. 8.11(*c*)

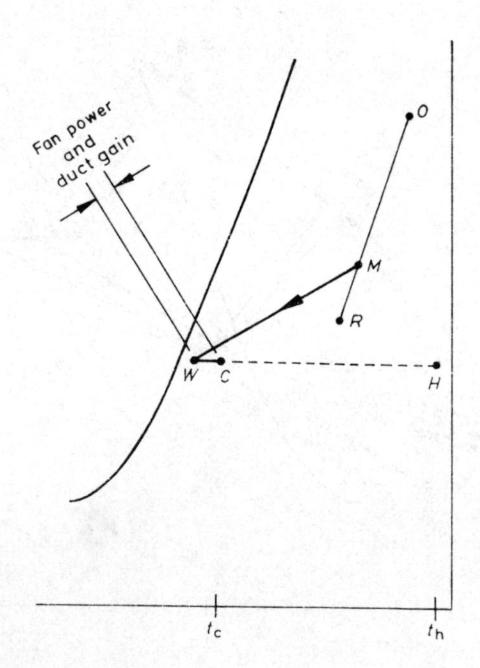

Fig. 8.11(*d*)

8.9 The load on air-water systems

These systems are dealt with in greater detail in Chapter 16 but, for the moment, it is worth commenting that they usually take some account of the diversity of cooling load. Just how much depends on how they are designed; for instance, re-heat on the air side can nullify some of the drop in sensible load caused by a fall in outside air temperature. The essence of such systems is that, by having individual chilled water cooler coils in units, either induction or fan coil, in each conditioned space, reductions in sensible load can be dealt with by throttling the flow of chilled water through such local coils.

One other point: the refrigeration load is not the sum of sensible gain, the load on the primary air cooler coil, the latent gain and the fan power. It must be remembered that the primary air delivered to the units has a sensible cooling capacity that must be regarded as offsetting part of the sensible gain. This point is clarified in section 16.4.

8.10 Booster coolers

One method adopted in an attempt to avoid the use of wasteful re-heat is to let all the latent load be handled by the fresh air. Dehumidified fresh air is then mixed with recirculated air and sensibly cooled by a number of zonal booster coolers, regulated from local thermostats. No re-heat is involved during the summer months. Booster heaters are usually sequenced with the booster coolness. Figure 8.12 illustrates the system.

EXAMPLE 8.9. A room has sensible and latent heat gains of 60 kW and 4·8 kW when conditioned at 22°C dry-bulb, 50 per cent saturation with an outside condition of 28°C dry-bulb and 19·5°C wet-bulb (sling). If the sort of plant illustrated in Fig. 8.12(*a*) is used, calculate the loads on the primary and booster cooler coils. Suppose a total of 50 people are present and the minimum fresh air requirement is 0·01 m^3/s per person.

Assume that the minimum practical temperature to which air at state *B* can be sensibly cooled is 16°C, and allow 3°C to cover fan power and heat gains to ducts.

Answer

To remove a sensible heat gain of Q_S kW, an air quantity m_s kg dry air/second must be supplied at state *S*. If *c* is the humid specific heat, the following equation applies.

$$Q_S = m_s c(t_r - t_s) \tag{8.1}$$

The same mass flow of supply air must have a moisture content of g_s in order to offset latent gains of Q_L kW. Hence,

$$Q_L = m_s(g_r - g_s)\,2454 \tag{8.2}$$

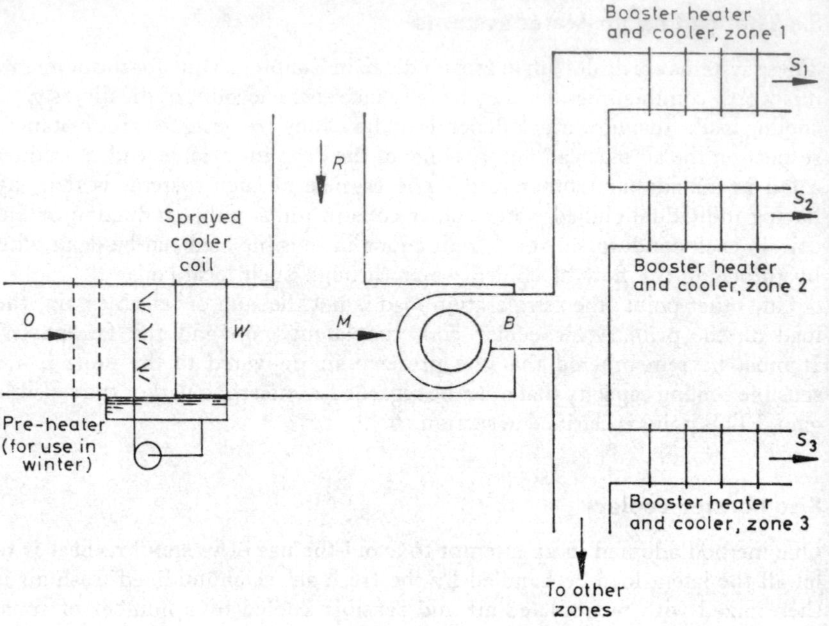

Fig. 8.12(*a*)

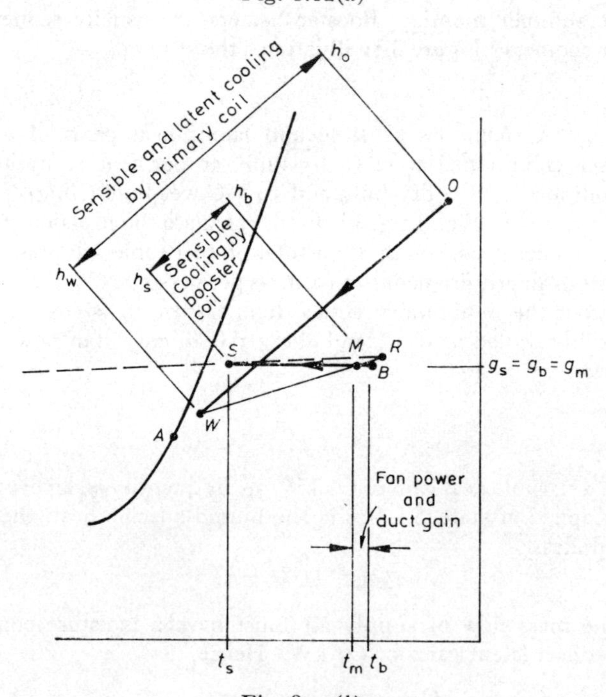

Fig. 8.12(*b*)

Since t_s is stipulated, m_s and so g_s can be calculated. State S is thus known and, because $g_s = g_m$, if the mass of fresh air handled, m_o, is fixed, so also is the ratio m_o/m_s.

Thus,
$$\frac{m_o}{m_s} = \frac{(g_r - g_m)}{(g_r - g_w)} \tag{8.3}$$

The only unknown here is g_w and so the performance of the primary air cooler coil can be assessed.

By eqn. (8.1)—

$$m_s = \frac{60}{1{\cdot}026 \times (22-16)} = 9{\cdot}75 \text{ kg/s}$$

By eqn. (8.2)—

$$g_s = g_r - \frac{4{\cdot}8}{9{\cdot}75 \times 2454}$$
$$= 0{\cdot}00837 - 0{\cdot}0002$$
$$= 0{\cdot}00817 \text{ kg/kg}$$
$$= g_m$$

Minimum fresh air requirement $= 50 \times 0{\cdot}01 = 0{\cdot}5 \text{ m}^3/\text{s}$

$$\therefore m_o = \frac{0{\cdot}50}{0{\cdot}8674} = 0{\cdot}576 \text{ kg/s}$$

$$\therefore \frac{m_o}{m_s} = \frac{0{\cdot}576}{9{\cdot}750} = 0{\cdot}0591$$

By eqn. (8.3)—

$$g_w = 0{\cdot}008366 - \frac{(0{\cdot}008366 - 0{\cdot}008166)}{0{\cdot}0591}$$
$$= 0{\cdot}004976 \text{ kg/kg.}$$

This is clearly too low a value of moisture content for the air leaving a cooler coil (the dew point is about 4°C, and this would mean an expensive cooler coil with possible frosting problems) and so the fresh air quantity must be increased.

Assume that a minimum practical moisture content, off the primary cooler coil, is 0·007 kg/kg. This will then determine a new and larger fresh air quantity, by eqn. (8.3)

$$\frac{m_o}{9{\cdot}750} = \frac{(0{\cdot}008366 - 0{\cdot}00816)}{(0{\cdot}008366 - 0{\cdot}0070)}$$

$$m_o = 9{\cdot}75 \times \frac{0{\cdot}0002}{0{\cdot}001366}$$

$$1{\cdot}43 \text{ kg/s.}$$

There is now some choice about the exact temperature of W. This is decided on by reference to a psychrometric chart and the assumption of a practical value for a contact factor.

If $t_w = 10°C$ is chosen, the apparatus dew point is about 7·8°C by construction and

$$\beta \simeq \frac{28° - 10°}{28° - 7{\cdot}8°} = \frac{18°}{20{\cdot}2°} = 0{\cdot}89$$

This is satisfactory.

The relevant enthalpies are (from tables or a chart):

$$h_o = 55{\cdot}36 \text{ kJ/kg}$$

$$h_w = 27{\cdot}70 \text{ kJ/kg}$$

$$h_r = 43{\cdot}39 \text{ kJ/kg}$$

$$h_m = \frac{1{\cdot}43}{9{\cdot}75} \times 27{\cdot}70 + \frac{8{\cdot}33}{9{\cdot}75} \times 43{\cdot}39$$

$$= 4{\cdot}06 + 37{\cdot}0$$

$$= 41{\cdot}06 \text{ kJ/kg.}$$

$$t_m = \frac{1{\cdot}43}{9{\cdot}75} \times 10° + \frac{8{\cdot}32}{9{\cdot}75} \times 22°$$

$$= 1{\cdot}46° + 18{\cdot}8°$$

$$= 20{\cdot}26°\text{C}$$

$$\therefore t_b = 23{\cdot}26°\text{C.}$$

$$\text{Primary coil cooling load} = 1{\cdot}43(55{\cdot}36 - 27{\cdot}70)$$

$$= 1{\cdot}43 \times 27{\cdot}66$$

$$= 39{\cdot}55 \text{ kW.}$$

$$\text{Secondary coil cooling load} = 9{\cdot}75 \times 1{\cdot}026 \times (23{\cdot}26° - 16°)$$

$$= 72{\cdot}6 \text{ kW.}$$

$$\text{Total cooling load} = 112{\cdot}15 \text{ kW.}$$

This load should now be analysed and checked.

$$\text{Sensible heat gain} = 60 \text{ kW}$$

$$\text{Latent heat gain} = 4{\cdot}8 \text{ kW}$$

Fan power, etc:

$$= 9{\cdot}75 \times 1{\cdot}026 \times 3° = 30 \text{ kW}$$

$$\text{Fresh air load} = 1{\cdot}43 \times (55{\cdot}36 - 43{\cdot}39) = 17{\cdot}12 \text{ kW}$$

$$111{\cdot}92 \text{ kW.}$$

8.11 Diversification of load

As was indicated in Example 8.5, if reductions in the sensible gain to a conditioned space are dealt with by elevating the supply air temperature through the use of re-heat, no corresponding reduction in refrigeration load will occur. All the air supplied is cooled to a fixed dew point and the false load imposed by the re-heat equals the reduction in sensible gain to the room. If methods are adopted such as the use of face and by-pass dampers, which raise the supply air temperature as the sensible gain diminishes without wasting any cooling, then the refrigeration load will not necessarily involve the sum of the maximum sensible gains to two or more rooms conditioned.

If a plant, such as that discussed in the last section, which separates the de-humidifying and fresh-air load from the main bulk of the sensible load, is used, then reductions in sensible gain are dealt with by reducing the secondary sensible cooling, rather than by cancelling cooling with re-heat.

8.12 Load diagrams

In multi-roomed buildings, with several orientations, diagrams can be made which give a useful picture of load variations.

The sensible load imposed on the room itself is composed of four major contributions:

(i) Transmission gain (denoted by T) due to the air-to-air temperature difference.

(ii) Electric lights (L).

(iii) People (P).

(iv) Solar (S).

The roles these four constituents play and the significance of their relative magnitudes are illustrated in the following numerical example:

A room with a west-facing window has maximum sensible heat gain at 5 pm in June, as follows:

		W
Glass, $16\ m^2 \times 2{\cdot}8\,(U\text{-value}) \times (28° - 22°)$	=	269
Wall, $30\ m^2 \times 1{\cdot}7\,(U\text{-value}) \times (28° - 22°)$	=	306
Solar gain, $16\ m^2 \times 215\ W/m^2 \times 0{\cdot}97$ (double glazing)	=	3335
Electric lights, 500 watts	=	500
People, 8×90 W/person	=	720
Infiltration, $\dfrac{140\ m^3 \times \frac{1}{2}\ \text{air change per hour}}{0{\cdot}8674\ m^3/kg \times 3600} \times 1026 \times (28° - 22°)$	=	138
	Total =	5268

This may be split up as follows:

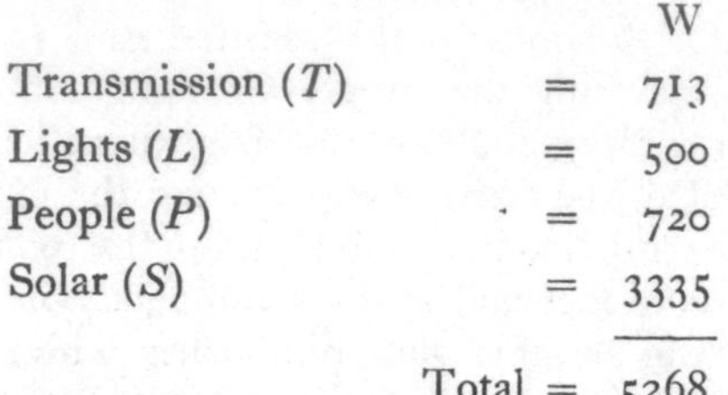

		W
Transmission (T)	=	713
Lights (L)	=	500
People (P)	=	720
Solar (S)	=	3335
Total	=	5268

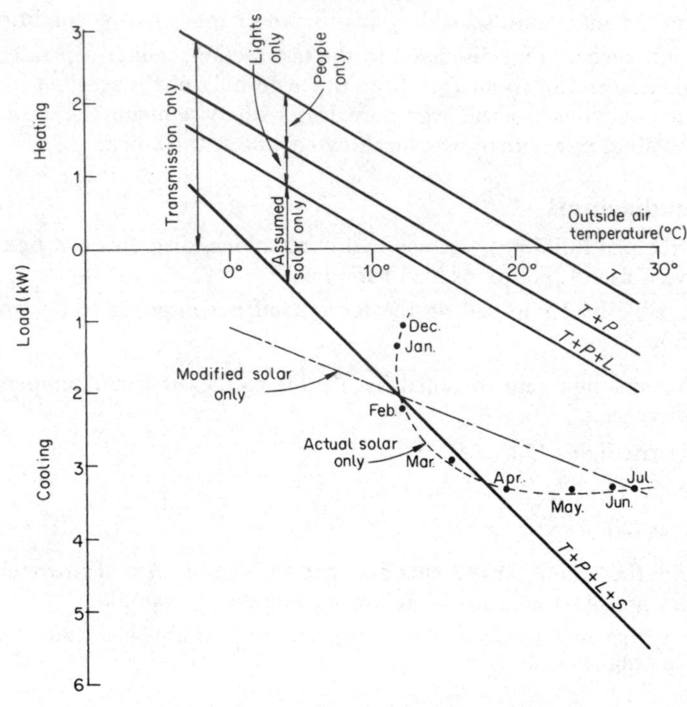

Fig. 8.13 Load diagram—western orientation

The above figures assume that the maximum outside air temperature occurs at 5 pm in June.

When the outside temperature, t_o, equals the room temperature, t_r, the transmission gains are zero, and when $t_o = 0$°C, the transmission figure becomes a loss of 2620 W.

If heating/cooling load in the room is used as the ordinate and t_o is used as the abscissa, a straight line can be drawn representing the contribution of T to the total room load. This line will pass through a value of zero load at $t_o = 22$°C, as Fig. 8.13 shows.

The contributions due to L and P are constant and can, therefore, be drawn parallel to T, offset by 500 W and 720 W respectively.

The way in which a line representing the variation in the solar contribution, S, should be drawn requires a little thought.

The load diagram is intended to represent maximum loads at different outside air temperatures. It is not intended to show diurnal variations in load but, rather, seasonal fluctuations. We can, therefore, consider the maximum solar gain through a west-facing window for different months of the year, and the corresponding mean monthly maximum temperatures provide the rest of the information necessary to plot the solar load line:

Month	Dec.	Jan.	Feb.	Mar.
Solar gain W/m^2	63	85	139	183
Solar load W	1008	1360	2222	2928
Mean monthly maximum temperature (°C)	12·2°	11·7°	12·2°	15·6°

Month	Apr.	May	June	July
Solar gain W/m^2	208	215	215	215
Solar load W	3330	3335	3335	3335
Mean monthly maximum temperature (°C)	19·4°	23·9°	26·7°	27·8°

It can be seen that the curve which results (shown as a broken line in Fig. 8.13) is unsatisfactory. One reason is that a direct correspondence between outside air temperature and solar gain is not to be expected. A further reason is that mean monthly maximum temperatures are not indicative of winter conditions. Neither, for that matter, are mean monthly minima (*see* section 5.10). To solve the difficulty, it is suggested that the minimum solar gain figure (for December) be associated with the winter design value for outside air temperature (0°C say). This can then locate the winter end of the solar contribution line, 3335 W and 28°C already locating the summer end of it.

Although clear skies, hence high solar intensities, are associated with cold weather in winter time, cloud cover prevails for a good deal of the time. This makes an accurate forecast of solar heat gain in winter rather open to question. For this reason, the single straight line shown in Fig. 8.13 for the solar contribution, is often accepted as reasonable.

The picture is a little more complicated for a room with southwards facing windows. The values of T, P and L are similar to other cases, but maximum solar heat gains through a window with a southern aspect occurring in February or October, at noon, it is unreasonable to associate them with the summer design value for outside air temperature.

In fact, the solar heat gains through single, internally shaded, glazing at noon, for a southerly orientation are:

Month	Dec.	Jan.	Feb.	Mar.	Apr.	May	June	July
Solar gain W/m²	193	208	227	215	186	142	127	142
Mean monthly maximum temperature (°C)	12·2°	11·7°	12·2°	15·6°	19·4°	23·9°	26·7°	27·8°

Figure 8.14 shows a plot of T, L, P and S for the same room as before, but with a southern window. The peak solar gain is now 16 m² × 227 W/m² × 0·97, and equals 3525 W. If this occurs in February on a warmish, clear, day, the corresponding mean monthly maximum temperature is 12·2°C. At this temperature the net value for $T+L+P$ is about 70 W and is a heat gain. So, $T+L+P+S$ is 3595 W when t_o is 12·2°C.

In March, if the mean monthly maximum temperature is again assumed to occur at noon, t_o is 15·6°C and $T+L+P$ is 480 W.

$S = 16 \times 215 \times 0{\cdot}97 = 3335$ W, and

$T+L+P+S = 3815$ W.

In April, t_o is 19·4°C and $T+L+P$ is 910 W.

$S = 16 \times 186 \times 0{\cdot}97 = 2885$ W, and

$T+L+P+S = 3795$ W.

In May, t_o is 23·9°C and $T+L+P$ is 1470 W.

$S = 16 \times 142 \times 0{\cdot}97 = 2202$ W, and

$T+L+P+S = 3672$ W.

In June, t_o is 26·7°C and $T+L+P$ is 1800 W.

$S = 16 \times 127 \times 0{\cdot}97 = 1973$ W, and

$T+L+P+S = 3773$ W.

The heat gain in December by solar radiation is almost the same as that occurring in April; if it is assumed that on a cold, clear day the outside air temperature is 0°C, then the solar gain in April can be transferred so as to take place at a temperature of 0°C and can provide a point representing a minimum winter solar heat gain. This point can be used with the solar gain

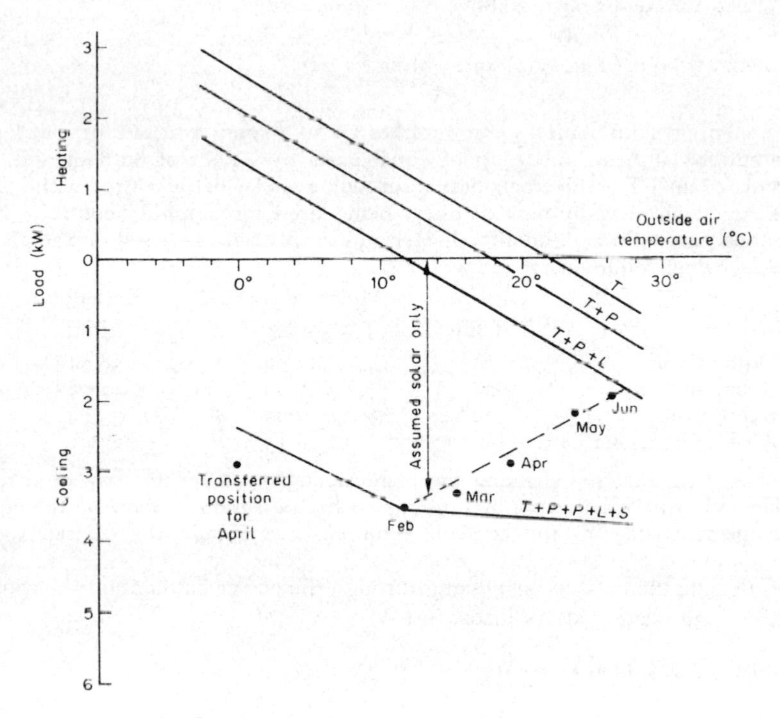

Fig. 8.14 Load diagram—southern orientation

points for February and June to provide the 'dog-leg' line shown in the figure for $T+L+P+S$.

In a similar manner load diagrams for eastern and northern orientations may be treated on their merits.

EXERCISES

1. Air enters an air washer at conditions of 32°C dry-bulb and 20·7°C wet-bulb, the flow rate being 4·72 m^3/s, measured at entry. If the air leaves the washer at 11°C dry-bulb, 10·3°C wet-bulb, using the following data, find—(*a*) the weight of moisture removed in kg/s, (*b*) the change of enthalpy.

Saturated vapour pressure at 20·7°C	24·41 mbar
Saturated vapour pressure at 10·3°C	12·52 mbar
Total barometric pressure	1013·25 mbar
Specific heat of dry air	1·007 kJ/kg °C
Specific heat of water vapour	1·84 kJ/kg °C
Latent heat of evaporation at 0°C	2501 kJ/kg

Constant for psychrometric equation for sling wet-bulb temperatures above 0°C = $6 \cdot 66 \times 10^{-4}$ °C^{-1}

Answers: (*a*) 0·0162 kg/s, (*b*) 29·13 kJ/kg.

2. An air-conditioning plant supplies air to a room at a rate of 1·26 kg/s, the supply air being made up of equal parts by weight of outdoor and re-circulated air. The mixed air passes through a cooler battery fitted with a by-pass duct, face-and-by-pass dampers being used for humidity control. The air then passes through an after-heater and supply fan. In a test on the plant, the following conditions were measured:

	Dry-bulb (°C)	Moisture content g/kg	Enthalpy kJ/kg
Outdoor air	27°	12·30	58·54
Room air	21°	7·86	41·08
Supply air	16°	7·07	33·99
Air leaving cooler battery	8°	6·15	23·51

Make neat sketches showing the plant arrangement and the psychrometric cycle and identify corresponding points on both diagrams. Show on the plant arrangement diagram the controls required to maintain the desired room condition.

Calculate the mass of air passing through the cooler battery and the cooler battery and heater battery loads, in kW.

Answers: 0·965 kg/s, 25·4 kW, 5·46 kW.

3. An air-conditioning system, containing an air washer using chilled water, supplies 2·36 m³/s to a room and extracts 80 per cent, by weight, of the air from the room, the balance being made up by outside air. Part of the air extracted from the room by-passes the washer and the remainder mixes with outside air before entering the washer. The conditions in the system are as follows:

	D.B.	Satn.	g/kg	kJ/kg	m³/kg
Room air	22·5°C	50%	8·632	44·58	0·8488
Supply air	15°C	76%	8·123	35·63	0·8265
Outside air	30°C	46%	12·560	62·29	0·8758
Washer leaving	11°C	90%	7·375	29·66	0·8141

Calculate:

(*a*) The mass of by-pass air to maintain these conditions.

(*b*) The temperature, moisture content and enthalpy of the air entering the washer.

(*c*) The washer cooling load in kW of refrigeration.

(*d*) Illustrate the cycle on a psychrometric sketch and show the system layout by means of a line diagram.

Answers: (*a*) 0·248 kg/s, (*b*) 24·15°C, 9·48 g/kg, 48·4 kJ/kg, (*c*) 4·65 kW of refrigeration.

4. The air in a room is to be maintained at 22°C dry-bulb and 50 per cent saturation by air supplied at a temperature of 12°C. The design conditions are as follows:

Sensible heat gain	6 kW
Latent heat gain	1·2 kW
Outside condition	32° dry-bulb, 24° wet-bulb (sling).

The ratio of recirculated air to fresh air is fixed at 3:1 by weight. The plant consists of a direct expansion cooler battery, an after-heater and a constant-speed fan. Allowing 1°C rise for fan power etc., calculate:

(i) The supply air quantity in m^3/s and its moisture content in g/kg.

(ii) The load on the refrigeration plant in kW of refrigeration.

(iii) The cooler battery contact factor.

Answers: (i) 0·478 m^3/s, 7·53 g/kg, (ii) 11·9 kW, (iii) 0·855.

5. A room is air conditioned by a system which maintains 22°C dry-bulb with 50 per cent saturation inside when the outside state is 28°C dry-bulb with 19·5°C wet-bulb (sling), in the presence of sensible and latent heat gains of 44 kW and 2·9 kW respectively. Fresh air flows over a cooler coil and is reduced in state to 10°C dry-bulb and 7·046 g/kg. It is then mixed with recirculated air, the mixture being handled by a fan, passed over another cooler coil and sensibly cooled to 13° dry-bulb. The air is then delivered to the conditioned room.

If the fresh air is to be used for dealing with the whole of the latent gain and if the effects of fan power and duct heat gain are ignored, determine the following:

(*a*) The amount of fresh air handled in m^3/s at the outside state and in kg/s.

(*b*) The amount of air supplied to the conditioned space in m^3/s at the supply state and in kg/s.

(*c*) The dry-bulb temperature, enthalpy and moisture content of the air handled by the fan.

(*d*) The load, in kW, involved in cooling and dehumidifying the outside air.

(*e*) The load in kW of refrigeration on the sensible cooler coil.

Answers: (*a*) 0·770 m^3/s, 0·887 kg/s, (*b*) 3·91 m^3/s, 4·77 kg/s (*c*) 19·8°C, 40·46 kJ/kg, 7·773 g/kg, (*d*) 24·8 kW, (*e*) 37·0 kW.

9

The Fundamentals of Vapour Compression Refrigeration

9.1 Basic concepts

If a liquid is introduced into a vessel in which there is initially a vacuum and whose walls are kept at a constant temperature it will at once evaporate. In the process the latent heat of vaporisation will be abstracted from the sides of the vessel. The resulting cooling effect is the starting point of the refrigeration cycle, which is to be examined in this chapter.

As the liquid evaporates, the pressure inside the vessel will rise until it reaches a certain maximum value for the temperature—the saturation vapour pressure (*see* section 2.8). After this, no more liquid will evaporate and, of course, the cooling effect will cease. Any further liquid introduced will remain in liquid state in the bottom of the vessel. If we now remove some of the vapour from the container, by connecting it to the suction of a pump, the pressure will tend to fall, and this will cause more liquid to evaporate. In this way, the cooling process can be rendered continuous. We need: a suitable liquid, called the *refrigerant*; a container where the vaporisation and cooling take place, called the *evaporator*; and a pump or fan to remove the vapour, called, for reasons which will appear later, the *compressor*.

The system as developed so far is obviously not a practical one because it involves the continuous consumption of refrigerant. To avoid this it is necessary to convert the process into a cycle. To turn the vapour back into a liquid it must be cooled with whatever medium is on hand for the purpose. This is usually water or air at a temperature substantially higher than the temperature of the medium being cooled by the evaporator. The vapour pressure corresponding to the temperature of condensation must, therefore, be a good deal higher than the pressure in the evaporator. The required step-up in pressure is provided by the pump acting as a *compressor*.

The liquefaction of the refrigerant is accomplished in the *condenser*, which is, essentially, a container cooled externally by air or water. The hot high-pressure refrigerant gas from the compressor is conveyed to the condenser and liquefies therein. Since there is a high gas pressure in the condenser, and the liquid refrigerant there is under the same pressure, it is easy to complete the cycle by providing a needle valve or other regulating device for injecting liquid into the evaporator. This essential component of a refrigerant plant is called the *expansion valve*.

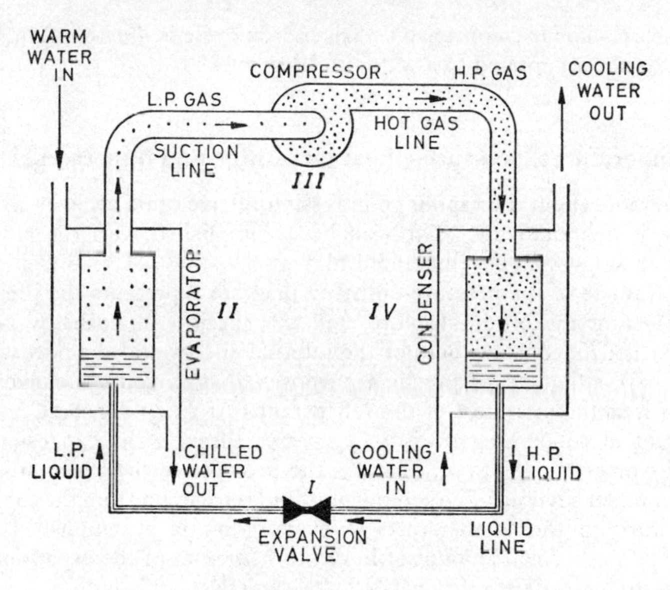

Fig. 9.1 Basic vapour compression cycle applied to a water chiller

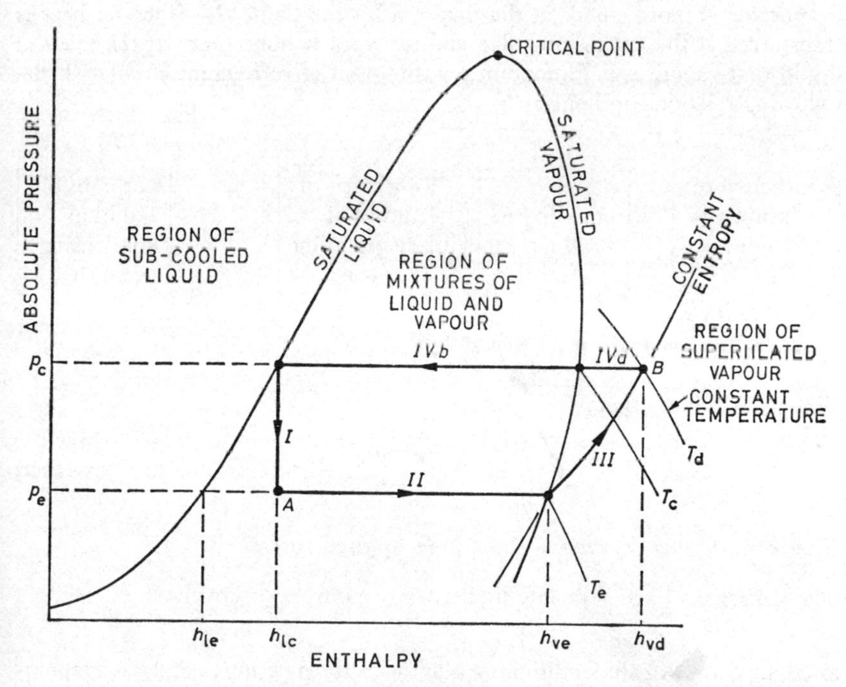

Fig. 9.2 Simple saturation refrigeration cycle on pressure-enthalpy diagram

This basic vapour compression refrigeration cycle is illustrated in Fig. 9.1 where it is shown applied to a water-chilling set.

9.2 Temperatures, pressures, heat quantities and flow rates

To learn more about the vapour compression refrigeration cycle we must now examine it quantitatively. This can be accomplished with the aid of the *pressure-enthalpy diagram* illustrated in Fig. 9.2.

The ordinate of the pressure-enthalpy diagram represents the pressure of the refrigerant in kN/m^2 absolute, and the abscissa its enthalpy in kJ/kg. Enthalpy is defined as the sum of the internal energy and the flow work (*see* eqn. (2.12)), although for practical purposes, it is sometimes convenient to think of it as the total heat of the refrigerant. At *I* (*see* Fig. 9.1), hot liquid refrigerant at condensing pressure p_c passes through the expansion valve, where its pressure falls to p_e, which is the pressure in the evaporator and at the compressor suction. The cooling of liquid refrigerant from the condensing temperature to the temperature of evaporation is accomplished by the vaporisation of a small amount of liquid downstream of the expansion valve. Vapour produced in this way is known as *flash gas.*

The state of the mixture of liquid refrigerant and vapour entering the evaporator is represented on the diagram by the point *A*. Since no heat is transferred at the expansion valve and no work is done there, if the mass of liquid that vaporises is f kilogram per kilogram of refrigerant circulated, the following relationship holds:

$$\begin{bmatrix}\text{enthalpy of} \\ \text{mixture en-} \\ \text{tering eva-} \\ \text{porator}\end{bmatrix} = \begin{bmatrix}\text{enthalpy of} \\ \text{flash gas}\end{bmatrix} f + \begin{bmatrix}\text{enthalpy of} \\ \text{liquid at} \\ \text{evaporating} \\ \text{pressure}\end{bmatrix} (1-f) = \begin{bmatrix}\text{enthalpy of} \\ \text{liquid at} \\ \text{condensing} \\ \text{pressure}\end{bmatrix}$$

$$= h_{ve}f + h_{le}(1-f) = h_{lc} \qquad (9.1)$$

Whence it follows that:

$$f = \frac{h_{lc}-h_{le}}{h_{ve}-h_{le}} \qquad (9.2)$$

The term *dryness fraction* is sometimes applied to f.

The subscripts l and v denote liquid and vapour, respectively.

EXAMPLE 9.1. An air conditioning plant uses Refrigerant 12 and has evaporating and condensing temperatures of 0°C and 35°C respectively. What will be the mass of flash gas per kilogram of refrigerant circulated?

Answer

Referring to Table 9.1, which gives some of the properties of saturated Refrigerant 12, the enthalpies we require are seen to be as follows:

Saturated vapour at 0°C, h_{ve} = 187·53 kJ/kg

Saturated liquid at 35°C, h_{lc} = 69·55 kJ/kg

Saturated liquid at 0°C, h_{le} = 36·05 kJ/kg

Therefore

$$f = \frac{69{\cdot}55 - 36{\cdot}05}{187{\cdot}53 - 36{\cdot}05} = 0{\cdot}221$$

TABLE 9.1 *Some properties of Refrigerant 12.*

Temperature °C	Absolute Pressure kN/m²	Gas volume m³/kg	Liquid enthalpy kJ/kg	Gas enthalpy kJ/kg	Gas entropy kJ/kg K	
0	308·6	0·0554	36·05	187·53	0·6966	Saturated
5	362·6	0·0475	40·69	189·66	0·6943	Saturated
35	847·7	0·0206	69·55	201·45	0·6839	Saturated
40	960·7	0·0182	74·59	203·20	0·6825	Saturated
5	308·6	0·0564	—	190·77	0·7081	Superheated
40	847·7	0·0212	—	205·21	0·6959	Superheated
45	847·7	0·0218	—	208·96	0·7078	Superheated
50	847·7	0·0224	—	212·72	0·7196	Superheated

Thus 22·1 per cent of the refrigerant in circulation will vaporise at the expansion valve before entering the evaporator.

Referring again to Fig. 9.2, line II represents the vaporisation of the remaining liquid at constant temperature and the transfer of heat from the water being chilled to the refrigerant in the evaporator. The *refrigerating effect* per kilogram of refrigerant in circulation is given by the following formula:

$$\text{refrigerating effect} = (h_{ve} - h_{le})(1 - f) \qquad (9.3)$$

$$= (h_{ve} - h_{lc}) \qquad (9.4)$$

It was common practice to measure amounts of refrigeration in *tons of refrigeration.* One ton of refrigeration (abbreviation: TR) is the amount of cooling produced by one U.S. ton of ice in melting over a period of 24 hours.

Since an American ton is 907·2 kg and the latent heat of fusion of water amounts to 334·9 kJ/kg, we have

$$1 \text{ TR} = \frac{907{\cdot}2 \times 334{\cdot}9}{24 \times 3600} = 3{\cdot}516 \text{ kW} \tag{9.5}$$

If the cooling required is X kW of refrigeration, it follows from eqn. (9.5) that the rate of refrigerant circulation necessary is given by the following equation:

$$m = \frac{X}{h_{ve} - h_{1c}} \text{ kg/s} \tag{9.6}$$

where m = the mass of refrigerant circulated.

EXAMPLE 9.2. If the plant of Example 9.1 has a capacity of 352 kW of refrigeration, what mass of Refrigerant 12 must be circulated per second? What is the volumetric rate of flow under suction conditions?

Answer

From Table 9.1 we have

Saturated vapour at 0°C, $h_{ve} = 187{\cdot}53$ kJ/kg.

Saturated liquid at 35°C, $h_{1c} = 69{\cdot}55$ kJ/kg.

Then, using eqn. (9.6) we have:

$$m = \frac{352}{187{\cdot}53 - 69{\cdot}55} = 2{\cdot}98 \text{ kg/s}$$

or

$$2{\cdot}98/352 = 0{\cdot}00846 \text{ kg/s kW.}$$

Coming to the second part of the question, from Table 9.1 it will be seen that the specific volume of Refrigerant 12 vapour when saturated at 0°C is 0·0554 m^3/kg. Hence, the volumetric rate of flow will be

$$2{\cdot}98 \times 0{\cdot}0554 = 0{\cdot}165 \text{ m}^3\text{/s}$$

or

$$0{\cdot}165/352 = 0{\cdot}00042 \text{ m}^3\text{/s kW}$$

The process labelled III in Fig. 9.2 represents the compression of the refrigerant vapour from pressure p_e to p_c. If compression can be regarded as adiabatic and reversible, the process line on the diagram is one of constant entropy. An *adiabatic* process is one during which there is no heat transfer (*see* section 2.17). Change in *entropy* is defined as the quantity of heat crossing the boundary of a reversible system, divided by the absolute temperature of the system. Change in entropy is given by the following formula:

$$\Delta s = \int \frac{dQ}{T} \tag{9.7}$$

where s is entropy in kJ/kg K, Q is the heat quantity in kJ/kg, and T the temperature in kelvin.

To give some examples, the increase in entropy during vaporisation is simply the latent heat of that process divided by the absolute temperature. Also, if the specific heat of a liquid c_1 remains constant during heating or cooling then the change in entropy will be:

$$\Delta s = s_2 - s_1 = \int_{T_1}^{T_2} \frac{dQ}{T} = c_1 \int_{T_1}^{T_2} \frac{dT}{T} = c_1 \log_e \frac{T_2}{T_1} \tag{9.8}$$

Notice that a reduction in the temperature of the liquid gives rise to a negative value of Δs; in other words, entropy diminishes with cooling and increases with heating. In the case of an ideal gas with constant specific heats the change in entropy is given by the following equations:

$$\Delta s = s_2 - s_1 = c_v \log_e \frac{T_2}{T_1} + (c_p - c_v) \log_e \frac{V_2}{V_1}$$

$$= c_p \log_e \frac{T_2}{T_1} - (c_p - c_v) \log_e \frac{p_2}{p_1} \tag{9.9}$$

In these expressions c_v and c_p are the specific heats of the gas at constant volume and constant pressure respectively, in J/kg °C, p is the absolute pressure in N/m^2 and V is the volume in m^3/kg.

Reverting to process III in Fig. 9.2: as the vapour is compressed, it generally rises in temperature and, finally, leaves the machine as a superheated gas at the condition represented on the diagram by point B. The work of compression, W_c, is given by

$$W_c = m(h_{vd} - h_{ve}) \text{ kW}. \tag{9.10}$$

where m is the mass of refrigerant circulated in kg/s.

EXAMPLE 9.3. Using the data of Examples 9.1 and 9.2, find the work of isentropic compression.

Answer

Reference to a pressure-enthalpy diagram or to tables giving the properties of Refrigerant 12 shows that the entropy of the saturated vapour entering the compressor at 0°C and 308·6 kN/m^2 is 0·6966 kJ/kg K. Superheated gas at the discharge pressure of 847·7 kN/m^2 absolute has an entropy of 0·6966 kJ/kg K, when the temperature is 40·29°C. Further reference to the diagram or tables indicates that refrigerant gas at 40·29°C and 847·7 kN/m^2 absolute pressure has an enthalpy of 205·43 kJ/kg. Using these values in eqn. (9.10) we find that

$$W_c = 2{\cdot}98(205{\cdot}43 - 187{\cdot}53)$$
$$= 53{\cdot}34 \text{ kW}$$

or $$= 53{\cdot}34/352 = 0{\cdot}15 \text{ kW/kW of refrigeration.}$$

On entering the condenser, the gas loses its superheat and latent heat of evaporation—a process represented in Fig. 9.2 by the lines IVa and IVb, at constant pressure—until it is again entirely in liquid state. The total amount of heat rejected to the condenser is given by the following equation:

$$\text{heat rejected to condenser in kW} = m(h_{vd} - h_{lc}) \qquad (9.11)$$

EXAMPLE 9.4. Again, using the data of the previous examples in this chapter, calculate the total amount of heat dissipated at the condenser.

Answer

Using eqn. (9.11), we have, for the heat rejected:

$$2{\cdot}98(205{\cdot}43 - 69{\cdot}55) = 404{\cdot}92 \text{ kW}$$
$$404{\cdot}92/352 = 1{\cdot}15 \text{ kW/kW of refrigeration.}$$

In the preceding sections several simplifying assumptions have been made. The cycle described is, therefore, called the *simple saturation cycle.* In practice, the liquid entering the expansion valve is usually several degrees cooler than the condensing temperature. The gas entering the compressor, on the other hand, is several degrees warmer than the temperature of evaporation. Furthermore, there are pressure drops in the suction, discharge and liquid pipelines and the compression process is not truly isentropic. The actual power required to drive the compressor is somewhat greater than W_c on account of frictional losses. All these factors have to be taken into account in a more exact quantitative treatment of the subject.

9.3 Coefficient of performance

The coefficient of performance of a refrigeration machine is the ratio of the energy removed at the evaporator (refrigerating effect) to the energy supplied to the compressor. Thus, we have the formula:

$$COP = \frac{h_{ve} - h_{lc}}{h_{vd} - h_{ve}} \qquad (9.12)$$

The best possible performance giving the highest *COP* would be obtained from a system operating on a Carnot cycle. Under such conditions the refrigeration cycle would be thermodynamically reversible, and both the expansion and compression processes would be isentropic. A Carnot cycle is shown on a *temperature-entropy diagram* in Fig. 9.3.

Point A indicates the condition of the liquid as it leaves the condenser. Line AB represents the passage of the refrigerant through an expansion engine, the work of which helps to drive the refrigeration machine. During the expansion process the pressure of the refrigerant drops from p_c to p_e, its temperature is reduced from T_c to T_e and part of the refrigerant vaporises. The expansion process, being reversible and adiabatic, takes place at constant entropy, with the result that line AB is vertical.

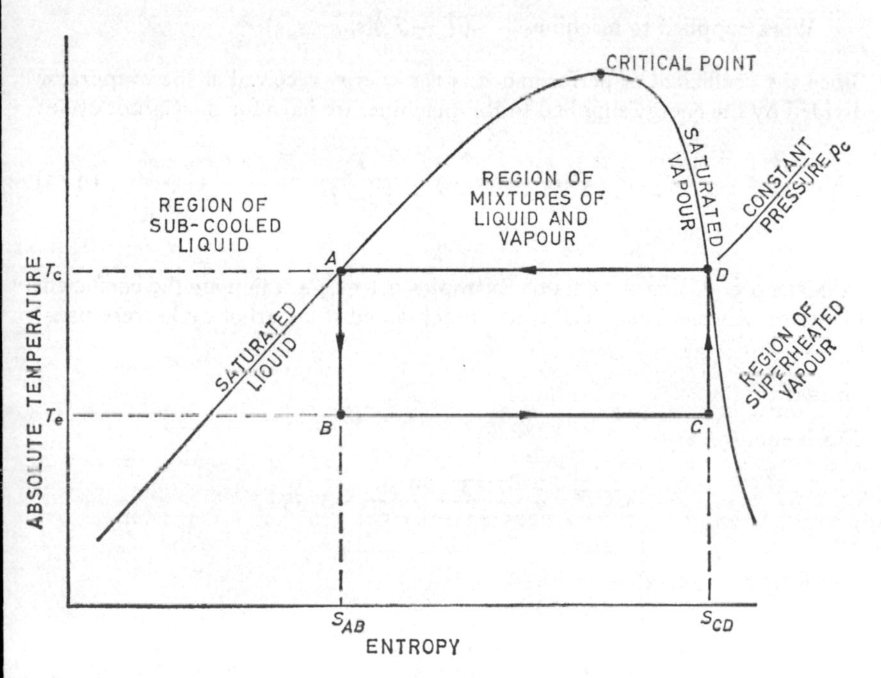

Fig. 9.3 Carnot refrigeration cycle on temperature-entropy diagram

During the next part of the cycle, refrigerant is evaporated at constant pressure p_e and temperature T_e, this being represented on the diagram by the line BC. The process comes to an end at point C, where the refrigerant still consists of a liquid-vapour mixture but the proportion of liquid is small.

The mixture then enters another engine, the compressor, which is driven by some external source of power supplemented by the output of the expansion engine already described. In the compressor, the refrigerant is increased in pressure from p_e to p_c, the temperature rises from T_e to T_c and the remaining liquid evaporates, so that the refrigerant leaves the machine as a saturated vapour—point D on the diagram. The compression process is adiabatic, reversible and, therefore, isentropic.

The final part of the cycle is shown by the line DA, which represents the liquefaction of the refrigerant in the condenser at constant pressure P_c and temperature T_c.

Since an area under a process line on a temperature-entropy diagram represents a quantity of energy (*see* eqn. (9.7)) the quantities involved in the Carnot refrigeration cycle are as follows:

Heat rejected at condenser	$T_c(s_{CD}-s_{AB})$
Heat received at evaporator	$T_e(s_{CD}-s_{AB})$
Work supplied to machines	$(T_c-T_e)(s_{CD}-s_{AB})$

Since the coefficient of performance is the energy received at the evaporator divided by the energy supplied to the machine, we have for the Carnot cycle:

$$COP\,(\text{Carnot}) = \frac{T_e}{T_c-T_e} \tag{9.13}$$

EXAMPLE 9.5. Using the data of Examples 9.1 to 9.4, calculate the coefficient of performance. What COP could be obtained if a Carnot cycle were used?

Answer

From eqn. (9.12)—

$$COP = \frac{187{\cdot}53-69{\cdot}55}{205{\cdot}43-187{\cdot}53} = 6{\cdot}59$$

and from eqn. (9.13)—

$$COP\,(\text{Carnot}) = \frac{0+273}{35-0} = 7{\cdot}8$$

It will be seen from this example that the simple saturation cycle under consideration had a coefficient of performance which was 80·9 per cent of that attainable with a Carnot cycle. Comparisons of this kind are sometimes used as an index of performance (*see* Table 9.2, for example).

For comparison with Fig. 9.2 the simple saturation cycle is illustrated on a temperature-entropy diagram in Fig. 9.4. The excursion of the process line into the superheat region and the irreversible method of expansion are largely responsible for the reduced coefficient of performance as compared to that theoretically attainable with a Carnot cycle.

The coefficient of performance of a refrigeration system operating on a Carnot cycle can be deduced from the changes in the enthalpy of the refrigerant as it passes through the cycle. The process is illustrated in Fig. 9.5.

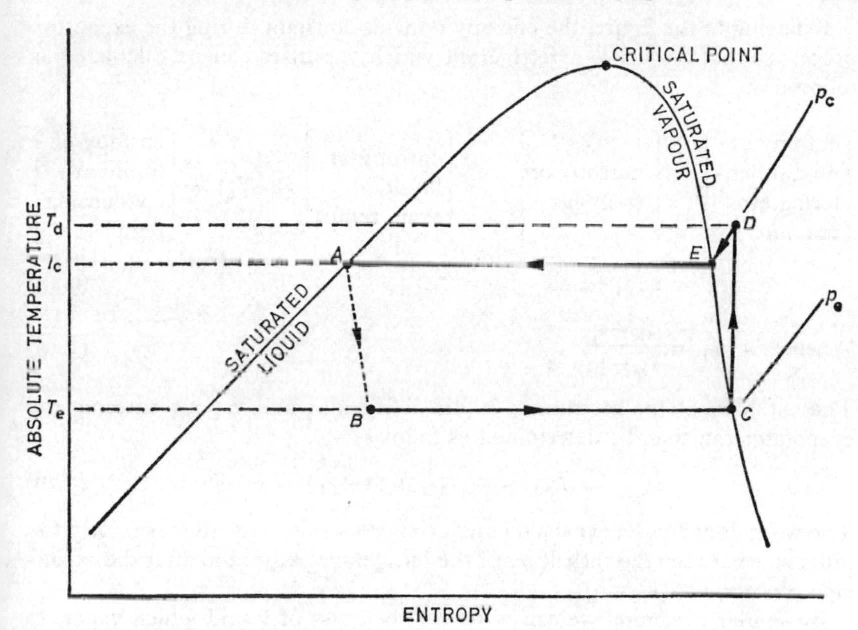

Fig. 9.4 Simple saturation refrigeration cycle on temperature-entropy diagram

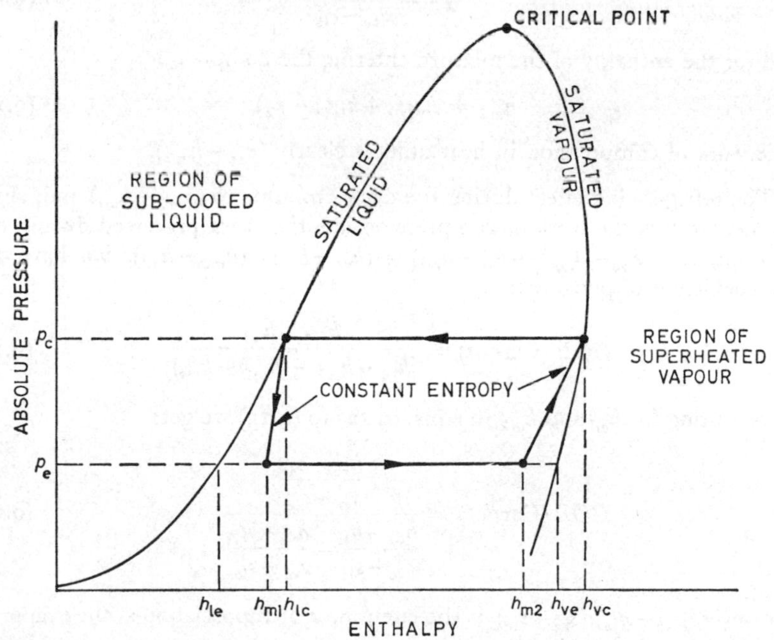

Fig. 9.5 Carnot refrigeration cycle on pressure-enthalpy diagram

Referring to the figure, the entropy remains constant during the expansion process so that the mass of refrigerant which vaporises can be calculated as follows:

$$\left\{\begin{matrix}\text{entropy of}\\ \text{mixture en-}\\ \text{tering eva-}\\ \text{porator}\end{matrix}\right\} = \left\{\begin{matrix}\text{entropy of}\\ \text{flash gas}\end{matrix}\right\} f_1 + \left\{\begin{matrix}\text{entropy of}\\ \text{liquid at}\\ \text{evap. temp.}\end{matrix}\right\}(1-f_1) = \left\{\begin{matrix}\text{entropy of}\\ \text{liquid at}\\ \text{condensing}\\ \text{temp.}\end{matrix}\right\}$$

$$= s_{ve}f_1 + s_{le}(1-f_1) = s_{lc} \tag{9.14}$$

Whence $$f_1 = \frac{s_{lc}-s_{le}}{s_{ve}-s_{le}} \tag{9.15}$$

The enthalpy of the mixture of liquid refrigerant and vapour entering the evaporator can then be determined as follows—

$$h_{ml} = h_{ve}f_1 + h_{le}(1-f_1) \tag{9.16}$$

The work done by the expansion engine expressed in heat units is clearly the difference between the enthalpies of the refrigerant before and after the expansion process, i.e. $(h_{lc}-h_{ml})$.

By similar reasoning we can write for the mass of liquid which vaporises during the compression process *CD*:

$$f_2 = \frac{s_{vc}-s_{le}}{s_{ve}-s_{le}} \tag{9.17}$$

and for the enthalpy of the mixture entering the compressor:

$$h_{m2} = h_{ve}f_2 + h_{le}(1-f_2) \tag{9.18}$$

The work of compression in heat units is clearly $(h_{vc}-h_{m2})$.

The refrigerating effect during the cycle amounts to $(h_{m2}-h_{ml})$ and, since the net work is the work of compression less the work produced during expansion, i.e. $(h_{vc}-h_{m2})-(h_{lc}-h_{ml}) = (h_{vc}-h_{lc})-(h_{m2}-h_{ml})$, we have for the coefficient of performance:

$$COP\ (\text{Carnot}) = \frac{h_{m2}-h_{ml}}{(h_{vc}-h_{lc})-(h_{m2}-h_{ml})} \tag{9.19}$$

Substituting for h_{ml} and h_{m2} in eqns. (9.12) to (9.15) we get:

$$COP\ (\text{Carnot}) = \frac{\dfrac{h_{ve}-h_{le}}{s_{ve}-s_{le}}}{\dfrac{h_{vc}-h_{lc}}{s_{vc}-s_{lc}} - \dfrac{h_{ve}-h_{le}}{s_{ve}-s_{le}}} \tag{9.20}$$

However $(h_{ve}-h_{le})/(s_{ve}-s_{le})$, is the latent heat of vaporisation at the evaporating pressure, divided by the difference in entropy between vapour and liquid under evaporating conditions. This is clearly equal to T_e. Similarly $(h_{vc}-h_{lc})/(s_{vc}-s_{lc})$, is equal to T_c and eqn. (9.20) reduces to eqn. (9.13).

From eqns. (9.6), (9.10) and (9.12) a useful relation between kW/kW and *COP* can be derived.

$$m = \frac{X}{h_{ve}-h_{lc}} \tag{9.6}$$

$$W_c = m(h_{vd}-h_{ve}) \tag{9.10}$$

$$\frac{W_c}{X} = \frac{(h_{vd}-h_{ve})}{(h_{ve}-h_{lc})}$$

but

$$COP = \frac{(h_{ve}-h_{lc})}{(h_{vd}-h_{ve})} \tag{9.12}$$

therefore

$$W_c = \frac{1}{COP} \text{ kW/kW of refrigeration} \tag{9.21}$$

9.4 Pressure-volume relations

The gas being compressed in a refrigeration machine can sometimes be regarded as an ideal one where pressures and volumes are related as follows:

$$p_e V_e^n = p_d V_d^n \tag{9.22}$$

Three cases can then be distinguished:

(i) The case of reversible adiabatic compression, in which the process is isentropic and we have for the value of n—

$$n = c_p/c_v = \gamma \tag{9.23}$$

Here c_p and c_v are the specific heats of the refrigerant gas at constant pressure and constant volume, respectively.

(ii) Reversible but non-adiabatic compression in which there is heat transfer between the refrigerant and its surroundings. Reciprocating compressors with cylinder cooling may approach this case and in such circumstances n will be less than γ.

(iii) Irreversible but adiabatic compression, in which there are thermal effects due to gas friction and turbulence but no actual heat transfer between the refrigerant and its enclosing surfaces. Centrifugal machines approach this type of compression, in which case n will be greater than γ.

When $n \neq \gamma$ the process is called *polytropic compression*. With polytropic compression the value of n depends upon the process—it is *not* a property of the refrigerant. During polytropic compression the entropy of the gas increases in case (iii) and diminishes in case (ii). Thus, the process departs from the simple saturation cycle which we have hitherto considered.

If the compression is reversible, as in cases (i) and (ii), the work done on 1 kg mass of refrigerant in Nm during steady flow is given by the following expression, where pressures are in N/m^2 absolute and volumes are in m^3.

$$\text{Work done} = \int_{p_e}^{p_d} V\mathrm{d}p \tag{9.24}$$

Using eqn. (9.22) we have:

$pV^n = c$ (where c is a constant), or

$V = c^{1/n}p^{-1/n}$

Thus,

$$\text{work done} = c^{1/n}\int_{p_e}^{p_d} p^{-1/n}\mathrm{d}p$$

$$= \frac{n}{(n-1)}\left\{p^{(n-1)/n}\right\}_{p_e}^{p_d} c^{1/n}$$

$$= \frac{n}{(n-1)}\left\{\left(\frac{p_d}{p_e}\right)^{(n-1)/n} - 1\right\}p_e V_e \tag{9.25}$$

It should be noted that the work done on 1 kg mass of refrigerant in Nm is here equal to the head in metres against which the compressor is working. Developing the expression further by using eqn. (9.6), we get for the work of compression, in W, on m kg of refrigerant

$$W_c = \frac{mn}{(n-1)}\left\{\left(\frac{p_d}{p_e}\right)^{(n-1)/n} - 1\right\}p_e V_e$$

$$= \frac{p_e V_e}{(h_{ve}-h_{lc})}\cdot\frac{n}{(n-1)}\left\{\left(\frac{p_d}{p_e}\right)^{(n-1)/n} - 1\right\}X \tag{9.26}$$

If the compression process is the isentropic one described as case (i) above, the equation becomes:

$$W_c = \left(\frac{p_e V_e}{h_{ve}-h_{lc}}\right)\left(\frac{\gamma}{\gamma-1}\right)\left\{\left(\frac{p_d}{p_e}\right)^{(\gamma-1)/\gamma} - 1\right\}X \tag{9.27}$$

In case (iii), where the process is irreversible but adiabatic, eqn. (9.24) still gives the fluid output in Nm/kg of refrigerant, or the head in metres against which the machine is working, but it no longer represents the required input energy to the fluid.

This exceeds the quantity

$$\int_{p_e}^{p_d} V\mathrm{d}p$$

and is calculated from the following expression—

$$W_c = \frac{p_e V_e}{h_{ve} - h_{lc}} \frac{\gamma}{\gamma - 1} \left\{ \left(\frac{p_d}{p_e} \right)^{(n-1)/n} - 1 \right\} X \tag{9.28}$$

Comparing this with eqn. (9.27) we see that

$$\frac{\text{theoretical work for case (i)}}{\text{theoretical work for case (iii)}} = \frac{\{(p_d/p_e)^{(\gamma-1)/\gamma} - 1\}}{\{(p_d/p_e)^{(n-1)/n} - 1\}} \tag{9.29}$$

This ratio is called the isentropic efficiency.

The quantity

$$\frac{n}{(n-1)} \frac{(\gamma - 1)}{\gamma} \tag{9.30}$$

is called the polytropic efficiency and that represented by eqn. (9.25) is called the polytropic head. Polytropic head has to be divided by the polytropic efficiency to give the power required by case (iii) in Nm/kg of gas pumped.

Other forms of eqns. (9.25) to (9.28) are possible. Thus, for an ideal gas we can write RT_e in place of $p_e V_e$, R being the particular gas constant for the refrigerant.

It should be noted that although eqn. (9.10) is applicable to cases (i) and (iii), it cannot be applied to case (ii), since the quantity $h_{vd} - h_{ve}$ does not, in that case, represent all the work of compression. Instead, we must write for case (ii)—

$$W_c = m\{(h_{vd} - h_{ve}) + h_j\} \tag{9.31}$$

Here, h_j is the heat lost to the jacket in J/kg of refrigerant in circulation. Its value can be calculated from

$$h_j = \left(\frac{n}{n-1} - \frac{\gamma}{\gamma - 1} \right) \left\{ \left(\frac{p_d}{p_e} \right)^{(n-1)/n} - 1 \right\} p_e V_e \tag{9.32}$$

The temperature of the hot gas leaving the compressor can be calculated from the formula:

$$T_d = T_e \left(\frac{p_d}{p_e} \right)^{(n-1)/n} \tag{9.33}$$

EXAMPLE 9.6. A refrigerant which behaves as an ideal gas has a molecular mass of 64·06 and a specific heat ratio of 1·26 (equal to c_p/c_v). If the compression ratio is 3·119 and the refrigerating effect is 322 kJ/kg with evaporation at 4·5°C, find the theoretical kW/kW, the enthalpy gain during compression and the temperature of the discharged gas for the following cases:

(i) isentropic compression,

(ii) polytropic compression with $n = 1{\cdot}22$,

(iii) polytropic compression with $n = 1{\cdot}30$.

Answer

(i) Writing RT_e in place of $p_e V_e$ and setting $R = 8314{\cdot}66/64{\cdot}06$, we obtain from eqn. (9.27),

$$W_c = \frac{8314{\cdot}66 \times 277{\cdot}5 \times 1{\cdot}26}{64{\cdot}06 \times 322 \times 1000 \times 0{\cdot}26}(3{\cdot}119^{(0{\cdot}26/1{\cdot}26)} - 1)$$

$$= 0{\cdot}1438 \text{ kW/kW}.$$

Using eqns. (9.6) and (9.10),

$$h_{vd} - h_{ve} = 322 \times 0{\cdot}1438$$

$$= 46{\cdot}3 \text{ kJ/kg}.$$

Finally, we have from eqn. (9.33),

$$T_d = 277{\cdot}5 \times 3{\cdot}119^{(0{\cdot}26/1{\cdot}26)}$$

$$= 351{\cdot}5 \text{ K},$$

which is equivalent to 78·5°C.

(ii) Here we proceed as in case (i) but we use eqn. (9.26) in place of eqn. (9.27).

$$W_c = \frac{8314{\cdot}66 \times 277{\cdot}5 \times 1{\cdot}22}{64{\cdot}06 \times 322 \times 1000 \times 0{\cdot}22}(3{\cdot}119^{(0{\cdot}22/1{\cdot}22)} - 1)$$

$$= 0{\cdot}1417 \text{ kW/kW}.$$

Using eqns. (9.6) and (9.31),

$$h_{vd} - h_{ve} = 0{\cdot}1417 \times 322 - h_j$$

$$= 45{\cdot}6 - h_j.$$

Equation (9.32) gives the jacket loss:

$$h_j = \left(\frac{1{\cdot}22}{0{\cdot}22} - \frac{1{\cdot}26}{0{\cdot}26}\right)\left(3{\cdot}119^{(0{\cdot}22/1{\cdot}22)} - 1\right)\left(\frac{8314{\cdot}66 \times 277{\cdot}5}{64{\cdot}06}\right)$$

$$= 5825 \text{ J/kg}.$$

Therefore,

$$h_{vd} - h_{ve} = 45{\cdot}6 - 5{\cdot}82$$

$$= 39{\cdot}78 \text{ kJ/kg}.$$

The temperature of the discharged gas will be

$$T_d = 277 \cdot 5 \times 3 \cdot 119^{(0 \cdot 22/1 \cdot 22)}$$
$$= 341 \text{ K},$$

which is equivalent to 68°C.

(iii) Proceeding as for case (i) but using eqn. (9.28) instead of eqn. (9.27)

$$W_c = \frac{8314 \cdot 66 \times 277 \cdot 5 \times 1 \cdot 26}{64 \cdot 06 \times 322 \times 1000 \times 0 \cdot 26}(3 \cdot 119^{(0 \cdot 3/1 \cdot 3)} - 1)$$
$$= 0 \cdot 1622 \text{ kW/kW}.$$

From eqns. (9.6) and (9.10),

$$h_{vd} - h_{ve} = 0 \cdot 1622 \times 322$$
$$= 52 \cdot 25 \text{ kJ/kg}.$$

The temperature of the gas leaving the compressor will be

$$T_d = 277 \cdot 5 \times 3 \cdot 119^{(0 \cdot 3/1 \cdot 3)}$$
$$= 361 \cdot 5 \text{ K},$$

which is equivalent to 88·5°C.

Summarising the results,

	Case (i)	Case (ii)	Case (iii)
n	1·26	1·22	1·30
kW/kW	0·1438	0·1417	0·1622
Enthalpy gain kJ/kg	46·3	39·78	52·25
Discharge gas temperature °C	78·5°	68·0°	88·5°

9.5 Refrigerants

The comparative performance of refrigerants commonly used in the vapour compression cycle is given in Table 9.2, the values of which were all obtained by calculation, using the methods of this chapter. Actual operating conditions will therefore be somewhat different, owing to the effects of the factors mentioned at the end of section 9.2.

In general, it has been assumed that the vapour enters the compressor in a saturated condition at 5°C. In refrigerants 113 and 114, however, where saturated suction gas would result in condensation during compression, enough superheat has been assumed to ensure saturated discharge gas. This superheat has not been counted as part of the refrigerating effect.

A condensing temperature of 40°C has been taken for all refrigerants.

TABLE 9.2 *Comparative performance of refrigerants evaporating at 5°C condensing at 40°C.*

Refrigerant Number	Name of refrigerant	Suction Temperature °C	Evaporating Pressure bar	Condensing Pressure bar	Compression Ratio	Refrigerating Effect kJ/kg	Specific Vol. of Vapour m^3/kg	Compressor Displacement litre/s kW	Power in kW per kW of Refrigeration	% Carnot Cycle Efficiency
718	Water	5°	0·009	0·074	8·46	2370·0	147·0	62·0	0·1355	92·9
11	Trichloromonofluoromethane	5°	0·496	1·747	3·52	157·0	0·332	2·12	0·1395	90·2
717	Ammonia	5°	5·160	15·55	3·01	1088·0	0·243	0·214	0·1456	86·4
114	Dichlorotetrafluoroethane	12·7°	1·062	3·373	3·18	106·2	0·122	1·14	0·1484	84·8
12	Dichlorodifluoromethane	5°	3·626	9·607	2·65	115·0	0·047	0·409	0·1502	83·8
113	Trichlorotrifluoroethane	10·4°	0·188	0·783	4·16	129·5	0·652	5·03	0·1511	83·3
22	Monochlorodifluoromethane	5°	5·838	15·34	2·63	157·8	0·040	0·255	0·1518	82·9
502	An azeotropic mixture	5°	6·678	16·77	2·51	101·0	0·026	0·259	0·1631	77·1

The performances have been calculated on the assumption that compression is isentropic.

A number of factors have to be taken into consideration when the thermodynamic characteristics listed in the table are being evaluated. For example, if operating pressures are high, then the materials comprising the refrigeration system will be heavy and the equipment expensive. The refrigerant containers will also be heavy, and this will increase the cost of transport. Sub-atmospheric operating pressures, on the other hand, mean that any leakage will result in air entering the system.

In reciprocating compressors, the displacement (litre/s kW) should be low so that the required performance can be achieved with a small machine. For a centrifugal compressor, on the other hand, large displacements are desirable in order to permit the use of large gas passages. The reduced frictional resistance to such passages improves the compressor efficiency.

The power required to drive the compressor is obviously very important because it affects both the first cost and the running cost of the refrigeration plant.

Although not listed in the table, the critical and freezing temperatures of a substance have also to be noted when assessing its suitability for use as a refrigerant.

There are, of course, factors other than thermodynamic ones which have to be taken into consideration when choosing a refrigerant for a particular vapour compression cycle. These include the heat transfer characteristics, dielectric strength of the vapour, inflammability, toxicity, chemical reaction with metals, tendency to leak and leak detectability, behaviour when in contact with oil, availability and, of course, the cost.

9.6 Supplementary worked examples

EXAMPLE 9.7. Saturated liquid Refrigerant 12 is expanded by a throttling process to a pressure corresponding to a temperature of 2°C. Determine the increase in entropy and the dryness fraction after throttling. The specific heat of liquid R12 can be taken as 0·96 kJ/kg °C and the latent heat at 2°C as 150 kJ/kg. Condensation is at 37°C.

Answer

(i) Referring to section 9.2 we have—

$$f = \frac{h_{1c} - h_{1e}}{h_{ve} - h_{1e}} \qquad (9.2)$$

Now $h_{1c} - h_{1e}$ is the difference between the enthalpies of saturated liquid at condensing and evaporating pressures and must be equal to the product of the specific heat of the liquid c_1 and the difference between the condensing and evaporating temperatures. That is,

$$h_{1c} - h_{1e} = c_1(T_c - T_e) \qquad (9.34)$$

The denominator of eqn. (9.2) is clearly equal to the latent heat of vaporisation at the evaporating temperature (symbol h_{fg}). That is,

$$h_{ve} - h_{le} = h_{fg} \tag{9.35}$$

Then, substituting in eqn. (9.2) from (9.34) and (9.35) we get—

$$f = \frac{c_1(T_c - T_e)}{h_{fg}} \tag{9.36}$$

Inserting values,

$$f = \frac{0{\cdot}96 \times (310 - 275)}{150} = 0{\cdot}224$$

(ii) The net gain in entropy per kilogram of refrigerant subjected to the throttling process will equal the gain in entropy on account of the vaporisation of f kg of *liquid* at the evaporating temperature *less* the loss of entropy resulting from the cooling of 1 kg of liquid fron T_c to T_e. Thus,

change in entropy on account of vaporisation (eqn. (9.7)) $= f \cdot \dfrac{h_{fg}}{T_e} = 0{\cdot}224 \times \dfrac{150}{275} = 0{\cdot}1222$ kJ/kg K

change in entropy on account of cooling (eqn. (9.8)) $= c_1 \log_e \dfrac{T_e}{T_c}$

$$= 0{\cdot}96 \times \log_e \frac{275}{310} = -0{\cdot}1152$$

Net gain $= +0{\cdot}0070$ kJ/kg K

We can check this result by referring to tables giving the properties of Refrigerant 12 and reasoning that

$$\Delta s = (1 - f)s_{le} + f s_{ve} - s_{lc}$$

Any discrepancy occurs because the specific heat of the liquid does not, in fact, remain constant during cooling.

EXAMPLE 9.8. Water is used in a simple saturation vapour compression refrigeration cycle and the evaporating and condensing temperatures and absolute pressures are 4·5°C with 0·008424 bar and 38°C with 0·066240 bar respectively. Assume that water vapour behaves as an ideal gas with $c_p/c_v = 1{\cdot}322$ and calculate the discharge temperature if compression is isentropic. Find also the kW/kW if the refrigerating effect is 2355 kJ/kg.

Answer

(i) From eqn. (9.33) we have

$$T_d = 277{\cdot}5\,(0{\cdot}066240/0{\cdot}008424)^{(0.322/1.322)}$$

$$= 459\text{ K or } 186^{\circ}\text{C.}$$

(ii) Since the molecular mass of water is 18·02 and $T_e = 277{\cdot}5$ K, we have

$$p_e V_e = \frac{8314{\cdot}66 \times 277{\cdot}5}{18{\cdot}02 \times 1000} = 128\text{ kJ}$$

Using this in eqn. (9.27)

$$\frac{W_c}{X} = \left(\frac{128 \times 1{\cdot}322}{2355 \times 0{\cdot}322}\right)\left[\left(\frac{0{\cdot}066240}{0{\cdot}008424}\right)^{(0{\cdot}322/1{\cdot}322)} - 1\right]$$

$$= 0{\cdot}1454\text{ kW/kW of refrigeration.}$$

EXERCISES

1. A refrigeration machine works on the simple saturation cycle. If the difference between the enthalpies of saturated liquid at the condensing and evaporating pressures is 158·7 kJ/kg and the latent heat of vaporisation under evaporating conditions is 1256 kJ/kg, find the dryness fraction after expansion. Calculate the refrigerating effect.

Answers: 0·1265 and 1097·3 kJ/kg.

2. Calculate the displacement of a compressor having 176 kW capacity if the refrigerating effect is 1097 kJ/kg and the volume of the suction gas is 0·2675 m³/kg. Assuming a volumetric efficiency of 75 per cent, what cylinder size will be needed if the speed is to be 25 rev/s and there are to be 6 cylinders with equal bore and stroke?

Answers: 0·0429 m³/s and 78·6 mm.

3. Find the change in entropy when a liquid refrigerant whose specific heat is 4·71 kJ/kg °C cools from 35·5°C to 2°C. If 12·64 per cent liquid then vaporises and the latent heat of that process is 1256 kJ/kg, what further change in entropy occurs? State the total change in entropy per kg of refrigerant for the cooling and partial vaporisation.

Answers: −0·544 kJ/kg K, +0·577 kJ/kg K and +0·033 kJ/kg K.

4. A 4-cylinder 75 mm bore × 75 mm stroke compressor runs at 25 rev/s and has a volumetric efficiency of 75 per cent. If the volume of the suction gas is 0·248 m³/kg and the machine has an operating efficiency of 75 per cent, what power will be required on a simple saturation cycle when the difference between the enthalpies of the suction and discharge gases is 150 kJ/kg? If the

refrigerating effect is 1087 kJ/kg what is the output of the machine in kW of refrigeration? State the coefficient of performance.

Answers: 20 kW, 108·7 kW of refrigeration and $COP = 7{\cdot}25$.

5. A refrigeration system works on a Carnot cycle with saturated vapour leaving the compressor at 37·8°C. If the evaporation takes place at 4·4°C and the specific heat of the liquid is 0·952 kJ/kg °C, calculate the dryness fraction of the refrigerant leaving the expansion engine. The latent heat of vaporisation at 4·4°C may be taken at 149·3 kJ/kg. What would be the dryness fraction if the refrigerant were throttled instead of expanded isentropically through an engine?

Answers: 0·20 and 0·21.

6. In Table 9.2 the kW/kW for Refrigerant 12 is given as 0·1502. Check this result using eqn. (9.27) and taking the value of $\gamma = 1{\cdot}13$.

Answer: 0·150 kW/kW of refrigeration.

10

Cooler Coils and Air Washers

10.1 Distinction between cooler coils and air washers

The evaporative cooling of air by adiabatic saturation (*see* sections 3.5 and 3.6) may be improved by lowering the temperature of the feed water, but little improvement results if all the feed water is evaporated. The picture changes, however, if the rate of injection of feed water is increased so that not all of it is evaporated, but some falls into a sump and is recirculated. When the rate of water flow is speeded up its temperature has an increasingly important influence on the heat exchange with the airstream. For example, if an infinitely large quantity of water were circulated at a temperature t_w, then the air would leave the spray chamber also at a temperature t_w, in a saturated condition.

In air washers used for cooling and dehumidification processes (*see* section 3.4), the quantities of water circulated are very large compared with the quantity that could be totally evaporated. The temperature of the water thus plays a major part in determining the state of the air leaving the washer. Figure 10.1 illustrates the psychrometric considerations.

Direct cooling of this sort is not necessarily the most effective method. True contra-flow heat exchange, or even a reasonable approximation to it, cannot be easily obtained in an air washer—with the possible exception of some capillary washers. On the other hand, when chilled water is circulated through a cooler coil of more than four rows a good realisation of contra-flow is obtained. This permits a closer approach between the leaving air temperature and the entering water temperature with cooler coils using chilled water or brine.

In comparison with the cooler coil the washer suffers from a number of other disadvantages:

(i) It is more bulky.
(ii) Corrosion is a greater risk.
(iii) Maintenance is more expensive.
(iv) It uses an open chilled water circuit, thus causing the deposition of scale, rust, slime, etc. in the water chiller (evaporator), reducing its heat transfer efficiency and increasing the cost of the refrigeration plant.

It is true that the washer allows the air to be humidified but this may also be achieved by using a sprayed cooler coil (*see* section 10.10). Steam injection is worthy of consideration for humidification since it dispenses with the need for extensive waterworks.

There is one point strongly in favour of air washers and sprayed cooler coils: the presence of the large mass of water in the sump tank and spray

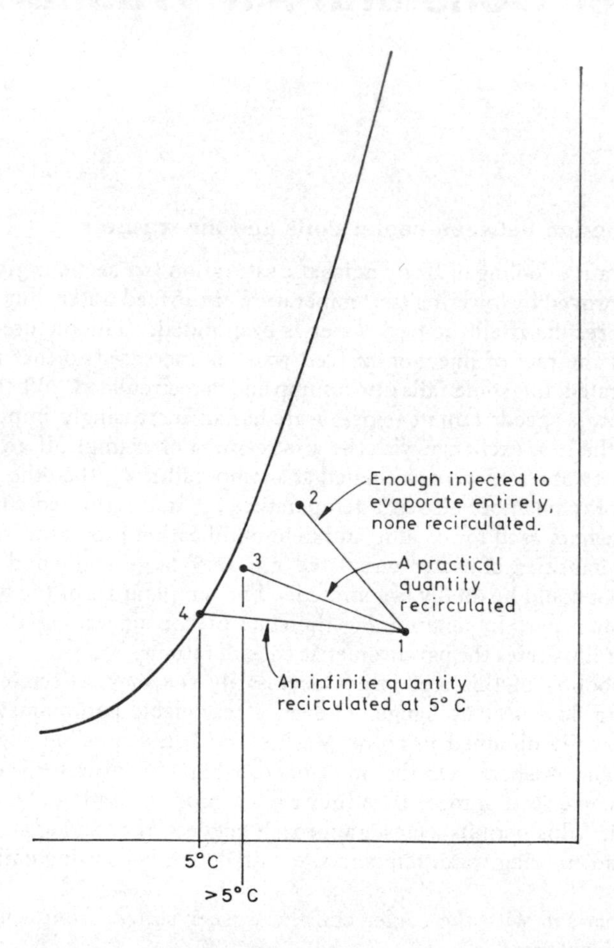

Fig. 10.1

chamber gives a thermal inertia to the system, smoothing out fluctuations in the state of the air leaving the coil or washer, and adding stability to the operation of the automatic controls.

However, in spite of these favourable aspects, the air washer is somewhat out of fashion because the cooler coil is more efficient, and because of the four points mentioned.

10.2 Cooler coil construction

A cooler coil is not merely a heater battery fed with chilled water or into which cold, liquid refrigerant is pumped. There are two important points of difference: first, the temperature differences involved are very much less for a cooler coil than for a heater battery, and secondly, moisture is condensed from the air on to the cooler coil surface. With air heaters, water entering and leaving at 85°C and 70°C may be used to raise the temperature of an airstream from 0° to 35°C, resulting in a log mean temperature difference of about 59° for contra-flow heat exchange. With a cooler coil, water may enter at 7°C and leave at 13°C in reducing the temperature of the airstream from 26°C to 11°C, a log mean temperature difference of only 7·6°C with contra-flow operation. The result is that much more heat transfer surface is required for cooler coils and, as will be seen in section 10.3, it is important that contra-flow heat exchange be obtained. The second point of difference, that dehumidification occurs, means that the heat transfer processes are more involved in cooler coils.

There are three forms of cooler coil: chilled water, direct expansion, and chilled brine. The first and third types make use of the sensible heat absorbed by the chilled liquid as it is circulated inside the finned tubes of the coil to effect the necessary cooling and dehumidification of the airstream. The second form has liquid refrigerant boiling within the tubes, and so the heat absorbed from the airstream provides the latent heat of evaporation for the refrigerant.

Chilled water coils are usually constructed of externally finned, horizontal, tubes, so arranged as to facilitate the drainage of condensed moisture from the fins. Tube diameters vary from 10 to 20 mm, and copper is the material commonly used, with copper or aluminium fins. Copper fins and copper tubes generally offer the best resistance to corrosion, particularly if the whole assembly is electro-tinned after manufacture. Fins are usually of the plate type, although spirally-wound and circular fins are also popular. Cross-flow heat exchange between the air and the cooling fluid occurs for a particular row but, from row to row, contra- or parallel-flow of heat may take place, depending on the way in which the piping has been arranged. Figure 10.2(*a*) illustrates this. Because of the improved heat transfer rate and the fact that the leaving air temperature may be reduced to below the value of the leaving fluid temperature, contra-flow connexion is essential for chilled water coils. In direct expansion coils, since the refrigerant is boiling at a constant temperature the surface temperature is more uniform and there is no distinction between parallel and contra flow, the logarithmic mean temperature difference being the same. However, with direct-expansion cooler coils a good deal more trouble has to be taken with the piping in order to ensure that a uniform distribution of liquid refrigerant takes place across the face of the coil. This is achieved by having a ' distributor ' after the expansion valve, the function of which is to divide the flow of liquid refrigerant into a number of equal streams. Pipes of equal resistance join the downstream side of the distributor to the coil so that the liquid is fed uniformly over the depth and height of the coil. It may be necessary to feed the coil from both sides if it is very wide. The limitations imposed by the need to secure effective distribution of the liquid refrigerant throughout the coil tend to discourage the use of very large direct-expansion cooler coils. Control problems exist.

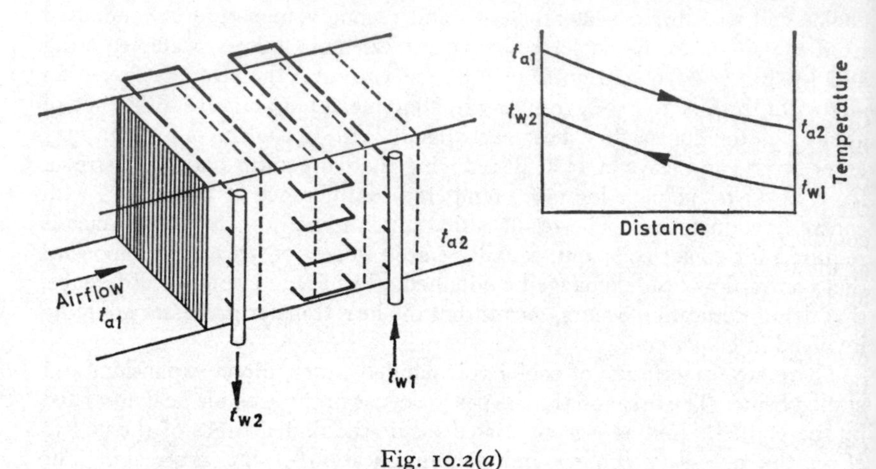

Fig. 10.2(*a*)

This arrangement shows contra-flow for the rows of a four-row cooler coil. Cross-flow heat exchange occurs in individual rows.

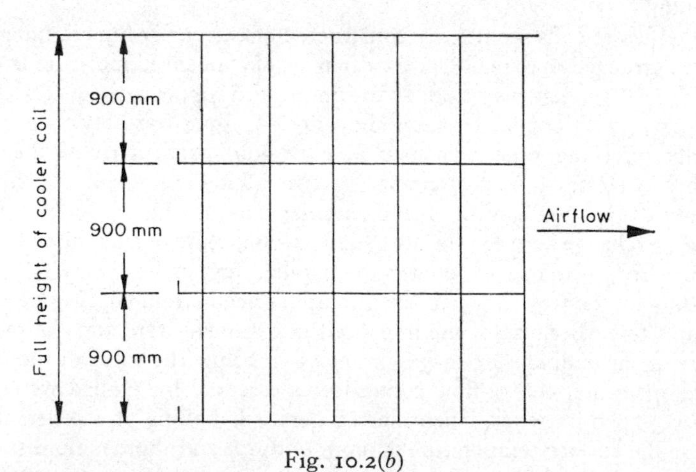

Fig. 10.2(*b*)

All cooler coils should be divided into sections (*see* Fig. 10.2(*b*)) so that horizontal condensate collection trays run across their full width and depth. The maximum desirable vertical distance between trays is 900 mm. If this is exceeded there is a tendency under conditions of high latent load for the condensate to flood the lower portion of the coil, diminishing both the airflow and the heat transfer rate. Trays should, of course, be drained to waste. A consequence of the distribution and condensate collection aspects is that wide, short, coils are cheaper than narrow, tall, ones.

Water velocities are usually between 1 and 2 m/s, at which values the coils are self-purging of air. Fin spacings vary between about 300 and 500 per metre.

10.3 Parallel and contra-flow

Consider a 4-row coil in stylised form, as shown in Fig. 10.3(*a*). Air enters the first row at state *O*. Water enters at a temperature t_{wa} and, its temperature rising as it absorbs heat from the airstream, leaves at a higher value. Denote the mean surface temperature of the first row by t_{w1}. If the first row had a contact factor of unity, the air would be conditioned to saturation at a state 1′, as shown in Fig. 10.3(*b*). In fact, the contact factor β is less than unity and the air leaves at a state denoted by 1. This is now the entry state for the second row, the mean surface temperature of which is t_{w2}. In a similar way, the exit state from the second row is 2, rather than 2′. This reasoning is applied to the succeeding rows, from which the leaving states of the airstream are denoted by 3 and 4. A line joining the state points 0, 1, 2, 3 and 4 is a concave curve and represents the change of state of the air as it flows past the rows under parallel-flow conditions. A straight line joining the points 0 and 4 indicates the actual overall performance of the coil. This condition line, replacing the condition curve, cuts the saturation curve at a point *A* when produced. By the reasoning adopted when flow through a single row was considered, one may regard the temperature of *A* as the mean coil surface temperature for the whole coil. It is denoted by t_{sm}.

Similar considerations apply when contra-flow is dealt with, but the result is different. A convex condition curve is obtained by joining the points 0, 1, 2, 3 and 4. Referring to Fig. 10.4 it can be seen that, this time, air entering the first row encounters a mean coil surface temperature for the row of t_{w1} and that this is higher than the mean surface temperatures of any of the succeeding rows. As air flow through the four rows occurs, progressively lower surface temperatures are met. The tendency of the state of the air is to approach the apparatus dew points, 1′, 2′, 3′ and 4′, each at a lower value than the point preceding it, as it passes through the rows. The condition curve thus follows a downward trend as flow proceeds.

The result of this is that, by comparison with the case of parallel flow, a lower leaving air temperature is achieved, greater heat transfer occurs, and the coil is more efficient, if it is piped up for contra-flow operation.

10.4 Contact factor

A psychrometric definition of this was given in section 3.4. Such a definition is not always useful—for example, in a cooler coil for sensible cooling only—and so it is worth considering another approach, in terms of the heat transfer involved, that is in some respects more informative though not, perhaps, so precise.

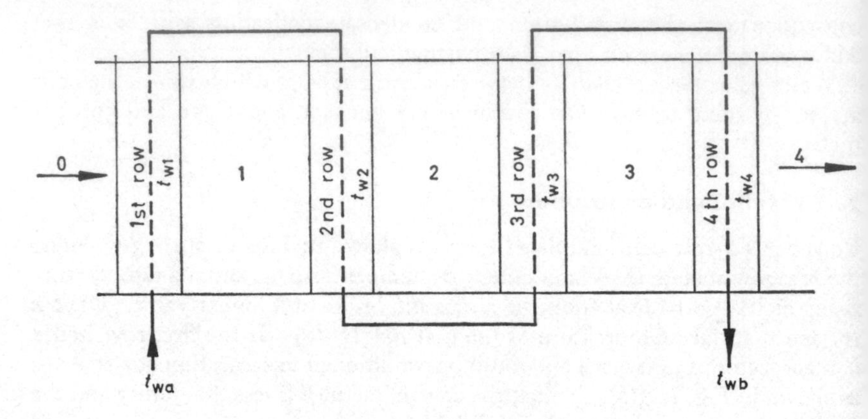

Fig. 10.3(*a*) Parallel flow

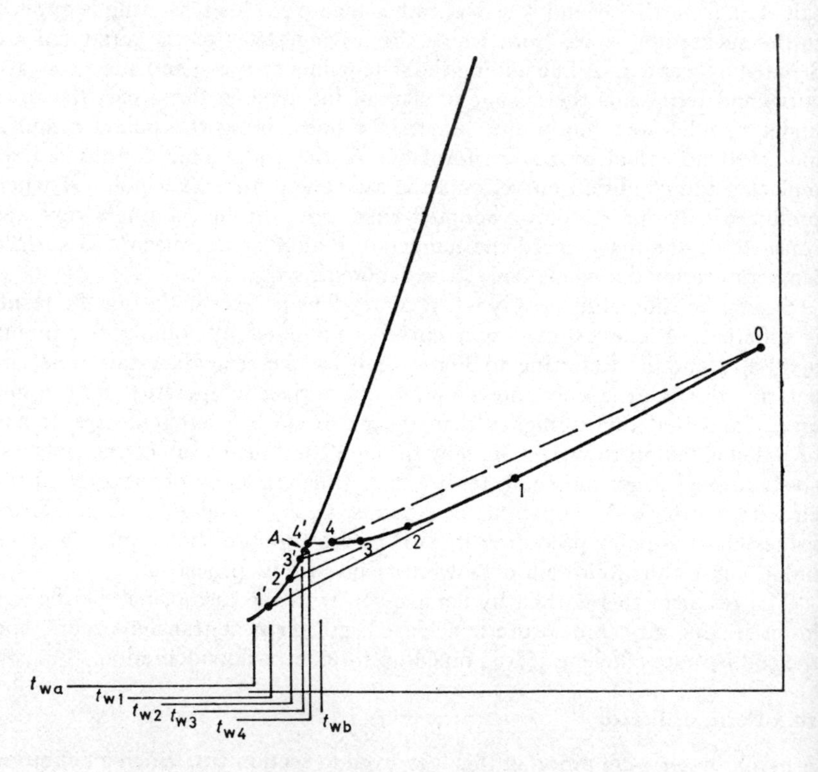

Fig. 10.3(*b*) Parallel flow

The points O, 4 and A are in a straight line. A is the apparatus dew point and its temperature is the mean coil surface temperature, t_{sm}.

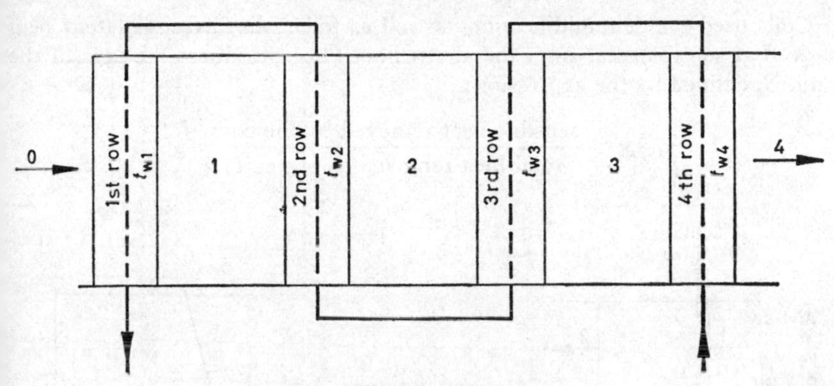

Fig 10.4(*a*) Contra-flow

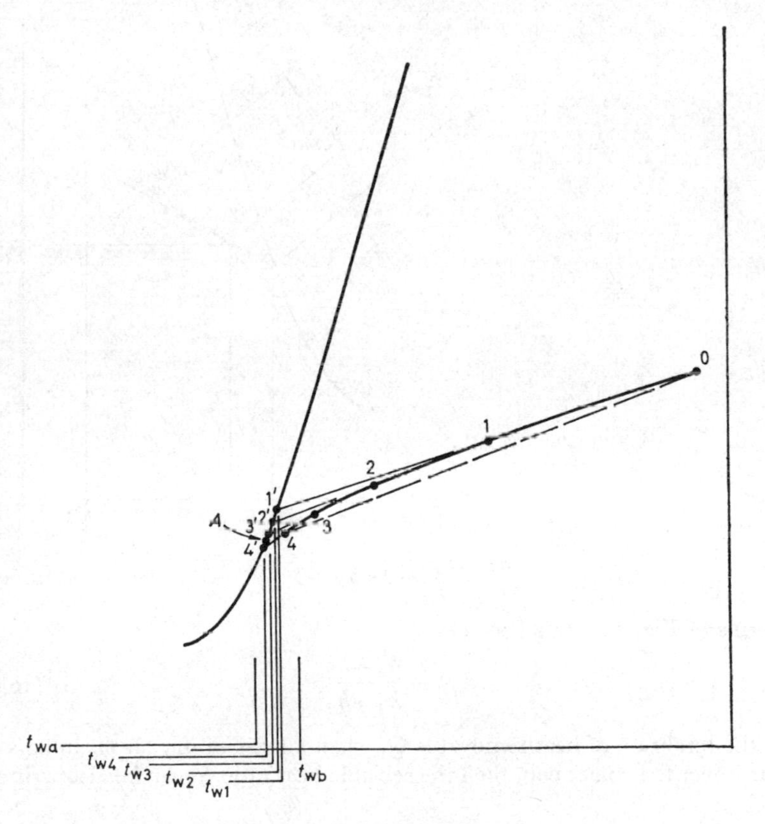

Fig. 10.4(*b*) Contra-flow

The points *O*, 4 and *A* are in a straight line. *A* is the apparatus dew point and its temperature is the mean coil surface temperature, t_{sm}.

Coils used for dehumidification as well as for cooling remove latent heat as well as sensible heat from the airstream. This introduces the idea of the ratio S, defined by the expression:

$$S = \frac{\text{sensible heat removed by the coil}}{\text{total heat removed by the coil}}.$$

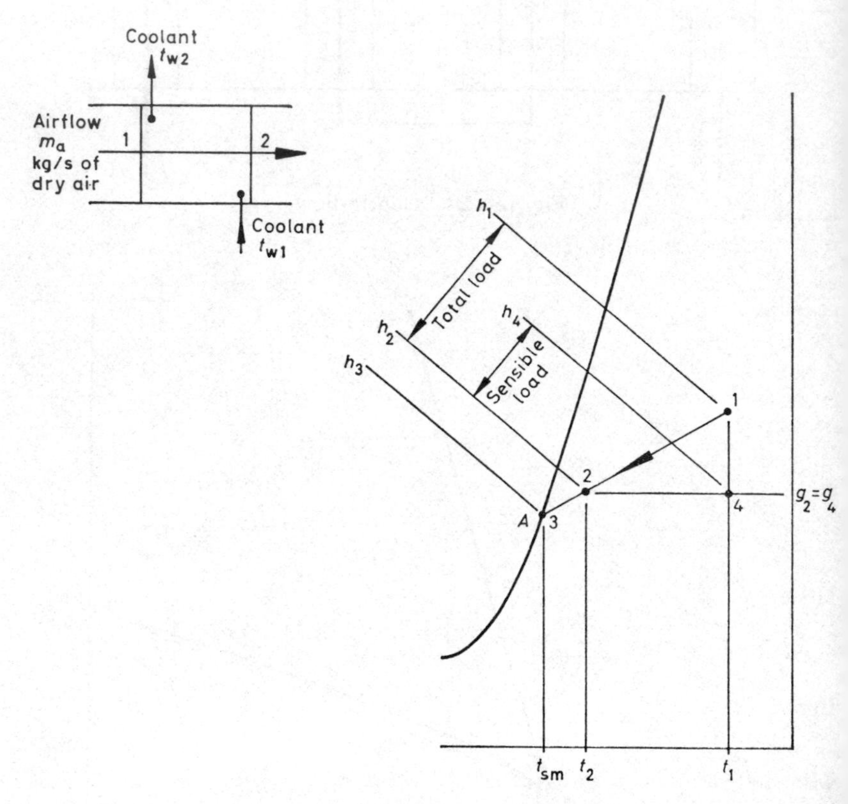

Fig. 10.5

In terms of Fig. 10.5 this becomes

$$S = \frac{h_4 - h_2}{h_1 - h_2} \tag{10.1}$$

If the total rate of heat removal is Q_t when a mass of dry air m_a in kg/s is flowing over the cooler coil, then the sensible heat ratio S, can be also written as

$$S = \frac{m_a c(t_1 - t_2)}{Q_t} \tag{10.2}$$

where c is the humid specific heat of the airstream.

Equation (6.12) gives an expression for sensible heat exchange which, when modified to the approximate form of eqn. (6.13), allows the expression for S to be re-written as

$$Q_t = \frac{1{\cdot}25 \times (\text{flow rate in m}^3/\text{s}) \times (t_1 - t_2)}{S} \tag{10.3}$$

The rate of transfer of sensible heat, Q_s, can also be considered in terms of the resistance R_a, of the surface film on the air side of the cooler coil

$$Q_s = \frac{A_t}{R_a} \times (LMTD) \tag{10.4}$$

where A_t is the total surface area of the coil in square metres and $LMTD$ is the logarithmic mean temperature difference between the airstream and the mean coil surface temperature t_{sm}.

$$LMTD = \frac{(t_1 - t_{sm}) - (t_2 - t_{sm})}{\log_e (t_1 - t_{sm})/(t_2 - t_{sm})} \tag{10.5}$$

Thus, eqn. (10.4) becomes

$$Q_s = \frac{A_t}{R_a} \cdot \frac{(t_1 - t_2)}{\log_e (t_1 - t_{sm})/(t_2 - t_{sm})} \tag{10.6}$$

and

$$1{\cdot}25 \times (\text{m}^3/\text{s}) \times (t_1 - t_2) = \frac{A_t}{R_a} \cdot \frac{(t_1 - t_2)}{\log_e (t_1 - t_{sm})/(t_2 - t_{sm})}$$

Hence,

$$\log_e (t_1 - t_{sm})/(t_2 - t_{sm}) = \frac{A_t}{R_a} \cdot \frac{1}{1{\cdot}25 \times (\text{m}^3/\text{s})} \tag{10.7}$$

But

$$(\text{m}^3/\text{s}) = A_f \, . \, v_f$$

where A_f is the face area of the coil in square metres and v_f is the face velocity in m/s.

Thus, eqn. (10.7) becomes

$$\log_e (t_1 - t_{sm})/(t_2 - t_{sm}) = \frac{A_t}{A_f} \cdot \frac{1}{R_a 1{\cdot}25 v_f}$$

$$= k.$$

k is a constant for a given coil and a given face velocity. It must be remembered, however, that it takes no account of the heat transfer through the water film on the inside of the tubes. Hence, one can write

$$\frac{(t_2 - t_{sm})}{(t_1 - t_{sm})} = e^{-k}$$

In approximate terms,

$$(1-\beta) = \frac{(t_2-t_{sm})}{(t_1-t_{sm})}$$

hence,

$$(1-\beta) = e^{-k}.$$

If r is the number of rows of the coil and if A_r is the total surface area per row, then $A_t = A_r r$ and

$$k = \frac{A_r}{A_f} \cdot \frac{r}{R_a 1 \cdot 25 v_f}$$

An expression for the contact factor now emerges:

$$\beta = 1 - \exp\left(-\frac{A_r}{A_f} \cdot \frac{r}{1 \cdot 25 R_a v_f}\right) \quad (10.8)$$

It is important to appreciate that the contact factor of a cooler coil is independent of the state of the air or the temperature of the coolant, provided that the ratio of air mass flow to water mass flow remains constant.

EXAMPLE 10.1. A four-row coil with a face velocity of 2·5 m/s has a contact factor of 0·85. Calculate the contact factor for the following cases:

(*a*) Face velocity 3·0 m/s, four rows.
(*b*) Face velocity 2·0 m/s, four rows.
(*c*) Face velocity 2·5 m/s, six rows.
(*d*) Face velocity 2·5 m/s, two rows.

Assume that changes in the face velocity have no significant influence on the value of R_a.

Answer

From the information given, and from eqn. (10.8) we can write

$$0 \cdot 85 = 1 - \exp\left(-\frac{A_r}{A_f} \cdot \frac{1}{R_a} \cdot \frac{4}{1 \cdot 25 \times 2 \cdot 5}\right)$$

$$0 \cdot 15 = \exp\left(-1 \cdot 28 \times \frac{A_r}{A_f R_a}\right)$$

$$\log_e 0 \cdot 15 = \frac{A_r}{A_f R_a} \times (-1 \cdot 28)$$

$$\frac{A_r}{A_f R_a} = \frac{-1 \cdot 8971}{-1 \cdot 28} = 1 \cdot 482$$

This constant can now be used in eqn. (10.8) as follows:

(*a*)
$$\beta = 1 - \exp\left(-1{\cdot}482 \times \frac{4}{1{\cdot}25 \times 3{\cdot}0}\right)$$
$$= 1 - \exp(-1{\cdot}581) = 0{\cdot}80.$$

(*b*)
$$\beta = 1 - \exp\left(-1{\cdot}482 \times \frac{4}{1{\cdot}25 \times 2.0}\right)$$
$$= 1 - \exp(-2{\cdot}37) = 0{\cdot}91.$$

(*c*)
$$\beta = 1 - \exp\left(-1{\cdot}482 \times \frac{6}{1{\cdot}25 \times 2{\cdot}5}\right)$$
$$= 1 - \exp(-2{\cdot}85) = 0{\cdot}94.$$

(*d*)
$$\beta = 1 - \exp\left(-1{\cdot}482 \times \frac{2}{1{\cdot}25 \times 2{\cdot}5}\right)$$
$$= 1 - \exp(-0{\cdot}95).$$
$$= 0{\cdot}61.$$

10.5 The U-value for cooler coils

Heat transfer to a cooler coil involves three stages: heat flows from the air-stream to the outer surface of the fins and pipes, it is then transferred through the metal of the fins and the wall of the piping and, finally, it passes from the inner walls of the tubes through the surface film of the cooling fluid to the main stream of the coolant.

The heat transfer coefficient on the air side, h_a, expressed in W/m^2 °C depends largely on the velocity of massflow of the airstream. For cooler coils having a staggered tube arrangement and the usual disposition of fins (300 fins/m), the values in Table 10.1 are typical.

TABLE 10.1

v_f (m/s)	1·0	1·5	2·0	2·5	3·0	3·5
h_a (W/m^2°C):	35	45	54	62	69	75

(Reproduced by kind permission from "Air Conditioning, Heating and Ventilating")

The figures quoted above are applied to the total external surface area of the cooler coil, not just to the face area, but they can be modified to do so if this is convenient. The values must be increased by dividing by the value of S (*see* eqn. 10.1), if the coil is dehumidifying as well as cooling.

If a coil has bare tubes, the value of the thermal resistance of the metal is negligible. All practical cooler coils, however, have fins, and under these circumstances the resistance is not to be ignored. The thermal resistance of metal-finned tubes is approximately 0·0009 to 0·005 m^2 °C/W, applied to the total net external surface area.

Heat transfer coefficients for the flow of heat through the surface film of cooling fluid within the tubes depends on whether chilled water or evaporating refrigerant is used. The presence of a coating of dirt or scale inside the tubes also has some bearing, although this is more relevant to shell-and-tube evaporators and to shell-and-tube condensers than it is to cooler coils. In condensers receiving cooling water from a cooling tower it is customary to allow a value of 0·000176 m² °C/W to cover this effect. With evaporators, which usually handle clean water in a closed circuit with a cooling coil, the fouling factor taken is usually 0·00088 m² °C/W. Needles to say, if it is a refrigerant that is being evaporated or dondensed, the tubes should be very clean indeed on the refrigerant side, hence no allowance is made for a fouling factor on this side.

TABLE 10.2

Water velocity m/s	Approximate internal tube diameter			
	15 mm		25 mm	
	5° C	15°C	5°C	15°C
0·4	0·00064	0·00050	0·00069	0·00064
0·5	0·00053	0·00042	0·00057	0·00053
0·75	0·00037	0·00029	0·00042	0·00036
1·00	0·00028	0·00022	0·00033	0·00027
1·25	0·00023	0·00019	0·00027	0·00022
1·50	0·00021	0·00017	0·00023	0·00020
1·75	0·00018	0·00015	0·00020	0·00018
2·00	0·00016	0·00014	0·00018	0·00016
2·25	0·00015	0·00013	0·00017	0·00015
2·50	0·00014	0·00012	0·00016	0·00014

The above values of thermal resistance are in m² °C/W and are applied to the internal surface area of the tube. To apply them directly to the external surface area they must be multiplied by the ratio of external to internal surface area.

(Reproduced by kind permission from "Air Conditioning, Heating and Ventilating")

A considerable amount of experimental evidence has been accumulated for the heat transfer occurring when fluids pass through pipes. It appears both from this and from theoretical considerations, that the major factor of influence is the velocity of flow through the tubes. Temperature plays a minor role, as Table 10.2 shows.

A *U*-value can now be defined for a cooler coil:

$$\frac{1}{U_t} = (R_a + R_m + R_w) \tag{10.9}$$

The sub-script 't' denotes that the U-value is applied to the total external surface area of the cooler coil. This is necessary since some authorities choose to refer the U-value to the face area. R_m denotes the thermal resistance of the metal (tube wall and fins) and R_w denotes the resistance of the film of cooling fluid, be it chilled water or evaporating refrigerant.

An expression for the total heat transfer through the air film, the metal of the tubes and fins, and the coolant film, can now be written:

$$Q_t = U_t \times A_t \times (LMTD) \qquad (10.10)$$

The value of the logarithmic mean temperature difference to be used in this expression is not to be confused with that given by eqn. (10.5), which referred to heat exchange between the airstream and a mean coil surface temperature, t_{sm}. With eqn. (10.10) the heat exchange is between air entering at a temperature t_1 and leaving at temperature t_2, and coolant entering at temperature t_{w1} and leaving at temperature t_{w2}.

Thus,

$$LMTD = \frac{(t_1 - t_{w2}) - (t_2 - t_{w1})}{\log_e (t_1 - t_{w2})/(t_2 - t_{w1})} \qquad (10.11)$$

Combining eqns. (10.10) and (10.11) we have—

$$Q_t = U_t \times A_t \times \frac{(t_1 - t_{w2}) - (t_2 - t_{w1})}{\log_e (t_1 - t_{w2})/(t_2 - t_{w1})} \qquad (10.12)$$

EXAMPLE 10.2. A cooler coil is capable of cooling and dehumidifying air from 28°C dry-bulb, 19·5°C wet-bulb (sling), 55·36 kJ/kg, 0·8674 m^3/kg and 10·65 g/kg, to 12°C dry-bulb, 11·3°C wet-bulb (sling), 32·41 kJ/kg, 0·8178 m^3/kg and 8·062 g/kg. 4·75 m^3/s enters the coil, and chilled water is to be used to effect cooling with a temperature rise of 5·5°C. The coil offered by the manufacturers has the following relevant data:

Number of rows	6
Face velocity	2·64 m/s
Fins/metre	316
Fin width	37·5 mm, in the direction of airflow.
Face dimensions	1·5 m wide × 1·2 m high.
Tubes	15 mm O.D., 13·6 mm I.D., staggered in pitch, in the direction of airflow, at 3·75 mm centres.

Calculate a value for the entering water temperature, assuming all the external coil surface is wet with condensate.

Answer

Plotting the states of the entering and leaving air on a psychrometric chart shows that the contact factor is 0·906 and that the mean coil surface temperature is 10·35°C.

$$Q_t = \frac{4{\cdot}75 \times (55{\cdot}36 - 32{\cdot}41)}{0{\cdot}8674}$$

$$= 125{\cdot}5 \text{ kW}.$$

By eqn. (6.12), the sensible heat transfer is

$$Q_s = \frac{4{\cdot}75 \times (28-12) \times 358}{(273+28)}$$

$$= 90{\cdot}4 \text{ kw}$$

$$S = Q_s/Q_t$$

$$= 0{\cdot}72.$$

Number of fins $1{\cdot}5 \times 316 = 474$ over a width of $1{\cdot}5$ m.

Gross surface area of fins $(2 \times 474) \times (0{\cdot}0375 \times 1{\cdot}2)$ per row.

$$= 42{\cdot}65 \text{ m}^2 \text{ per row.}$$

Cross-sectional area of tubes $(\pi \times 0{\cdot}015^2)/4$ per tube.

$$= 0{\cdot}0001767 \text{ m}^2 \text{ per tube.}$$

Number of tubes $= (1{\cdot}2/0{\cdot}0375)$

$$= 32 \text{ tubes per row.}$$

Net surface area of fins $42{\cdot}65 - 32 \times 0{\cdot}0001767 \times 474 \times 2$

$$= 42{\cdot}65 - 5{\cdot}36$$

$$= 37{\cdot}29 \text{ m}^2 \text{ per row.}$$

External perimeter of one tube $(\pi \times 0{\cdot}015)$

$$= 0{\cdot}0471 \text{ m.}$$

External surface of 32 tubes $= 0{\cdot}0471 \times 1{\cdot}5 \times 32$

$$= 2{\cdot}26 \text{ m}^2 \text{ per row.}$$

(This ignores the fact that a very small part of the tube is shielded by the thickness of each fin.)

Total net external surface area $= (37{\cdot}29 + 2{\cdot}26) \times 6$

$$= 39{\cdot}55 \times 6$$

$$= 237{\cdot}3 \text{ m}^2.$$

Internal perimeter of one tube $= 0{\cdot}0136\,\pi$

$$= 0{\cdot}04275 \text{ m.}$$

Total net internal surface area $= 0{\cdot}04275 \times 1{\cdot}5 \times 32 \times 6$

$$= 12{\cdot}3 \text{ m}^2.$$

$$\frac{\text{External surface area}}{\text{Internal surface area}} = 237{\cdot}3/12{\cdot}3$$

$$= 19{\cdot}3.$$

The water splits into 32 distinct paths and then flows through the six rows of the coil in series. Hence, the mean velocity of waterflow inside the tubes is easily found:

Cross-sectional internal area of one tube $= (\pi \times 0{\cdot}0136^2)/4$

$= 0{\cdot}0001453 \text{ m}^2$.

Mass flow of water $= 125{\cdot}5 \text{ kW}/(4{\cdot}2 \text{ kJ/kg °C} \times 5{\cdot}5°)$

$= 5{\cdot}44 \text{ kg/s}$.

Volumetric flow of water $= 5{\cdot}44$ litre/s

$= 0{\cdot}00544 \text{ m}^3/\text{s}$.

Mean velocity of water flow $= 0{\cdot}00544/(0{\cdot}0001453 \times 32)$

$= 1{\cdot}17 \text{ m/s}$.

From Table 10.1, $h_a = 64 \text{ W/m}^2 \text{ °C}$, for sensible heat transfer only on the air side and a face velocity of airflow of 2·64 m/s.

For total heat transfer, $h_a = 64/S$

$= 64/0{\cdot}72$

$= 89 \text{ W/m}^2 \text{ °C}$.

Hence the resistance of the air-side film is

$$R_a = 1/h_a = 0{\cdot}011236 \text{ m}^2 \text{ °C/W}.$$

Take the resistance of the metal tube and finning as $0{\cdot}0035 \text{ m}^2 \text{ °C/W}$.

The value of R_w, the thermal resistance of the water film is found, from Table 10.2 by interpolation, to be $0{\cdot}00022 \text{ m}^2 \text{ °C/W}$ at 10°C and applied to the internal surface area of the pipe.

Then, applied to the external surface area of the cooler coil,

$$R_w = 19{\cdot}3 \times 0{\cdot}00022$$

$$= 0{\cdot}00425 \text{ m}^2 \text{ °C/W}.$$

By eqn. (10.9),

$$1/U_t = 0{\cdot}011236 + 0{\cdot}0035 + 0{\cdot}00425$$

$$= 0{\cdot}01899 \text{ m}^2 \text{ °C/W}$$

$$U_t = 52{\cdot}65 \text{ W/m}^2 \text{ °C}.$$

By eqn. (10.10),

$$LMTD = Q_t/(U_t \times A_t)$$

$$= 125\,500/(52{\cdot}65 \times 237{\cdot}3)$$

$$= 10{\cdot}03\text{°C}.$$

By eqn. (10.11),

$$10 \cdot 03 = \frac{(28-t_{w2})-(12-t_{w1})}{\log (28-t_{w2})/(12-t_{w1})}$$

Because $t_2 = t_{w1}+5 \cdot 5^\circ$,

$$10 \cdot 03 = \frac{(22 \cdot 5-t_{w1})-(12-t_{w1})}{\log (22 \cdot 5-t_{w1})/(12-t_{w1})}$$

Hence,

$$\log (22 \cdot 5-t_{w1})/(12-t_{w1}) = 10 \cdot 5/10 \cdot 03 = 1 \cdot 046$$

Thus, $$(22 \cdot 5-t_{w1})/(12-t_{w1}) = 2 \cdot 85$$

and so $t_{w1} = 6 \cdot 32$°C and $t_{w2} = 11 \cdot 82$°C.

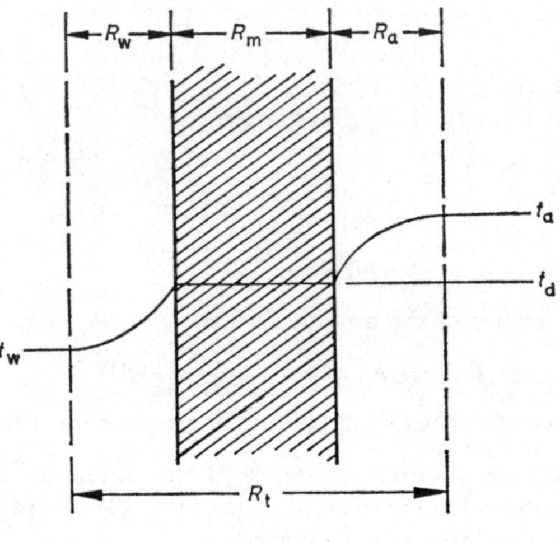

Fig. 10.6

10.6 Sensible cooling

No part of the external surface of a cooler coil may be at a temperature which is less than the dew point of the airstream if dehumidification is to be avoided. There is thus a minimum permissible value for the temperature of the chilled water or evaporating refrigerant inside the coil. Figure 10.6 shows the temperature variation across the wall of a tube through which chilled water is flowing. If t_a is the dry-bulb temperature of the airstream and t_d is its dew point, then we can write a set of equations expressing the heat transfer taking place:

$$Q_t = (t_a - t_d)/R_a \quad (10.13a)$$

$$= (t_d - t_w)/R_w \quad (10.13b)$$

$$= (t_a - t_w)/R_t \quad (10.13c)$$

Regardless of the presence of finning, the outside surface of the tube must not, at any place, be less than the dew point temperature. If the thermal resistance of the metal of the tube wall itself is ignored, there is no change of temperature across the wall and the gradient is as shown in Fig. 10.6. If fins are present, a lot more heat is removed from the airstream than if the tubes are bare. This has the effect of raising the temperature of the outer surface, both for the fins and for the tube. The lowest external surface temperature is then appreciably greater than the inner surface temperature, and the gradient is more complicated than as shown in Fig. 10.6. Under these circumstances only eqns. (10.13a) and (10.13c) apply. (Because the inner wall surface temperature no longer equals t_d.)

Combining the two relevant equations, we can obtain an expression for the minimum permissible coolant temperature:

$$t_w = t_a - (t_a - t_d)(R_t/R_a) \tag{10.14}$$

EXAMPLE 10.3. If the coil in Example 10.2 is used to cool 4·75 m³/s at 28°C dry-bulb, 19·5°C wet-bulb (sling) to 21°C dry-bulb, without dehumidification taking place, what is the value of the minimum permissible chilled water temperature?

Answer

The dew point of the entering airstream is 14·9°C. The lowest dry-bulb temperature of the airstream is 21°C, and so this will be the critical value to use for t_a in eqn. (10.14). From Example 10.2,

$$\begin{aligned} R_t &= R_a + R_m + R_w \\ &= 1/64 + 0{\cdot}0035 + 0{\cdot}00425 \\ &= 0{\cdot}015625 + 0{\cdot}0035 + 0{\cdot}00425 \\ &= 0{\cdot}0234 \text{ m}^2\ {}^\circ\text{C/W}. \end{aligned}$$

Then, by eqn. (10.14):

$$\begin{aligned} t_w &= 21 - (21 - 14{\cdot}9)(0{\cdot}0234/0{\cdot}0156) \\ &= 21 - 9{\cdot}15 \\ &= 11{\cdot}85^\circ\text{C}. \end{aligned}$$

Figure 10.7 illustrates the psychrometric changes. It is seen that if the condition curve 1–2 is produced to intersect the saturation curve, an apparatus dew point of 14·9°C is obtained. This implies that the contact factor, β, is

$$\begin{aligned} \beta &= (28 - 21)/(28 - 14{\cdot}9) \text{ approximately}, \\ &= 0{\cdot}535. \end{aligned}$$

This is a good deal less than the value of 0·906 obtained in Example 10.2 and it appears to contradict the statement in section 3.4 that, for a given ratio of air to water flow, the contact factor is independent of the state of the air or the temperature of the coolant. In fact, it is wrong to assume that the intersection of the condition curve with the saturation line gives the mean coil surface temperature for the case of sensible cooling. A moment's thought verifies this: if the temperature (t_4) at the intersection were the mean coil surface temperature, then it would imply the existence of a lower surface temperature, and this, in its turn, being less than the dew point of the entering

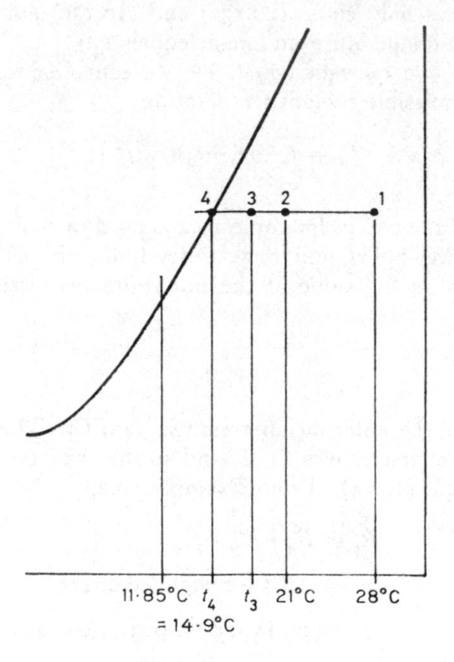

Fig. 10.7

airstream (also t_4), would result in dehumidification. Thus, the mean coil surface temperature cannot easily be determined by construction on a psychrometric chart unless, of course, the value of the contact factor is already known. It can, however, be calculated.

EXAMPLE 10.4. Determine the mean coil surface temperature for the coil in Example 10.3 (*a*) by calculation and, (*b*) by construction on the psychrometric chart, given that the contact factor is 0·906.

Answer

(*a*) Adopting the notation of Fig. 10.7, in which the mean coil surface temperature is denoted by t_3, the following relationships hold.

By eqn. (10.4) for the flow of heat through the air side of the cooler coil

$$Q_s = (A_t/R_a) \times (LMTD)$$

and, from Example 10.3,

$$A_t/R_a = 237{\cdot}3/0{\cdot}015625 = 15180\ °C/W.$$

$$\text{Heat lost by the airstream} = \frac{4{\cdot}75 \times (28-21) \times 358}{(273+28)}$$

$$= 39{\cdot}55\ \text{kW}.$$

Hence,

$$39\,550 = 15\,180 \times (LMTD)$$

$$(LMTD) = 2{\cdot}61°C.$$

But, by eqn. (10.5),

$$(LMTD) = \frac{(28-21)}{\log_e (28-t_3)/(21-t_3)}$$

Hence, $\log_e (28-t_3)/(21-t_3) = 7/2{\cdot}61 = 2{\cdot}69$

and $(28-t_3)/(21-t_3) = 14{\cdot}73.$

Thus, $t_3 = 20{\cdot}4°C.$

(*b*) Plotting points 1 and 2 on a psychrometric chart is straightforward, but use must be made of the approximate equation for contact factor to evaluate t_3:

$$0{\cdot}906 = (28-21)/(28-t_3)$$

$$t_3 = 20{\cdot}3°C.$$

For a given removal of sensible heat, the flow rate determines the temperature rise suffered by the water. This, in turn, affects the value of the mean coil surface temperature. So it does not follow that the six-row coil in the example will give the required leaving air state, if the rate of water flow is kept constant. In fact it will not and, if the rate is kept constant, the number of rows required to effect the heat removal specified must be reduced.

EXAMPLE 10.5. Using the data in Examples 10.3 and 10.4, calculate the number of rows necessary for cooling $4{\cdot}75\ m^3/s$ from 28°C to 21°C, without dehumidification, if the rate of water flow is to remain at $5{\cdot}44$ kg/s.

Answer

Water temperature rise $= 39{\cdot}55/4{\cdot}2 \times 5{\cdot}44$

$= 1{\cdot}73°C$

Entry water temperature $= 11{\cdot}85°C$ (*see* Example 10·3)

Exit water temperature $= 13{\cdot}58°C$

Entry air temperature $= 28°C$

Exit air temperature $= 21°C$

Hence,

$$(LMTD) = \frac{(28-13\cdot58)-(21-11\cdot85)}{\log_e (28-13\cdot58)/(21-11\cdot85)}$$
$$= 5\cdot27/\log_e 1\cdot576 = 5\cdot27/0\cdot45$$
$$= 11\cdot59^\circ C.$$

If A_r is the area of one row of the cooler coil and r is the number of rows, then

$$Q_s = U \times A_r \times r \times (LMTD), \text{ and}$$
$$1/U = 1/64 + 0\cdot0035 + 0\cdot00425$$
$$= 0\cdot0234 \text{ m}^2 \text{ °C/W}$$
$$U = 42\cdot73 \text{ W/m}^2 \text{ °C.}$$

(This should be compared with the value of $52\cdot65$ W/m² °C obtained for a wet coil. It demonstrates that U-values are larger if the coil surface is wet with condensate.)

Hence, since the value of A_r is $237\cdot3/6 = 39\cdot55$ m²,

$$r = 39\,550/(42\cdot73 \times 39\cdot55 \times 11\cdot59)$$
$$= 2\cdot02 \text{ rows.}$$

Clearly, if the duty specified is to be achieved, two rows must be used, the rate of water flow being somewhat throttled in order to obtain a leaving air dry-bulb of 21°C.

10.7 The maximum temperature for a completely wet coil

To secure an optimum U-value it may be necessary to ensure that the whole of the external surface area of a cooler coil is wet with condensate. Equation (10.14) provides the answer to this. Instead of using the lowest value of the dry-bulb temperature (t_a) of the airstream, the highest value must be used, that is, at the entering condition. This ensures that even at the beginning of the coil condensation will be occurring. As the airstream penetrates deeper into the coil it encounters lower surface temperatures, since the coil will, of course, be connected for contra-flow heat transfer.

EXAMPLE 10.7. If the coil in Example 10.2 must be completely wetted by condensate, what is the highest permissible chilled water temperature?

Answer

The critical value of t_a is now 28°C, hence, by eqn. (10.14),

$$t_w = 28-(28-14\cdot9)(0\cdot01899/0\cdot011236)$$
$$= 28^\circ - 22\cdot1^\circ$$
$$= 5\cdot9^\circ C.$$

This, being a leaving water temperature, implies that the coil in Example 10.2 is not entirely wet with condensate. Hence the value of t_a chosen should have been modified and the answer obtained is not strictly correct.

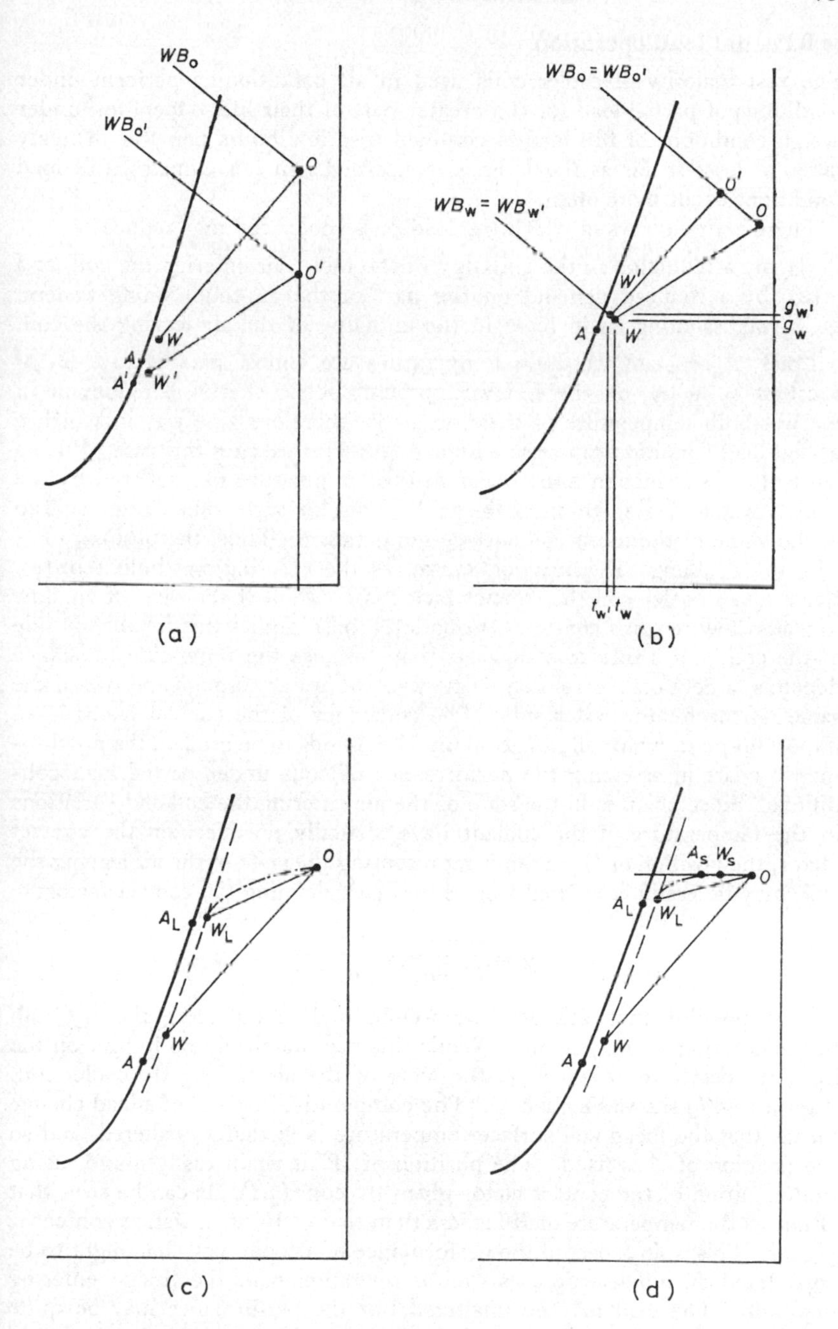

WB_o
$WB_{o'}$
O
O'
W
A
A'
W'
(a)
$WB_o = WB_{o'}$
$WB_w = WB_{w'}$
O'
O
W'
A
W
$g_{w'}$
g_w
$t_{w'}$ t_w
(b)
O
A_L
W_L
W
A
(c)
A_s W_s
O
A_L
W_L
W
A
(d)

Fig. 10.8

10.8 Partial load operation

The vast majority of cooler coils used in air conditioning perform under conditions of partial load for the greater part of their life. Operation under design conditions of full load is confined to a few hours per year in many cases, at least so far as the U.K. is concerned. In hot climates, full load conditions occur more often.

There are two ways in which the load on a cooler coil may reduce:

(i) by a reduction in the enthalpy of the moist air entering the coil, and
(ii) by a reduced demand on the part of the air-conditioning system, necessitating an increase in the enthalpy of the air leaving the coil.

Lines of constant wet-bulb temperature are almost parallel to lines of constant enthalpy, on the I.H.V.E. psychrometric chart. A reduction in the wet-bulb temperature of entering air is, therefore, one way in which a partial load condition can arise. Figure 10.8(*a*) illustrates this case. With a fall in load a reduction must occur in the temperature rise suffered by the chilled water passing through the coil. This, in turn, must mean a drop in the value of the mean coil surface temperature. Thus, the position of A falls to A', along the saturation curve, as the entering wet-bulb reduces. For a given cooler coil the contact factor is constant if the ratio of air flow to water flow remains constant. Equation (10.8) implies this for the air side of the coil, and Table 10.2 suggests that, because the water-side resistance depends largely on the velocity of flow of the water through the tubes, the same is true for the water side. The constancy of the contact factor is an important point which allows geometrical methods to be used on the psychrometric chart in assessing the performance of coils under partial load conditions. Since changes in the state of the air entering the coil and variations in the temperature of the coolant have virtually no effect on the contact factor, the position of W', a point representing the state of the air leaving the coil, may be easily calculated from eqn. (3.1), defining the contact factor:

$$\beta = \frac{h_o - h_w}{h_o - h_a} = \frac{h_{o'} - h_{w'}}{h_{o'} - h_{a'}}$$

It is possible that although the wet-bulb does not alter, the dry-bulb of the entering air may reduce. While this has practically no impact on the load, it does have an effect on the state of the air leaving the cooler coil. Figure 10.8(*b*) shows such a case. The comparative absence of a load change means that the mean coil surface temperature is virtually unaltered, and so the position of A is fixed. The position of W' is again easily found, using the definition of the contact factor given by eqn. (3.1). It can be seen that although the temperature of W' is less than that of W, its moisture content is higher. This is an aspect of the performance of a cooler coil that ought to be considered during selection, as well as operation with the design entering dry-bulb. The load may be unaltered but the performance may be quite unsatisfactory under such reduced dry-bulb conditions if this has not been taken account of.

As was discussed in Chapter 8, a reduction in the sensible or latent heat gains in the conditioned space may require the supply of air at a state different from its design value. For example, if the latent heat gains fall off (the sensible heat gains remaining unaltered), the moisture content of the supply air must be elevated if the relative humidity in the room is to be kept constant. (This would be coupled with the supply of air at the same dry-bulb temperature, achieved by means of a re-heater, as described in § 8.3.) There are two ways in which the state of the air leaving a cooler coil using chilled water may be altered: (i) by varying the rate of water flow and (ii) by changing the temperature of the chilled water flowing on to the coil.

If the state of the air entering a coil remains fixed but the rate of flow of the chilled water is reduced, the temperature rise suffered by the water increases. The mean coil surface temperature therefore goes up and the position of A rises up the saturation curve. At first, when this happens, the value of the contact factor stays virtually unaltered, little variation in the value of R_w taking place. So it is possible to draw a set of condition lines O–W, all having the same contact factor, β. In due course a limiting point, A_L, is reached. Any further reduction in the flow rate of the water causes such an increase in the value of R_w that the value of β cannot any longer be regarded as constant. The states of the air leaving the coil can be joined by a broken line from W to W_L, as shown in Fig. 10.8(*c*). Above the limiting state, W_L, the value of β falls away and the locus of the air-leaving state is represented in the figure by the chain-dotted line which runs from W_L to O. This last part of the locus is never parallel to a line of constant moisture content since, the entering water temperature being assumed constant, some small amount of dehumidification always takes place.

A variation in the temperature of the chilled water flowing on to a coil, the flow rate of water being constant, produces a condition rather similar to the one just considered. A progressive increase in the value of the temperature of the chilled water causes the mean coil surface temperature to rise and A moves up the saturation curve to a limiting position at A_L, as in Fig. 10.8(*d*). Before this point the contact factor may be determined by geometrical methods on the psychrometric chart but afterwards it may not, although the value of β remains constant. As was mentioned in section 10.6, a chilled water temperature is eventually reached which results in the coil executing sensible cooling only. Under these conditions, air leaves the coil at state W_s and the locus from this point to O is along a line of constant moisture content.

10.9 The performance of a wild coil

When no control is exercised over either the temperature or the flow rate of the coolant, a coil is termed 'wild'. Such an operation may be quite satisfactory, and the method may have economic advantages, since the expense of three-way valves and thermostats is avoided. It is not quite true to assume that no control exists over the temperature of the coolant since, in a chilled water coil, the water chiller produces water at a nominally constant temperature. Figure 10.8(*a*) gives a picture of what happens: as the entering wet-bulb drops, the load on the coil falls, and the mean coil surface temperature reduces, the position of A moving down the saturation curve. However, this will not continue indefinitely. The value of the temperature of the water produced by the chiller is a limiting value. Under no-load conditions the temperature of A equals this value. So, provided the dew point of the air supplied to the conditioned space may be permitted to drop to such a low value, and that there is no objection to the refrigeration plant which produces the chilled water operating, the wild coil is acceptable. The economic question that must be answered is: will the saving in capital cost, by not using control valves and thermostats, be offset by the extra running cost of the refrigeration plant?

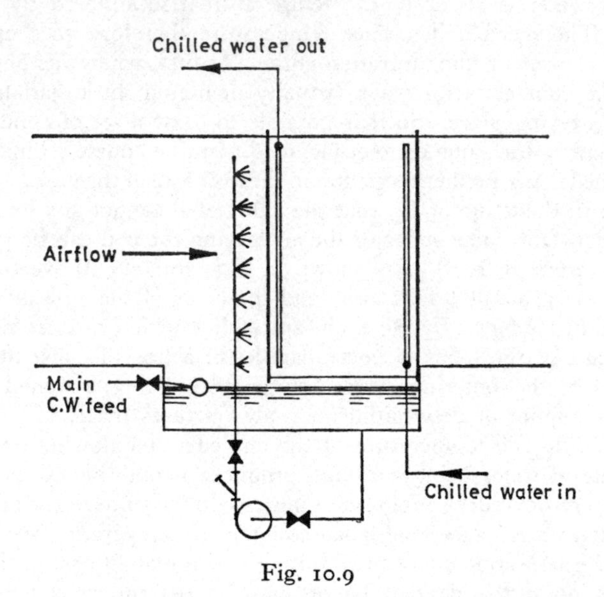

Fig. 10.9

10.10 Sprayed cooler coils

A sprayed cooler coil is a coil, fed with chilled water or liquid refrigerant in the usual way, positioned over a recirculation tank, with a bank of stand pipes and spray nozzles located a short distance from its upstream face. Mains water is fed to the tank through a ball-valve to make good any evaporative losses that may occur. Water is drawn from the tank by a pump and is delivered through the nozzles on to the face of the coil. This water then falls down the fins into the tank and is largely recirculated. Figure 10.9 illustrates this.

The prime function of the sprayed cooler coil is to provide humidification for operation of the air-conditioning plant in winter. A cooler coil offers a very large surface area to the airstream passing over it, and it is by wetting this very large area that the spray nozzles achieve humidification. The nozzles, therefore, should not have a very large pressure drop across them, since this is associated with an atomisation effect, but should be placed close enough to the face of the coil to ensure that all the water delivered by them goes on to the coil. Only one bank of nozzles is necessary and, since this faces downstream, the humidifying efficiency associated with an atomised spray is small: of the order of 50 per cent. On the other hand, if atomisation is not relied on, but the surface of the coil is properly wetted instead, a better humidifying efficiency can be obtained. Only tests carried out by the manfacturer in the works or by the user on site, can provide reasonably accurate

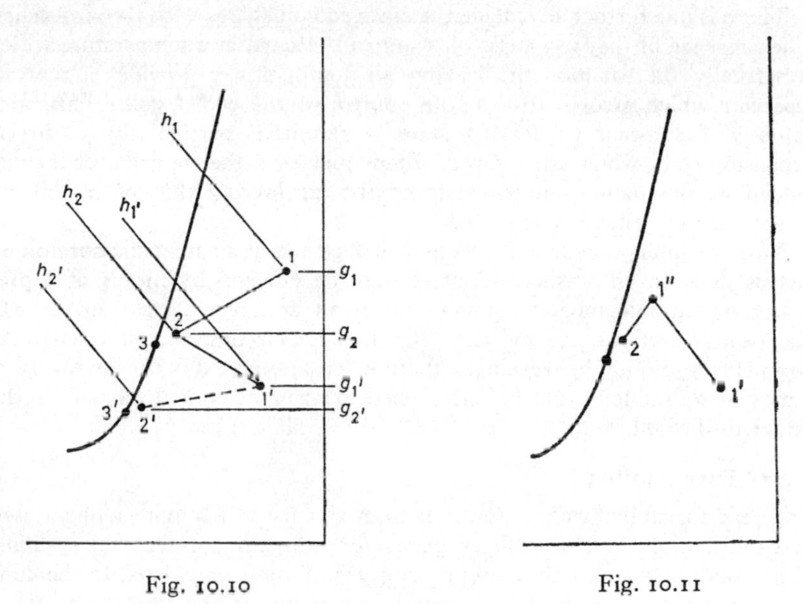

Fig. 10.10

Fig. 10.11

data as to the efficiencies likely. But, the contact factor of the coil itself provides some clue. If the coil were completely wetted throughout its depth, then it would perhaps be reasonable to take the humidifying efficiency as approaching the value of the contact factor. Since the coil is not completely wetted, the actual value will be less than this; a practical value is 80 per cent.

An effective distribution of water is about 0·82 litres per second on each square metre of face area of the coil. The pressure drop across the nozzles should be about 42 kN/m^2. If more pressure than this is lost little is gained in effectiveness, but a penalty is paid in extra running costs for the pump.

When chilled water flows through the inside of the coil and spray water is circulated over the outside of the coil, accurate control of dew point is possible for a wide range of entering air conditions. Figure 10.10 illustrates how this may be achieved. If the coil is not sprayed, and is chosen to cool and dehumidify air from state 1 to state 2 under design conditions in summer, it will produce a state 2′ if the entering air state is 1′, during autumnal or spring weather. This has a moisture content which is too low if control over dew point is necessary. If a sprayed coil is used, however, summer operation will, for all practical purposes, be the same as before but when the entering air state is 1′ it is now possible to produce a leaving state 2. For purposes of illustration the process can be regarded as one of adiabatic humidification from 1′ to 1″, followed by cooling and dehumidification to 2. This is shown in Fig. 10.11. So, as with an air washer, the sprayed cooler coil, when properly controlled can execute a process of cooling and humidification, as typified by the change 1′ to 2.

There is one further advantage the sprayed coil shares with the air washer. The presence of the large mass of water in the tank, at a temperature which is virtually the same as the leaving air temperature, provides a thermal reservoir which gives inertia to the control of the cooler coil. This is of value if a constant air leaving state is required, particularly in direct-expansion coils, which are sprayed. Inadequacies in the control over the coil output or fluctuations in the state of the air leaving the cooler coil are smoothed out. Stability is added.

During winter, of course, the wetted surface acts as an adiabatic humidifier, just as does an air washer. Control may be effected by means of a pre-heater or variable mixing dampers for fresh and recirculated air, as was discussed in sections 3.10 and 3.11. Under such circumstances the refrigeration plant would not be working, although, as is remarked in the next section, it may be desirable to run the pump used to circulate chilled water from the chiller to the coil.

10.11 'Free cooling'

The need for chilled water, even in winter, can arise with more sophisticated air-conditioning designs such as perimeter induction or fan coil systems, which achieve their conditioning by the use of water circulated to the conditioned spaces as well as air. Another instance is the treatment of the interior zones of office blocks which suffer no heat loss in winter but, instead, have a year-round heat gain from electric lighting and people. Such interior zones may use local subsidiary sensible cooler coils in their supply ducts. Chilled water could, of course, be provided by the refrigeration plant, even in the winter. However, it is often convenient and economical to take advantage of the fact that a sprayed cooler coil can, under suitable outside wet-bulb conditions (approximately below 4°C) provide the chilled water necessary.

Referring to Fig. 10.12(*a*): if air at state 1, with a wet-bulb temperature t'_1, flows over a sprayed cooler coil, and if the temperature of the water flowing through the inside of the tubes has also a value t'_1, then adiabatic saturation occurs and the change of state of the air is up a wet-bulb line. The air leaves at state 2—the exact position of which along the wet-bulb line depends on the humidifying efficiency—with the same wet-bulb temperature but with a dry-bulb temperature t_2, which is less than its entering value, in accordance with eqn. (2.22). Assuming that wet-bulb lines are parallel to lines of constant enthalpy, no heat exchange takes place and the water flowing inside the tubes leaves at the same temperature t'_1, at which it entered.

Referring to Fig. 10.12(*b*): if water is fed to the tubes at a constant temperature t_{w1}, less than t'_1, then heat transfer will take place from the air to

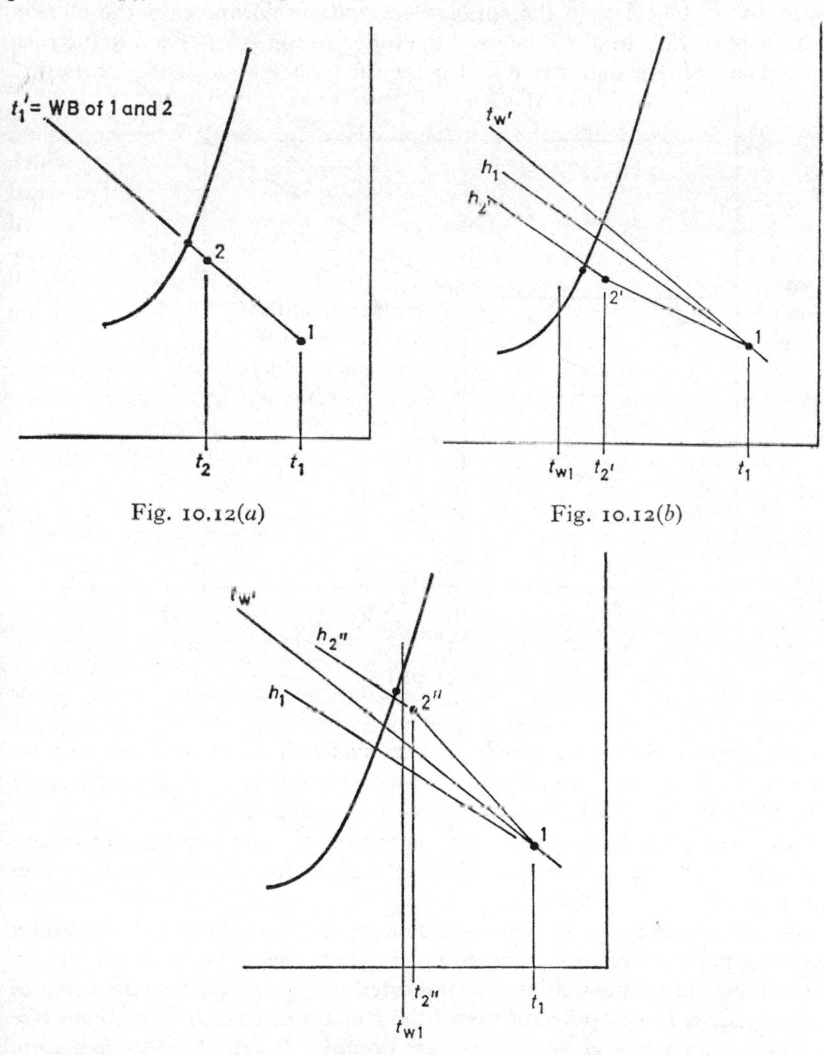

Fig. 10.12(*a*)

Fig. 10.12(*b*)

Fig. 10.12(*c*)

the water inside the tubes and adiabatic saturation will not occur. The cooling load will be the fall in enthalpy of the air flowing over the outside of the tubes, namely, $h_1 - h_2$, in kJ/kg of dry air. This must equal the heat gain to the water flowing inside the tubes, which will leave the coil at a temperature t_{w2}, equal to $t_{w1}+(h_1-h_2)/m_w$, where m_w is the mass flow of water inside the tubes in kg of water per kg of dry air flowing outside the tubes.

Referring to Fig. 10.12(*c*): in a similar way, if t_{w1} is greater than t'_1, the wet-bulb temperature of the entering airstream, heat will flow from the water inside the tubes to the spray water and air flowing over the outside of the tubes. This heat will be used to evaporate spray water and to increase the enthalpy of the airstream, unbalancing the process of adiabatic saturation.

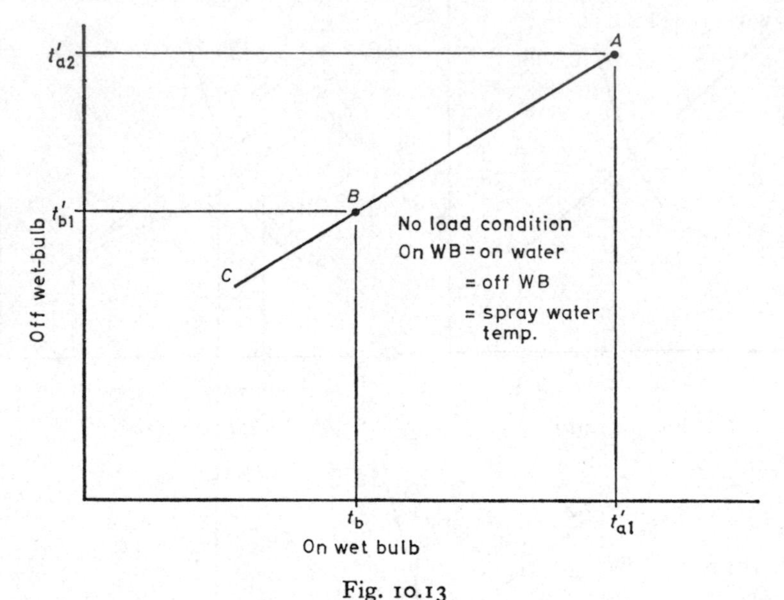

Fig. 10.13

This is the case of 'free cooling'. The sprayed coil is now acting as a heater battery and humidifier, and the water flowing inside the tubes is suffering a drop in temperature as it gives up energy to the airstream.

Free cooling also occurs with dry cooler coils but the heat transfer is very much less, the outer surface not being wet and, hence, no evaporative cooling occurring.

Although some rough idea of free cooling performance can be obtained by joining a point representing the summer design operation with that taking place under no-load conditions, as indicated in Fig. 10.13, a straight line of this sort tends to be optimistic about the amount of free cooling obtainable. In the figure, point *A* represents the summer design state of operation. Air enters the sprayed coil at a wet-bulb temperature t'_{a1} and leaves at a wet-bulb temperature t'_{a2}. To effect this heat exchange water at a temperature t_{w1} enters the coil and leaves at a value of t_{w2}. A heat balance establishes the value of t_{w2}. The no-load point of operation is at the point *B*. Here, air enters at a wet-bulb temperature t'_{b1}, adiabatic saturation takes place, and the air leaves at the same temperature. There is therefore no change in the water temperature. Between the points *A* and *B* a straight line is probably a reasonably accurate representation of the behaviour of the coil. Beyond the point *B*, say at *C*, where free cooling is occurring, this is no longer true. As was remarked, the graph is optimistic.

10.12 Direct-expansion coils

The heat transfer principles already discussed, apply equally to direct-expansion cooler coils, with little modification. The outstanding difference is that the value of the inside film resistance is changed; a liquid is boiling within the tubes of a direct-expansion coil whereas the fluid flowing through a chilled water coil suffers an increase of sensible heat. If the symbol R_r replaces R_w in eqn. (10.9), a *U*-value can be calculated and used as it was for chilled water coils. Table 10.3 gives values for R_r, in m^2 °C/W, in terms of the capacity of the refrigerant flowing through the tubes in any particular circuit, and in terms of the outside diameter (OD) of the tubes.

TABLE 10.3

Cooling capacity in kW/circuit	12 mm OD	15 mm OD	18 mm OD
1·75	0·00070	—	—
2·46	0·00047	—	—
2·81	—	0·00070	—
4·21	—	0·00044	0·00070
5·26	—	0·00035	—
5·98	—	—	0·00047

The temperature of the refrigerant inside the coil is substantially constant, except near the end of each circuit, where all the liquid refrigerant has been boiled to a gas and some superheating occurs. This is usually about 5°C but, being a sensible process, it does not represent a large amount of heat exchange compared with the latent heat exchanges that have occurred earlier in the circuit. An allowance is sometimes made for this effect when calculating the logarithmic mean temperature difference, by assuming that the outlet gas temperature is 1 or 2 degrees higher than the inlet liquid temperature.

Direct-expansion coils differ from chilled water cooler coils in one important respect: performance depends not only on the considerations of air and coolant flow mentioned, but also on the point of balance achieved between the characteristic of the coil's performance and the characteristic of the performance of the condensing set to which the coil is connected. This aspect of the matter is dealt with in section 12.5.

10.13 Air washers

A diagrammatic picture of an air washer is given in Fig. 10.14. Essentially, a washer consists of a spray chamber in which a dense cloud of finely divided spray water is produced by pumping water through nozzles. On its path through the chamber the air first passes an array of deflector plates or a perforated metal screen, the purpose of which is to secure a uniform distribution of air flow over the cross-section of the washer and to prevent any moisture accidentally being blown back up the duct. Within the spray

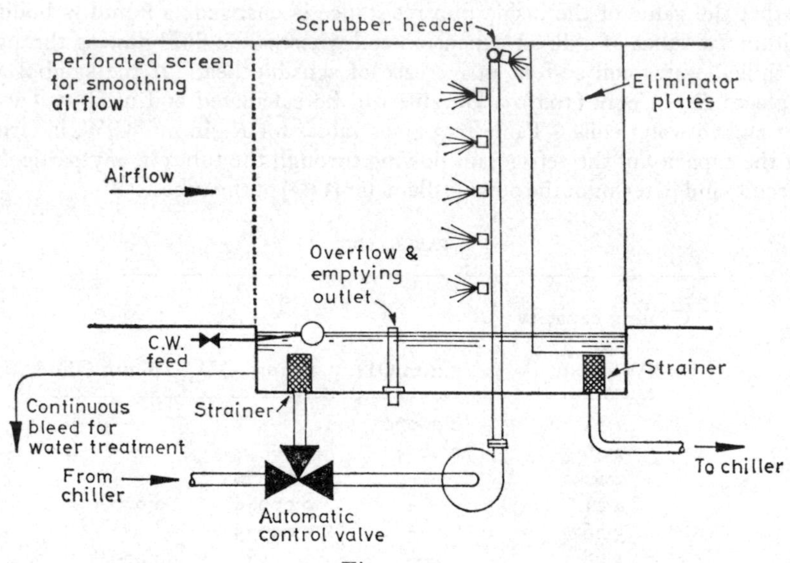

Fig. 10.14

chamber itself, the nozzles which atomise the water are arranged in banks, usually one or two in number, but occasionally three. The efficiency of the washer depends on how many banks are used and which way they blow the spray water, upstream or downstream. The banks consist of stand pipes with nozzles mounted so as to give an adequate cover of the section of the chamber. The arrangement is usually staggered, as shown in Fig. 10.15. Finally, the air leaves the washer by passing through a bank of eliminator plates, the purpose of which is primarily to prevent the carry-over of unevaporated moisture beyond the washer into the ducting system. A secondary purpose is to improve the cleaning ability of the washer by offering a large wetted surface on which the dirt can impinge and be washed away. In general, washers are poor filters; they should not be relied on alone to clean the air.

There is another type of washer, the capillary sort (Fig. 10.16), which achieves much the same ends (of humidification or dehumidification) by using wetted cells instead of banks of spray nozzles. The cells contain vast numbers of glass fibres, arranged parallel to one another and to the direction of air and water flow. The intimate contact between the air and the water

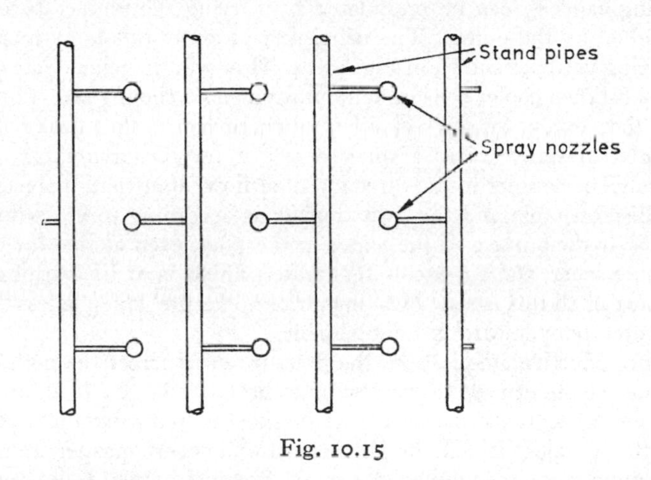

Fig. 10.15

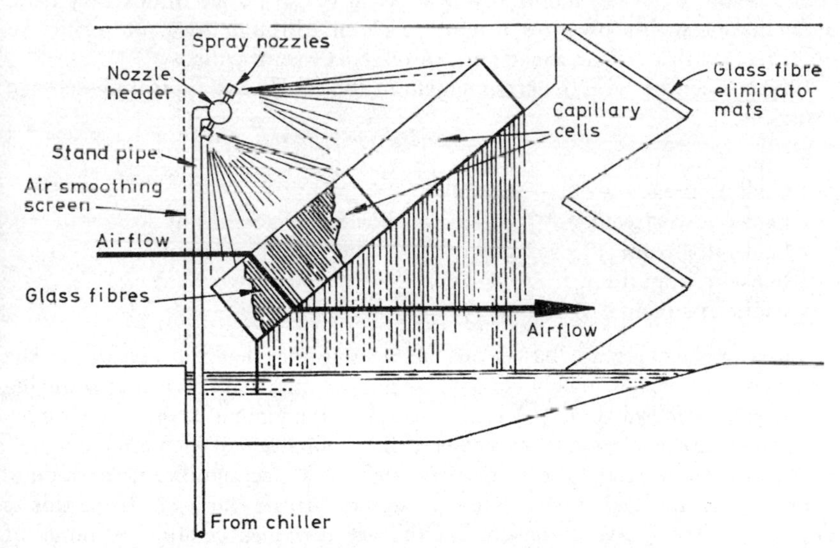

Fig. 10.16 Capillary air washer arranged for parallel-flow of air and water

Contra-flow is preferred for air-conditioning. Pumps, bleeds, overflows, etc., as for Fig. 10.14.

which this produces gives rise to a high humidifying efficiency, of the order of 98 per cent. It is claimed that this sort of filter cleans the air better than the more conventional type. Mats of glass fibres are used on the downstream end of the washer, to act as eliminators.

Provided the temperature of the water circulated is less than the dew point of the entering air, a washer can dehumidify, and its cooling and dehumidifying capacity can be modulated by varying the temperature of the water handled by the pump. The usual way of doing this is by means of a 3-port mixing valve, as shown in Fig. 10.14. However, in general, air washers are less useful than cooler coils in dehumidifying and cooling air. The reason for this is that, except for some capillary washers, contra-flow does not occur. As a droplet of water leaves a spray nozzle it travels for a short distance either parallel or counter to the direction of airflow, after that, it starts falling to the collection tank, and the heat transfer is according to cross-flow conditions. Near the surface of the water in the tank, even at the downstream end of the washer, the air encounters water which is at its warmest. The consequence of all this is that a leaving air temperature which is less than the leaving water temperature is unobtainable.

To secure effective atomisation, the pressure drop across the nozzles used in the spray-chamber type of washer must be fairly large. If an attempt is made to modulate the capacity of the washer by reducing the water flow through the nozzles, it will be unsuccessful, because proper atomisation ceases below a certain minimum pressure. Typical nozzle duties vary from 0·025 litres/s with a pressure drop of 70 kN/m^2 through a 2·5 mm orifice to 0·3 litres/s with a drop of 280 kN/m^2 through a 6·5 mm orifice.

Efficiencies vary with the arrangement of the banks, and typical percentage values are:

1 bank downstream	50	For face velocities of airflow over the section of 2·5 m/s.
1 bank upstream	65–75	
2 banks downstream	85–90	
2 banks upstream	92–97	
2 banks in opposition	90–95	
3 banks upstream	>95	

These efficiencies are based on the assumption that the section of the spray chamber is adequately covered by nozzles and that these are producing a properly atomised spray. A typical nozzle arrangement is 45/m^2 yielding a water flow rate of about 2 litres/s over each m^2 of section of each bank.

As with cooler coils, the velocity of airflow is also significant: a value of 2·5 m/s over the face area is usually chosen. Little variation from this is desirable, either way, if the eliminators are to be successful. A range of 1·75 m/s to 3·75 m/s has been suggested.

Good practice in the design and use of air washers for air conditioning requires a water quantity of 0·11 litres/s per nozzle and 22 nozzles per m^2 of cross-sectional area in each bank, with a mean air velocity of 2·5 m/s. Two banks, blowing upstream, are adequate, so the total circulated water quantity should be 0·44 litres/s m^2 of cross-section.

EXERCISES

1. (*a*) Define the term 'contact factor' as applied to cooler coils and state under what conditions it remains constant.

(*b*) Explain how the performance of a chilled water cooler coil varies under conditions of (i) varying water flow rate through the tubes, and (ii) varying water flow temperature.

(*c*) How is apparatus dew point, as applied to chilled water cooler coils, related to the flow and return temperatures of the chilled water passing along the tubes?

2. A sprayed cooler coil is chosen to operate under the following conditions:

Air on: 28°C dry-bulb, 20·6°C wet-bulb (sling), 28·3 m^3/s.
Air off: 12°C dry-bulb, 8·062 g/kg dry air.
Chilled water on: 5·5°C.
Chilled water off: 12·0°C.

You are asked to calculate (*a*) the contact factor, (*b*) the design cooling load and (*c*) the water flow rate through the coil.

Answers: 0·87, 868 kW; 31·8 kg/s.

3. An air-conditioning system comprises a chilled water cooler coil, re-heater and supply air fan, together with a system of distribution ductwork. Winter humidification is achieved by steam injection directly into the conditioned space. It is considered economic, from an owning and operating cost point of view, to arrange that a constant minimum quantity of outside air is used throughout the year. The fixed quantity of outside air is 25 per cent by weight of the total amount of air supplied to the room. Using the psychrometric chart provided and the data listed below, calculate

(*a*) The contact factor of the cooler coil.
(*b*) The design load on the cooler coil in summer.
(*c*) The load on the cooler coil under winter design conditions.
(*d*) The winter re-heat load.

Design Data

Sensible heat gain in summer	25 kW
Latent heat gain in summer	0·3 kW
Sensible heat loss in winter	4·4 kW
Outside state in summer	28°C dry-bulb, 19·5°C wetbulb.
Outside state in winter	−1°C saturated.
Inside state in summer and winter	21°C dry-bulb, 50 per cent saturation.
Supply temperature in summer design conditions	12·5°C dry-bulb.
Temperature rise due to fan power and duct heat gains	0·5°C

Assume that the ratio of latent to total heat removal by the cooler coil is halved under winter design conditions. Take the specific heat of dry air as 1·012 kJ/kg °C.

Answers: 0·83, 36·85 kW, 20·05 kW, 37·2 kW.

4. (*a*) Air at a constant rate and at a state of 25·5°C dry-bulb, 17·9°C wet-bulb (sling), flows on to a cooler coil and leaves it at 11°C dry-bulb, 10·7°C wet-bulb (sling), when the coil is supplied with a constant flow rate of chilled water at an adequately low temperature. Calculate the state of the air leaving the coil if the load is halved when the state of the air entering the coil changes to 20°C dry-bulb with 14·2°C wet-bulb (sling). Make use of the psychrometric chart provided.

(*b*) State briefly how the contact factor of a cooler coil alters when (i) the number of rows is reduced and (ii), the face velocity is reduced, assuming that the ratio of air mass flow to water mass flow is constant.

(*c*) Derive an expression for the minimum chilled water temperature which may be used with a sensible cooler coil, in terms of the dry-bulb temperature and the dew point of the airstream and the relevant thermal resistances.

Answers: (*a*) 10·7°C dry-bulb, 10·2°C wet-bulb, 30·02 kJ/kg, (*b*) (i) diminishes and (ii) increases.

5. Moist air at 28°C dry-bulb and 20·6°C wet-bulb flows over a 4-row cooler coil, leaving it at 9·5°C dry-bulb and 7·107 g/kg. Using the psychrometric chart provided, answer the following:

(*a*) What is the contact factor?

(*b*) What is the apparatus dew point?

(*c*) If the air is required to offset a sensible heat gain of 2·4 kW and a latent heat gain of 0·3 kW, in the space being conditioned, calculate the weight of dry air which must be supplied to the room in order to maintain 21°C dry-bulb therein.

(*d*) What is the percentage saturation in the room?

(*e*) If the number of rows of the coil is decreased to 2, the rate of waterflow, the rate of airflow and the apparatus dew point remaining constant, what temperature and humidity will be maintained in the room?

Ignore any temperature rise due to duct friction or heat gain.

Answers: (*a*) 0·925, (*b*) 8·1°C, (*c*) 0·204 kg/s, (*d*) 49 per cent, (*e*) 28°C dry-bulb and 40 per cent saturation.

BIBLIOGRAPHY

1. M. Fishenden and O. A. Saunders. *An Introduction to Heat Transfer*. Clarendon Press, Oxford, 1950.
2. W. H. Carrier, R. E. Cherne, W. A. Grant and W. H. Roberts. *Modern Air Conditioning, Heating and Ventilating*. Pitman, 1959.
3. L. C. Bull. Coils for cooling and dehumidifying air, *J.I.H.V.E.*, April 1959; Cooling coil performance, *ibid.*, February 1960.
4. J. K. Hardy, K. C. Hales and G. Mann. *The Condensation of Water on Refrigerated Surfaces*. H.M.S.O., 1951.
5. D. Q. Kern. *Process Heat Transfer*. McGraw-Hill, 1950.
6. W. F. Stoecker. *Refrigeration and Air Conditioning*. McGraw-Hill, 1958.
7. M. A. Ramsey. How to figure cooling coils—1, *Air Conditioning, Heating and Ventilating*, November, 1963.

11

The Rejection of Heat from Condensers and Cooling Towers

11.1 Methods of rejecting heat

All heat gains dealt with by an air-conditioning system must be rejected at the condenser. To accomplish this, the condenser must be water-cooled, evaporative-cooled or air-cooled.

Except in the comparatively rare cases where a supply of lake or river water may be drawn on, or the even rarer cases where mains water is available, the water used by a water-cooled condenser must be continuously recirculated through a cooling tower. Figure 11.1 illustrates what occurs. Water is pumped through the condenser and suffers a temperature rise of $(t_{w2}-t_{w1})$ as it removes the heat rejected by the refrigerant during the process of condensation. The cooling water then flows to the top of a cooling tower (an induced-draught type is illustrated) whence, in falling to a catchment tank at the bottom of the tower, it encounters airflow. Contra-flow heat exchange occurs, the water cooling by evaporation from t_{w2} to t_{w1} and the air becoming humid in the process. The heat lost by the water constitutes a gain of enthalpy to the air. The water is then ready for recirculation to the condenser at the desired temperature. Some water is necessary from the mains, to make good the evaporative losses from the system but, since 1 kg of water liberates about 2400 kJ during a process of evaporative cooling (whereas 1 kg rising through 5 degrees absorbs only 21 kJ), the amount of make up required is quite small, being of the order of 1 per cent of the circulation rate in this respect. There is some loss also due to the carry over of droplets by the emergent airstream but, even allowing for this, the loss is still little more than 1 per cent.

Evaporative condensers operate in a similar fashion but more directly. In Fig. 11.2 it can be seen that the pipe coils of the condenser are directly in the path of the air and water streams and that evaporative cooling takes place directly on the outer surface of the tubes.

With air-cooled condensers, no water is used at all. The condenser consists of a coil of finned tube, and over this air is drawn or blown by means of a propeller fan or, occasionally, a centrifugal fan. As with cooling towers and evaporative condensers, several fans may be used.

The evaporative condenser is probably the most efficient way of rejecting heat to the atmosphere, if it is properly designed. This is reflected in the typical air quantities required:

0·03 to 0·06 m^3/s kW for evaporative condensers,
0·04 to 0·08 m^3/s kW for cooling towers and
0·14 to 0·20 m^3/s kW for air-cooled condensers.

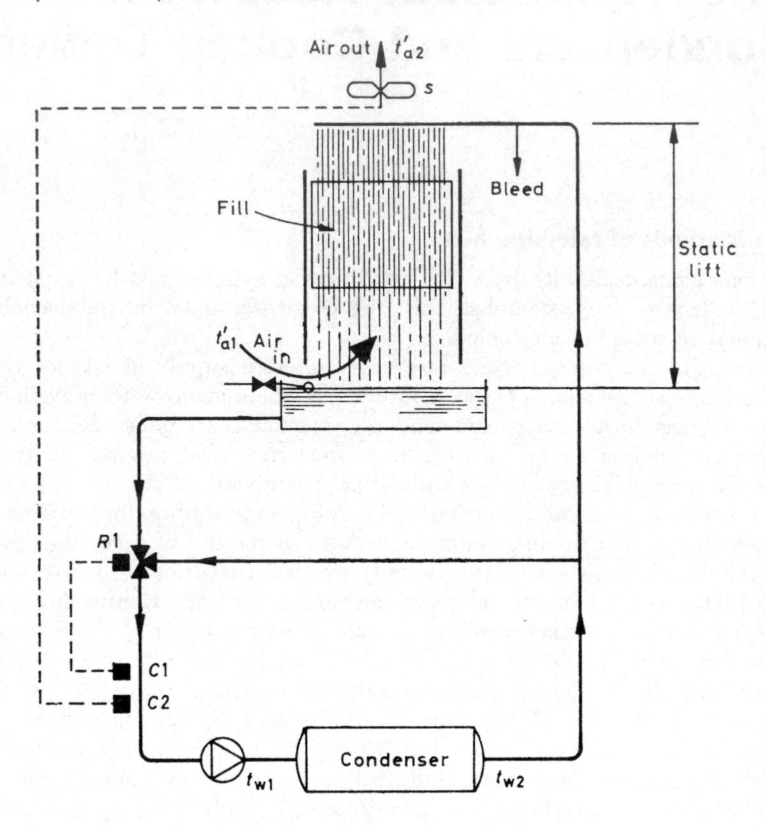

Fig. 11.1 Induced-draught cooling tower

*C*1 is an immersion thermostat controlling the temperature of cooling water flowing to the condenser through the agency of motorised mixing valve *R*1.

*C*2 is an immersion thermostat which cycles the fan and switches it off when adequate water cooling can be achieved by natural draught. t'_{a1} and t'_{a2} are inlet and outlet air wet-bulb temperatures.

11.2 Types of cooling towers

Cooling towers may be broadly classified as (i) induced draught, (ii) forced draught and (iii) cross flow.

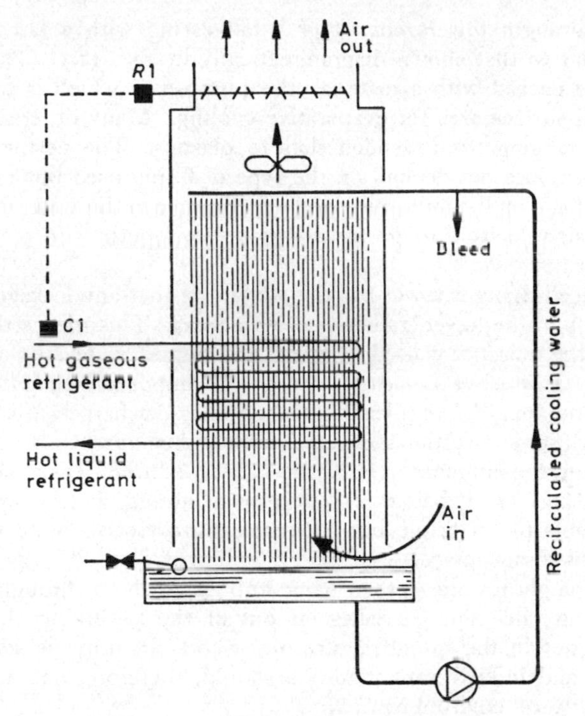

Fig. 11.2 Evaporative condenser
C1 is a pressure controller which keeps refrigerant condensing at a constant pressure through the agency of the motorised modulating dampers *R1*.

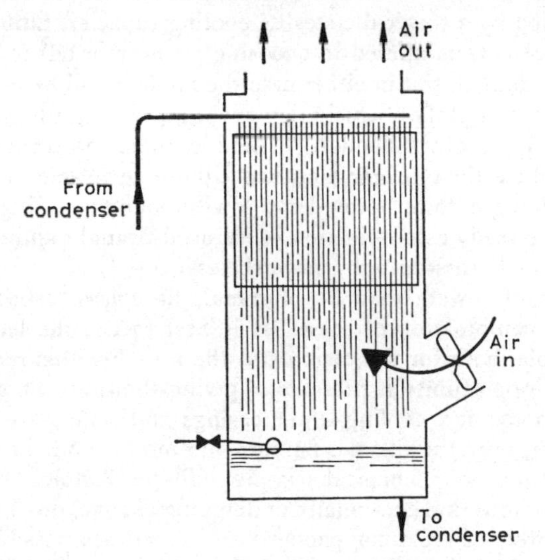

Fig. 11.3 Forced-draught cooling tower

Induced-draught towers consist of a tall casing with a fan at the top, rather similar to that shown diagrammatically in Fig. 11.1. The casing of the tower is packed with a material, the purpose of which is to provide a large wetted surface area for evaporative cooling. Many different materials are in use, ranging from wooden slats to plastics. The performance of a cooling tower does not depend on the type of filling used but rather on its arrangement and on the uniformity and effectiveness of the water distribution. A constant air velocity through the filling is also important in securing good performance.

The induced-draught tower has the advantage that any leakage will be of ambient air into the tower, rather than vice-versa. This means that there is less risk of the nuisance caused by water and humid air leaking outwards to the vicinity than arises in other towers. An important advantage is that humid air and spray (if any) leaving the tower is discharged at a fairly high velocity, with some directional effect. There is thus much less risk of short-circuiting, and the influence of wind pressure is reduced. It is true, however, that the fan and water, being in the stream of humid air leaving the tower, are more liable to corrosion, but adequate anti-corrosive protection should minimise this disadvantage.

Forced-draught towers, as their name implies, push air through the tower. Fig. 11.3 illustrates this. Leakage is out of the tower but the fans and motors are not in the humid airstream. Short-circuiting is something of a problem, and forced-draught towers should, therefore, not be used in a restricted physical environment.

Although the height of a tower is of importance in producing a long path for contra-flow evaporative heat exchange to take place, the cross-section of the tower is also important because the larger its value, for a given height, the greater will be the wetted surface area available. It follows that the volume occupied by a tower dictates its cooling capacity, rather than height alone. A choice is thus offered in choosing a tower: a tall tower or a short tower may be adopted, but in either case the volume will be about the same. Because of their unsightly obtrusion on skylines, cooling towers are frowned on by architects and town planners. There is, thus, sometimes considerable pressure to reduce their height, and low silhouette towers are chosen.

The cross-draught tower provides a low silhouette, as Fig. 11.4 shows. Such towers are really a combination of induced-draught and cross-draught, but are designated cross-draught, nevertheless.

The casings of towers may be of metal, fibreglass, asbestos, wood or concrete, or a combination of these. It is best to use the least corrodable material possible, both for the casing and the fill. For this reason, and also because of the opportunity it presents of giving the tower an architecturally harmonious anonymity, cooling-tower casings and ponds are often erected by the builders, only the fill, the fan and the motor, and the water piping, being installed by the mechanical services' sub-contractors.

The design of towers is continually undergoing change, much of it effective. As a result, towers of great compactness are sometimes possible. One such type of tower uses a rotating water header coaxial with the induced-draught fan. This, coupled with the use of a cellulose fill, rather like corrugated cardboard, which presents a very large surface area, produces a uniform distribution of water and constancy of airflow.

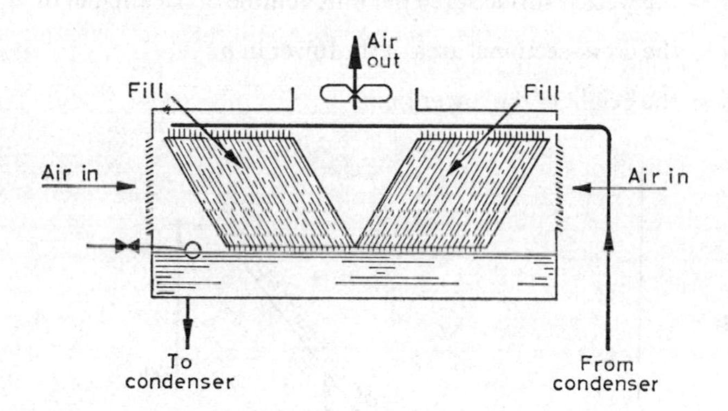

Fig. 11.4 Cross-draught cooling tower

11.3 Theoretical considerations

A full theoretical treatment yielding a design complete in every detail is not possible. A theoretical approach takes the design only so far; beyond this, the design can be completed by assigning empirical values to the unsolved variables.

A heat balance must be struck between the water and the air:

$$m_w(t_{w1}-t_{w2}) = m_a(h_{a2}-h_{a1}) \tag{11.1}$$

where m_w = mass flow of water in kg/s,

t_{w1} = inlet water temperature,

t_{w2} = outlet water temperature

m_a = mass flow of air in kg of dry air/s,

h_{a1} = enthalpy of the air at inlet in kJ/kg dry air,

h_{a2} = enthalpy of the air at outlet in kJ/kg dry air.

Although m_w decreases by evaporation as it flows through the tower, it is regarded here as a constant for simplicity. It can be shown that the enthalpy difference between the water and the air at any point in the tower is the force promoting heat and mass transfer, and the following expression can be derived:

$$Z = \frac{2{\cdot}61 m_w(t_{w1}-t_{w2})}{ksa\Delta h_m} \tag{11.2}$$

where k = the coefficient of vapour diffusion in kg water/s m^2 for a unit value of Δh_m,

Δh_m = the 'mean driving force', in kJ/kg,

s = the wetted surface area per unit volume of packing, in m^{-1},

a = the cross-sectional area of the tower in m^2,

Z = the height of the tower in m.

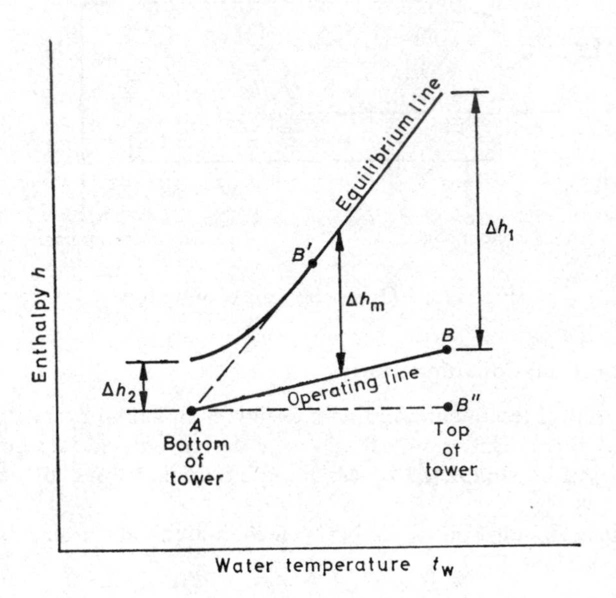

Fig. 11.5

Using eqn. (11.1), a complementary form of eqn. (11.2) exists:

$$Z = \frac{2{\cdot}61 m_a(h_{a2} - h_{a1})}{ksa\Delta h_m} \tag{11.3}$$

The 'volume transfer coefficient', ks, is independent of atmospheric conditions and can be expressed by re-arranging eqn. (11.2).

In the vicinity of the water in the tower there is a thin film of saturated air, at a temperature t_w, the same as the temperature of the water. This air has an enthalpy h_w which is greater than that of the ambient air, h_a. Thus, for any given value of t_w, the water temperature, there exists an enthalpy difference, Δh, between the enthalpy of the film, h_w, and the enthalpy of the ambient air:

$$\Delta h = h_w - h_a \tag{11.4}$$

The mean value of this is Δh_m and is termed the mean driving force.

Figure 11.5 shows this diagrammatically. The equilibrium line represents the variation in the value of h_w with respect to the water temperature, t_w. The operating line shows the variation of the enthalpy of the air, with respect to t_w.

The driving force, promoting heat transfer at any value of t_w, is the difference given by eqn. (11.4). Since heat exchange occurs through the height of the tower, the value of t_w reduces as it falls through the tower. Thus, Δh_2 represents the driving force at the bottom of the tower and Δh_1 at the top. The mean value occurs at some intermediate position.

The operating line is straight, being determined by eqn. (11.1). Hence, the slope of the line depends on the ratio of water flow to air flow, m_w/m_a. When m_w/m_a is zero, the driving force is zero and the tower is infinitely tall. Such a condition would arise if the operating line were AB', tangential to the equilibrium curve, because at the point of tangential touching, Δh is zero. The converse case is AB'' parallel to the abscissa. Here the tower is not tall but, to secure the wetted surface area, it must be of large cross-section. It is evident that increasing the value of m_a with respect to m_w, for a given tower, increases the mean driving force and produces a bigger tower capacity.

A typical value of m_w/m_a is 1·0.

EXAMPLE 11.1 Assuming a water-to-air mass-flow ratio of 1·0, calculate the air quantity likely to be handled by a cooling tower used to cool water from 32° to 27°C, for a refrigeration plant having a coefficient of performance of 4.

Answer

Heat rejected at the condenser, per kW of refrigeration

$$= 1{\cdot}25 \text{ kW}$$

Water flow rate through the condenser

$$= 1{\cdot}25/(4{\cdot}2 \times 5)$$

$$= 0{\cdot}0596 \text{ kg/s}$$

$$= 0{\cdot}06 \text{ litres/s}$$

Hence the air flow rate through the cooling tower is also 0·0596 kg/s.

Assuming an induced-draught tower, the fan handling saturated air at 20°C wet-bulb, then the humid volume is 0·8497 m^3/kg and the air flow rate is

$$0{\cdot}0596 \times 0{\cdot}8497 = 0{\cdot}0506 \text{ m}^3\text{/s for each kW of refrigeration.}$$

11.4 Evaporative condensers

Whereas a condenser/cooling tower arrangement requires a system of water distribution piping, this is almost entirely absent in the evaporative condenser. Only enough water need be circulated to ensure that the outside surface of the condenser coils is completely wet. The heat exchange is solely latent, and less water is required in circulation than is necessary with a condenser/cooling tower, where a sensible heat exchange occurs in the shell-and-tube condenser.

The evaporative condenser thus is more compact and is cheaper. It suffers from lack of flexibility, and oil return and other problems demand that the condenser should not be too far from the compressor. A cooling tower, on the other hand, may be located miles away, the condenser then being adjacent to the compressor.

Scaling on the tubes of an evaporative condenser may be something of a problem, particularly if a high condensing temperature is used.

11.5 Air-cooled condensers

Unlike evaporative condensers, air-cooled condensers have a capacity which is related to the dry-bulb temperature of the ambient air, rather than to its wet-bulb temperature. If working condenser pressures are not to become excessively high, making the plant expensive to run, large condenser surface areas must be used. This has set a limit on the practical upper size of air-cooled condensers. Their use in air-conditioning has been commonly confined to plants having a capacity of less than 70 kW of refrigeration, although they have been used for duties as high as 250 kW, in temperate climates.

A 22-degree difference between the entering dry-bulb temperature and the condensing temperature is often consistent with the avoidance of excessively large condenser surface areas. Because of the absence of water piping and the consequent simplicity of operation, air-cooled condensers are increasing in popularity.

EXAMPLE 11.2. Assuming 0·15 m^3/s kW, and a difference of 22 degrees between the ambient dry-bulb and the condensing temperature of R.12, determine (*a*) the condensing pressure and (*b*) the leaving air temperature for an installation using an air-cooled condenser in London.

Answer

(*a*) The design outside dry-bulb temperature for London is about 28°C (*see* section 5.11).

At 50°C, the condensing pressure of R.12 is 1·22 MN/m^2 absolute. This is a manageable pressure.

(*b*) To cool at a rate of 1 kW in the evaporator, about 1·25 kW must be rejected at the condenser, assuming a coefficient of performance of 4. All this heat is rejected into the airstream and so the leaving air temperature will be

$$t = 28° + \frac{1{\cdot}25}{0{\cdot}15} \times \frac{(273+28)}{358}$$

$$= 35°\text{C}.$$

If it had been contemplated that R.22 be used, the condensing pressure would have been embarrassingly high: at 50°C it is 1·96 MN/m^2 absolute.

11.6 Automatic control

It is usually desirable to maintain a reasonably stable condensing pressure, particularly when the refrigeration plant has to run throughout the year in a non-tropical climate. The 'temperate' winter experienced in the U.K. can result in condensing pressures which are too low for the plant to operate properly, unless steps are taken to limit the rate at which heat is rejected in the condenser.

Figure 11.1 shows a method adopted for the cooling tower. The rate of water flow through the condenser is kept constant but its temperature is varied by means of a 3-port mixing valve, *R*1. In winter, when low wet-bulbs prevail, the water leaving the tower is very much less than the 27°C desired for t_{w1} in normal summer operating conditions. The immersion thermostat, *C*1, may have a set point lower than 27°C, say 22°C±1°, so that some fall in condensing head may occur under reduced wet-bulb conditions. This is desirable, up to a point, because the power per kW reduces as the condensing head falls. Below 18°C is usually undesirable, with reciprocating compressors.

Instead of a 3-port valve, a 2-port valve may be used to throttle the rate of flow of cooling water. This must *never* be controlled from water temperature but directly from condensing pressure.

It is common practice to arrange that the cooling tower fan is cycled on-off, in addition to the above-mentioned devices. This saves running cost and avoids over-cooling the water unnecessarily. Induced-draught cooling towers, because they are taller than the cross-draught type, are capable, to some extent, of cooling by natural draught when their fans are switched off. Thus, an immersion thermostat *C*2, with a set point of, say, 25°C±2°C, might be used, when, the temperature falling from 27°C (as the ambient wet-bulb decreased), the fan would remain on until water at 23°C was coming off the tower. It would then be switched off. Unless the ambient wet-bulb, or the load, was low enough to permit the tower to produce water at 23°C by natural draught, the fan would cycle and the flow temperature would vary between 23°C and 27°C. All this time the valve *R*1 would have its by-pass port fully closed. With the fan off, if the load or the ambient wet-bulb continued to fall, the cooling water temperature to the condenser would be maintained between 21°C and 23°C, through the modulation of the mixing valve.

The pressure drop across *R*1, when fully open to by-pass, should be about 25 per cent of the static lift (*see* Fig. 11.1).

Evaporative condensers have their capacity controlled by means of a condenser pressure-sensing element, *C*1. This modulates the flow of air passing over the wetted coils by means of motorised dampers, as shown diagrammatically in Fig. 11.2.

It is customary to have auxiliary contacts on the damper motor control so that, when the dampers are fully closed, the fan is switched off. A cruder form of control, not acceptable in temperate climates when the refrigeration plant has to run throughout the winter, is just to cycle the fan. Some economy in running costs can be effected, if thought necessary, by arranging to switch off the pump when the outside dry-bulb temperature is low enough to permit the condenser to reject its heat without the aid of evaporation.

Air-cooled condensers are controlled in a similar way to evaporative condensers, but there is a difficulty that is worth mentioning. They tend to be used for less important purposes and, hence, are asked to reject heat throughout the winter. Since they are frequently installed on roofs and in exposed

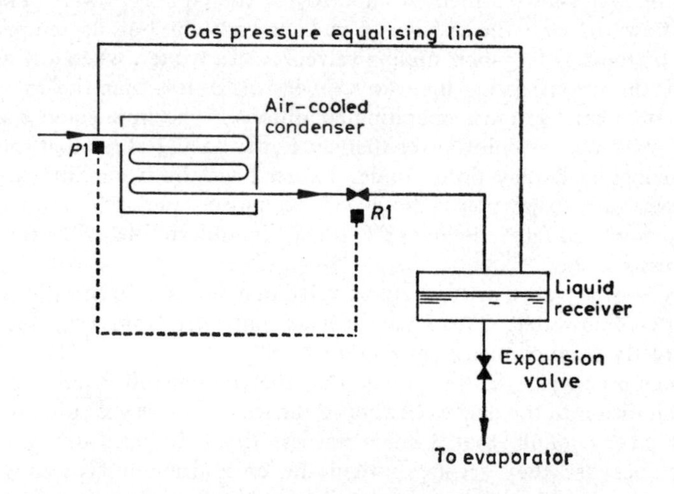

Fig. 11.6

positions, they are very sensitive to the effects of the weather. Even if the fan is off and motorised dampers shut off the face of the coil entirely, the cooling effect of the wind on the casing is considerable, especially at low air temperatures. Further, motorised dampers do not provide a hermetic seal, and some leakage is inevitable. This means that the normal methods of control are likely to be ineffective, unless the condenser is well protected and shielded from the weather.

An alternative control (Fig. 11.6) is sometimes used. There are several variations of this, some probably patented, but in principle the intention is to reduce the capacity of the condenser by reducing the surface area available for heat transfer. The liquid refrigerant is backed-up inside the condenser by restricting its flow to the liquid receiver. This reduces the area of the condenser through which heat is transferred from the condensing vapour to the environment and thus varies the rate of heat rejection in a way that can be automatically controlled from condensing pressure. As the pressure falls (sensed by $P1$), the motorised valve $R1$ reduces the outflow of liquid to the receiver. The size of the liquid receiver needs to be increased somewhat so that the refrigerant demands of the system can be met, even though $R1$ is fully closed for a short while. It is also necessary to have a gas-pressure equalising line between the condenser and the receiver so that an adequate pressure is maintained on the upstream side of the expansion valve.

11.7 Practical considerations

Three major problems have to be faced with cooling tower installations:

(i) Corrosion.

(ii) Scale formation.

(iii) Protection against freeze up.

The answer to the corrosion problem is two-fold: first, a tower constructed from materials that are likely to be adequately resistant to corrosion for the expected life of the tower should be selected; secondly, due thought should be given to the use of water treatment.

Continuous evaporation means that the concentration of dissolved solids in the tank of the cooling tower will increase. Unless steps are taken to nullify the effects of this, in due course some scale or sludge formation is inevitable. This will reduce the performance of the cooling water system by being deposited on the heat transfer surfaces in the condenser and on the fill of the tower. Remedial action takes the form of allowing an adequate continuous bleed from the tower (usually arranged in the discharge pipe, as shown in Fig. 11.1, so that water is not wasted when the pump is off) coupled with some form of water treatment. If the bleed rate is high enough, it may, under certain circumstances, be sufficient to minimise scale formation without the use of water treatment.

Bleed rates are often quite high. The rate desirable may be related to the hardness of the water, expressed in parts per million $CaCO_3$. For example, water which has a hardness of 100 ppm might need a continuous bleed equal to 100 per cent of the evaporation rate. At 200 ppm, it might require to be 200 per cent of the evaporation rate, and at 300 ppm, to be ten times the evaporation rate. This can sometimes mean a formidable consumption of water.

The Metropolitan Water Board currently demand that a break tank be interposed between the cooling tower (or other device, using mains water for cooling) and their mains. The break tank must be large enough to provide 4 hours or half a day's supply of cooling water, in the event of a failure of the mains supply. They have three reasons for insisting on this:

(*a*) A degree of stand-by is offered to the user.
(*b*) A break is necessary between the mains in order to prevent back-siphonage and contamination of the Board's water supply.
(*c*) Fluctuations in demand are averaged; thus, the Board's mains can be sized for average, rather than peak demands.

There is a fourth possible reason why such a tank might be required. In a large office block (for instance), if the cooling tower supply is taken from the cold water storage tank used for supplying water to lavatories, there is reduced stand-by water available for lavatory flushing purposes. If the mains supply failed, the cooling tower would use this water, to the detriment of the hygiene in the building.

EXAMPLE 11.3. Estimate the capacity of the cooling-tower break tank necessary for a 720 kW air-conditioning installation. The hardness of the mains water is 100 ppm and the plant runs for an 8-hour working day.

Answer

Assuming a coefficient of performance of 4, the heat rejected in the cooling tower is, therefore,

$$720 \times 1{\cdot}25 = 900 \text{ kW}$$

Taking an approximate value of 2450 kJ/kg as the latent heat of water, then the evaporation rate will be 0·367 kg/s.

For the hardness quoted, a continuous bleed of 100 per cent is probably required. Thus, the break tank must have a capacity of 11000 litres.

To prevent freeze-up, two precautions are necessary: first, the provision of an immersion heater with an in-built thermostat in the tank, and secondly, the protection of all exposed pipework which contains water with weather-proofed lagging. Below the static water line (when the pump is off), such pipework should be traced with electric cable, thermostatically controlled to keep the water above 0°C. As an alternative, the tower may be drained in winter, if the refrigeration plant is off. But such drainage is a tedious maintenance chore and may not be acceptable to a user. In any case, operation of a cooling tower in winter requires considerable forethought.

The tank itself should be of sufficient capacity to contain the water flowing into it when the pump starts, without overflowing.

Vortices of air bubbles entrained in the water may form at the outflow from the cooling tower tank. Up to 10 per cent of the volume of the outflow may be air, in bad cases. The formation of vortices is thus very undesirable.

It can be prevented by having a large enough outflow diameter, or a greater depth of water above the outflow, or by the suitable positioning of baffles (*see* Fig. 11.7).

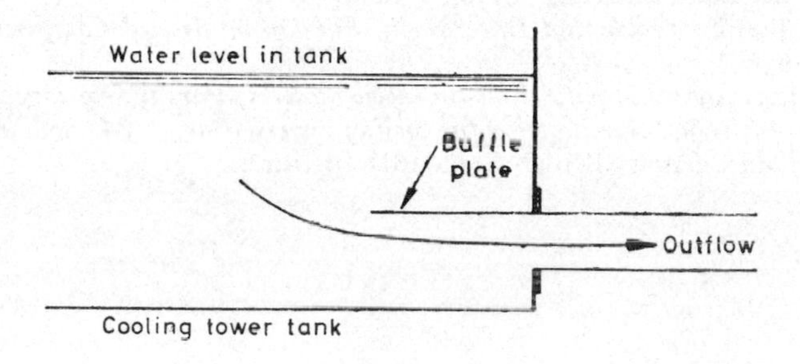

Fig. 11.7

Finally, a screen should be provided around the outflow pipe to filter out the larger pieces of debris which collect in the tank of any cooling tower. The size of the holes should be just smaller than the internal diameter of the water tubes in the condenser. Holes that are too small may be undesirable because the screen may clog up and frequently stop water flow. Birds seem to favour cooling towers in some districts and their droppings and feathers may prove a hazard. Frequent maintenance inspections should resolve this problem.

EXERCISES

1. Compare the merits of a forced convection evaporative condenser with those of a shell-and-tube water cooled type.

2. Compare briefly the relative merits of a shell-and-tube condenser with cooling tower, and an evaporative condenser. Make a sketch of an evaporative condenser, indicating the main components.

3. (*a*) With the aid of neat sketches, distinguish between cooling towers and evaporative condensers. Name two other methods of rejecting the heat from the condenser of a refrigeration plant used for air conditioning.

(*b*) Discuss the practical limitations of the four methods mentioned in part (*a*) of this question. Give emphasis to application, running cost, maintenance and noise.

BIBLIOGRAPHY

1. J. Jackson. *Cooling Towers.* Butterworth Scientific Publications, 1951.
2. W. H. McAdams. *Heat Transmission*, pp. 285–292. McGraw-Hill, 1942.

3. D. F. Denny and G. A. J. Young. *The Prevention of Vortices and Swirl at Intakes*. British Hydrodynamics Research Association, Publication SP583.
4. S. A. Littman. Should we specify water treatment? *Air Conditioning, Heating and Ventilating*, October 1964, 74–78.
5. A.S.H.R.A.E. *Guide and Data Book, Fundamentals and Equipment*, 1963, 623–636.
6. B.S. 4485: 1969: *Specification for cooling towers, Part 1, Glossary of Terms.*
7. B.S. 4485: 1969: *Specification for cooling towers, Part 2, Methods of test and acceptance testing*. British Standards Institute.

12

Refrigeration Plant

12.1 The expansion valve

In general terms, the function of the expansion valve is to reduce the pressure in the system between the high value in the condenser and the low value in the evaporator so that a corresponding low temperature may be obtained in the evaporator, where it is wanted. Its secondary function depends on the way in which it is controlled. In air conditioning, a ' thermostatic expansion valve ', as it is called, is in common use. The secondary function of this valve is to meter the flow of liquid refrigerant to the evaporator so that the gas leaving it will be slightly superheated. This is necessary because the compressor is designed to pump a gas, not a liquid. If liquid entered the compressor damage would be done.

Figure 12.1 is a diagram of a thermostatic expansion valve. Suppose that the spring, *S*, is slackened so that it exerts no force on the underside of the diaphragm *D*. The space above *D*, the capillary tube and the sensitive phial *P*, is filled with a measured amount of the same refrigerant that has been used to charge the rest of the system. *P* is clamped to the outside of the pipe carrying the gas leaving the evaporator, and therefore, it is at the same temperature as the refrigerant inside the outlet pipe—ignoring any heat transfer with the environment. If the refrigerant leaving the evaporator is in the dry saturated or the wet state, and if there is some liquid refrigerant inside the phial, then the saturation pressure inside *P* will be the same as the saturation pressure in the evaporator. Since *P* is connected to the upper side of *D* and the evaporator is connected to the under side of *D*, the diaphragm will be in equilibrium. This will persist, regardless of whether the vapour leaving the evaporator is dry, slightly wet, or very wet. There is thus considerable risk that liquid will enter the compressor.

If *S* is adjusted so that it exerts an upward force on *D*, the valve will partly close, the diaphragm being attached to the valve spindle. Less refrigerant will enter the evaporator and, for a given load, the vapour leaving the evaporator will be drier. A further increase in the upward force exerted by the spring will eventually reduce the flow of refrigerant so much that all the liquid will be boiled off before it reaches the outlet of the evaporator. At outlet it will then be in a superheated state.

It follows from this that the surplus force exerted by the spring determines the amount of superheat in the gas leaving the evaporator and that the amount of superheat may be adjusted by setting the spring.

EXAMPLE 12.1 Refrigerant 12 evaporates at 2°C in a direct-expansion cooler coil. What must be the surplus pressure exerted by the spring in the thermostatic expansion valve if 8 degrees of superheat are to be maintained at the outlet from the coil?

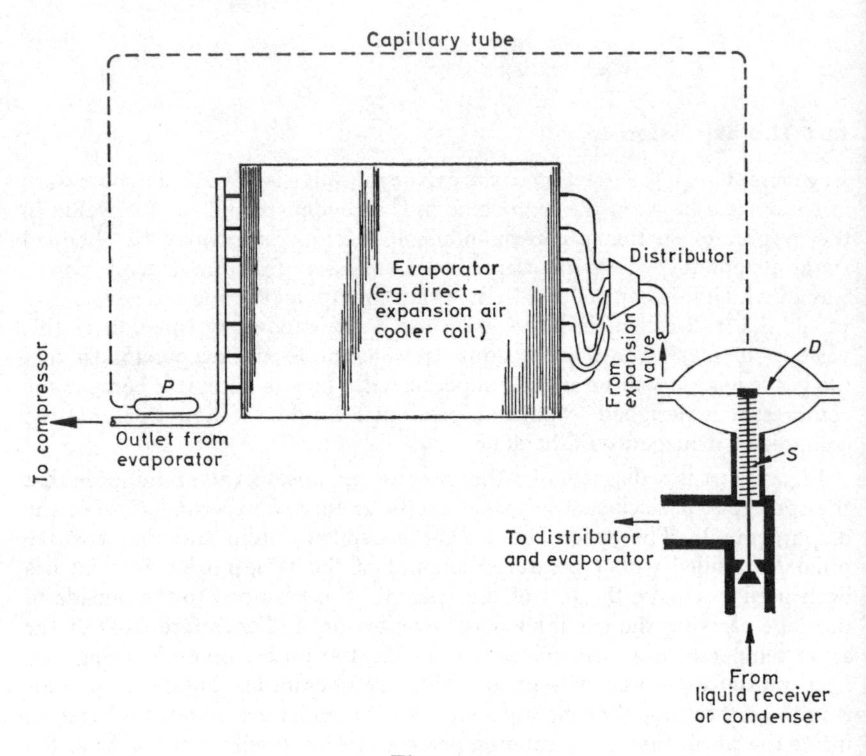

Fig. 12.1

Answer

From tables of the properties of saturated Refrigerant 12, it is found that, at 2°C, the saturation vapour pressure is 330 kN/m^2 absolute and that at 10°C (=2°+8°) it is 424 kN/m^2 absolute.

The spring must therefore be adjusted until it exerts a force corresponding to 424−330 or, 94 kN/m^2.

The pressure drop across an expansion valve depends on the rate of flow of refrigerant and the valve size. Expansion valves must be chosen to match the capacity of the evaporator, and they are usually rated in terms of a particular condensing temperature and one or more evaporating temperatures.

To make sure that the under side of the diaphragm senses a pressure equal to that in the outlet pipe near the phial, a pressure equalising line is run from the outlet to the under side of the diaphragm.

In a similar way, a pressure drop occurs in the liquid line between the condenser outlet and the inlet port of the expansion valve. If this results in a pressure in the line which is less than the saturation vapour pressure corresponding to the particular temperature of the refrigerant, some of the liquid will flash to gas, in order to reduce the temperature to the saturation value, before the liquid enters the expansion valve. This represents a waste of refrigerating effect. It may be dealt with by arranging for the liquid to be sub-cooled below the condensing temperature, to the extent of 2 degrees or so.

It was observed, in the case of chilled water cooling coils, that the mean surface temperature fell when the load diminished. A similar effect occurs with direct-expansion cooler coils fitted with thermostatic expansion valves. Under conditions of full-load operation the expansion valve is in an equilibrium position and the liquid refrigerant is completely evaporated a short distance before it emerges from the cooler coil. It acquires the correct amount of superheat in travelling over this short distance. If the load falls, the liquid is required to absorb less heat and so the place where it is completely evaporated is further along, closer to the outlet from the coil. It thus suffers a reduction in superheat, which the sensitive phial, strapped to the outside of the outlet pipe, notices and corrects by making the expansion valve partly close and reduce the amount of refrigerant flowing. A bigger pressure drop now occurs across the expansion valve, because it is partly shut. If the condensing head is kept constant, this means that the evaporating pressure, and so the evaporating temperature, are less than they were for the full-load condition. Hence, the mean coil surface temperature falls as the load reduces.

It should be clear from the above that a thermostatic expansion valve is not used to vary the capacity of the evaporator. If it is desired to change the state of the air leaving a direct-expansion cooler coil, or to keep the state of the air leaving the coil constant, then other controls must be used, in addition to the expansion valve.

12.2 Evaporators

The evaporator is the place in the refrigeration circuit where heat is removed from the substance being cooled, air or water in the case of air conditioning. Liquid refrigerant within the evaporator absorbs heat from the air or water and, in doing so, boils. To effect this boiling, two distinct types of evaporator are in use:

(i) the flooded evaporator, used mostly for water-chilling, and

(ii) dry, expansion evaporators, used for both water-chilling and for air-cooling.

Water-chillers may take several forms. For example, they may be shell-and-tube heat exchangers, submerged coils in an open tank of water, or Baudelot coolers, as illustrated in Fig. 12.2. In (*a*) and (*b*), the evaporator is entirely enclosed, and a feed and expansion tank, elsewhere in the system, permits it to be filled with water. If a freeze-up occurs, expensive damage may be done to the tubes in the heat exchanger. In (*c*) and (*d*), the water

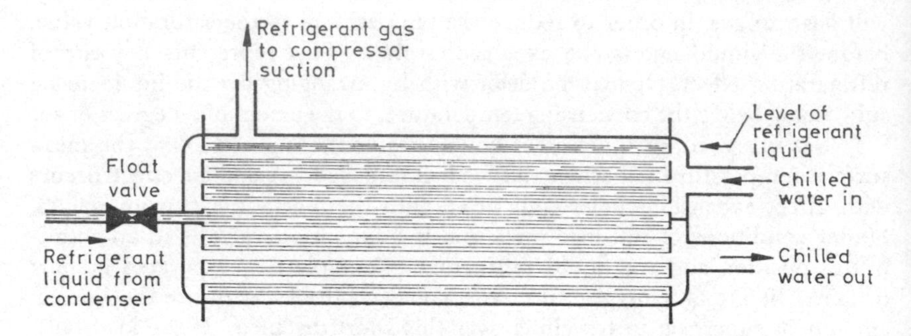

Fig. 12.2(*a*) Flooded shell-and-tube evaporator

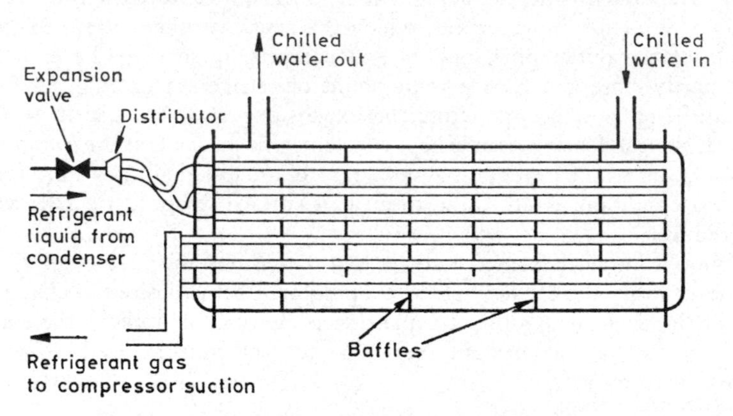

Fig. 12.2(*b*) Dry shell-and-tube evaporator

circuit is open and the water storage vessel constitutes its own feed and expansion tank. Heat transfer from the submerged coils in Fig. 12.2(*c*) is promoted by employing a water agitator, whereas the Baudelot cooler improves matters by arranging for water to flow downwards from a header tank, over the pipes containing the boiling refrigerant. These last two types are rather expensive and bulky and for this reason are out of favour to-day. Their virtues lie in their freedom from risk of damage should a freeze-up occur and in their ease of control. The large mass of water stored in the tanks provides a cold thermal reservoir and permits the compressor to be cycled on-off without capacity control. They have the further advantage, particularly in the case of the Baudelot cooler, that a lower chilled water temperature may be obtained than is possible with either of the shell-and-tube types of water-chiller. Whereas the latter are best used to produce water at a temperature of 6°C

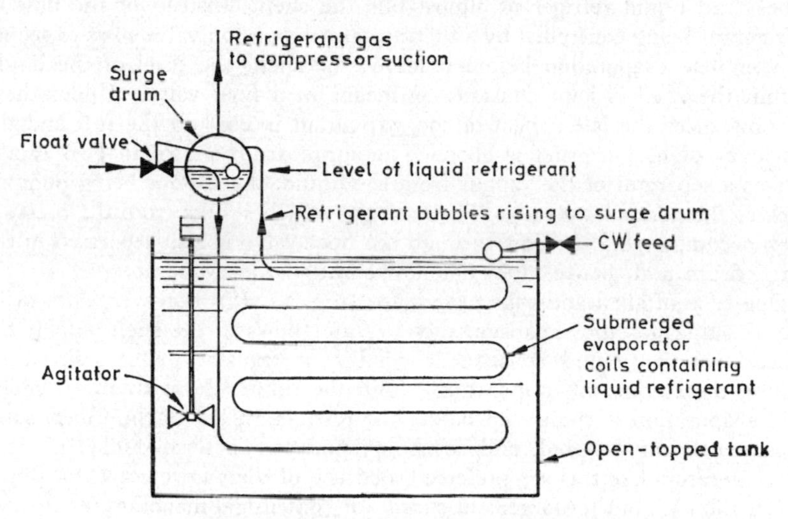

Fig. 12.2(*c*) Submerged evaporator

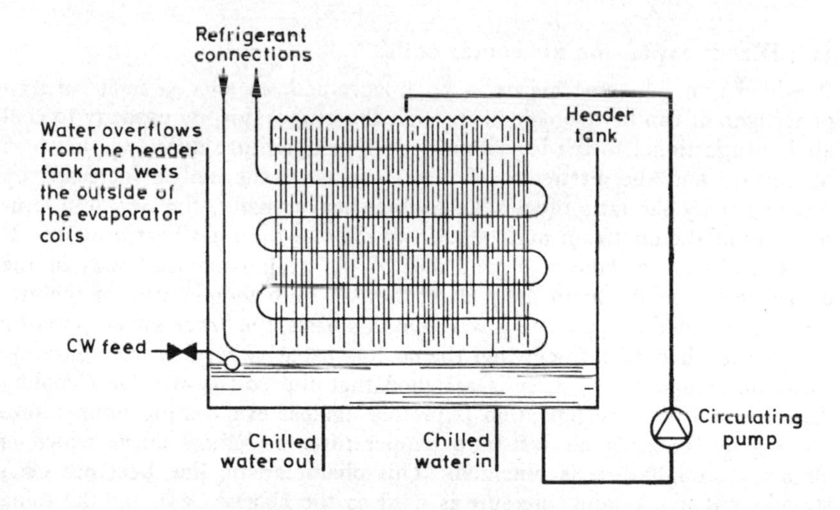

Fig. 12.2(*d*) Baudelot cooler

or above, a Baudelot cooler can be made to give water at 2°C, or even lower. Ice can actually form on such submerged coils without serious risk. However, for the reasons mentioned earlier, they are seldom used to-day in commercial air conditioning.

Flooded evaporators (Figs. 12.2(*a*) and (*c*)), as their name implies, have liquid covering the whole of their heat transfer surfaces. For example, in the shell-and-tube evaporator water flows in multiple passes through the tubes, and liquid refrigerant almost fills the shell. Instead of the flow of refrigerant being controlled by a thermostatic expansion valve so as to secure its complete evaporation before it leaves the shell, the level of the liquid within the shell is kept constant by means of a float valve. Under these circumstances the latent heat of the refrigerant is used to the full and the efficiency of heat transfer is good. The surge drum shown in Fig. 12.2(*c*) acts as a separator of the vapour from the liquid, the vapour being pumped back to the compressor only. The flash gas which results from the pressure drop accompanying the flow through the float valve is also separated in the surge drum and, hence, does not flow through the evaporator.

One type of shell-and-tube evaporator (Fig. 12.2(*b*)) however, does make use of a thermostatic expansion valve. Water flows in the shell outside the tubes, turbulence and consequent good heat transfer being assisted by baffles, the refrigerant evaporating within the tubes. As with an air-cooler coil, evaporation is completed before the refrigerant leaves the tubes, some superheat being ensured, and so the evaporation is termed dry.

Evaporators like this are preferred, because of their lower cost, for duties below 140 kW, but for larger duties and for centrifugal machines the flooded evaporator is the cheaper, and is in common use.

12.3 Direct expansion air cooler coils

Provided that, by some means as yet unspecified, an adequate rate of flow of refrigerant can be maintained through the evaporator, its capacity to cool air is proportional to the logarithmic mean temperature difference between its surface and the airstream. If the *U*-value of the coil is increased by dividing it by the ratio of sensible to total heat transfer, the dry-bulb temperature of the airstream may be used to express the total heat transfer. If the *U*-value is not known, a good idea of the proportional capacity of the cooler coil may be obtained by using the wet-bulb temperature of the airstream, provided that the coil is wet. On this basis, the lower the evaporating temperature becomes, the greater the cooling capacity for any fixed entering wet-bulb temperature. If it is assumed that the coil is wet, and cooling capacity in kW of refrigeration is plotted against evaporating temperature for a given entering air wet-bulb temperature, a shallow curve which is almost a straight line is obtained. This characteristic line becomes even straighter if evaporating pressure is used as the abscissa. If, for the same cooler coil, air enters at a different wet-bulb temperature, then a new characteristic curve is obtained, parallel to the old one. Figure 12.3 illustrates this in terms of the example which follows.

EXAMPLE 12.2. A direct expansion air cooler coil is selected to give air at a leaving wet-bulb of 10°C when the entering wet-bulb is 17°C. Under these design conditions, when using refrigerant 22, the load is 100 kW of refrigeration and the evaporating temperature is 1°C. Assuming a wet coil, construct a pair of characteristic lines, in terms of cooling capacity and evaporating temperature, which express the performance of the coil for the design entering wet-bulb value, and for an entering wet-bulb of 13°C.

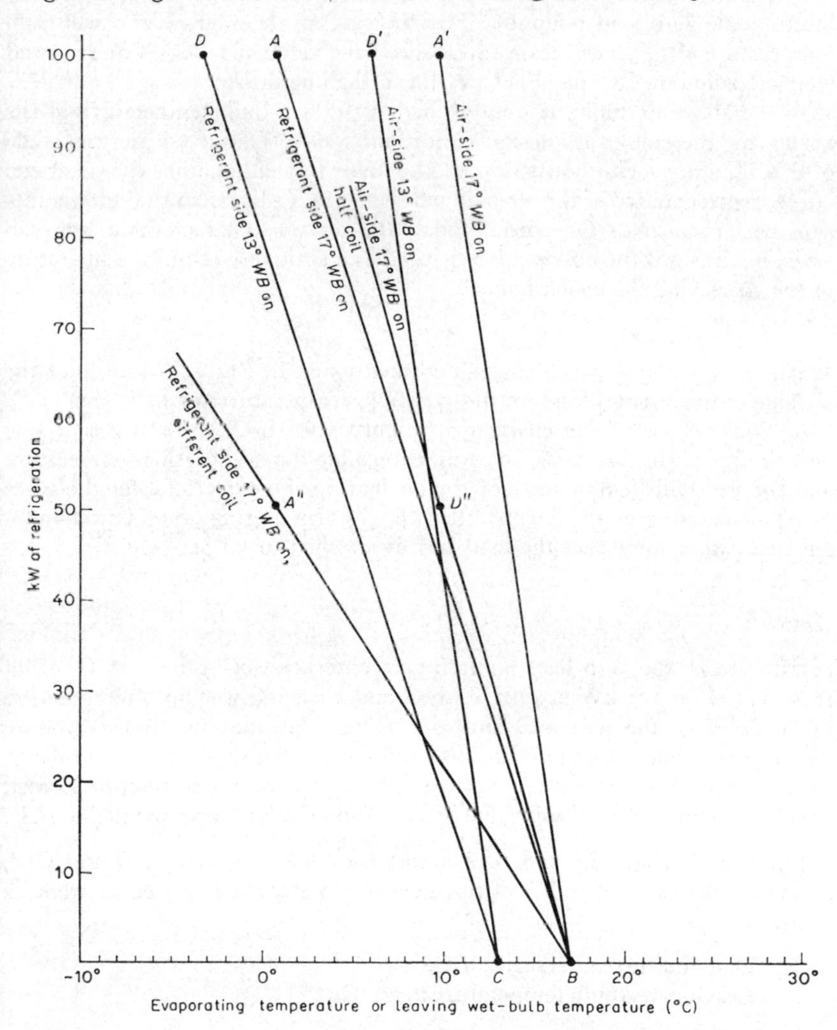

Fig. 12.3 Evaporating temperature or leaving wet-bulb temperature

Answer

For the design case, two points on the characteristic can be immediately established: point A is for an evaporating temperature of 1°C and the given load of 100 kW, and point B is for the case of zero load. If the evaporating temperature is 17°C and the entering wet-bulb is also 17°C, then the load is zero, it being assumed that the difference between these two temperatures is an indication of the heat transfer. The characteristic line for the design case is then the straight line joining *A* to *B*.

The performance characteristic for the second case can be drawn by locating the zero load point *C*. For the case of air entering at a wet-bulb temperature of 13°C, this is on the abscissa at a value of 13°C. The required characteristic line *CD*, parallel to *BA*, can then be drawn.

In the above example, it was given that the wet-bulb temperature of the air leaving the coil under design conditions was 10°C, but no use was made of this. In fact, such information can be made to yield another characteristic curve, representative of the air-side performance, rather than the refrigerant-side performance so far considered. The air-side characteristic line can easily be drawn if the abscissa is regarded as a scale of wet-bulb temperature of the air leaving the cooler coil.

EXAMPLE 12.3. (*a*) For the cooler coil mentioned in Example 12.2, plot the air-side characteristics for entering wet-bulb temperatures of 17°C and 13°C.

(*b*) Making use of the characteristic curves for both the refrigerant-side and air-side performances of the coil, determine the evaporating temperature and the wet-bulb temperature of the air leaving the cooler coil for the cases of (i) air entering at 17°C wet-bulb and, (ii) air entering at 13°C wet-bulb. In each case assume that the load has diminished to 50 per cent.

Answer

(*a*) Locate *B*, the zero load point for an entering wet-bulb of 17°C. Plot the point *A′* at 100 kW of refrigeration and a leaving wet-bulb temperature of 10°C. Join the points *B* and *A′*. Then the line *BA′* is the air-side characteristic line for an entering wet-bulb temperature of 17°C. Similarly, for an entering wet-bulb temperature of 13°C, the zero load point is located at *C* (with a leaving wet-bulb value of 13°C) and *CD′* is drawn parallel to *BA′*.

(*b*) Refer to Fig. 12.3 and read across the lines *BA*, *BA′*, *CD* and *CD′*, at an ordinate value of 50 kW of refrigeration to give the required answers, as follows:

(i) Evaporating temperature: 9°C.
Leaving wet-bulb temperature: 13·5°C.

(ii) Evaporating temperature: 5°C.
Leaving wet-bulb temperature: 9·5°C.

Several important conclusions can be drawn from this example. First, a reduction in the value of the entering wet-bulb temperature causes a fall in evaporating temperature and a drop in the value of the wet-bulb temperature of the air leaving the coil. Secondly, if the load is reduced, the entering wet-bulb remaining unaltered, the leaving wet-bulb and evaporating temperature rise. These conclusions cannot be generalised because no indication is given of the manner in which the flow of refrigerant is manipulated in order to give the specific loads. A proper appreciation of this can be obtained only by considering the behaviour of the cooler coil in relation to its associated condensing set. Before dealing with this aspect, it is necessary to examine one further way in which the air- and refrigerant-side characteristics of direct-expansion cooler coils may alter.

A cooler coil changes physically into a different coil if its surface area or U-value alters. The air-side and refrigerant-side characteristics which may have been plotted are then not applicable. This is conveniently illustrated by considering a cooler coil which is split into two equal sections, and over

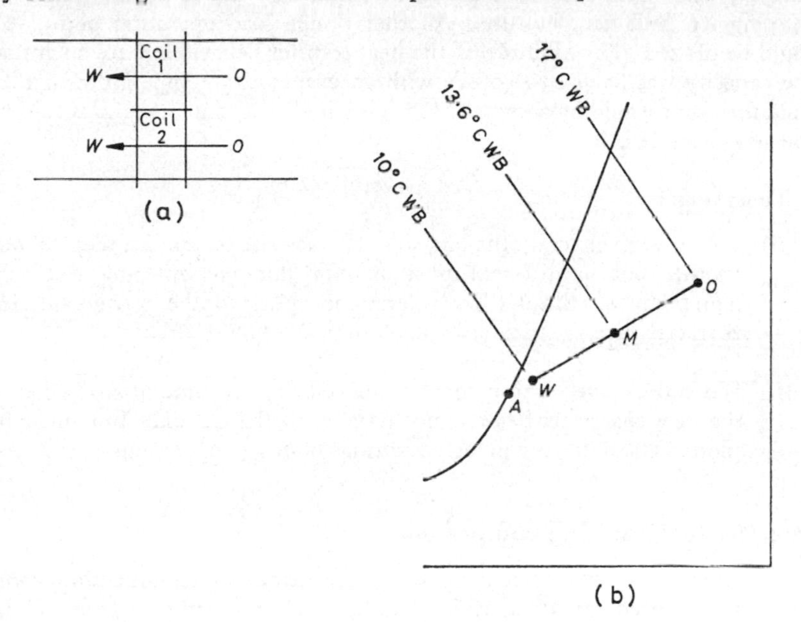

Fig. 12.4

which the airflow is in parallel, as in Fig. 12.4(*a*). Figure 12.4(*b*) shows the psychrometry involved. Air flowing over a section of the coil in use enters at a state O and leaves at a state W. If one section of the coil is taken out of service, say by closing a solenoid valve in the liquid line upstream of the expansion valve, then air flowing over this section is unchanged in state. It leaves at the same state at which it entered, that is, state O. The other section of the coil, however, is still able to change the state of the air from O to W. Thus, the state of the air leaving the coil as a whole is M, a mixture of equal proportions by weight of air at state O with air at state W. The mass of conditioned air having been halved, the cooling load has also halved. Although the air-side and refrigerant-side characteristics for the section that is operating are unaltered, a new air-side characteristic can be drawn for the ' new ' coil, which consists of an operative half and an inoperative half. This is shown by the line BD'' in Fig. 12.3, the point D'' being located by the co-ordinates 50 kW of refrigeration and 10°C leaving wet-bulb. The important point to observe is that although this revised characteristic is for an entering wet-bulb of 17°C, it is not parallel to BA', also for 17°C. In a similar way, if the U-value were modified, a new characteristic would result. Although it cannot be shown by the example of two equal coil sections, the refrigerant-side characteristic also alters its slope, if the coil changes. Suppose the coil just considered were in one section only and that either its U-value or its area were altered, in some unspecified way, so that, in fact, it became a different coil. The zero load point B would stay the same for the same entering wet-bulb temperature. Another design load operating point, A'', could be plotted. For example, if the heat transfer behaviour were such that the capacity was halved to 50 kW with an evaporating temperature of 1°C. then the point would be shown in Fig. 12.3 at A'' and the line $A''B$ would be the new characteristic.

To sum up:

(i) For any given coil, the air-side characteristics are parallel to one another but in different positions for different entering wet-bulb temperatures. Similar considerations apply to the refrigerant-side characteristics.

(ii) If a coil changes with respect to air velocity, U-value or surface area the new characteristics are not parallel to the old ones but must be plotted afresh for the given conditions of design operation.

12.4 The reciprocating compressor

It is customary when considering the performance of reciprocating compressors, to do so for a particular condensing temperature. That is, the performance of a condensing set is what is really being dealt with. As the condensing pressure against which the compressor has to deliver the gaseous refrigerant rises, more work is expended per kW of refrigeration, and the refrigeration capacity of the condensing set reduces. On the other hand, as suction pressure at inlet to the cylinders of the compressor rises, less work need be done to secure the same refrigerating effect. Furthermore, with an increase in suction pressure, the density of the gas entering the cylinders also increases and so, for a given swept volume, the mass of refrigerant handled by the compressor becomes greater. A picture is thus formed of the characteristic behaviour of a condensing set. Figure 12.5 illustrate this.

A compressor has a rising capacity characteristic; the higher the suction pressure the greater the capacity. The characteristic lines for different condensing temperatures are virtually straight, particularly if suction pressure is used as the abscissa. If the saturation temperature corresponding to suction pressure is used, superheat being ignored, some slight curvature results. Although straight lines can be used, the inaccuracy resulting being accepted, it is better to draw the lines as curves if the data is available for this purpose. It usually is, since compressor characteristics cannot be readily guessed. Reference must be made to the manufacturers for the information needed to plot the characteristic.

The capacity of a compressor, for a given condensing temperature, depends directly on the mass of refrigerant being pumped. Thus, if the speed of rotation or the number of cylinders used is altered, the refrigeration capacity of the compressor will change in direct proportion. This fact enables a family of characteristics to be plotted, provided, of course, that the initial information is to hand from the manufacturers.

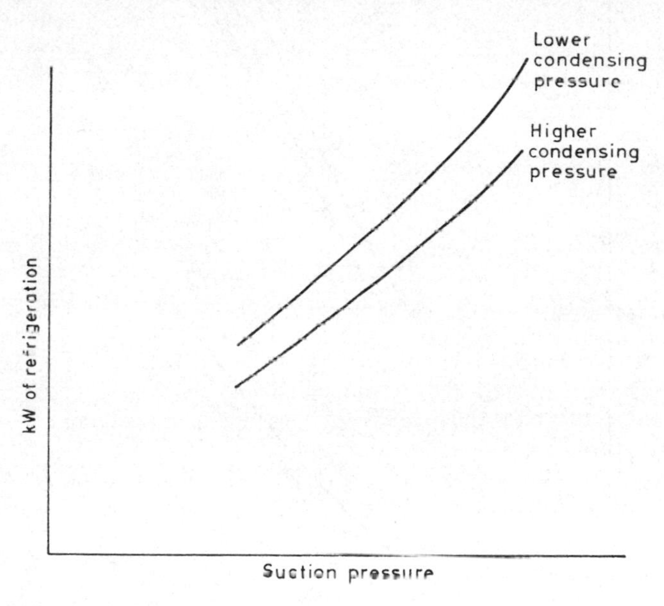

Fig. 12.5

EXAMPLE 12.4. A compressor using R22 has a performance as follows:

Suction temperature °C	−10°	−5°	0°	+5°
Suction pressure kN/m²	219	261	309	362
Capacity in kW (condensing at 35°C)	65	84·7	108	135
Capacity in kW (condensing at 40°C)	57·9	77·3	99·8	126·5

(*a*) Plot the two capacity characteristic curves for the system condensing at 35°C and 40°C.

(*b*) If two of the compressor's four cylinders are unloaded, plot the capacity characteristic for the remaining pair when condensation occurs at 40°C.

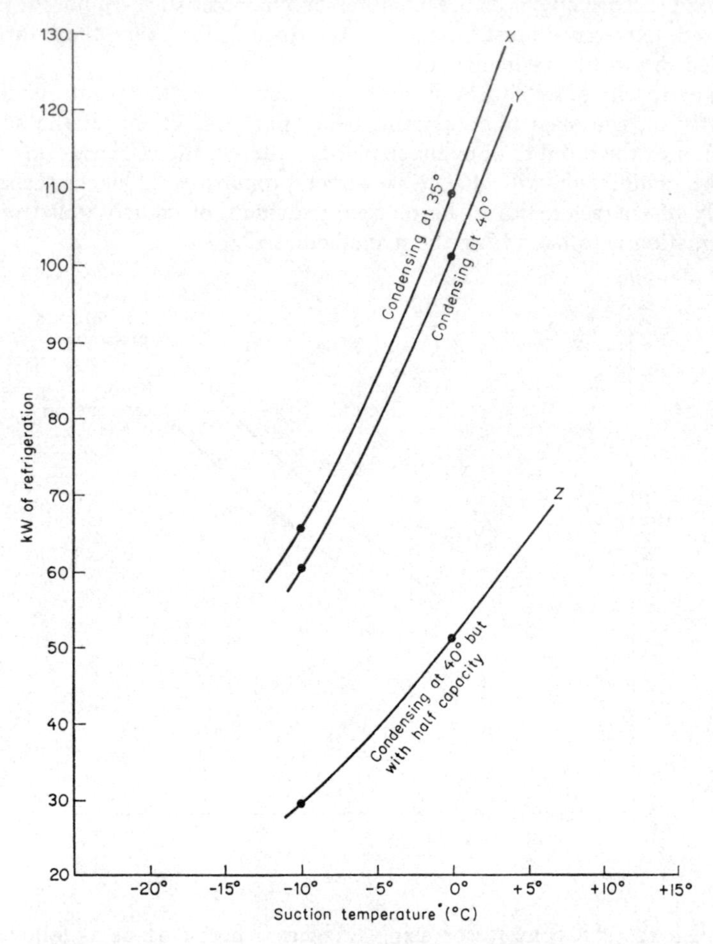

Fig. 12.6

Answer

Either temperature or pressure may be used for the abscissa. In this case, choose temperature.

(*a*) Figure 12.6 shows the result of plotting the given information, by means of the two curves X and Y.

(*b*) Since half the cylinders are unloaded, the pumping capacity of the compressor is also halved. Hence for any given suction temperature the capacity is reduced by 50 per cent. Thus, curve Z in Fig. 12.6 can be drawn, every point on which has an ordinate value exactly half that of curve Y.

In Chapter 9 reference was made to the theoretical volumetric flow rate of refrigerant and to the theoretical mass flow rate, in m^3/s kW and kg/s kW, respectively. The calculations were based on the assumption that the full amount of the volume of the cylinder swept by the piston was available for pumping the refrigerant gas. In reality, the actual weight of gas pumped is less than the theoretical weight, hence giving rise to a concept of *volumetric efficiency*, which is defined as

$$\frac{\text{the actual volume of fresh gas}}{\text{the swept volume}}$$

measured at the suction pressure.

The actual volume of fresh gas is always different from the swept volume for the following reasons:

(*a*) Because the cylinder walls are warmer than the refrigerant gas leaving the evaporator, the gas within the cylinder expands and resists the flow of gas entering.

(*b*) The density of the gas within the cylinder is less than that about to enter because of the pressure drop which accompanies flow through the suction valve.

(*c*) Some clearance volume must be left at the top of the cylinder, otherwise there would be a danger of the piston damaging the top of the cylinder as the stroke increased, with wear on the crankshaft. The small amount of vapour trapped in this clearance volume (about 1·0 to 0·5 mm deep) at the top of the stroke, re-expands as the piston moves downward and reduces the volume available for the entry of fresh gas.

Unlike the internal combustion engine, the compression ratio does not depend on the ratio of the swept volume to the total cylinder volume but, instead, it is expressed as the ratio of the absolute condensing pressure to the absolute suction pressure. The value of the compression ratio will thus have an effect on the volumetric efficiency. Approximate values are as given in Table 12.1 which quotes the actual volumetric efficiency.

TABLE 12.1

Compression ratio	Volumetric efficiency
2·0	78
3·0	74
4·0	70
5·0	66
6·0	62

12.5 The interaction of the characteristics of evaporators and condensing sets

When an evaporator is coupled to a condensing set the point of rating is established by plotting the characteristics of the two on the same set of co-ordinates, using pressure or temperature as the abscissa and kW of refrigeration as the ordinate. These two characteristics refer to the refrigerant-side performance and their intersection gives the pressure or temperature at which evaporation will take place and the capacity that will be obtained, in tons of refrigeration.

EXAMPLE 12.5. The condensing set shown by curve Y in Fig. 12.6 is connected to the evaporator shown by curve BA in Fig. 12.3.

(*a*) Determine the evaporating temperature and the refrigeration capacity.

(*b*) If the answer obtained in part (*a*) is for the compressor running at 6 rev/s, at what speed must the compressor be run in order to obtain a capacity of 100 kW of refrigeration? What will the evaporating temperature then be?

Answer

Curves Y and BA are plotted, as shown in Fig. 12.7, when it will be seen that they intersect at P, the point of rating.

(*a*) The co-ordinates of P provide the required answers:
Evaporating temperature = +0·5°C.
Capacity = 103·4 kW of refrigeration.

(*b*) Because the capacity of a compressor is directly proportional to the mass of refrigerant pumped and, hence, to the speed at which it runs, at the same suction state, the revised running speed is

$$n_2 = 6 \times \frac{97{\cdot}5}{103{\cdot}4}$$

$$= 5{\cdot}65 \text{ rev/s.}$$

The evaporating temperature at 100 kW on the coil characteristic is 1°C.

If the wet-bulb temperature of the air entering the coil falls, then the evaporating temperature and the capacity also fall, as was remarked in section 12.3. The new point of rating can be established by plotting the new coil characteristic on the same co-ordinate system as that showing the performance of the condensing set.

EXAMPLE 12.6. Determine the evaporating pressure and the refrigerating effect for the conditions of operation used in Example 12.5 (the compressor running at 6 rev/s), if the entering wet-bulb is reduced to 13°C.

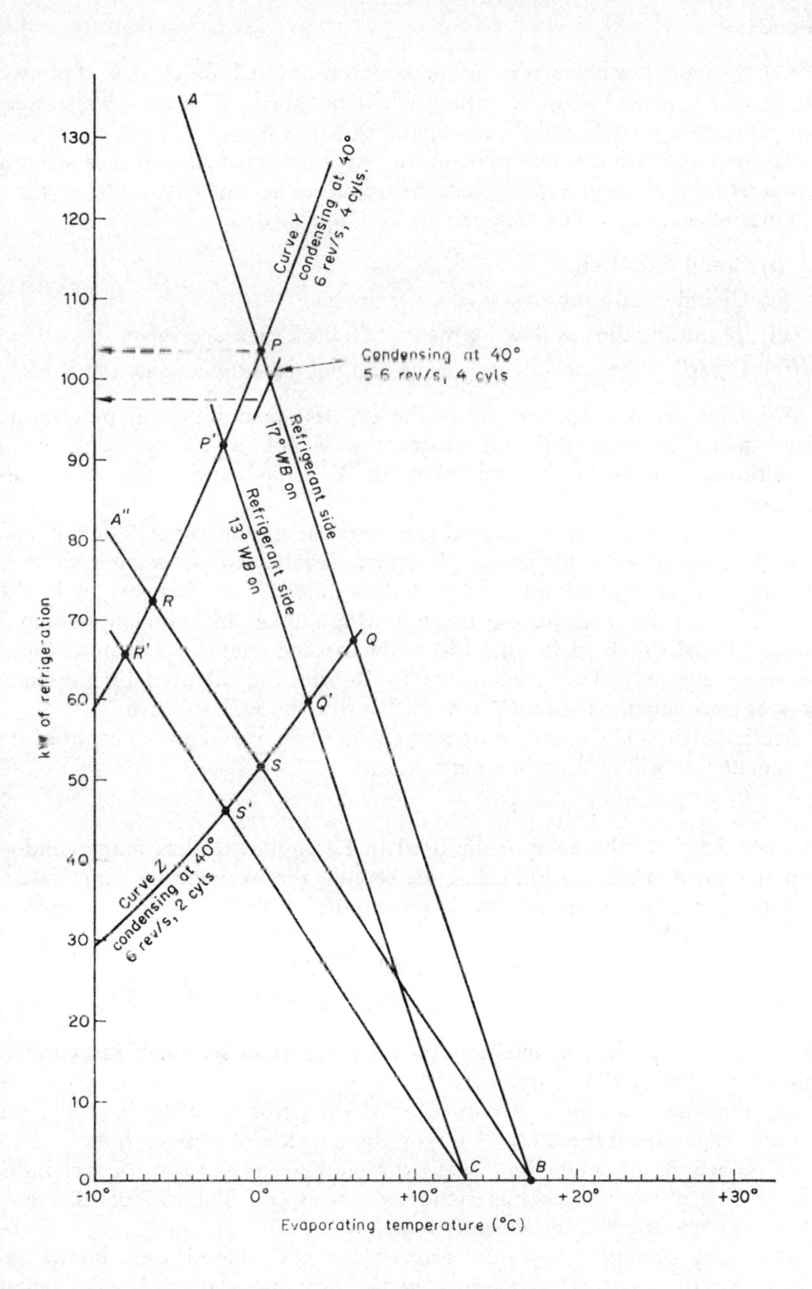

A
P
Curve Y
condensing at 40°
6 rev/s, 4 cyls.
Condensing at 40°
5·6 rev/s, 4 cyls
P'
17° WB on
Refrigerant side
13° WB on
Refrigerant side
A''
R
R'
Q
Q'
S
S'
Curve Z
condensing at 40°
6 rev/s, 2 cyls
C
B
kW of refrigeration
Evaporating temperature (°C)

Fig. 12.7

Answer

If a new evaporator characteristic for an entering wet-bulb of 13°C is plotted on Fig. 12.7, a new point of rating, P', is obtained. The new evaporating temperature is $-1{\cdot}7$°C and the new capacity is 92 kW.

To meet the reduced load without an excessive reduction in evaporating temperature, the compressor may be reduced in capacity. There are a number of ways in which this can be accomplished:

(i) Speed reduction.
(ii) Cylinder unloading.
(iii) Throttling the gas flow by means of a back pressure valve.
(iv) Diverting the gas flow by means of a hot gas by-pass valve.

Although speed reduction offers the greatest flexibility and permits an exact match between load and capacity, it is not used very much in air conditioning because of the cost involved. Variable-speed induction motors are expensive.

The commonest method adopted is to vary the number of cylinders in use. This is accomplished by lifting the suction valve off its seating so that, although gas is induced into the cylinders unloaded in this way, it is also expelled from the them during the following stroke, and no compression is done. Virtually no work is done and so the method offers an economic way of reducing capacity. The pressure of the lubricating oil used by the compressor is usually the hydraulic agency that lifts the suction valve.

Methods (iii) and (iv) are, more strictly, used for varying the output of the evaporator, as will be seen in what follows.

EXAMPLE 12.7. If the compressor used in Example 12.6 has four cylinders and if two of these are unloaded, determine the evaporating temperature and the capacity for (*a*) an entering wet-bulb of 17°C and (*b*) an entering wet-bulb of 13°C.

Answer

On Fig. 12.7 plot the characteristic for the compressor with half its cylinders unloaded. This is shown by curve Z.

(*a*) With an entering wet-bulb of 17°C the point of rating is Q and the system evaporates at about 6·3°C, giving about 67 kW of refrigeration.

(*b*) When the air enters the direct expansion cooler coil at 13°C wet-bulb, intersection of the two characteristics occurs at Q'. The evaporating temperature is about 3·7°C and the duty is 59 kW.

If the evaporating temperature drops below 0°C, there is a distinct possibility that the surface temperature of the evaporator may also fall below 0°C. With water chillers this will result in the formation of ice, and a risk of expensive damage. With direct-expansion air-cooler coils, frost will form on the external finned surface. In both these cases the process is an accelerating one. The formation of ice or frost reduces the heat transfer and the load drops still further, the evaporating temperature following suit. When frost forms, the reduction in heat transfer causes a tendency for the amount of super-heat to drop. This could result in wet vapour entering the compressor. There is a risk of damage to the compressor and valves if it is handling liquid instead of gas.

If the pair of coil sections considered in section 12.3 is coupled to a common compressor, it is easy to see what happens when one of the coil sections is taken out of the circuit by being valved-off.

EXAMPLE 12.8. If the cooler coil considered in Example 12.7 is divided into two equal parts, airflow over them being in parallel (*see* Fig. 12.4) and if the two coil sections are coupled with the compressor used in Example 12.7, determine the following:

(*a*) the evaporating temperature and duty if one coil is taken out of circuit by being valved-off, all four cylinders of the compressor continuing to pump,

(*b*) the evaporating temperature and duty if two of the cylinders of the compressor are unloaded when one section of the cooler coil is shut off.

Answer

It is necessary to plot the characteristic for a single section of the cooler coil. This was done on Fig. 12.3 (curve *BA*″) for an entering wet-bulb of 17°C, but no characteristic for the performance of the compressor was shown at the time. It is now possible to re-plot curve *BA*″ on Fig. 12.7 for an entry wet-bulb of 17°C.

(*a*) For an entering wet-bulb of 17°C, the intersection of characteristics for the half-evaporator and the four-cylinder compressor occurs at the point *R*. The evaporating temperature is −6·2°C and the duty is 72 kW. (If the entering wet-bulb were 13°C, the evaporating temperature would be −8°C and the duty would be 65·5 kW.)

(*b*) For an entering wet-bulb of 17°C the intersection takes place at *S*, for the half-evaporator and the compressor using only two cylinders. The evaporating temperature is 0·5°C and the duty is precisely halved to 51·7 kW (If the entering wet-bulb fell to 13°C, then evaporation would be at −1·7°C and the capacity would be 46 kW.)

Two points of interest emerge. First, halving the surface area of the evaporator but doing nothing to reduce the capacity of the compressor gives rise to a reduced evaporating temperature but does not give half the duty. (In this illustration the reduced duty of 72 kW is 70 per cent of the design duty.) Secondly, if the compressor is unloaded to match the reduced coil capacity, a proper balance can be obtained.

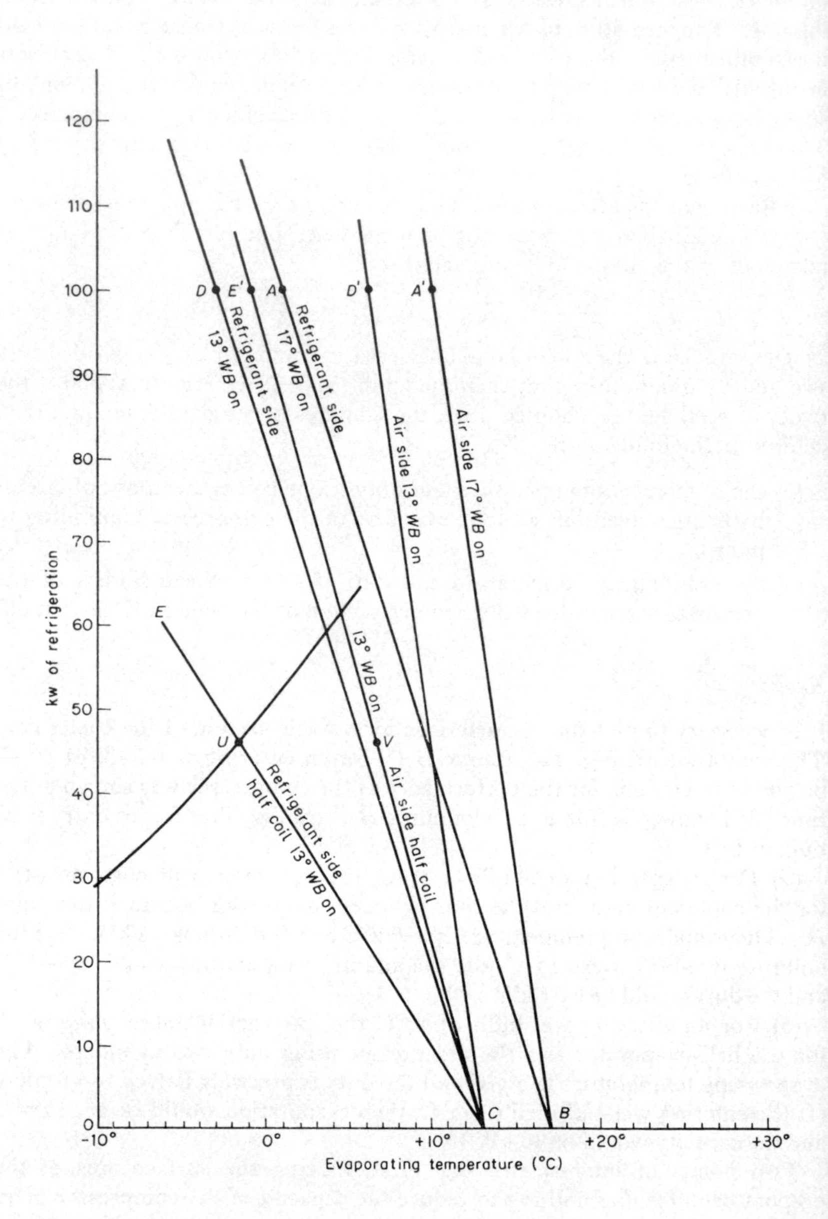

kw of refrigeration
Evaporating temperature (°C)
D
E'
A
D'
A'
Refrigerant side 13° WB on
Refrigerant side 17° WB on
Air side 13° WB on
Air side 17° WB on
E
U
V
13° WB on
Air side half coil
Refrigerant side half coil 13° WB on
C
B
120
110
100
90
80
70
60
50
40
30
20
10
0
-10°
0°
+10°
+20°
+30°

Fig. 12.8

EXAMPLE 12.9. Determine the wet-bulb temperature of the air leaving the direct-expansion cooler coil referred to in Example 12.8, the evaporating temperature, and the duty, if the entering wet-bulb falls to 13°C, half the coil being valved-off and the compressor having two of its cylinders unloaded to match.

Answer

The appropriate characteristics of performance are plotted (*see* Fig. 12.8) for the unloaded compressor, the refrigerant-side of the full coil with an entering wet-bulb of 17°C and the air-side of the full coil with an entry wet-bulb of 17°C. These two coil lines are denoted by the letters *BA* and *BA'*, respectively. The coil characteristics for an entering wet-bulb temperature of 13°C are parallel to *BA* and *BA'* but start from the point *C*, on the abscissa, at 13°C. They are denoted by *CD* and *CD'*.

The half-coil lines can now be plotted from these, in the manner described earlier, starting from the point *C*. They are shown in the figure by *CE* for the refrigerant-side and by *CE'* for the air-side.

The refrigerant-side characteristic intersects the characteristic for the unloaded compressor at the point *U*. This yields the information that the evaporating temperature is about −1·8°C and that the duty is about 46 kW of refrigeration. At the same duty, the point *V* is located on the air-side characteristic which, as can be seen, gives a leaving wet-bulb temperature of about 6·5°C.

A loss of pressure occurs in the suction pipe between the outlet from the evaporator and the suction valve on the compressor. This means that the evaporating pressure is not the same as the suction pressure. The point is illustrated in Fig. 12.9.

If there were no friction in the suction pipe, the evaporating pressure and the suction pressure would be identical. The intersection of the two characteristics at the point 1 shows this. The evaporating pressure is 505·5 kN/m^2 absolute, corresponding to a saturation temperature of 0°C for R22. If the pressure drop due to friction in the suction pipe is, say, 17 kN/m^2, then evaporation will occur at the point 3 and the gas will enter the suction valve at the point 2. The duty will be 100 kW of refrigeration instead of 103·4 kW. The suction pressure will be 497 kN/m^2 absolute and the evaporating pressure will be 514 kN/m^2, corresponding to the saturation temperatures of −1°C and +1°C, read from the points 2 and 3 in the figure. Since the capacity of the compressor is directly proportional to the mass flow of refrigerant gas and since the frictional loss is also related to the mass flow of gas, another characteristic can be drawn showing the reduced performance resulting from the frictional losses in the suction pipe. This is shown by the broken line in Fig. 12.9. It is only true, of course, for one particular piping arrangement and so it is more usual to give compressor performances without taking this into account.

The above remarks lead naturally to the use of a back-pressure valve for varying the duty. A back-pressure valve is only a device which varies the frictional loss in the suction pipe. There are several ways in which this can be used to advantage. A variation in the temperature of air recirculated

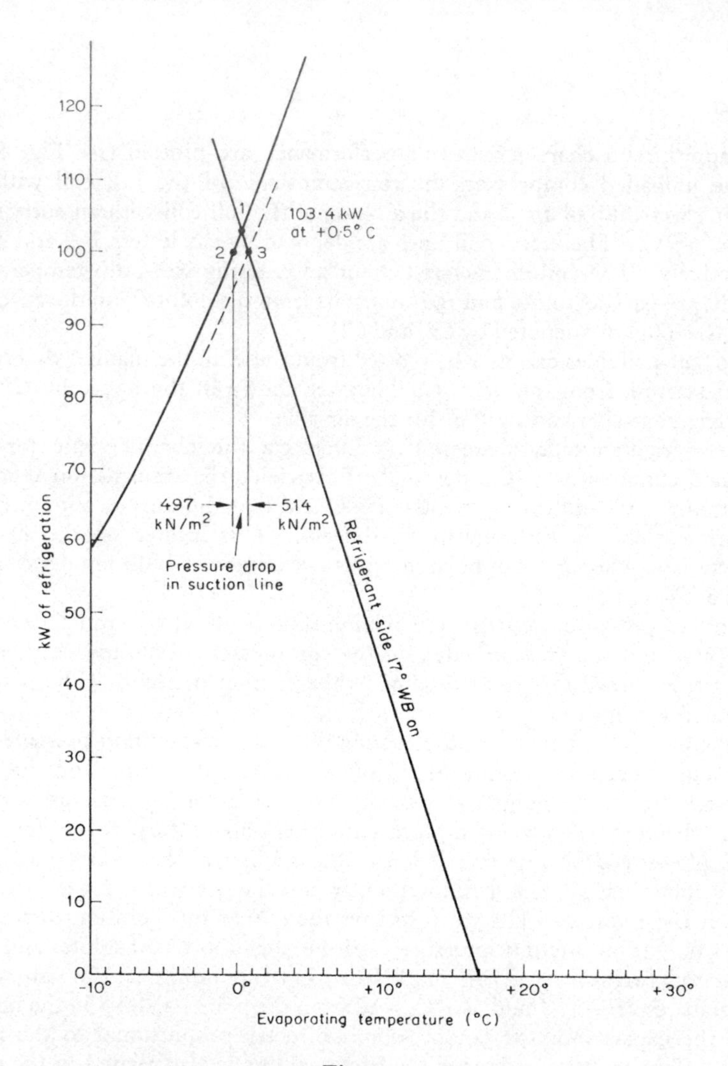

Fig. 12.9

from a conditioned space can be used to send a signal to the valve instructing it to vary the flow of gas in the suction pipe. A fall in the temperature of the recirculated air would ask the valve to throttle the flow of gas and so

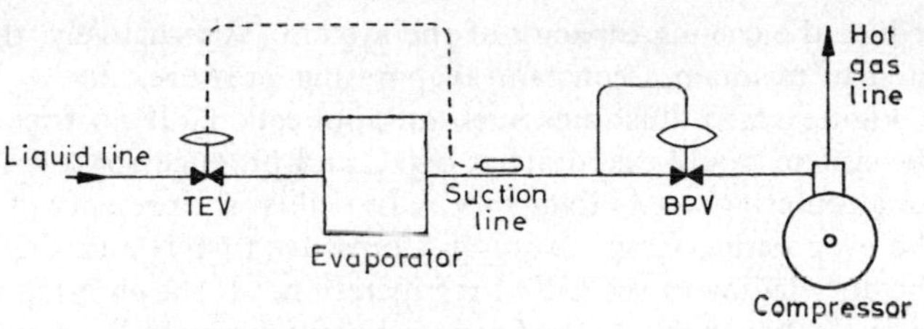
Hot gas line
Liquid line
TEV
Evaporator
Suction line
BPV
Compressor

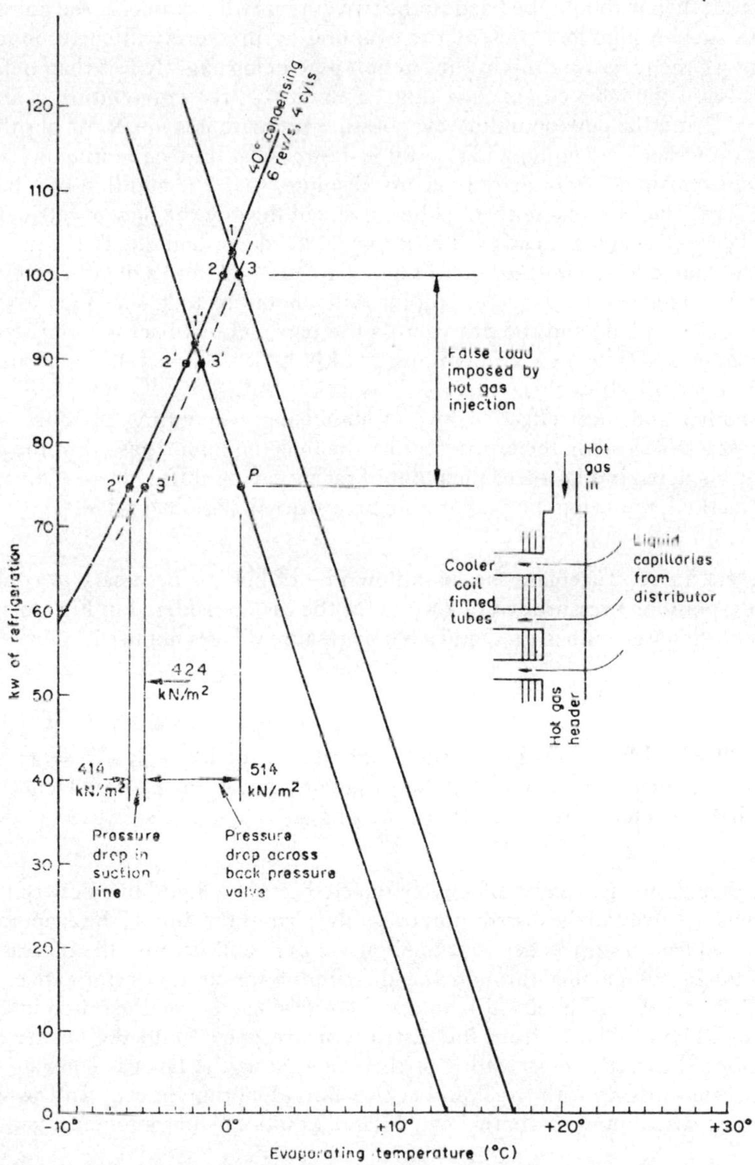
40° condensing 6 rev/s, 4 cyls
kw of refrigeration
Evaporating temperature (°C)
False load imposed by hot gas injection
Hot gas in
Liquid capillaries from distributor
Cooler coil finned tubes
Hot gas header
424 kN/m²
414 kN/m²
514 kN/m²
Pressure drop in suction line
Pressure drop across back pressure valve

Fig. 12.10

reduce the cooling capacity of the system. Alternatively, the valve could be used to maintain a constant evaporating pressure.

Figure 12.10 illustrates such an application. If no friction were present, the system would evaporate at 0·5°C and produce 103·4 kW of refrigeration, for an entering wet-bulb of 17°C. In reality, the presence of friction increases the evaporating temperature and pressure to 1°C and 514 kN/m^2 absolute, the duty falling to 100 kW of refrigeration. If the entering wet-bulb reduces to 13°C, then without pipe friction the two curves will intersect at the point 1′. With suction pipe loss present the evaporating pressure will correspond to point 3′, the pressure drop in the suction pipe being slightly less than before, because of the reduced gas flow rate. At point 3′, the temperature is about −1·2°C and the corresponding evaporating pressure is 479 kN/m^2 absolute. The duty is only about 89·2 kW. If it is desired that the evaporating pressure should remain constant at 514 kN/m^2 absolute, then the addition of a back-pressure valve can ensure this. The valve will throttle the flow of refrigerant until evaporation occurs at 514 kN/m^2 absolute, corresponding to the point P on the characteristic for the coil in Fig. 12.10, at +1°C and a duty of just over 75 kW. The loss in the suction piping will amount to 10 kN/m^2 (424 kN/m^2 minus 414 kN/m^2) and the drop across the partly closed back-pressure valve will be 90 kN/m^2 (514 kN/m^2 minus 424 kN/m^2). The points 2″, 3″ and P in the figure illustrate this.

Another and most effective way of stabilising evaporating pressure is to impose a false load on the evaporator by the injection of hot gas. For the case just considered in Fig. 12.10 the point of rating can be shifted from 3″ to 3, by this method, returning the evaporating pressure to its original value.

EXAMPLE 12.10 Calculate the mass flow rate of hot gas necessary to control the evaporating pressure at 514 kN/m^2 for the case considered in Fig. 12.10 if the entering wet-bulb is 13°C and a back pressure valve is not used.

Answer

The false load to be imposed on the evaporator is 100 kW − 75 kW = 25 kW. Liquid refrigerant at 514 kN/m^2 has a latent heat of 205 kJ/kg. Hence the required mass flow rate is 25/205 = 0·122 kg/s.

Although hot gas can be effectively injected into the liquid line between the expansion valve and the distributor, this only permits the duty of the evaporator to be reduced from 100 per cent to about 60 per cent, because the increasing amount of gas passing through the distributor upsets its performance. A much better way of injection is to use a hot gas header, as illustrated in Fig. 12.10. The capillaries from the distributor are brazed into the header and feed liquid directly to each tube of the evaporator, the hot gas entering the header and mixing with the liquid at the individual tube inlets. In this way the best distribution of both hot gas and liquid can be obtained over the cooler coil and the turn-down of the cooling capacity is limited only by the quality of the valve used to modulate the hot gas flow.

Instead of controlling cooler coil capacity from evaporating pressure (for which a self-acting valve can be used) it is possible to use a motorised or pneumatic valve, with a bellows seal in place of a gland. The capacity of the evaporator can then be modulated, with integral action or other control refinement if desired, from the temperature of the air leaving the coil.

The use of hot gas valves is not confined to air cooler coils; some centrifugal chillers use them to improve the turn-down of their inlet guide vanes.

12.6 The condenser

The rejection of heat from condensers was discussed in Chapter 11. The treatment here is confined to water-cooled condensers of the shell-and-tube type.

Hot superheated gas usually enters at the top of the shell and is de-superheated and condensed to a liquid at high pressure and temperature by coming in contact with horizontal tubes which convey the cooling water. Some sub-cooling of the liquid below the saturation temperature corresponding to the pressure prevailing inside the shell also occurs and is desirable. The condensed liquid is usually only enough to fill about the lower fifth of the shell, but the shell must be large enough to contain the full charge of refrigerant when repair or maintenance work is carried out on other components of the system. There is thus no need for a separate liquid receiver.

The capacity of a condenser to reject heat depends on the difference of temperature between the condensing refrigerant and the cooling water. Thus, its capacity may be increased by raising the condensing temperature or by increasing the rate of water-flow. Raising the condensing temperature has the side effect of lowering the capacity of the compressor, as has been observed earlier.

Some choice is open to the designer of a refrigeration system: a small condenser may be chosen, giving a high condensing temperature, or vice-versa. The first choice will result in a motor of larger horsepower to drive the compressor, with consequent increased running costs. The alternative will give lower running costs, but the size and capital cost of the condenser will be greater.

A typical choice of water flow rate is about 0·06 litres/s for each kW of refrigeration, rising in temperature from 27°C to 32°C, as it flows through the condenser.

12.7 Piping and accessories

From a thermodynamic point of view, the only essential substance in the system is the refrigerant, anything else is undesirable. Of the other substances which may find their way into a system, only oil is necessary, and this only because the compressor requires a lubricant for its proper operation.

Thus, air, water and dirt are not wanted, and every possible step must be taken to ensure that they are absent. Their total absence may be impossible but care in the proper choice of materials, and in assembly and charging, can minimise their presence.

The piping used must be of a material which will not be attacked by the refrigerant (for example, copper is acceptable with the fluorinated hydrocarbons but not with ammonia) and, apart from having the necessary structural properties, it must be clean and dehydrated, prior to use. The same holds true for all fittings, valves, and other components used.

Piping must be large enough to permit the flow of the refrigerant without undue pressure drop and, most important, it must permit the oil in the system to be carried to the place where it is wanted, that is, to the crankcase. Inevitably, some oil is discharged with the high-pressure refrigerant from the compressor, into the condenser. When the refrigerant is in the liquid state, it mixes well with the oil, if it is a fluorinated hydrocarbon, and both are easily carried along. When the refrigerant is a gas, however, mixing does not occur. The oil turns into a mist and the pipes must be sized to give a refrigerant gas velocity high enough to carry over the lubricant. It is clear from this that a problem may arise under conditions of partial load, when the system is not pumping the design quantities of gas. Oil usually collects on the walls of the piping and drains down into the low points of the system. For this reason, proper oil return to the crankcase can be assisted by pitching the pipes. An oil separator between the compressor and the condenser is also useful, particularly in air-conditioning applications where large variations of load occur.

An oil pressure failure switch is essential. This is interlocked with the compressor motor starter so that if the pressure falls below a minimum value, the compressor is stopped, damage from lack of lubrication being prevented.

Water in a system can cause trouble in a number of ways and to minimise this a dryer is sometimes inserted in the piping. This should not be regarded as a valid substitute for proper dehydration in the assembly and setting to work, but should be considered in addition to it.

Strainers should be fitted upstream of the expansion valve and any solenoid valve which may accompany it.

Sight glasses are essential. They should be placed in the liquid line between the condenser or liquid receiver and the expansion valve. Their purpose is to verify the presence of an adequate charge in the system and to monitor the state of the liquid entering the expansion valve.

Gauges are also essential although they are not always fitted on some of the modern packaged plant. They permit a proper check to be kept on the running of the plant and they are of assistance in commissioning the plant and in locating faults.

Gauge glasses are necessary, although they may not always be present. Their purpose is to establish the levels of the liquid in the condensers and receivers, this being of particular use in charging the system.

12.8 Charging the system

Before any charge of refrigerant is put into a system, it must be thoroughly pressure-tested for leaks on both the low and high side of the compressor. Anhydrous carbon dioxide should be used for this and all joints and connexions carefully inspected, soapy water being used for a bubble test. A watch on the pressure gauges will indicate if a serious leak is present. Following this, a small amount of R12 should be added to the system, and all joints, pipework and connexions gone over with a hallide torch. (This is a tracer technique for detecting leaks when fluorinated hydrocarbons are used. Any escaping refrigerant reacts with the flame of the torch to produce a green colour.)

The system is regarded as free from leaks if, after having been left under pressure for 24 hours, no variation is observed in the gauge readings, due attention being paid to variations in the ambient temperature, which will alter the gauge readings.

When the system is free from leaks, it must be dehydrated. This is accomplished by using an auxiliary vacuum pump (never the refrigeration compressor itself) to pull a hard vacuum on the entire system. Dehydration can be accomplished in this way only if liquid water in the system is made to boil, the vapour then being drawn off. At 20°C, the saturation vapour pressure is 2·337 kN/m^2 absolute. That is, a vacuum of about 99 kN/m^2. The application of external heat to the system may be of help in attaining the required amount of dehydration.

Following the above testing and dehydration procedure, the ancillary systems should be set to work. By this is meant the associated air handling and water handling plant, together with their automatic controls. It is not possible to charge and set to work a refrigeration system unless this is done. For example, unless the cooling-water pump is circulating an adequate amount of water through the condenser, the refrigeration plant will fail continually on its high-pressure cut-out.

The system should be charged from a cylinder or drum of refrigerant through a dehydrator. For any given system, if the design duty is to be obtained, only one weight of refrigerant is the correct amount for it to operate between the design condensing and evaporating pressures. For packaged plants of well-known size and performance, a measured weight of refrigerant should give the desired operation. However, for plants of a more bespoke character, sight glasses and gauge glasses may be used to advantage in assessing the necessary amount of the charge.

12.9 Centrifugal compressors

Reciprocating compressors running at about 24 rev/s have capacities of the order of 25 kW per cylinder for R12, and 35 kW per cylinder for R22 when used for conventional air-conditioning applications. The maximum number of cylinders on one machine is about 16, so this puts the maximum capacity of a piston machine at about 550 kW of refrigeration, excluding freak machines with large strokes and bores. Although one reciprocating compressor of this size is likely to be cheaper than a centrifugal machine of like capacity, the complication of controlling its output (cylinder unloading) may put it in an unfavourable light compared with a centrifugal compressor, which can have a modulating control exercised over its capacity. Centrifugal systems are available for capacities as low as 280 kW of refrigeration, but they come into their own, economically speaking, at about 500 kW.

Whereas a reciprocating compressor is a positive displacement device, a centrifugal compressor is not. If the flow of gas into a piston machine is throttled it will continue to pump, even though the amount delivered is small, provided that its speed is kept up by an adequate input of power to its crankshaft. There is no ' stalled ' condition. This is not so with a centrifugal compressor. The rotating impeller of a centrifugal compressor increases the pressure of the gas flowing through its channels by virtue of the centrifugal force resulting from its angular velocity. The velocity of the impeller is constant in a radial direction but the linear velocity in a direction at right angles to the radius of the impeller increases as the radius gets greater. The energy input to the gas, which is rotated within the impeller, thus increases towards the periphery of the wheel. This input of energy is what causes the gas to flow outwards through the impeller against the pressure gradient, that is, from the low pressure prevailing at the inlet eye to the high pressure existing at the periphery. The function of the impeller casing, or the volute, is to convert the velocity pressure of the gas leaving the wheel to static pressure, with as much efficiency as possible.

In addition to the radial movement imposed on the gas by the impeller, the gas stream tends to rotate relative to the impeller. This is illustrated in Fig. 12.11(*a*). On an absolute basis, any particular particle of gas will tend not to rotate but, since the wheel is rotating, the particle will appear to rotate relative to the wheel. The point P_1 is facing the convex side of an impeller blade initially but, later in the process of rotation, denoted by P_4, it is facing the concave side of the preceding impeller blade. The effect of this is to produce a circulatory movement of the gas within the impeller, as shown in Fig. 12.11(*b*). It can be seen that this circulatory movement assists the flow to the periphery of the wheel, produced by centrifugal force, on the convex side of a blade but retards it on the concave side. This effect introduces losses which can be minimised by using a wheel with narrow channels between the impeller blades.

For a given compressor, running at a given speed, the pressure-volume characteristic curve is virtually a straight line, as shown in Fig. 12.12, if no losses occur. Losses do occur, however. They are the circulatory losses just described, losses due to friction, and losses caused by the fact that the gas entering the impeller has to change direction by 90 degrees, as well as having a rotation imposed on it. These entry losses may be reduced by giving the gas a swirl before it enters the inlet eye of the impeller. There is a proper angle of swirl for each rate of gas flow, that is, for each load. Variable-inlet guide vanes are fitted on all modern centrifugal compressors. Their position is adjusted to suit variations in the load, which permits a modulating control of output with little alteration in efficiency. The intention is that the machine should operate at a design point which involves the minimum loss or, put another way, the maximum efficiency.

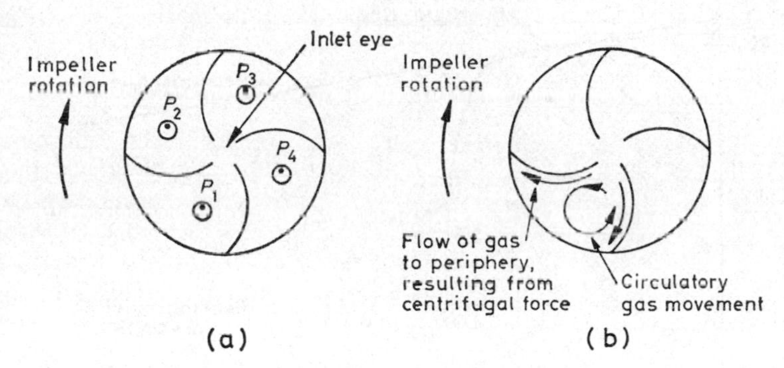

Fig. 12.11

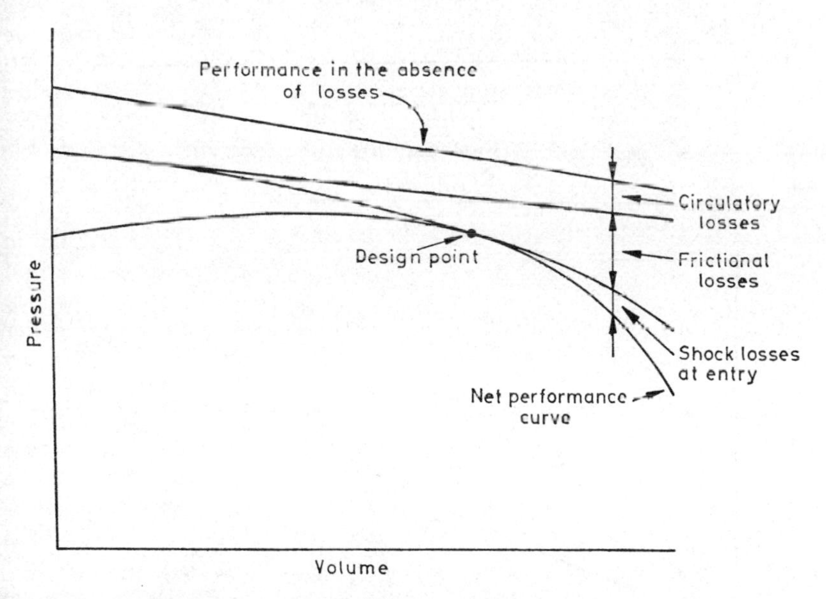

Fig. 12.12

A centrifugal impeller is designed to pump gas between the low suction pressure and the high condensing pressure. If the condensing pressure rises, the difference between these two pressures exceeds the design value and the compressor quite soon finds the task of pumping beyond its ability. Thus, whereas the reciprocating machine will continue to pump, but at a

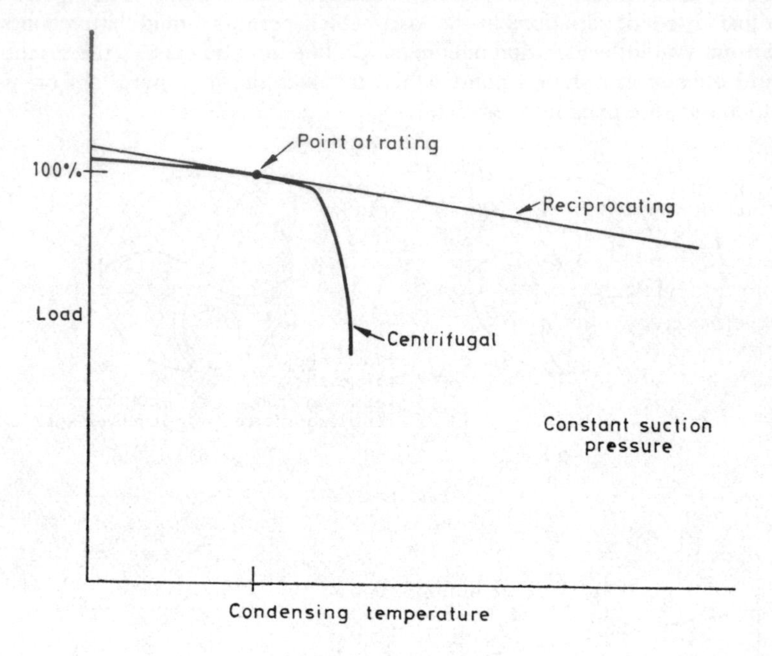

Fig. 12.13(*a*)

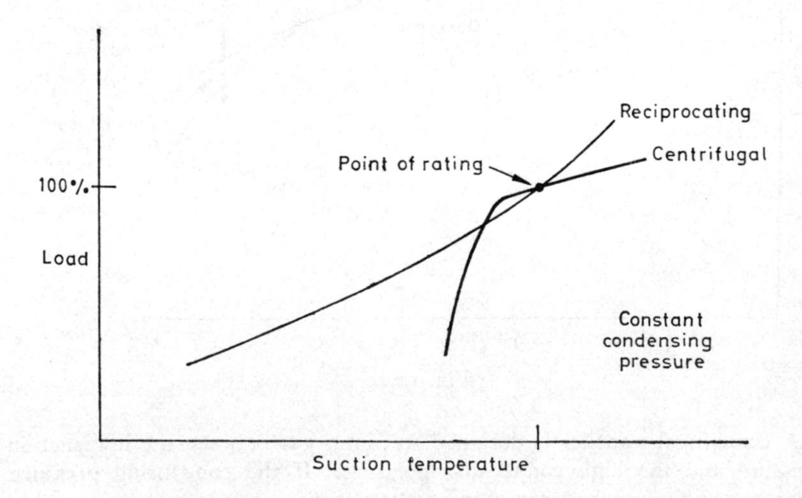

Fig. 12.13(*b*)

steadily reducing rate as the condensing pressure rises, the pumping capacity of the centrifugal compressor falls rapidly away. This is illustrated in Fig. 12.13(*a*). A similar behaviour is seen if the suction pressure is reduced, the condensing pressure being held at a constant value, as Fig. 12.13(*b*) shows.

This feature of the performance of the centrifugal machine gives rise to a phenomenon termed 'surging'. When the pressure difference exceeds the design pumping ability of the impeller, flow ceases and then reverses, because the high condensing pressure drives the gas backwards to the lower suction pressure. Pressure in the evaporator then builds up, and the difference between the high and low sides of the system diminishes until it is again within the ability of the impeller to pump. The flow of gas then resumes its normal direction, the pressure difference rises again, and the process repeats.

This oscillation of gas flow and the rapid variation in pressure difference which produces it is *surging*. Apart from the alarming noise which surging produces, the stresses imposed on the bearings and other components of the impeller and driving motor may result in damage to them. Surging continuously is most undesirable, but some surge is quite likely to occur from time to time, unless a very careful watch is kept on the plant. This is particularly likely with plants which operate automatically and are left for long periods unattended. Surge is likely to occur under conditions of light load (when the suction pressure is low) coupled with a high condensing temperature.

The proper use of inlet guide vanes can give a modulating control of capacity down to 15 per cent or even, it is claimed, to 10 per cent of design full load.

The high heads necessary for air-conditioning applications may be developed in two ways: either by running the impeller fast enough to give the high tip speed wanted or by using a multi-stage compressor. High tip speeds can also be obtained by using large-diameter impellers, but if diameters are excessively large, structural and other difficulties arise.

High speeds of rotation are usually produced by using speed-increasing gears which multiply the 48 rev/s normally obtainable from a two-pole induction motor. Another method adopted was to use a motor-generator set which multiplied the mains frequency by a factor of 6 or more. The motor actually driving the compressor could then be run at 285 rev/s, without the use of speed-increasing gearing. This method had the advantage that because the speed of rotation could be very much higher than gearing could produce, the diameter of the impeller might be considerably reduced and a more compact plant arrangement obtained. Development has unfortunately been difficult.

Although speed-increasing gear is noisy it has largely displaced two-stage compression. Figure 12.14(*a*) shows a diagrammatic arrangement of two-stage compression with an 'intercooler' added. The purpose of this is to improve the performance of the cycle by using two expansion valves, *A* and *B*, and feeding some of the low pressure gas at state 9, obtained after valve *A*, to the suction side of the second stage compressor. Figure 12.14(*b*) describes the cycle in terms of the temperature-entropy changes involved.

Hot, high-pressure liquid leaves the condenser at a state denoted by 5. The pressure and temperature of the fluid is dropped by passing it through the first expansion valve, *A*. Some liquid flashes to gas in the process, the flash gas being fed from the intercooler (merely a collection vessel) to the inter-

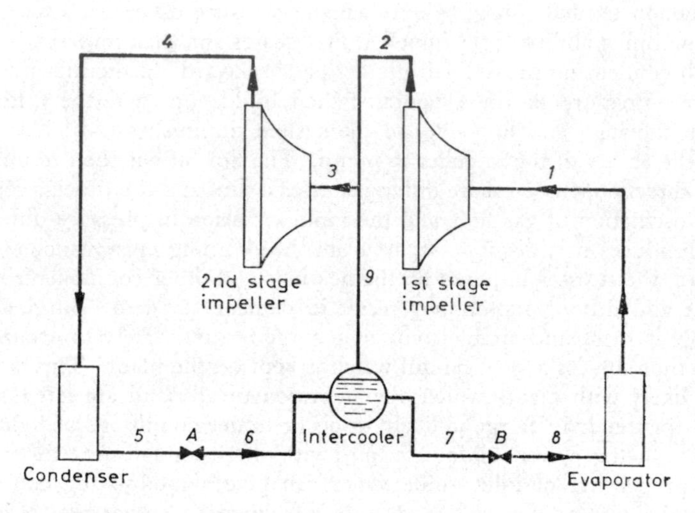

Fig. 12.14(*a*)

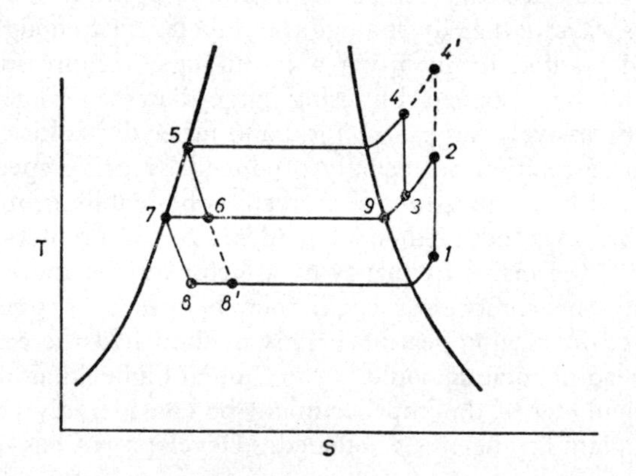

Fig. 12.14(*b*)

mediate stage of compression. The remaining liquid passes from the intercooler through the second expansion valve, *B*, and thence to the evaporator. Gas leaves the evaporator at state 1 and enters the first stage impeller, being compressed to state 2. When gas at state 2 mixes with gas at state 9, from the intercooler, it forms state 3 and enters the second stage impeller, leaving this at state 4. This then enters the condenser, and the process is repeated.

If the intercooler is not used, the cycle will follow the line 1–4′–5–8′–1. The refrigerating effect, represented by the area beneath 8′–1, is clearly less than the effect produced when an intercooler is used, represented by the area beneath 8–1.

12.10 The screw compressor

This is a positive displacement machine used with refrigerants at pressures above atmospheric for chilling water in air conditioning applications, usually with a water-cooled condenser. The compressor housing contains a pair of intermeshing screws, one with male lobes and the other female. The male screw is directly coupled to a semi-hermetic, 2-pole, squirrel-cage motor, although open drive arrangements are also possible. No suction or discharge valves are used. Instead, rotation of the screws exposes a space of increasing volume as the male lobe leaves the female, on each thread of the screw at the suction end of the housing, refrigerant gas flowing into the openings presented. For each thread the male lobe then starts to cross the opening and close it, further rotation compressing the trapped gas in the inter-lobe space. Compression is continuous, hot, high-pressure gas being discharged from the far end of the housing as inter-lobe spaces are offered to it. If each screw has four lobes and if the drive is at 48 rev/s, 192 inter-lobe spaces are shown to the suction and discharge openings per second and uniform gas flow and compression occurs. In fact, there are six female and four male lobes.

Smooth, modulating capacity control down to 10 per cent of design duty is achieved by means of a sliding valve in the bottom of the housing. As this is progressively opened suction gas escapes from the inter-lobe spaces and recirculates without being compressed. The compressor displacement is thereby reduced and the capacity diminished with little loss. The partial load characteristic performance is a shallow curve, extending from 100 per cent power input at full load to about 20 per cent input at 10 per cent duty.

Although some of the earlier screw machines, which used gear trains, produced a good deal of noise and vibration, modern versions that have overcome these problems are readily available and give satisfactory results. At present they are obtainable in the range from about 350 kW to 2000 kW of refrigeration. Because, like reciprocating compressors, they are positive displacement machines that will continue to pump gas provided sufficient power is fed into them, surging is impossible and they are therefore eminently suitable for heat-reclaim applications where condenser cooling water is required at the highest practical temperature under conditions of low cooling load. For such applications a double-bundle condenser is used, one bundle of tubes handling dirty water from the cooling tower in the usual way and the other being used to reject some or all of the heat to clean water circulated from heater batteries.

BIBLIOGRAPHY

1. W. H. Carrier, R. E. Cherne, W. A. Grant and W. H. Roberts. *Modern Air Conditioning, Heating and Ventilating*. Pitman, 1959.
2. W. F. Stoecker. *Refrigeration and Air Conditioning*. McGraw-Hill, 1958.
3. *Trane Air Conditioning Manual*. The Trane Company, La Cross, Wisconsin.
4. *Trane Refrigeration Manual*. The Trane Company, La Cross, Wisconsin.
5. M. A. Ramsey. Non-freeze design for coils—2, *Air Conditioning, Heating and Ventilating*, February 1962.
6. M. A. Ramsey. Control for all outside air—2, *Air Conditioning, Heating and Ventilating*, May 1964.
7. M. A. Ramsey. Refrigerant piping guide, *Air Conditioning, Heating and Ventilating*, March 1965, 71–86.

13

Automatic Controls

13.1 The principle of automatic control

In general, the design capacity of any component in an air-conditioning system will not match the load. The loads are continually changing, so the steady-state design condition is an unusual one. This means that the state maintained within the conditioned space will not stay constant—if the plant is left to run wild the capacity will exceed the load for most of the time.

Taking as a particular case the capacity of an air-heater battery installed in a plenum ventilation system, a fall in the heating load, associated, say, with a rise in the outside air temperature, will result in the battery output exceeding the load. If the battery does not have its capacity reduced, an increase in room air temperature will occur. Thus, if it is desired to keep the room air temperature at a constant value in the face of load fluctuations, the output of the battery must be appropriately varied. If this change of capacity is to be achieved automatically, several problems must be dealt with. First, the temperature in the room must be measured. Secondly, any variation in this temperature must be used to send a signal to the heater battery. Thirdly, the strength of the signal must be used to initiate a variation in the output of the battery and a means must be found of regulating the capacity of the battery in response to this signal. Finally, the time taken to achieve all this must not be so long that the load changes greatly in the meantime and the battery output becomes out of phase with it.

The need to measure a variation in the room temperature implies that the temperature cannot be kept at a fixed value; some change must occur in order that a variation in the load can be sensed. This deviation is inherent in many systems of control but is minimised with certain sophisticated systems described later.

The transmission of the signal from the measuring device can take several forms but its strength must be related to the measured deviation. Thus, the capacity regulating device at the battery can gauge its response according to signal strength.

The principle whereby an observed deviation in the load is made to give a corrective response is termed ' negative feedback ', and a system of controls making use of this principle is termed a ' closed loop '.

A closed-loop is not essential in the whole of an automatic-control system although elements of the system may be sub-systems which themselves use negative feedback. An open-loop system of this sort can give successful control: for example, a change in the value of the controlled variable (room temperature, say) may not be directly measured but, instead, an alteration in the variable which produces the load change (outside air temperature, say) may be measured. In the context of the illustration

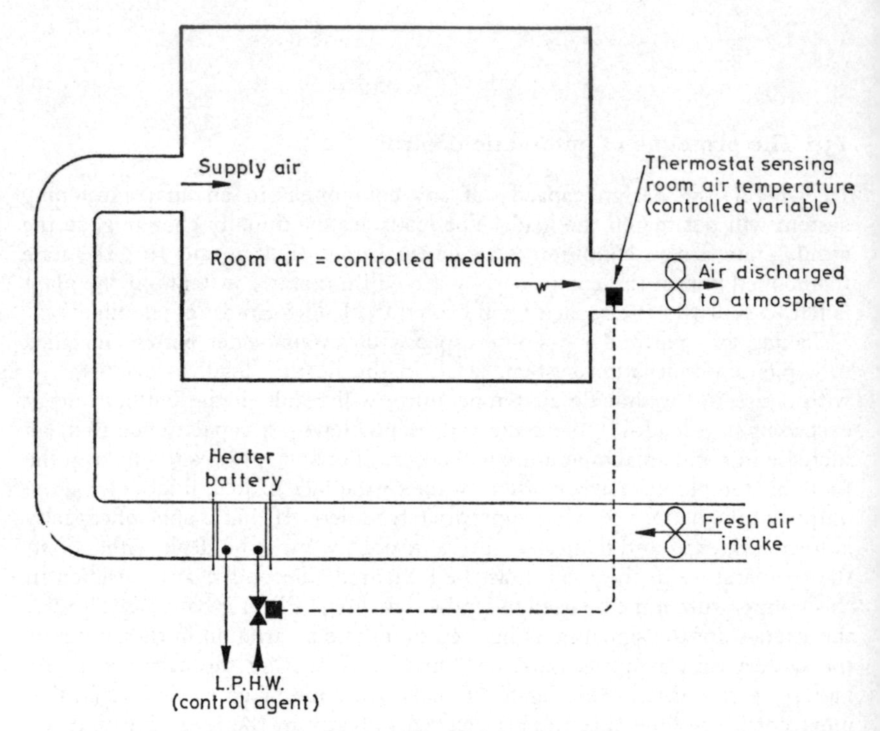

Fig. 13.1 Closed-loop system

making use of the heater battery in the plenum system, an open-loop system of automatic control could be adopted as follows. Fabric heat losses are directly proportional to the difference between the inside and outside air temperatures. Hence, ideally, there is a unique battery output for every outside air temperature that will give the correct inside air temperature without any deviation.

Summarising:

A closed-loop system (illustrated in Fig. 13.1) is one which measures a departure from the desired value of the controlled variable and feeds this information back to the device which regulates the capacity of the manipulated variable, so that corrective action may be taken.

An open-loop system does not make use of negative feedback from the controlled variable, but regulates the manipulated variable in some pre-arranged manner.

13.2 Definitions

For a discussion of the theory, operation and application of automatic controls to be developed, some agreement must be reached on the terms to be used. A British Standard (B.S. 1523: Part 1: 1967) exists which provides a glossary of terms used in automatic process control, but these terms differ in some respects from those in common use in both America and in the U.K. The terms in common use are defined below and, where appropriate, reference is made to the corresponding British Standard term.

1. *Controlled Variable* (B.S. Controlled Condition)
The quantity or physical property measured and controlled—for example, room air temperature.

2. *Controlled Medium*
The substance which has a physical property that is under control—for example, the air in a room.

3. *Manipulated Variable*
The physical property or quantity regulated by the control system in order to achieve a change of capacity which will match the change of load—for example, the flow rate of L.P.H.W. through a heater battery.

4. *Control Agent*
The substance whose physical property or quantity is regulated by the control system—for example, the L.P.H.W. fed to a heater battery.

5. *Desired Value* (also B.S.)
The value of the controlled variable which it is desired that the control system will maintain.

6. *Set Point* (B.S. Set Value)
The value on the scale of the controller at which it is set. For example, a thermostat may have its pointer set against the figure 21°C on the scale.

7. *Control Point*
The value of the controlled condition which the controller is trying to maintain. This is a function of the mode of control. For example, with proportional control and a set point of 21°C ± 2°C, the control point would be 23°C at full-load and only 21°C at 50 per cent load.

8. *Deviation* (also B.S.)
The difference between the set point and the measured value of the controlled condition, at any instant. For example, although the set point on the thermostat is 21°C, the measured value of the room air temperature may be 19°C. The deviation is then −2°C, at that instant of time.

9. *Offset* (also B.S.)
A sustained deviation caused by some inherent characteristic of the control system. For example, if the set point is 21°C and the measured room air temperature is at a steady value of 19°C for some period of time, then the offset is −2°C, for this period.

10. *Primary Element* (B.S. Measuring Unit)
That part of the controller which responds to the value of the controlled condition (B.S. Detecting Element) and gives a measured value of the condition (B.S. Measuring Element)—for example, a bimetallic strip type of thermostat.

11. *Final Control Element* (B.S. Correcting Unit)
This is the mechanism which alters the value of the manipulated variable in response to a signal initiated at the primary element, for example, a motorised valve.

12. *Automatic Controller* (B.S.)
A device which compares a signal from the detecting element with the set point and, hence, initiates a corrective action to reduce the deviation—for example, a room thermostat.

13. *Differential Gap* (also termed Differential)
This is the smallest range of values through which the controlled variable must pass for the final control element to move from one to the other of its two possible positions. For example, if a two-position controller has a set point of 21°C ± 1°C, then the differential gap is 2°C. *See* Fig. 13.2.

14. *Proportional Band* (also B.S.)
Also known as the throttling range, this is the range of values of the controlled variable which corresponds to the movement of the final control element between its extreme positions. For example, if a proportional controller has a set point of 21°C ± 2°C, then the proportional band is 4°C. *See* Fig. 13.3.

15. *Cycling* (B.S. Hunting)
This is a persistent periodic change in the controlled condition which is self-induced.

13.3 Measurement and lag

The accuracy with which the value of the controlled variable is measured is the basis of the accuracy of control exercised by the automatic system. Tight control over the condition cannot be expected if the primary element is only capable of coarse measurement. The crudity of a primary element can, in a sense, be summed up by referring to its measuring lag. Lag, in general, is a delay in the response of one part of a control system to a signal initiated by another. For example, in a thermostat used for controlling room air temperature, measuring lag can arise from several causes. A change in the sensible heat gain to the room (the load) causes a departure in the value of

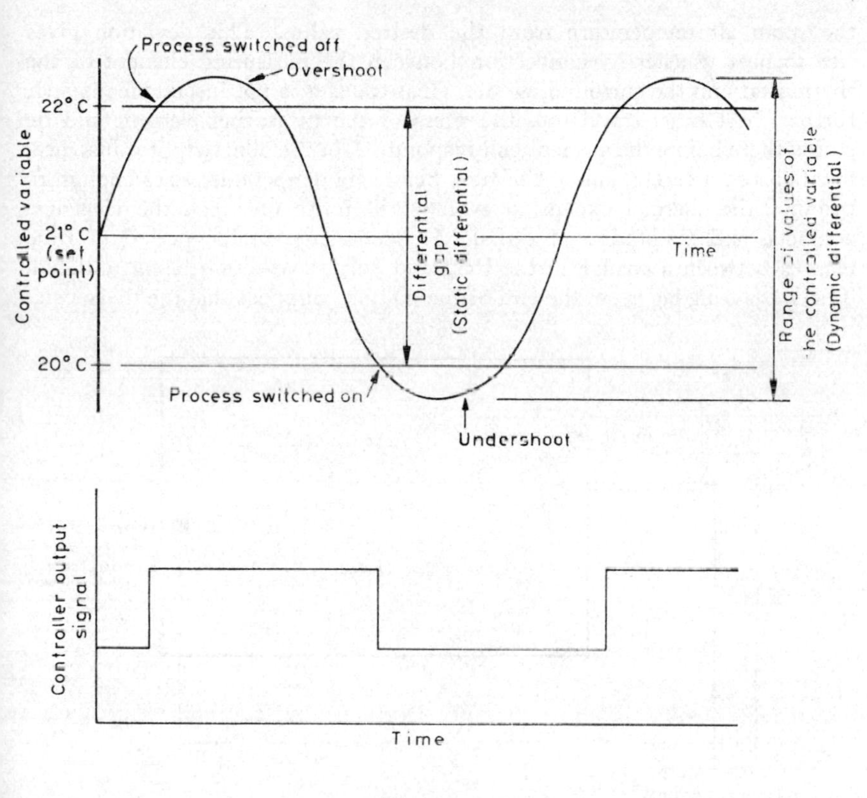

Fig. 13.2

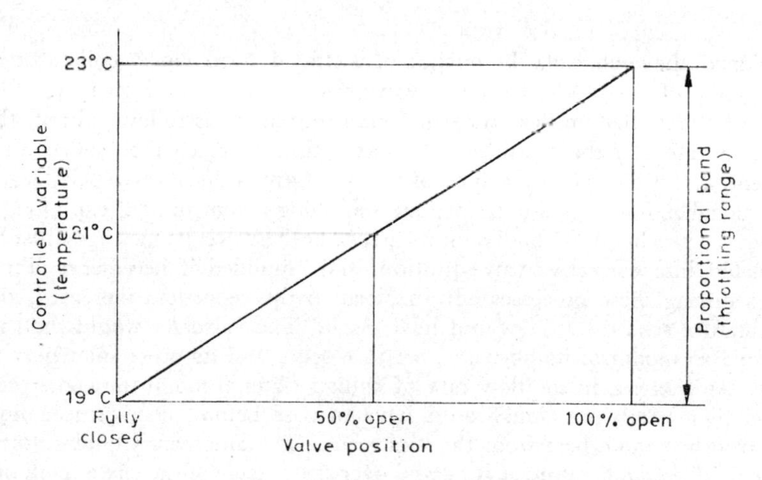

Fig. 13.3

the room air temperature from the desired value. This deviation gives rise to heat transfer by convection between the measuring element of the thermostat and the surrounding air. Heat transfer is not instantaneous and, further, heat must travel into the mass of the measuring element and be stored there before the element can respond. A bi-metallic strip, for instance, must absorb a certain amount of heat before its temperature rises enough to produce the thermal expansion which will make or break the electrical contacts, and so produce a signal. Hence, a measurable period of time elapses between a change in the load and corrective action being initiated. This measuring lag is not the end of the matter. Suppose that the thermostat

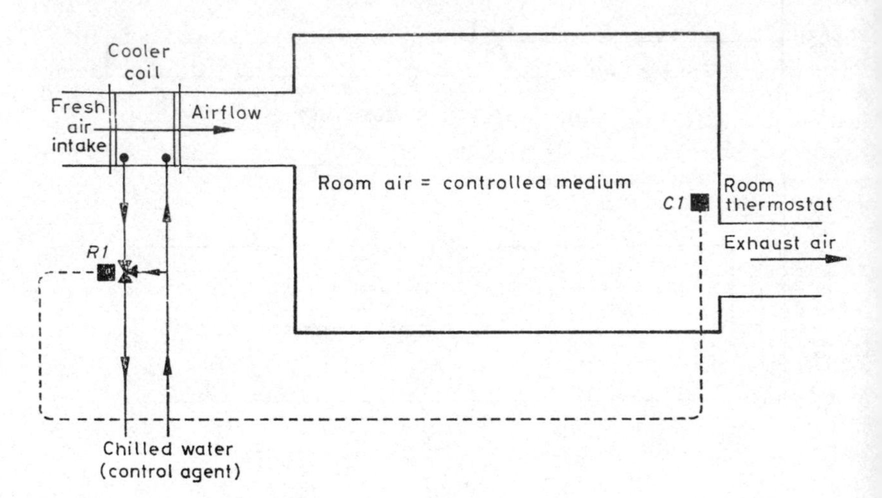

Fig. 13.4 Closed loop system

mentioned above controls the output of a chilled water cooler coil through the agency of a motorised mixing valve, as illustrated in Fig. 13.4. The chain of events that ensues after the measuring lag is as follows. First, the signal initiated by the thermostat $C1$ takes time to reach the final control element, $R1$. (This is an instance of what is termed distance-velocity lag—the time taken for a signal to be propagated along a control line, e.g. along a pneumatic pipeline.) Secondly, on its arrival at the valve, its message must be translated into corrective valve motion, and amplified if necessary. Time passes during these processes. If the load in the room had increased, the temperature sensed by $C1$ would have risen. The valve $R1$ would then be required to modulate its by-pass port to closure and its other inlet port to open. An increase in the flow rate of chilled water through the cooler coil would then result but time would have to pass before the increased flow rate absorbed more heat from the airstream. The airstream would suffer a gradual fall in temperature as it flowed over the outside of the cooler coil, but this reduction of supply air temperature would not be felt in the conditioned room until the air had travelled along the length of the supply duct to the delivery grille. Even then the story would not be completed. For the deviation to be countered the supply air must diffuse throughout the room, taking time in the process. A typical air-conditioning installation might handle 15 air changes an hour and so, on an average, 4 minutes will be required for one air change to be completed.

The total time lag of the system is the sum of all the individual lags in both the automatic control system and the system being controlled, including the measuring lag and the lag inherent in the process of diffusion of the air through the room.

Because of these lags, there is the possibility that the response of the controlled system may be out of phase with the corrective information fed back from the measuring element. Apart from the risk of instability which this situation implies, there is the fact that the change in the value of the controlled variable is always greater than the change indicated by the feedback. Thus, although a thermostat may have a differential gap of only 2 degrees, the temperature maintained in the room may fluctuate by 3 or 4 degrees, because an instantaneous system response is impossible.

13.4 Measuring elements

The function of the measuring element is to measure the value of the controlled variable and to produce a corresponding signal which may be used to achieve a reduction in the deviation, through the agency of the rest of the automatic control system. This signal, or feedback variable, may take the form of a force, a mechanical displacement, a pressure, or a difference of electrical potential or current.

The commonest elements used for temperature measurement are the following.

1. *Contact Thermometers.* These are mercury-in-glass thermometers with electrical contacts within the stem. A rise or fall in the height of the column of mercury makes or breaks an electrical circuit, thus providing a signal. A proper choice of thermometer with the appropriate degree of accuracy, gives a reliable, direct and precise, form of measurement, within the range −40°C (the freezing point of mercury) to 538°C (the softening point of glass). They are suitable for two-position control.

2. *Bi-metallic Strip Thermometers.* As the name implies, a pair of dissimilar metal strips is joined firmly together and their differing coefficients of thermal expansion cause the strips to bend together, making or breaking an electrical circuit, on change in temperature. Their range of suitability is from −180°C to 420°C.

3. *Fluid Expansion Phials.* A phial filled with a fluid having a suitable coefficient of thermal expansion is connected by a capillary tube to a bellows or to a Bourdon tube. The latter, which is a spirally-wound tube, increases its length and unwinds as the liquid expands. This unwinding is arranged to produce a rotational movement and so operate a proportional or other controller. If a bellows is used, a linear movement is produced as the bellows

fill upon liquid expansion. These devices are suitable for use in the range −45°C to 650°C. A word of caution: the amount of fluid used to fill the phial is critical and depends upon the temperature range of the application. An inadequate fill, or a use outside the range of temperature suitable for the fill, may cause the fluid to evaporate. The pressure then exerted will be a saturation vapour pressure, and the relationship between expansion and temperature will no longer be virtually linear. (Vapour pressure thermometers are often intentionally used; their range is −35°C to 300°C.)

A gas (as opposed to a vapour, *see* section 2.2) is also used instead of a liquid. The element then is a pressure thermometer and has a linear expansion-temperature relationship, according to Charles' Law, operating as it does at what is virtually constant volume. The range of use is −85°C to 540°C.

4. *Thermocouples.* When a pair of dissimilar metallic wires are joined together to form a loop, a current flows around the loop if the two junctions are at different temperatures. The magnitude and direction of the current depend on the temperature difference and the pairing of the metals. The electrical current (or the potential difference, if the loop is not closed) is used as the feedback variable. Depending on the metals used, the range of application varies from −260°C to 2600°C.

5. *Resistance Thermometers.* The electrical resistance of a conductor varies with its temperature, and this is used to provide an accurate measurement and feedback. The range of use, depending on the material used, is from −265°C to 650°C.

6. *Thermistors.* These are made of oxides of metals and give an inverse, exponential relationship between temperature and electrical resistance. Their response is non-linear but very sensitive up to 100°C.

As regards humidity measuring elements, there are two basic types:

(*a*) The mechanical type, employing a hygroscopic material such as human hair, parchment, etc., whose length varies as the ambient relative humidity changes.

(*b*) The electrical type which uses a hygroscopic salt, such as lithium chloride, the electrical resistance of which alters as the humidity fluctuates.

The first type produces a mechanical displacement which may be used for two-position control with electrical systems or proportional control with pneumatic systems. Its disadvantage is in the non-linearity of its response, except over the range of middle humidities.

The second type produces changes in electrical current of very small magnitude, which, when amplified electronically, may be used to achieve proportional control.

13.5 Types of system

There are five types in common use:

1. *Self-acting.* With this form of system the pressure, force or displacement, produced as a signal by the measuring element, is used directly as the source

of power at the final control element. For example, a temperature-sensitive, liquid-filled phial, produces a force which may act on a diaphragm to which the spindle of a modulating valve is attached. The force exerted by the expanding or contracting liquid moves the valve spindle and so its position is related to the temperature measured, without any external power supply or amplification. This form of control is simple, proportional in nature and usually has a fairly wide throttling range because the temperature sensed must change sufficiently to produce the force necessary for valve movement. (*See* Fig. 13.5(*a*).)

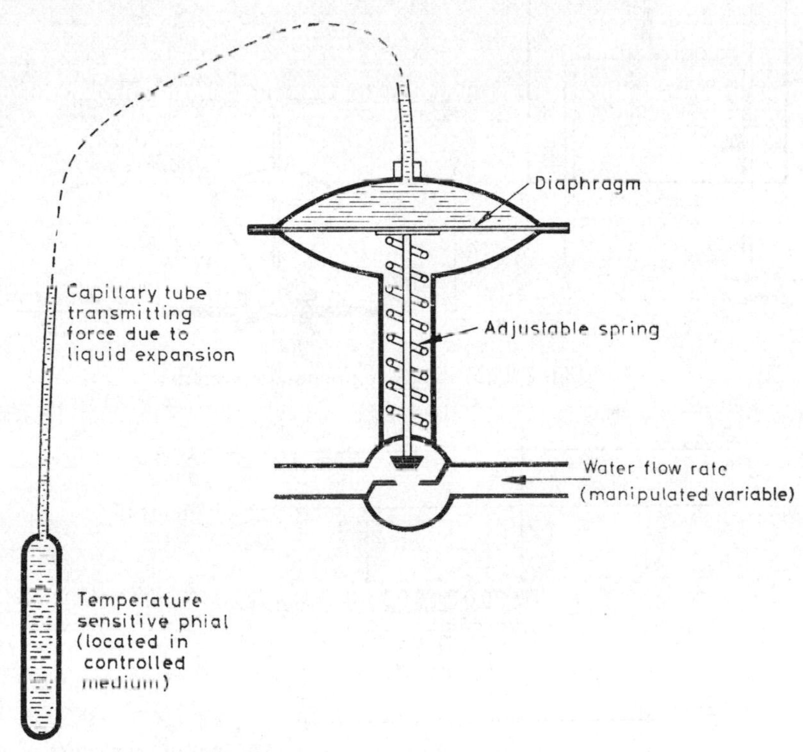

Fig. 13.5(*a*) Self-acting system

2. *Pneumatic.* Compressed air is piped to each controller and the controller reduces the pressure in a manner related to the value of the controlled variable that it senses. This is achieved by bleeding some air to waste. The reduced output, or the control pressure, is then transmitted to the relevant final control element, which is caused to move as the control pressure changes. (*See* Fig. 13.5(*b*).)

3. *Hydraulic.* This form of system is similar to the pneumatic, but oil, water, or other fluid is used to transmit the signals. Hydraulic control systems are used for higher force transmission than are pneumatics.

4. *Electrical.* Variations in voltage are used to transmit signals and to provide the current necessary for moving the final control elements. Normally, these systems are used with 24 volts. (*See* Fig. 13.5(*c*).)

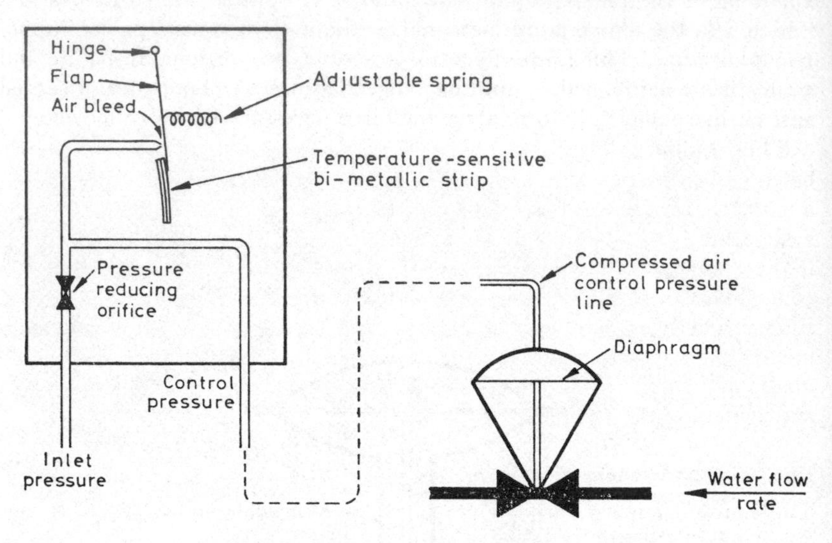

Fig. 13.5(*b*) Simple pneumatic system

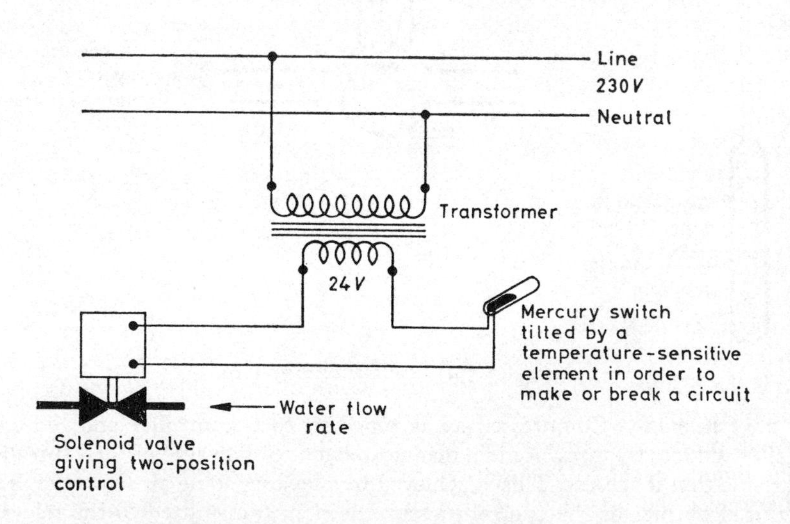

Fig. 13.5(*c*) Simple electrical system

5. *Electronic.* This system is similar to the electrical type, but much smaller signal strengths are transmitted by the sensing elements (usually resistance thermometers or thermocouples for temperature control). Elec-

tronic amplifiers change the magnitude of the signals to values suitable for actuating the final control elements.

13.6 Methods of control

There are three methods of control in common use in air-conditioning: two-position, proportional and, to a lesser extent, floating. All are capable of sophistication, aimed at improvement, whereby offset is reduced or stability enhanced.

The form of control chosen must suit the application. For example, one essential feature of floating control is that it works best with a system having a small thermal storage and a predictable response. If it is misapplied, the system may hunt as a result of excessive lag and may be difficult to set up if the response is not well known. It is also essential that any method of control chosen should be as simple as is consistent with the results desired. Elaboration may result in some controls being redundant, the system becoming temperamental and wearisome to commission. There is more to commissioning an air-conditioning installation than setting 21°C on a thermostat and switching the plant on.

13.7 Simple two-position control

There are only two values of the manipulated variable: maximum and zero. The sensing element switches on full capacity when the temperature falls to, say, the lower value of the differential and switches the capacity to zero when the upper value of the differential is reached. Figure 13.2 illustrates the way in which the controlled variable alters with respect to time when an air heater battery, say, is switched on and off in this way. Overshoot and undershoot occur because of the total lag. It can be seen that the range of values of the controlled condition exceeds the differential gap and that if the lag is large, poor control will result. This is not to denigrate the method; it is an excellent, simple and cheap form of control for the correct application.

13.8 Timed two-position control

An exact match between load and capacity does not generally exist for two-position control; for example, if the load is 50 per cent then, on average, the plant will be on for half the time and off for the other half. This would be so if the plant were on for 30 minutes in each hour, or if it were on for every alternate minute in each hour. But it is clear that a 30-minute on-off cycle would produce a much wider swing in the controlled variable than would a one-minute cycle. The lower the value of the room air temperature, then the longer the period for which the heater battery, say, must be on, in a small plenum heating installation. Thus, for large loads the ratio of 'on' to 'off' time would not be one to one but, say, 0·8 'on' to 0·2 'off'. Simple two-position control can be improved to obtain a timed variation of capacity, which gives a smaller differential gap, by locating a small heater element next to the temperature-sensitive element in the thermostat. If the ambient temperature is low, the heating element loses heat more rapidly and so takes a longer time for its temperature and the temperature of the adjacent sensitive element to reach the upper value of the differential gap which switches off the heater battery. Thus, the heater battery is on for longer periods when the ambient room air temperature is low.

13.9 Floating action

Floating control is so called because the final control element floats in a fixed position as long as the value of the controlled variable lies between two chosen limits. When the value of the controlled variable reaches the upper of these limits, the final control element is actuated to open, say, at a constant rate. Suppose that the value of the controlled variable then starts to fall in response to this movement of the final control element. When it falls back to the value of the upper limit, movement of the final control element is stopped and it stays in its new position, partly open. It remains in this position until the controlled condition again reaches a value equal to one of the limits. If the load alters and the controlled variable starts to climb again, then, when the upper limit is again reached, the final control element will again start to open and will continue to open until either it reaches its maximum position or until the controlled variable falls again to a value equal to the upper limit, when movement will cease. If the load change is such that the controlled variable drops in value to below the lower limit, say, then the final control element will start to close, and so on. Thus, the final control element is energised to move in a direction which depends on the deviation; a positive deviation gives movement of the element in one direction and a negative deviation causes movement in a reverse direction. There is a dead or floating band between the two limits which determines the sign of the deviation.

Figure 13.6 illustrates an example of floating control. A room is ventilated by a system as shown. The capacity of the L.P.H.W. heater battery is regulated by means of a motorised two-port valve, *R*1, controlled by floating action from a thermostat, *C*1, located in the supply air duct. There is negligible lag between *C*1 and *R*1. It is assumed that the variation in the state of the outside air is predictable and so the response rate of the controller can be properly chosen.

When the load is steady, the valve is in a fixed position, indicated by the line *AB* in Fig. 13.6(*b*), provided that the supply air temperature lies between the upper and lower limits, as indicated by the line *AA'*, in Fig. 13.6(*c*). Under this steady-state condition, the capacity of the heater battery matches the load. If the outside air temperature increases, but the valve position remains unchanged, the supply air temperature starts to rise, as shown by the line *A'B* in Fig. 13.6(*c*). When the air temperature in the supply duct attains a value equal to the upper limit (point *B*), the valve starts to close (point *B* in Fig. 13.6(*b*)). The rate of closure remains fixed, having been previously established during the initial process of setting up the controls, the battery output is continuously decreased, and the supply air temperature, after some overshoot, starts to fall. When the temperature reaches the upper

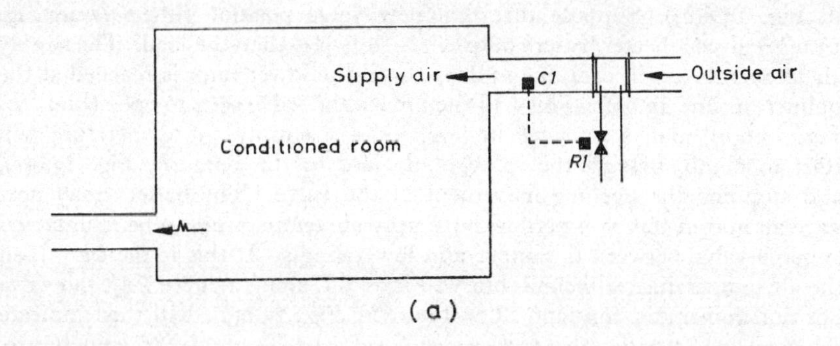

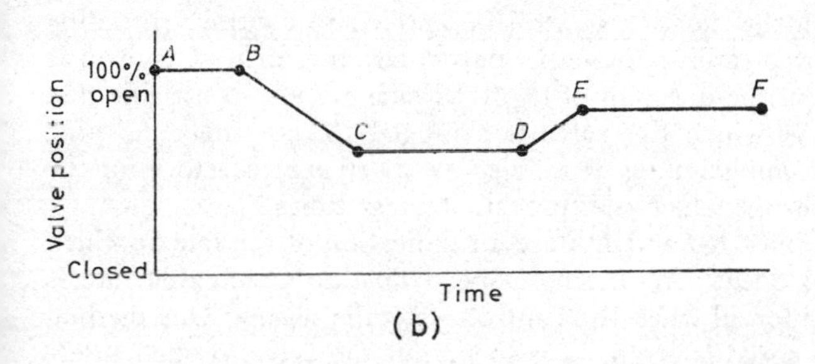

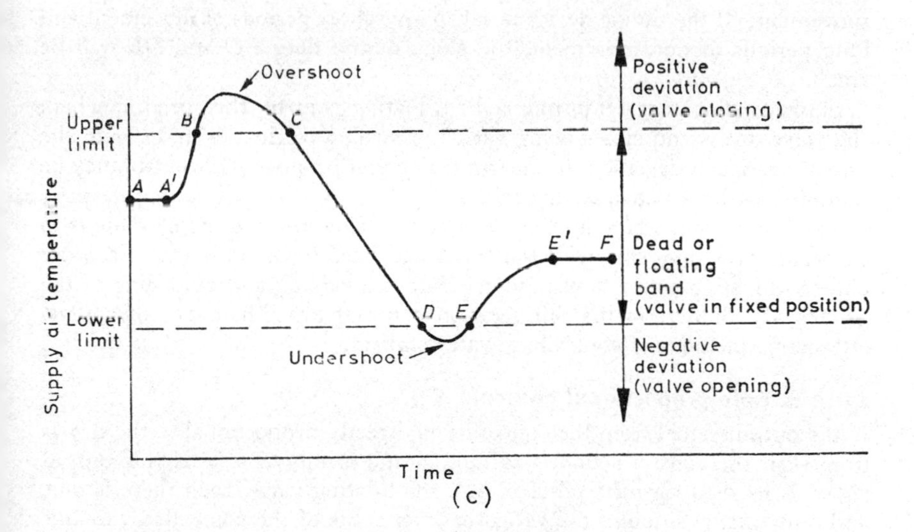

Fig. 13.6 Floating action

limit (point *C* in Fig. 13.6(*c*)), the movement of the valve stops (point *C* in Fig. 13.6(*b*)). Suppose that this new valve position (line *CD* in Fig. 13.6(*b*)) gives a heater battery output which is less than the load. The supply air temperature will continue to drop, until the lower limit is reached at the point *D* in Fig. 13.6(*c*). This will then make the valve start to open (line *DE*, Fig. 13.6(*b*)) and, after some undershoot, the supply air temperature will start to climb, passing the value equivalent to the point *E*, Fig. 13.6(*c*), and stopping the opening movement of the valve. The battery may now have an output that will permit the supply air temperature to be maintained at some value between the upper and lower limits. If this is the case, then the air temperature will climb but will level off, along a curve *EE'*, the valve position remaining constant. The line *EF*, Fig. 13.6(*b*), will then indicate the fixed valve position that gives a steady value of the supply air temperature, corresponding to the line *E'F* in Fig. 13.6(*c*).

This example is intended only to describe floating action. It is not intended as a recommendation that floating control be applied to control ventilating systems. In fact, for plenum systems in particular, the form of control is unsuitable; apart from the question of lag, load changes are so variable that although the speed at which the valve is arranged to open may be quite satisfactory during commissioning, it may prove most unsatisfactory for the other rates of load change which will prevail at other times.

The slopes of the lines *BC* and *DE* are an indication of the rate at which the valve closes and opens. With single-speed floating action this rate is determined once and for all when the control system is set up. One method whereby a choice of speed is arranged is to permit the valve to open intermittently with short adjustable periods of alternate movement and non-movement. If the timing device is set to give short periods of movement and long periods of non-movement, the slope of the lines *BC* and *DE* will be small.

Sophistication may be introduced to floating control, the speed at which the valve opens and closes being varied according to the deviation or to the rate of change of deviation, in the same way that proportional control may be modified, as described in section 13.11.

One application where floating control has been found to be of value is in controlling the temperature of the water delivered from the outlet of a water chiller. Floating action is used to regulate the loading and unloading of the groups of cylinders in the refrigeration compressor. There are, of course, alternative methods of controlling water chillers.

13.10 Simple proportional control

If the output signal from the controller is directly proportional to the deviation, then the control action is termed simple proportional. If this output signal is used to vary the position of a modulating valve, then there is one, and only one, position of the valve for each value of the controlled variable. Figure 13.7 shows this. Offset is thus an inherent feature of simple proportional control. Only when the valve is half open will the value of the controlled variable equal the set point. At all other times there will be a deviation; when the load is a maximum the deviation will be greatest in one direction, and vice-versa.

The terms direct action and reverse action are used to denote the manner in which the final control element moves in response to the signals it receives from the primary element. For example, suppose that a room suffers a heat gain and that this is offset by means of a fan-coil unit fed with chilled water, the output being regulated by means of a pneumatically actuated, two-port, modulating valve. If, when the room temperature rises, the controlling thermostat sends an increased air-pressure signal to the valve, it is direct

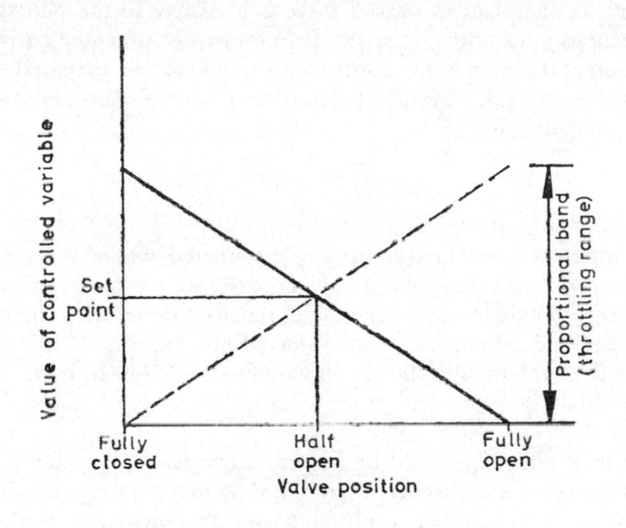

Fig. 13.7

acting; a signal of increased strength accompanies an increase in the value of the controlled variable. If this higher air pressure is delivered to the upper side of the diaphragm in the valve actuator, the valve will partly close, reducing the flow of chilled water through the coil of the unit. This is clearly the wrong response to the signal. So a pneumatic relay must be inserted in the compressed-air line between the controller (thermostat) and the regulator (valve) which will reverse the signal. A rise in the control pressure from the thermostat will then cause a fall in the pressure above the valve diaphragm. The controller has had its action reversed and the output of the fan coil unit will now be increased as the room temperature rises, the feedback from the thermostat giving rise to corrective action. The broken line in Fig. 13.7 shows this case of reversed action with chilled water.

If hot water were pumped through the same fan coil unit in winter, to deal with a heat loss, then a rise in room temperature would again cause a rise in the air pressure transmitted as a signal from the thermostat. This could then be permitted to act directly on the valve diaphragm, since corrective action would result. The pneumatic relay would then be kept out of the circuit and the thermostat would be direct acting, as shown by the full line in Fig. 13.7.

An alternative view of simple proportional control is obtained by introducing the concept of ' potential value ' of the controlled variable (due to Farrington). If there is not a match initially between load and capacity, the controlled variable will gradually change, approaching a steady value at which a match will prevail. This steady value, attainable exponentially only after an infinity of time, is the potential value of the controlled variable. Every time the capacity of the plant is altered there is a change in the potential value. Capacity variation can thus be spoken of in terms of potential correction, φ: if a deviation of θ occurs in the controlled variable then a potential correction of φ must be applied. Simple proportional control can consequently be defined by an equation:

$$\varphi = -k_p\theta \tag{13.1}$$

where k_p is the proportional control factor.

Thus, under a constant load condition, sustained deviation is the rule. This offset corresponds to a value of φ and when, after a load change, the controlled variable has responded to this correction, it settles down with some decaying overshoot and undershoot at a steady value. Such a steady value is the control point. Hence, even though the set point is unaltered (at the mid-point of the proportional band) the control point takes up a variety of values, depending on the load.

To minimise offset k_p should be large and increasing its value is known as increasing the control sensitivity, corresponding to narrowing the proportional band. For a given operating condition too much sensitivity (or too narrow a proportional band) can induce hunting, proportional control degenerating to two-position control.

13.11 Sophisticated proportional control

The offset which occurs with simple proportional control may be removed to a large extent by introducing reset. The set point of a proportional controller is altered in a way which is related either to the deviation itself or to the rate at which the deviation is changing.

A controller whose output signal varies at a rate which is proportional to the deviation is called a proportional controller with integral action.

When the output signal from a proportional controller is proportional to the rate of change of the deviation, the controller is said to have derivative action. Since derivative action is not proportional to deviation, it cannot be used alone but must be combined with another control action. Such a combination is then termed ' compound control action '.

Figure 13.8 illustrates some of the behaviours of different methods of control action. Deviation is plotted against time. Curve *A*, for proportional control plus derivative action, gives the least deviation; the controlled

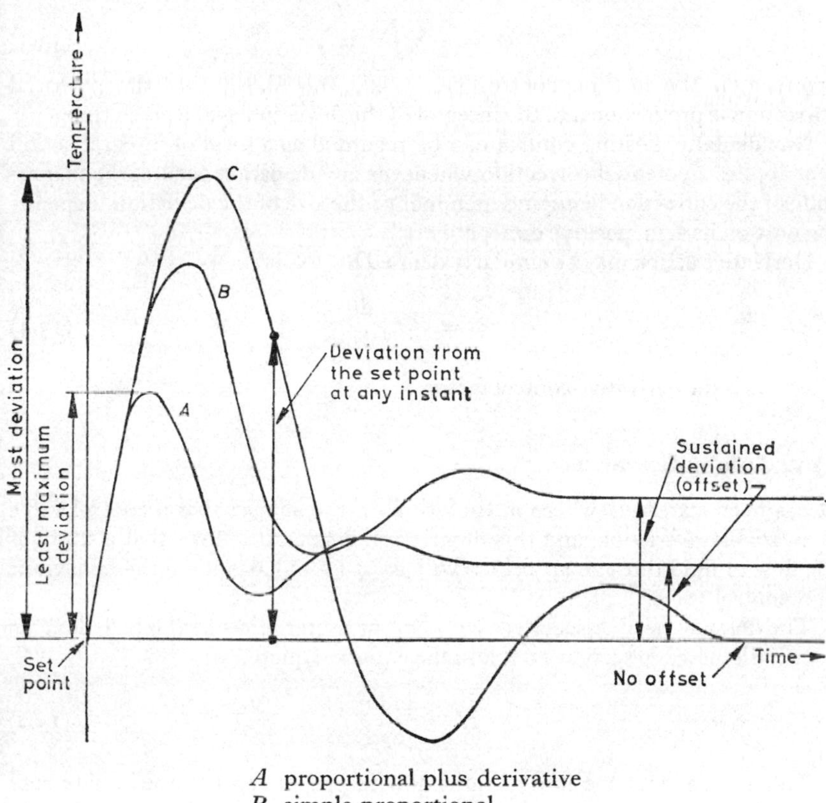

A proportional plus derivative
B simple proportional
C proportional plus integral

(*After* D. P. Eckman, *Automatic Process Control*)

Fig. 13.8

variable reaches a steady state quite rapidly but some offset is present. Curve *B* shows the case of simple proportional control; there is a larger maximum deviation than with curve *A* and the offset is greater. Curve *C* is for proportional plus integral control; the maximum deviation is greater than for the other two curves and the value of the controlled variable oscillates for some time before it settles down to a steady value without any offset at all.

Proportional plus derivative plus integral control is not shown. Its behaviour is similar to curves *A* and *C* but the controlled variable settles down more quickly. There is no offset. The method is most expensive.

In terms of potential correction, integral action may be defined by

$$\frac{d\varphi}{dt} = -k_i\theta \tag{13.2}$$

whence

$$\varphi = -k_i \int \theta \, dt \tag{13.3}$$

where k_i is the integral control factor and this shows that the potential correction is proportional to the integral of the deviation over a given time.

(Incidentally, floating control can be regarded as a form of integral action that applies a potential correction whenever any deviation occurs, the magnitude of the correction being independent of the size of the deviation, depending only on its sign, positive or negative.)

Derivative action may be similarly defined by

$$\varphi = -k_d \frac{d\theta}{dt} \tag{13.4}$$

where k_d is the derivative control factor.

13.12 Automatic valves

It has been stated elsewhere in literature on the subject that a control valve is a variable restriction, and this description is apt. It follows that a study of the flow of fluid through an orifice will assist in understanding the behaviour of a control valve.

The loss of head associated with air or water flow through a duct or pipe is discussed in section 15.1, and the equation quoted is:

$$H = \frac{4f\,lv^2}{2gd} \tag{15.2}$$

This implies that the rate of fluid flow is proportional to the square root of the pressure drop along the pipe and to the cross-sectional area of the pipe. We can therefore write a basic equation for the flow of fluid through any resistance, for example, a valve:

$$q = Ka\sqrt{\{2g(h_1-h_2)\}} \tag{13.5}$$

where q = volumetric flow rate in m^3/s,
a = cross-sectional area of the valve opening, in m^2,
h_1, h_2 = upstream and downstream static heads, in m of the fluid,
K = a constant of proportionality.

If it is assumed that the position of the valve stem, z, is proportional to the area of the valve opening, then

$$q = K_1 z\sqrt{\{2g(h_1-h_2)\}} \tag{13.6}$$

where K_1 is a new constant of proportionality.

Unfortunately, the picture is not as simple as this and the flow rate is not directly proportional to the position of the valve stem; the constants of proportionality are not true constants (they depend on the Reynold's number, just as does the coefficient f in eqn. (15.2)) and the area of the port opened by lifting the valve stem is not always proportional to the lift. A more realistic picture of the behaviour is obtained if the flow of fluid is considered through a pipe and a valve in series, under the influence of a constant difference of head across them. Figure 13.9 illustrates the case.

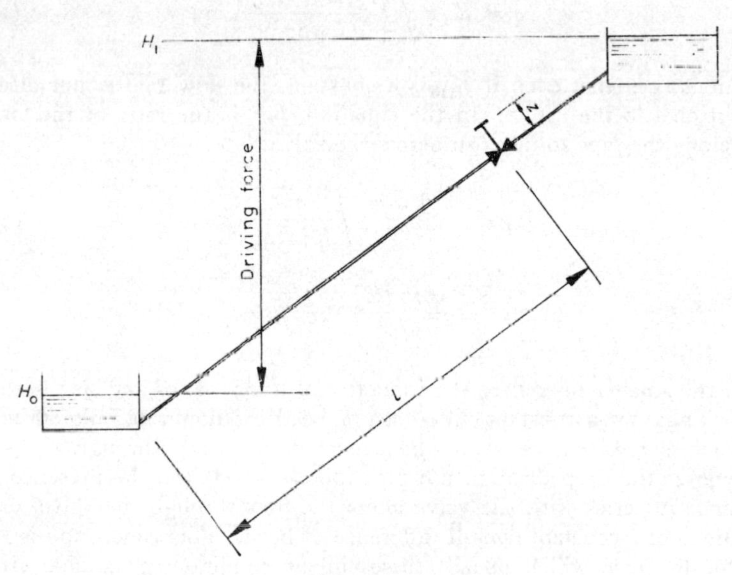

Fig. 13.9

The head loss in the pipe, plus that across the valve, must equal the driving force produced by the difference of head between the reservoir, H_1, and the sink, H_0.

The friction factor f (due to Fanning) is sometimes re-expressed as the Moody factor, f_m ($= 4f$). Using the Moody factor the head lost in the pipe is given by

$$h_1 = \frac{f_m \, lv^2}{2gd}$$

Also,
$$q^2 = \left(\frac{\pi d^2}{4}\right)^2 v^2$$

Hence,
$$h_1 = \frac{8 f_m \, lq^2}{\pi^2 g d^5}$$

The head lost across the valve is then $(H_1 - H_0 - h_1)$. Thus eqn. (13.6) for the flow through the valve can be re-written:

$$q = K_1 z[2g\{H_1 - H_0 - (8f_m\, lv^2/\pi^2 g d^5)\}]^{\frac{1}{2}}$$

$$q^2 = K_1 z^2 2g\{H_1 - H_0 - (8f_m\, lq^2/\pi^2 g d^5)\}$$

Write $\beta = (16 f_m\, l K_1^2/\pi^2 d^5)$ and the equation simplifies to

$$q = K_1 z \sqrt{\left\{\frac{2g(H_1 - H_0)}{(1+\beta z^2)}\right\}} \qquad (13.7)$$

So it is seen that even if K_1 is a constant, the flow rate is not directly proportional to the lift, z. In the equation, βz^2 is the ratio of the loss of head along the pipe to that lost across the valve:

$$\beta z^2 = \frac{8 f_m\, l q^2}{\pi^2 g d^5} \bigg/ \frac{q^2}{K_1^2 z^2 2g} \qquad (13.8)$$

$$= \frac{16 f_m\, l K_1^2}{\pi^2 d^5} \cdot z^2$$

For the smaller pipe sizes the influence of d^5 increases and βz^2 becomes large. Thus, for a constant difference of head the flow rate falls off as the pipe size is reduced, as would be expected, the drop through the valve reducing as the drop through the pipe increases. Hence, the presence of a resistance in series with the valve alters the flow through the valve, under conditions of a constant overall difference of head. This conclusion is valid also for the case which usually arises in air conditioning: a pipe circuit through which water is circulated by means of a centrifugal pump. Pumps of this sort have a fairly flat pressure-volume characteristic in any case, so the condition of constant overall head is approximately retained.

If a valve is to exercise good control over the rate of flow of the fluid passing through it, the ideal is that q/q_0 shall be directly proportional to z/z_0, where q_0 is the maximum flow and z_0 is the maximum valve lift (when the valve is fully closed, q and z are both zero). A direct proportionality between these two ratios is not attainable in practice. As has been remarked, the resistance of the rest of the controlled (piping) circuit has an influence on the flow rate, for a given valve lift.

The ratio of the pressure drop across the valve when fully open to the pressure drop through the valve and the controlled circuit, is termed the 'authority' of the valve. For example, if a valve has a loss of head of 5 metres when fully open and the rest of the piping circuit has a loss of 15 metres, then the authority of the valve is 0·25.

The effect of valve authority on the flow-lift characteristic of a valve can be seen by means of an example.

EXAMPLE 13.1 To simplify the arithmetic, divide eqn. (13.7) throughout by $K_1\sqrt{2g}$, choosing new units for q and z so that

$$q = z\sqrt{\{(H_1 - H_0)/(1+\beta z^2)\}}$$

Suppose that $H_1 - H_0 = 100$ units of head and that the valve lift, z, varies between 0 units (fully closed) and 1 unit (fully open). Assume also that the loss of head in the piping circuit connected to the valve is 50 units for maximum fluid flow, q_0 and maximum valve lift, z_0. Then the valve authority is 0·5.

Note that eqn. (13.8) showed βz^2 to equal h_l/h_v, where h_v is the head lost across the valve. Since the authority of the valve, α, is defined for full flow conditions by $h_v/(h_1+h_v)$ we can see that

$$\alpha = \frac{1}{1+\beta z^2} \qquad (13{\cdot}9)$$

Hence, since $z_0 = 1$ and $\alpha = 0{\cdot}5$, $\beta = 1$ for our given piping circuit and modified units. So the simplified version of eqn. (13.7) becomes

$$q = 10z\sqrt{\{1/(1+z^2)\}}$$

We can now compile the following table

z	0·1	0·3	0·5	0·7	0·9	1·0
z/z_0	0·1	0·3	0·5	0·7	0·9	1·0
q	0·995	2·873	4·472	5·734	6·690	7·071
q/q_0	0·141	0·406	0·632	0·811	0·946	1·000

If we increased the valve size and reduced the size of the piping circuit so that the head loss through the valve was 20 units and the loss through the piping 80 units, the authority would be 0·2. Then $\beta z^2 = 80/20 = 4$ and, since when fully open, $z = z_0 = 1$, β also equals 4. The simplified version of eqn. (13.7) now becomes

$$q = 10z\sqrt{\{1/(1+4z^2)\}}$$

and a new table may be calculated:

z	0·1	0·3	0·5	0·7	0·9	1·0
z/z_0	0·1	0·3	0·5	0·7	0·9	1·0
q	0·982	2·572	3·354	4·068	4·370	4·472
q/q_0	0·220	0·575	0·750	0·910	0·977	1·000

Figure 13.10(*a*) shows the results of these two calculations. It is clear that the greater the authority the nearer the characteristic approaches linearity, *for the control of fluid flow rate* and for this purpose between 0·2 and 0·5 is often chosen.

Generally speaking, an automatic control valve used in air conditioning is asked not just to regulate the flow of fluid, but to modulate the output of a heat exchanger such as a cooler coil or heater battery, neither of which has a linear flow rate-heat transfer characteristic. If a valve of the gate type is used, its characteristic is as shown in Fig. 13.10(*a*) and it is quite unsuitable for control purposes, being suited only for a shut-off application, because a small amount of valve lift results in a large increase in the rate of fluid flow.

The output of a heater battery, say, in terms of a varying flow rate of L.P.H.W. through it, is not linear. A typical characteristic is shown in Fig. 13.10(*b*). A small change in the flow rate produces a large change in the output at the lower end of the curve, and vice-versa at the upper end of the characteristic. It is only over the middle range of the output that any linearity exists. A gate valve would obviously be most unsuitable for control

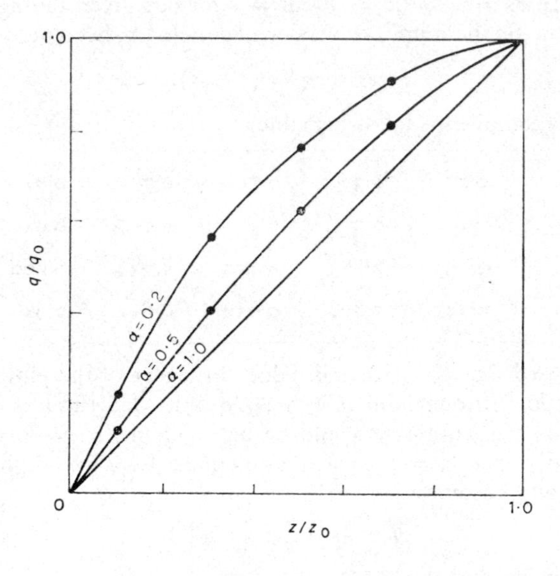

Fig. 13.10(*a*)

purposes here. To achieve proper control a valve is required which will give a very small increase in flow rate for a small lift at the bottom of its characteristic and a large increase in flow rate for a small lift at the top of its characteristic. Its characteristic would then counteract that of the heater battery, and the tendency would be towards a linear characteristic for battery output against valve lift.

A valve which behaves like this is termed an 'equal percentage' type. It is alternatively known as an 'increasing sensitivity', or a 'parabolic', or a 'logarithmic', or a 'characterised' type. It is said to have a characterised port. The valve sensitivity, $\Delta q/\Delta z$, is the slope of the characteristic curve and, or characterised valves, the sensitivity is a constant percentage of the flow ate. That is, its slope increases as the flow rate increases. So the valve passes nore fluid for a given increment of lift at the end of its movement in opening han it does at the beginning. This is what is desired. The required character-

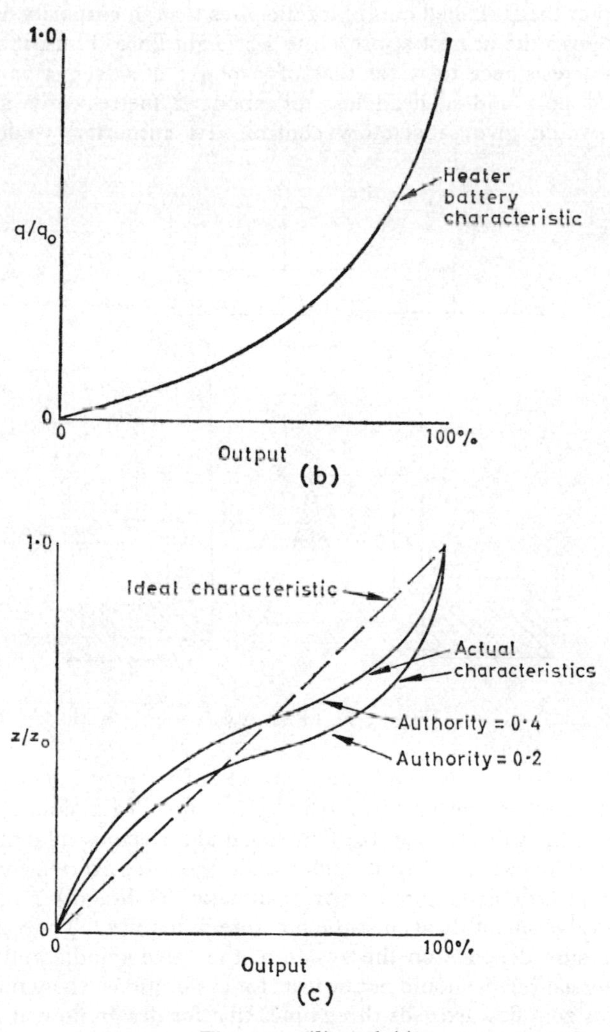

Fig. 13.10(*b*) and (*c*)

stic is achieved by using a piston type of valve (Fig. 13.11) that offers an area to flow which increases faster than the lift as the valve opens. Referring o the figure, the area for flow associated with a lift of z_2 is much greater han z_2/z_1 multiplied by the area associated with z_1.

When a valve of this type is used to regulate the output of an air heater battery (say) by modulating the flow of L.P.H.W. through it, the battery characteristic of output versus lift is a shallow S-shaped curve (Fig. 13.10(*c*)) which is much closer in appearance to a straight line than the characteristic a gate valve would give. The authority of the valve has a bearing on the shape and position of the S-shaped curve: it transpires that an authority of between 0·2 and 0·4 gives the nearest approach to a straight line. Thus, if the heater battery has a resistance to water flow of 5 metres of water, a valve with a characterised port and a head loss of about 2 metres of water when fully open would give satisfactory control. Its authority would be 2/7, or 0·29.

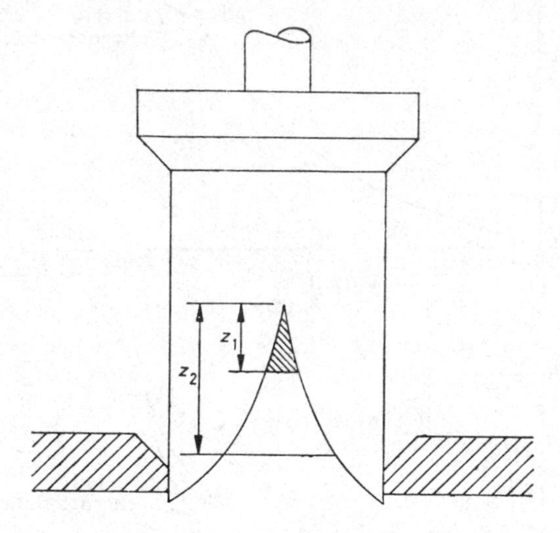

Fig. 13.11 Piston valve

The authority of a valve and the character of its port are not the only matters that affect its performance when piped up. In a simple throttling application, as the valve reduces the flow rate it absorbs more and more of the available pump pressure. A time comes when the force exerted by the valve actuator (its electric driving motor or pneumatic diaphragm) is not enough to close the valve smoothly against the pressure difference. The value of this shut-off pressure depends on the design of the valve spindle and actuator; most commercial valves should not be used for applications where the pressure difference for zero flow exceeds three times that for design flow, if good proportional control is desired. Furthermore, for free movement of the valve there must be some clearance between the plug or skirt and the seating and this means there is always a minimum flow rate for proportional control, beyond which control becomes on-off.

This latter effect gives rise to the concept of *rangeability* or *turn-down ratio*, the former being defined as 'the ratio of maximum controllable flow to minimum controllable flow ' and the latter as ' the ratio of maximum usable flow to minimum usable flow '. The slight difference covers the possibility that even though the maximum controllable flow may be 10 litres/s the valve may be in a hydraulic circuit where it is never asked to pass more than 7 litres/s. If the minimum flow in each case is 0·5 litres/s the rangeability would be 20 : 1 but the turn-down only 14 : 1.

Commercial valves can have rangeabilities of up to 30 : 1 and better made industrial valves up to 50 : 1. It must be remembered though that because of the non-linearity of a heat-exchanger output characteristic, a valve with a turn-down of 20 : 1 and a parabolic character may only control proportionally down to 15 or 20 per cent of the heat-exchanger maximum output.

It must be remembered that when assessing valve authority it is the pressure drop in the part of the hydraulic circuit where there is variable flow which is relevant.

The flow coefficient, A_v, is used when sizing valves, in the formula $q = A_v\sqrt{\Delta p}$, a version of eqn. (13.5) in which q is the flow rate in m^3/s and Δp the pressure drop in N/m^2. For a given valve, A_v represents the flow rate in m^3/s for a pressure drop of 1 N/m^2.

Three-port mixing valves present a problem: because each port of the valve does not necessarily give a linear relation between flow rate and valve position, constancy of flow is not provided for all port positions. The desirable performance is often that of a linear relation between the output of the controlled heat-exchanger and the valve lift for one port, whilst the other port gives constant total flow in the non-controlled circuit. Asymmetrical port characterisation goes some of the way to achieving this.

Valve positioners should be used whenever a valve (or a damper) is asked to operate in conditions of fluctuating external pressures, not related to its control function. Positioners are external mechanical links ensuring that there is only one valve position for each actuator position.

13.13 Automatic dampers

The dampers used in ducting systems for balancing purposes are analogous to the gate valves frequently used in heating systems for a similar reason. They have a relatively low pressure drop when fully open, a non-linear characteristic, and give poor regulation, although this may not matter unduly if enough time is available for achieving the regulation wanted. While this may then be tolerable for such dampers as are adjusted only during commissioning, it is quite intolerable for motorised modulating dampers which are used for achieving a controlled condition. Control dampers should be selected to have a fairly high pressure drop across them so that their characteristics may approach linearity.

A simple butterfly damper disturbs the airflow greatly when it is partly closed. This is overcome by using a multiple array of parallel-blade dampers or, better, of opposed-blade dampers. If the pressure drop across an array of dampers is kept artificially at a constant value, an inherent flow characteristic

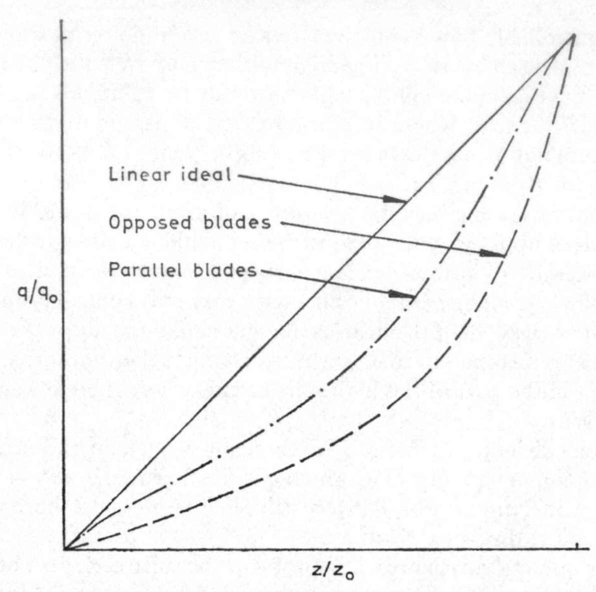

Fig. 13.12(*a*) Damper characteristics for constant pressure drop

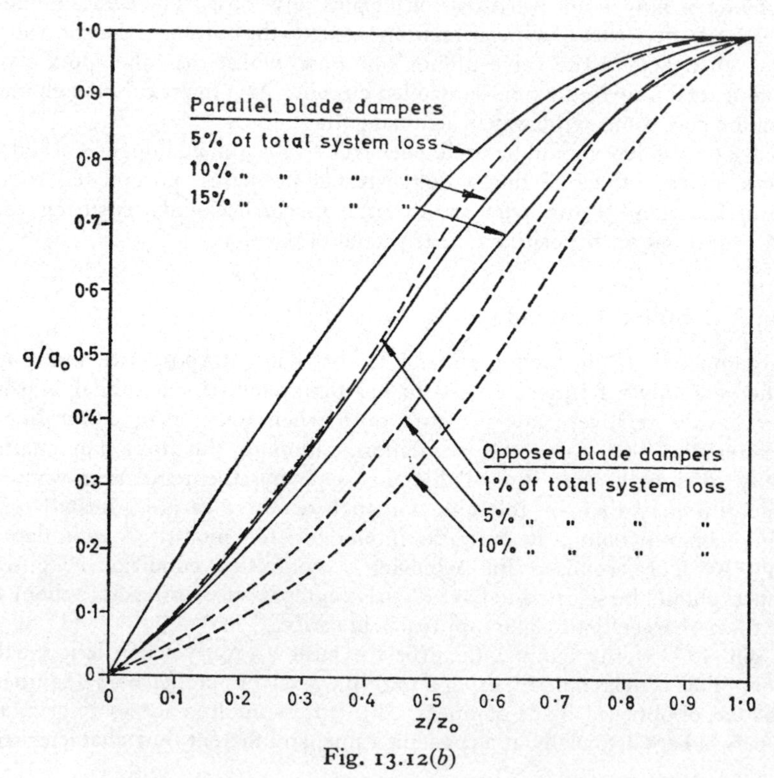

Fig. 13.12(*b*)

is obtained, as illustrated in Fig. 13.12(*a*). As with control valves, the resistance in the rest of the system determines the authority and so, for different authorities, the damper characteristic occupies different positions (*see* Fig. 13.12(*b*)). In fact, as a damper in an actual system closes, the pressure drop across it does not stay constant but increases until, when it is fully closed, the full value of the static pressure at fan discharge for zero volumetric flow will be acting across it (assuming it is on the delivery side of the fan). Thus, the characteristic curve in Fig. 13.12(*a*) for a constant pressure drop does not apply. As the damper partly closes the increased upstream pressure tends to counter the reduction in flow, and the characteristic shifts upwards.

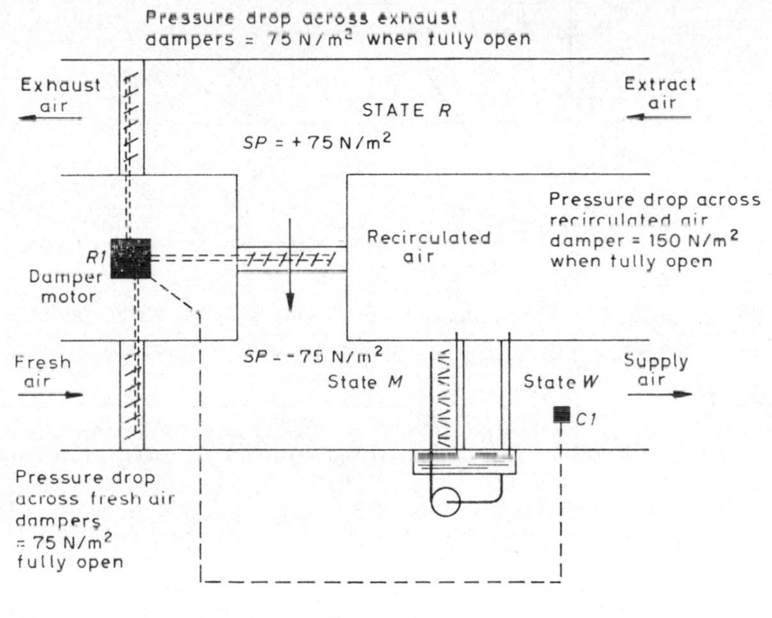

Fig. 13.13

(This effect is complicated if the point of operation of the system is near the top of a flat fan curve.) So one gets a curve something like that shown for a gate valve in Fig. 13.10(*a*). Hence, to combat this effect one selects a damper with a characteristic of an opposite nature, that is, with a characteristic appropriate to the application.

The opposed-blade damper has a characteristic which is furthest below the linear, and so this type of damper is the best choice when the pressure drop across the damper varies a good deal; a good application (if somewhat unusual in air conditioning) would be a straightforward throttling control over the airflow handled by a fan, from 100 per cent to 0 per cent. A more typical application would be that of face and by-pass dampers across a chilled water cooler coil, where the pressure drop across the face dampers alters as they are closed.

One of the commonest applications of motorised dampers in air-conditioning is when it is necessary to vary the mixing proportions of recirculated and fresh air (*see* section 3.11), as shown in Fig. 13.13. A thermostat *C*1, positioned in the ducting to sense condition *W*, would be used to vary the mixing propor-

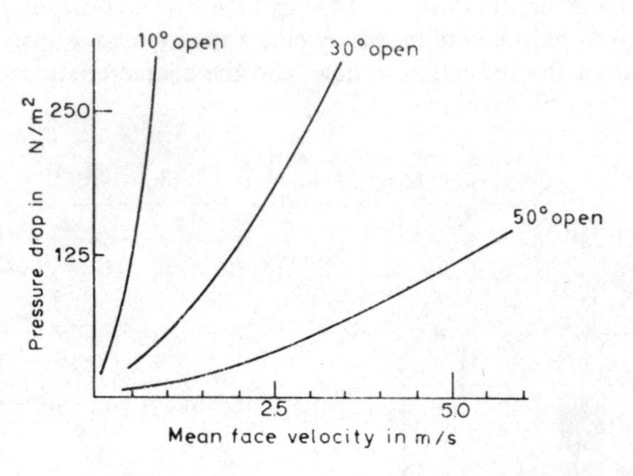

Fig. 13.14(*a*) Typical pressure drops for parallel blade dampers

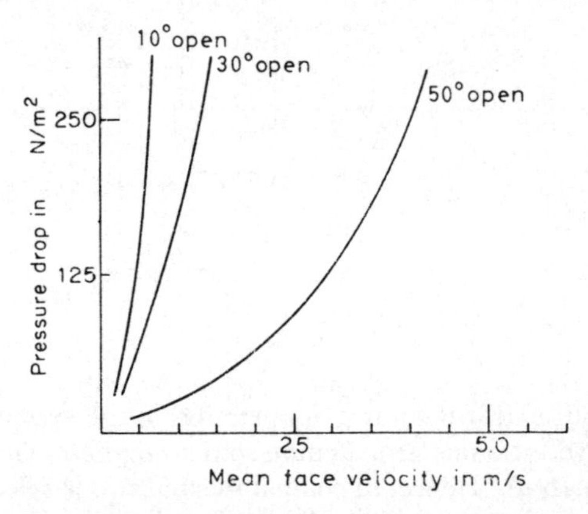

Fig. 13.14(*b*) Typical pressure drops for opposed blade dampers

tions of fresh and recirculated air in winter, so that adiabatic saturation over the sprayed cooler coil produced the required state *W*. The dampers would be moved by means of the damper motor (or group of damper motors) *R*1. Consider the variations of pressure which may occur as the mixing proportions alter. If the discharge air and the variable fresh-air dampers have been chosen to have a pressure drop across them of 75 N/m^2, then, when 100 per cent of fresh air is handled the static pressure at R is +75 N/m^2 and the static pressure at M is —75 N/m^2. The difference of pressure over the recirculation dampers is therefore 150 N/m^2. When the recirculation dampers are fully open, the discharge air and the variable fresh air dampers being fully closed, the pressure difference should still be 150 N/m^2 if the volumes handled by the supply and extract fans are not to vary. This suggests that since, virtually, constant pressure operation is the rule, parallel-blades are a better choice than opposed-blades, since they have a more nearly linear characteristic. A flat S-shaped curve can be obtained if the dampers are properly selected.

To secure an adequate pressure drop and characteristic it is necessary to size dampers with a fairly high velocity, and this may mean that their cross-section is less than the duct in which they are mounted. Figures 13.14(*a*) and (*b*) illustrate typical velocity-pressure drop relationships for parallel and opposed-blade dampers. To get a good pressure drop and characteristic the dampers are often arranged so that they are never more than 50° open. As an example, if the damper movement is restricted to 50°, then a face velocity of about 3 m/s is needed to secure a pressure drop of 125 N/m^2 across opposed-blade dampers, whereas about 5·5 m/s is required to get the same pressure drop with parallel-blade dampers. When securing a higher face velocity across the dampers, by reducing their cross-section, it is better to diminish the duct section gradually, by tapering, rather than to use blanking-off plates which give an abrupt contraction of the duct section.

13.14 Application

Although no two designs are ever alike, there are basic similarities in all designs, particularly for comfort conditioning. To illustrate the application of automatic control Fig. 13.15(*a*) shows a simple dew point plant and Fig. 13.15(*b*) the psychrometry.

Suppose the room is to have air supplied to it at a constant dew point throughout the year, the variation in latent heat gain being negligible. Suppose also that the temperature of the air in the room is to be kept at a nominally constant value.

Provided the wet-bulb of the outside air exceeds that of the air in the room, it is more economical in terms of the running cost of the refrigeration plant to use minimum fresh air. Hence *C1b* is a wet-bulb thermostat with a set point equal to the value of the room wet-bulb. It keeps the fresh-air dampers in the minimum position and the recirculation dampers in the fully open position so long as it senses a wet-bulb temperature in excess of its set point. Under these conditions *C1a*, a thermostat located after the sprayed cooler coil, maintains a nominally fixed dew point by varying the rate of flow of chilled water through the cooler coil by means of the mixing valve *R1a*. (Two points are worth observing at this juncture. First, *C1a* is generally an ordinary dry-bulb thermostat; how close the state of the air is to the dew point at exit from the sprayed coil depends on the contact factor of the coil—the larger the value of this the nearer will the leaving state of the air be to saturation and the more truly will *C1a* be controlling the dew point. Secondly, the criterion adopted in terming a valve a ' mixer ', rather than a ' diverter ', is the number of inlet ports. If a valve has two inlet ports and one outlet port, as has *R1a*, then it is a mixing valve. If a valve has one inlet port and

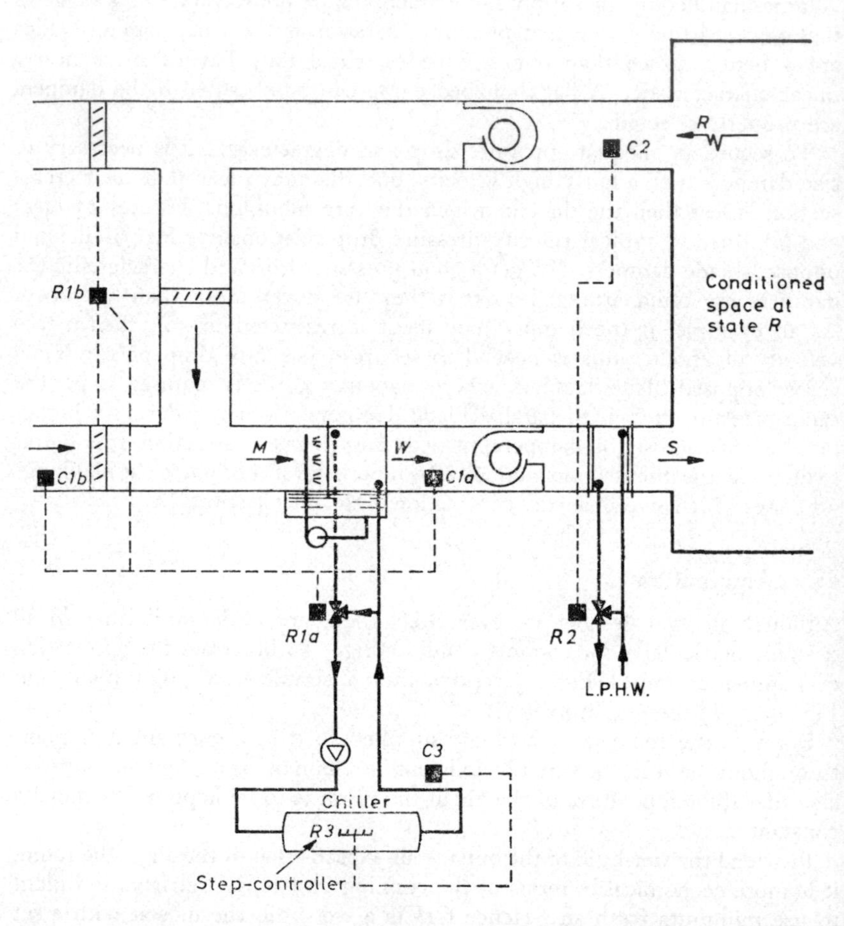

Fig. 13.15(*a*)

two outlet ports then it is a diverter valve. It is impossible to get any control at all with a mixing valve misused as a diverter valve. The out-of-balance pressures which arise when the valve disc is in anything but a dead central position result in a rapidly vibrating snap action by the disc, between its two extreme positions. (Some types of rotary plug valve may be used as diverters.)

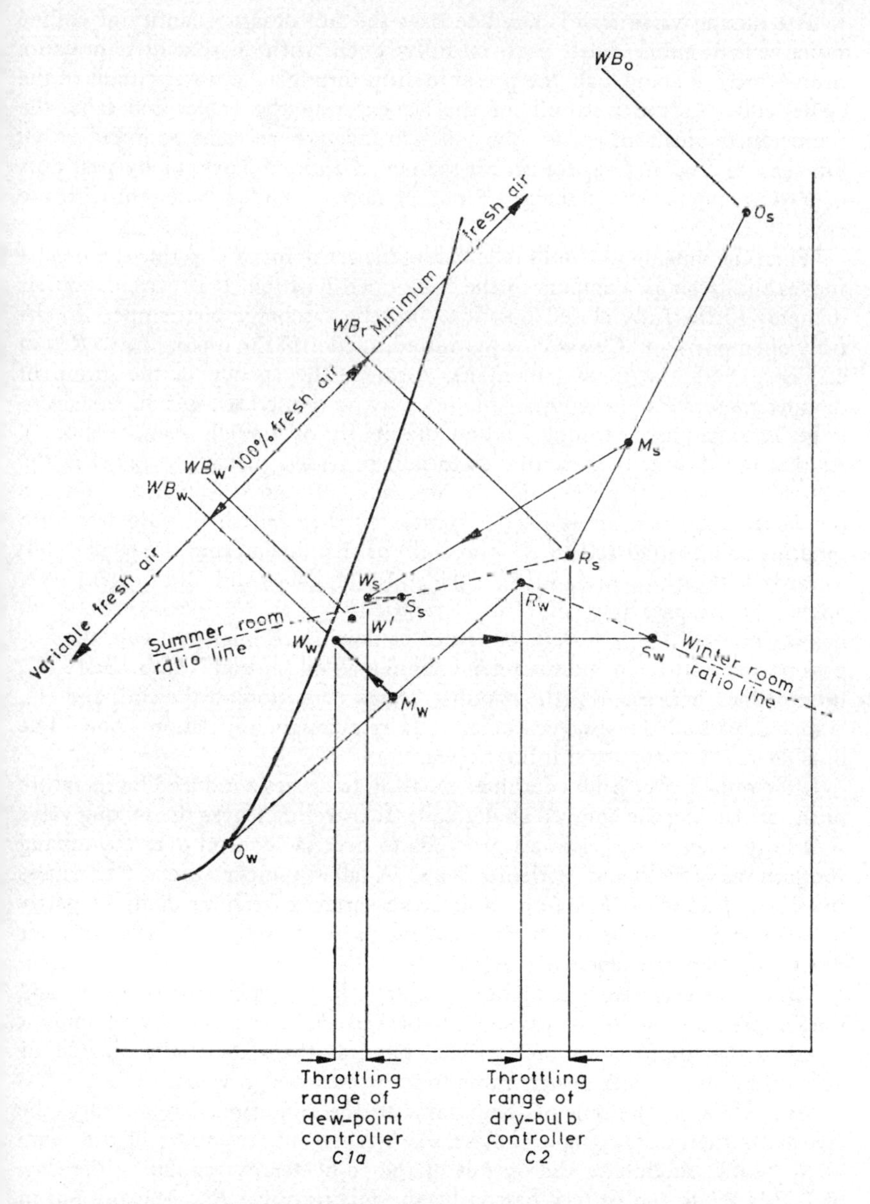

Fig. 13.15(*b*)

Subscript s denotes summer design
Subscript w denotes winter design

The mixing valve *R1a* is sized to pass the full design quantity of chilled water when either inlet port is fully open with a specified pressure drop which is about half the pressure drop through the water circuit of the cooler coil. As the wet-bulb of the air entering the cooler coil falls, the temperature of the air leaving the coil also falls (*see* sections 10.9 and 12.3). On sensing a fall in temperature off the coil, *C1a* modulates the by-pass port of *R1a* to open, thus reducing the rate of flow of chilled water through the coil.

When the outside wet-bulb falls below the set point of *C1b* this thermostat moves the fresh air dampers to the fully open position, the recirculation air dampers to the fully closed position, and the discharge air dampers to the fully open position. *C1a* is now permitted to control the mixing valve *R1a* in sequence with the mixing dampers, through the agency of the group of damper motors *R1b* (a group of motors may be needed if the dampers are so large as to require a torque beyond the ability of a single damper motor). As the outside wet-bulb continues to fall, provided that it lies between the wet-bulb of state R (WB_r) and the wet-bulb of state W (WB_w), then 100 per cent of outside air is used. With a further fall in outside wet-bulb (and hence a further fall in the wet-bulb of the air entering the cooler coil) towards WB_w, the cooling load on the coil diminishes and *C1a* progressively opens the by-pass port of *R1a*. Eventually, when the outside wet-bulb has a value of $WB_{w'}$, no chilled water is flowing through the coil, the by-pass port is fully open and the refrigeration load on the coil is zero. State W', intermediate between W_s (the summer design condition off the coil) and W_w is maintained off the sprayed cooler coil, by adiabatic saturation alone. The outside air dampers are still fully open.

If the outside wet-bulb continues to fall, *C1a* senses a reduced temperature of the air leaving the sprayed cooler coil. It therefore leaves the mixing valve *R1a* fully open to by-pass and proceeds to exercise control over the mixing proportions of fresh and recirculated air. A fall in temperature at *C1a* causes the damper motor(s) *R1b* to modulate the variable fresh-air dampers partly to close and the recirculated air dampers partly to open, the discharge air dampers also partly closing.

Thus, *C1a* exercises a sequencing control, in a proportional manner, and the temperature of the air leaving the sprayed cooler coil falls by an amount equal to the thermostat's proportional band as the state of the outside air alters from its summer design value to its winter design value.

Meanwhile, as the sensible heat gains to the conditioned room vary, the return air thermostat, *C2*, which senses a representative value of the room air dry-bulb, modulates the output of the re-heater by regulating the flow of L.P.H.W. to the battery passed by the mixing valve *R2*. When a fall in temperature is sensed by *C2*, the by-pass port of *R2* is modulated to closure. Thus, the room temperature falls by an amount approximately equal to the proportional band of *C2* as the sensible heat gains diminish from their summer design value and, passing through zero, approach their design value of sensible heat loss in winter.

The dampers would be sized to pass the appropriate air quantities, with pressure drops adequate for achieving a near-linear characteristic. The re-heater control valve *R*2 would be sized to pass the design quantity of L.P.H.W., when fully open, with a pressure drop of between one-third and one-half of the hydraulic loss through the heater battery.

Other controls are, of course, also needed. For example, *C*3 would be an immersion thermostat with a set point of, say, 6°C ± 1°C. It would be located in the chilled water outlet and would exercise a floating action over the step controller which unloaded the refrigeration compressor forming part of the water chilling set.

In formulating a scheme for the automatic control of an air-conditioning system it is essential that a schematic or flow diagram be prepared, after the fashion of Fig. 13.15(*a*) and that all relevant data be shown on it. Such relevant data would be, for example, all airflow and waterflow rates and all psychrometric states. It is also essential that a schedule of operation of the automatic controls be drawn up. This should include such information as the set points, proportional bands (or differential gaps) and location of all controllers (thermostats, humidistats, etc.), and the flow rates, pressure drops when fully open, location, size, etc., of all regulators (motorised valves, dampers, etc.). A short description of the action of each regulator on, say, rise in temperature, should also be included in the schedule. All controllers and regulators should be denoted in a systematic fashion, such as *C*1, *R*1, *C*2, *R*2 and so on.

It is also necessary that the sequence of operation of the components of the air-conditioning plant be stated, due care being paid to safety considerations, e.g. the chilled water pump must start before the refrigeration compressor does.

13.15 Fluidics

A frictional pressure drop occurs when air flows over a surface, pulling the airstream towards it. If the pressure at the surface is artificially increased slightly, by an external agency, the airstream can be persuaded to leave the surface and seek a more stable pattern of flow. The techniques of *fluidics* exploit this so-called ' Coanda ' effect to provide a simple manner of switching an airstream between two stable configurations by using small pressure differences, no moving mechanical parts being involved.

The resultant two-position control may be utilised in many ways, one of which, in the form of a variable volume air conditioning terminal unit, is illustrated in Fig. 16.19(*d*). Supply air from the central air handling plant provides at *P* a flow to the pilot fluidic switch where it continues along either path *A* or path *B*, to provide a bias pressure in the main, amplifying, fluidic switch at *C*, thereby causing the supply airflow to proceed along path *E* instead of *F*. The pressure drop along the surface of leg *E* causes a reduced pressure at tapping *G*, which in turn exerts a negative bias in the pilot switch at *B*, diverting the airstream from leg *A* to leg *B* and hence exerting a positive bias at *D*. This switches the supply airstream to path F. The process repeats and the supply airstream is delivered as alternating impulses through legs *E* and *F* with a frequency of about 120 Hz.

In the illustration these legs feed air to the conditioned room or back to the plant for recirculation, respectively and, for the case considered, a 50 per cent sensible heat gain would be dealt with by providing a 50 per cent supply airflow. If connections to a thermostat are made via tubes K and *Y*, the frequency and bias of the switching can be influenced so that the majority of the air is delivered to the room, or vice-versa. Static air pressure in the duct is the source of energy for the operation of the system which is thus self-acting and without moving parts.

EXERCISES

1. (*a*) Explain what is meant by (i) proportional control and (ii) offset. Give an example of how offset is produced as a result of proportional action.

(*b*) Show by means of a diagram with a brief explanation, the basic layout and operation of a simple electrical *or* a simple pneumatic proportional control unit, suitable for use with a modulating valve and a temperature-sensitive element producing mechanical movement.

2. (*a*) Explain briefly the basic principles by which the relative humidity in a room is controlled by an air-conditioning system using an air washer.

(*b*) What is the purpose of a low limit thermostat in a plenum ventilation system?

3. An air-conditioning plant comprises an air washer, after heater, and constant-speed fan. The weight of supply air is made up of 20 per cent of outside air and 80 per cent of recirculated air. To reduce the refrigeration load under conditions of maximum heat gain a duct is installed allowing part of the recirculated air to by-pass the washer instead of mixing with fresh air before the washer.

The conditioned space has a sensible heat gain in the summer and a loss in the winter. Describe the operation of an automatic control system for the plant that would maintain a constant temperature and humidity in the conditioned space throughout the year. Do not describe the mechanism of the actual control equipment.

BIBLIOGRAPHY

1. *Engineering Manual of Automatic Control.* Minneapolis-Honeywell Regulator Company, Minneapolis 8, Minnesota.
2. JOHN E. HAINES. *Automatic Control of Heating and Air Conditioning.* McGraw-Hill, 1961.
3. DONALD P. ECKMAN. *Automatic Process Control.* John Wiley, 1958.

4. G. H. FARRINGTON. *Fundamentals of Automatic Control.* Chapman & Hall, 1951.
5. *Damper Manual.* Johnson Service Company, Milwaukee, Wisconsin.
6. J. T. MILLER. *The Revised Course in Industrial Instrument Technology.* United Trade Press Ltd., 1964.
7. W. H. WOLSEY. A theory of 3-way valves. *J. Instn Heat. Vent. Engrs*, 1971, **39**, 35–51.

14

Vapour Absorption Refrigeration

14.1 Basic concepts

Following the procedure adopted in section 9.1, let a liquid be introduced into a vessel in which there is initially a vacuum, and let the walls of the container be maintained at a constant temperature. The liquid at once evaporates, and in the process its latent heat of evaporation is abstracted from the sides of the vessel. The resultant cooling effect is the starting point of the refrigeration cycle to be examined in this chapter, just as it was in the beginning of the vapour compression cycle considered in Chapter 9.

As the liquid evaporates the pressure inside the vessel rises until it eventually reaches the saturation vapour pressure for the temperature under consideration. Thereafter, evaporation ceases and the cooling effect at the walls of the vessel is not maintained by the continued introduction of refrigerant. The latter merely remains in a liquid state and accumulates in the bottom of the container. To render the cooling process continuous it is necessary, as we have already seen earlier, to provide a means of removing the refrigerant vapour as fast as it forms. In the vapour compression cycle this removal is accomplished by connecting the evaporator to the suction side of a pump. A similar result may be obtained by connecting the evaporator to another vessel containing a substance capable of absorbing the vapour. Thus, if the refrigerant were water, a hygroscopic material such as lithium bromide could be used in the absorber. The substance used to absorb the refrigerant vapour is called the ' carrier '.

In order to obtain closed cycles for both refrigerant and carrier the next stage in the process must be the release of the absorbed refrigerant at a convenient pressure for its subsequent liquefaction in a condenser. This is accomplished in the ' generator ', where heat is applied to the carrier-refrigerant solution and the refrigerant driven off as a vapour.

The absorber and generator together take the place of the compressor in the vapour-compression cycle. So far as the refrigerant is concerned, the rest of the absorption cycle is similar to the compression cycle, i.e. the vapour is liquefied in the condenser and brought into the evaporator through the expansion valve. As for the carrier, on leaving the generator it is, of course, returned to the absorber for another cycle.

In an absorption refrigeration system cooling water is required for both the condenser and the absorber.

The principal advantages of the absorption cycle over other refrigeration systems are that it can operate with low-grade energy in the form of heat (exhaust steam or high-pressure hot water) and that it has few moving parts. Theoretically, only a single pump is required, that needed for conveying the carrier-refrigerant solution from the low-pressure absorber to the comparatively high-pressure generator. In practice, two more pumps are frequently

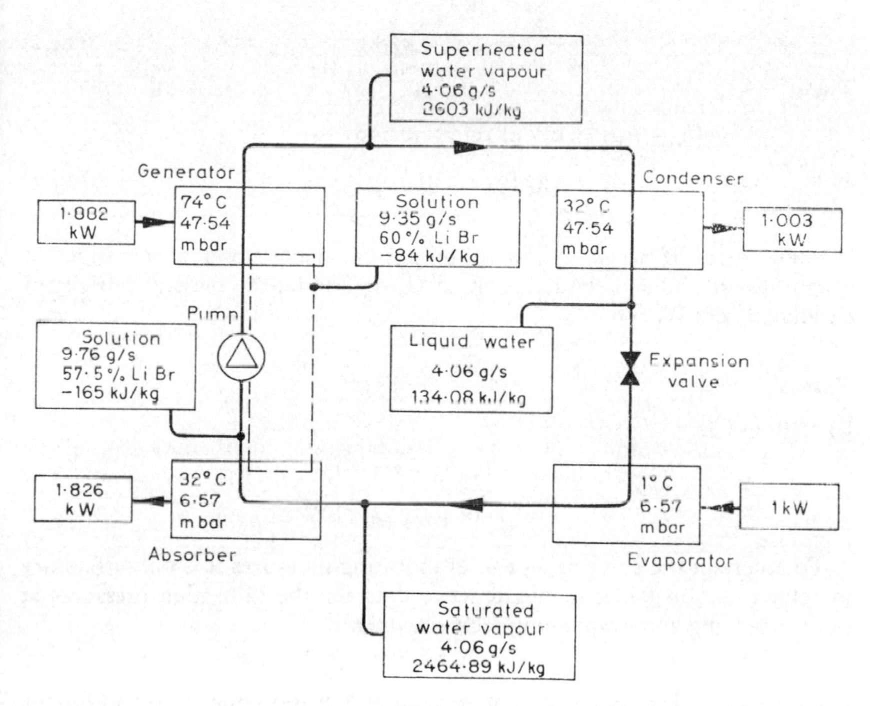

Fig. 14.1 Absorption refrigeration cycle using lithium bromide as carrier and water as refrigerant

used, one to recirculate solution over cooling coils in the absorber and another to recirculate the refrigerant over chilled water coils in the evaporator. The basic absorption refrigeration cycle without these refinements is illustrated in Fig. 14.1.

14.2 Temperatures, pressures, heat quantities and flow rates for the lithium bromide-water cycle

If it is assumed that the temperature of the liquid refrigerant (water) leaving the condenser is t_c and that the temperature of evaporation is t_e, then it is easy to calculate the mass of refrigerant, m_r, which has to be circulated per kW of refrigeration. This calculation depends on the further assumption

$$h_{ve} = 2463 + 1{\cdot}89t_e \text{ kJ/kg} \tag{14.1}$$

This is approximately true for the pressures and temperatures in common use in air-conditioning.

Similarly, for the enthalpy of the liquid leaving the condenser we can write (in the case of water)—

$$\begin{aligned} h_{lc} &= t_c c_l \\ &= 4{\cdot}19t_c \text{ kJ/kg} \end{aligned} \tag{14.2}$$

Then,

$$m_r(h_{ve} - h_{lc}) = \text{kW of refrigeration}$$

and

$$m_r = 1/(2463 - 2{\cdot}3t_e) \text{ kg/s kW} \tag{14.3}$$

EXAMPLE 14.1. If a vapour absorption system using water as a refrigerant evaporates at 1°C and condenses at 32°C, determine the mass of refrigerant circulated per kW.

Answer

By eqn. (14.3)—

$$\begin{aligned} m_r &= 1/(2463 - 2{\cdot}3 \times 1) \\ &= 0{\cdot}0004064 \text{ kg/s kW} \end{aligned}$$

To ascertain the condensing and evaporating pressures it is only necessary to refer to steam tables or hygrometric data for the saturation pressures at the condensing and evaporating temperatures.

EXAMPLE 14.2. Determine the condensing and evaporating pressures for the system used in Example 14.1.

Answer

For $t_c = 32$°C, the saturation vapour pressure, from steam tables, is

$$\begin{aligned} p_c &= 4{\cdot}754 \text{ kN/m}^2 \\ &= 47{\cdot}54 \text{ mbar.} \end{aligned}$$

For $t_e = 1$°C, the saturation vapour pressure is

$$\begin{aligned} p_e &= 0{\cdot}657 \text{ kN/m}^2 \\ &= 6{\cdot}57 \text{ mbar.} \end{aligned}$$

Alternatively, the values can be read from a graph such as that shown in Fig. 14.2.

Considering now the absorber-generator part of the system, it is assumed that the temperature in the absorber is t_a and that the generator is operating at a temperature t_g. Knowing these temperatures and pressures p_c and p_e,

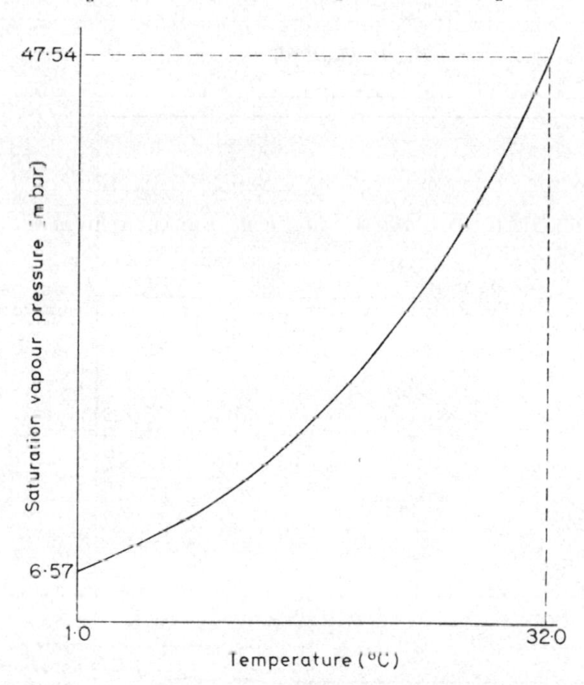

Fig 14.2 Saturation pressure of water vapour

the following data are obtained from tables or from a chart such as Fig. 14.3, which gives the properties of lithium bromide-water solutions:

Solution leaving absorber (at p_e and t_a)	concentration C_a% enthalpy h_a kJ/kg
Solution leaving generator (at p_c and t_g)	concentration C_g% enthalpy h_g kJ/kg

EXAMPLE 14.3. Determine the concentrations and enthalpies of the solution in the absorber and the generator, for Example 14.1.

Answer

Referring to Fig. 14.3, we read off the following:

At 0·657 kN/m^2 and 32°C, $C_a = 57{\cdot}5\%$ and $h_a = -165$ kJ/kg.

At 4·754 kN/m^2 and 74°C, $C_g = 60\%$ and $h_g = -84$ kJ/kg.

It is now possible to calculate the mass of solution that must be circulated to meet the needs of the cycle. If m_{sa} is the mass of solution leaving the absorber and m_{sg} the mass leaving the generator, both in kg/s kW, then for a mass balance the following must hold—

$$C_a m_{sa} = C_g m_{sg} \tag{14.4}$$

$$m_{sa} = m_{sg} + m_r \tag{14.5}$$

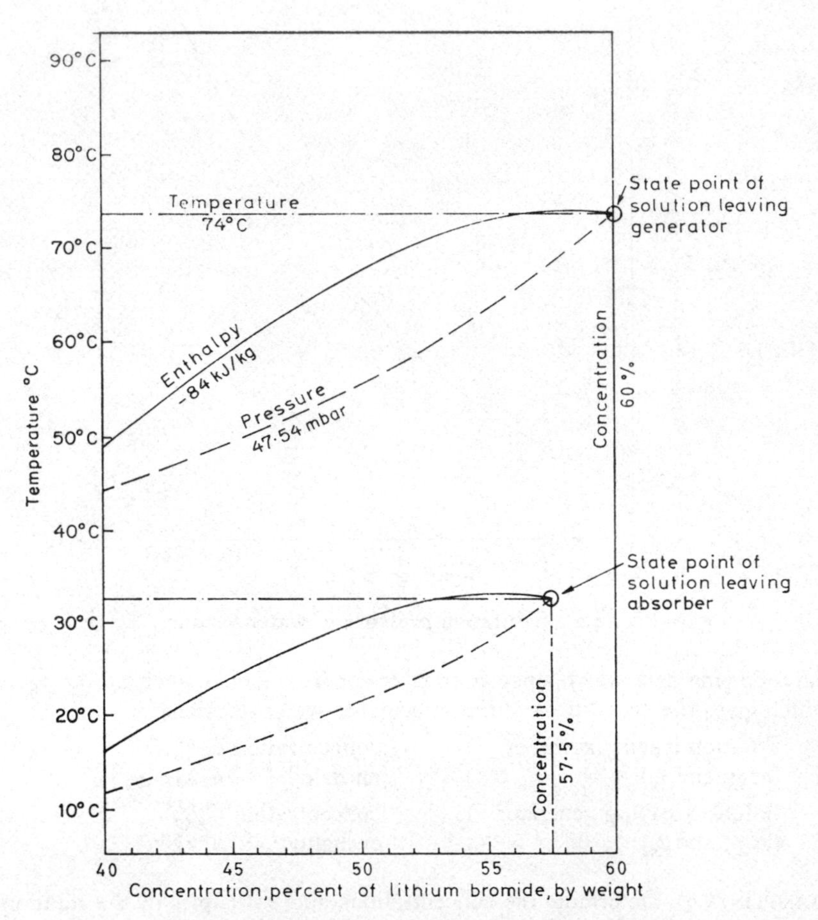

Fig. 14.3 Use of chart giving properties of lithium bromide-water

Substituting for m_{sa} from eqn. (14.5) in eqn. (14.4), we obtain

$$m_{sg} = \frac{C_a m_r}{C_g - C_a} \tag{14.6}$$

EXAMPLE 14.4. Using the data of the previous examples, determine the mass of solution that must be circulated in the absorber and in the generator.

Answer

From eqn. (14.6),

$$m_{sg} = \frac{57{\cdot}5 \times 0{\cdot}000406}{60 - 57{\cdot}5} = 0{\cdot}00935 \text{ kg/s kW}$$

and using eqn. (14.5),

$$m_{sa} = 0{\cdot}00935 + 0{\cdot}000406 = 0{\cdot}00976 \text{ kg/s kW.}$$

An equation similar to (14.1) can be used to determine h_{vg}, the enthalpy of the superheated water vapour leaving the generator,

$$h_{vg} = 2463 + 1{\cdot}89 t_g. \qquad (14.7)$$

EXAMPLE 14.5. Using the earlier data and eqn. (14.1), determine the enthalpy of the water vapour leaving the generator.

Answer

$h_{vg} = 2463 + 1{\cdot}89 \times 74 = 2603$ kJ/kg.

With the information now available it is possible to calculate the heat balance for the whole cycle.

In the absorber

$$\begin{bmatrix}\text{heat of}\\\text{entering}\\\text{water}\\\text{vapour}\end{bmatrix} + \begin{bmatrix}\text{heat of}\\\text{entering}\\\text{solution}\end{bmatrix} - \begin{bmatrix}\text{heat of}\\\text{leaving}\\\text{solution}\end{bmatrix} = \begin{bmatrix}\text{heat to be}\\\text{removed at}\\\text{absorber}\end{bmatrix}$$

$$m_r h_{ve} + m_{sg} h_g - m_{sa} h_a = H_a \text{ kW/kW of refrigeration} \qquad (14.8)$$

In the generator

$$\begin{bmatrix}\text{heat of}\\\text{leaving}\\\text{water}\\\text{vapour}\end{bmatrix} + \begin{bmatrix}\text{heat of}\\\text{leaving}\\\text{solution}\end{bmatrix} - \begin{bmatrix}\text{heat of}\\\text{entering}\\\text{solution}\end{bmatrix} = \begin{bmatrix}\text{heat to be}\\\text{supplied to}\\\text{generator}\end{bmatrix}$$

$$m_r h_{vg} + m_{sg} h_g - m_{sa} h_a = H_g \text{ kW/kW of refrigeration} \qquad (14{\cdot}9)$$

In the condenser

$$\begin{bmatrix}\text{heat of}\\\text{entering}\\\text{water}\\\text{vapour}\end{bmatrix} - \begin{bmatrix}\text{heat of}\\\text{leaving}\\\text{liquid}\end{bmatrix} = \begin{bmatrix}\text{heat to be}\\\text{removed at}\\\text{condenser}\end{bmatrix}$$

$$m_r h_{vg} - m_r h_{lc} = H_c \text{ kW/kW of refrigeration.} \qquad (14.10)$$

EXAMPLE 14.6. Using the data of preceding examples, calculate H_a, H_g and H_c. Check the heat balance.

Answer

From eqns. (14.8) and (14.1),

$$H_a = 0{\cdot}000406(2463+1{\cdot}89\times 1)+0{\cdot}00935\times(-84)-0{\cdot}00976\times(-165)$$
$$= 1{\cdot}0017-0{\cdot}7854+1{\cdot}6094$$
$$= 1{\cdot}826 \text{ kW/kW of refrigeration.}$$

From eqn. (14.9),

$$H_g = 0{\cdot}000406\times 2603+0{\cdot}00935\times(-84)-0{\cdot}00976\times(-165)$$
$$= 1{\cdot}057-0{\cdot}785+1{\cdot}610$$
$$= 1{\cdot}882 \text{ kW/kW of refrigeration.}$$

From eqns. (14.10) and (14.2),

$$H_c = 0{\cdot}000406(2603-4{\cdot}19\times 32)$$
$$= 1{\cdot}003 \text{ kW/kW of refrigeration.}$$

Summarising these results,

Heat removed		*Heat added*	
From absorber	1·826	To generator	1·882
From condenser	1·003	To evaporator	1·000
Total	2·829 kW/kW		2·882 kW/kW.

These two results are within 2 per cent.

14.3 Coefficient of performance and cycle efficiency

In Chapter 9 the coefficient of performance was defined as the ratio of the energy removed at the evaporator to that supplied at the compressor. In the absorption refrigeration cycle, the energy for operating the system is applied through the generator, hence the coefficient of performance may be defined as the ratio of the refrigerating effect to the energy supplied to the generator:

$$COP = 1/H_g \tag{14.11}$$
$$= \frac{m_r(h_{ve}-h_{lc})}{m_r h_{vg}+m_{sg}h_g-m_{sa}h_a}$$

The highest possible coefficient of performance would be obtained by using reversible cycles. Thus, the heat supplied per kilogram of refrigerant ($Q_g = H_g/m_r$) to the generator at temperature T_g might be used in a Carnot engine rejecting its heat to a sink at temperature T_c. The efficiency of this engine would be

$$(T_g - T_c)/T_g = W/Q_g$$

or

$$Q_g = (T_g W)/(T_g - T_c) \quad (14.12)$$

where W is the work done, i.e. the area $CDEF$ in Fig. 14.4.

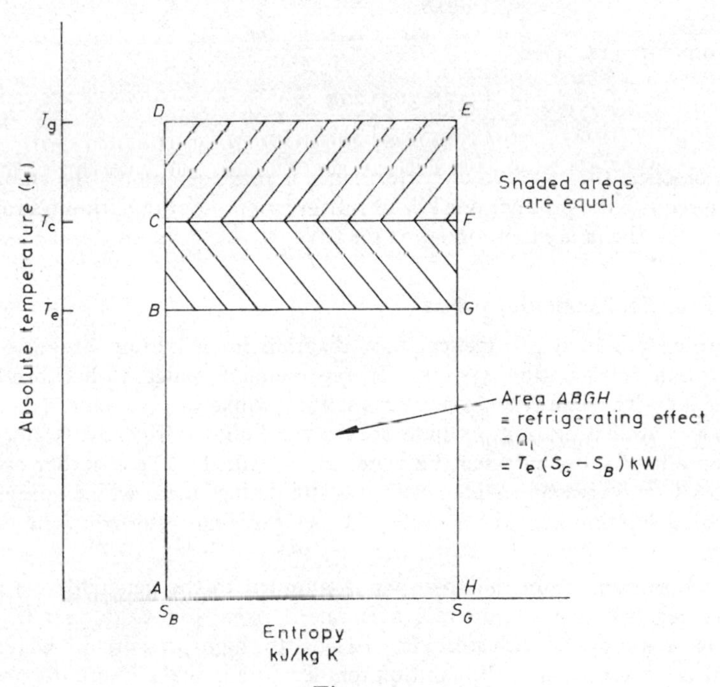

Fig. 14.4

If this work is used to drive a Carnot refrigerating machine then the work input to this, the area $BCFG$, equals W above, and the ratio of areas $BCFG$ and $ABGH$ is

$$W/Q_r = (T_c - T_e)/T_e$$

or

$$Q_r = (T_e W)/(T_c - T_e) \quad (14.13)$$

where Q_r is the refrigerating effect in kW.

Using eqns. (14.12) and (14.13) an expression for the maximum possible coefficient of performance is obtained:

$$COP_{max} = Q_r/Q_g$$

$$= \frac{(T_g - T_c)T_e}{(T_c - T_e)T_g} \quad (14.14)$$

EXAMPLE 14.7. Calculate the coefficient of performance for the cycle in Example 14.6 and compare this with the maximum obtainable.

Answer

From eqn. (14.11),

$$COP = \frac{1}{1{\cdot}8818} = 0{\cdot}53$$

From eqn. (14.14)—

$$COP_{max} = \frac{(347-305)\,274}{(305-274)\,347} = 1{\cdot}06$$

In practice, the criterion of performance is more commonly the amount of steam required to produce one kW of refrigeration. For the lithium bromide-water cycle the figure is around 0·72 g/s kW.

14.4 Practical considerations

Figure 14.5 is a more practical flow diagram for a lithium bromide-water absorption refrigeration system. It represents a water chiller producing water at 6·7°C. The evaporator works at 101·3 mbar, or 4·4°C, and is equipped with a recirculating pump which sprays the liquid refrigerant (water) over the bundle of tubes carrying the water to be chilled. The absorber operates at 40·6°C and is also provided with a recirculating pump which sprays concentrated solution over the bundle of tubes carrying water from the cooling tower.

Weak solution from the absorber is pumped to the generator via a heat exchanger, where its temperature is raised from 40·6°C to 71·1°C. The generator is supplied with steam at 1·841 bar (or high-pressure hot water leaving at 117·8°C) and in it the solution temperature is 104·4°C and the pressure 101·3 mbar. The heat exchanger effects an economy in operation since the cold solution is warmed from 40·6° to 71·1°C before it enters the generator for further heating to 104·4°C, and warm solution is cooled from 104·4° to 73·9°C before it enters the absorber for further cooling to 40·6°C.

In the example illustrated, cooling water from the tower or spray pond at 30°C first passes through the absorber and then the condenser, which it leaves at 39·4°C. The machine shown in the diagram consumes, at full load, from 0·70 to 0·72 g/s of steam per kW of refrigeration and the cooling water requirement is about 0·07 litre/s kW. If control is by varying the evaporation rate, steam consumption increases from about 0·72 to about 1·08 g/s kW, as the load falls from 50 to 10 per cent of full capacity. With solution control, on the other hand, there is a reduction in steam consumption from 0·72 to about 0·61 g/s kW as the load falls from 100 to 30 per cent and a small subsequent rise of 0·02 g/s kW to 0·63 g/s kW, when the load falls further through the range 30 to 10 per cent.

The output of the machine is controlled in two ways: (i) by varying the rate at which the refrigerant boils off in the generator, or (ii) by allowing some of the solution leaving the heat exchanger to by-pass the generator. The first method involves either modulating the heat supply to the generator or varying the flow of cooling water through the condenser.

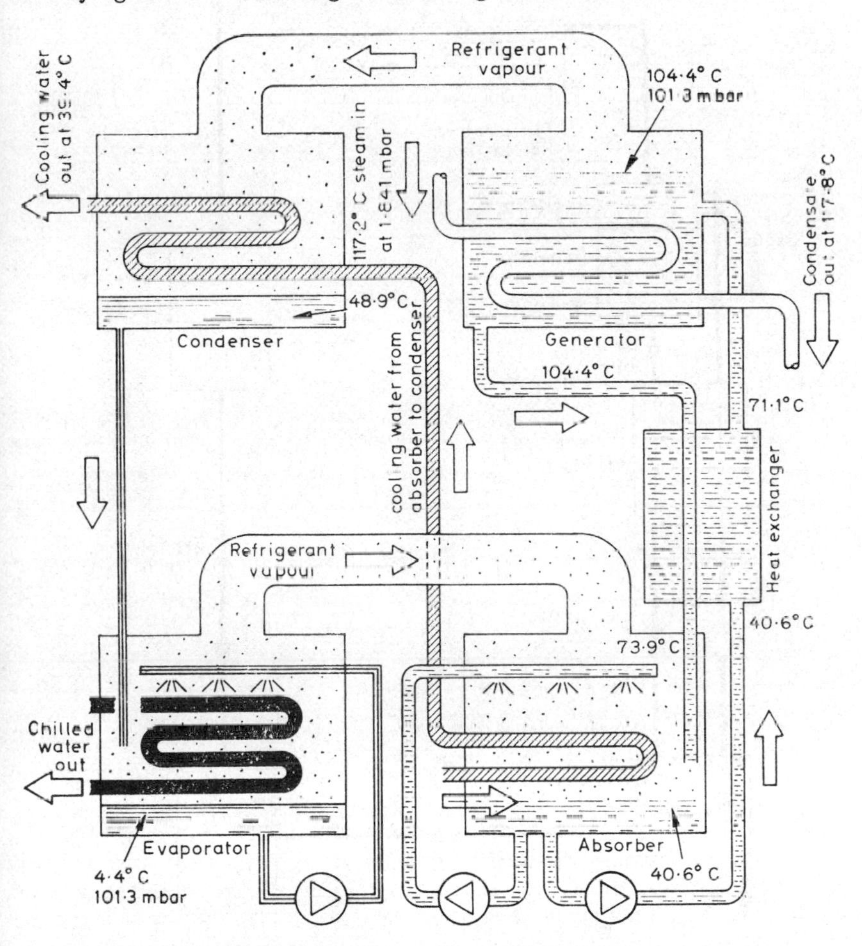

Fig. 14.5 Practical flow diagram for a lithium bromide-water system

14.5 Other absorption systems

The lithium bromide-water system is now almost the only one chosen for air-conditioning applications. An alternative that has been used extensively in industrial work and, to a small extent, in air conditioning, is the system in which ammonia is the refrigerant and water the carrier. Lower temperatures can be produced with this arrangement than are possible when water is used as a refrigerant, but there are difficulties. These arise because the carrier vaporises in the generator as well as the refrigerant, and the system therefore

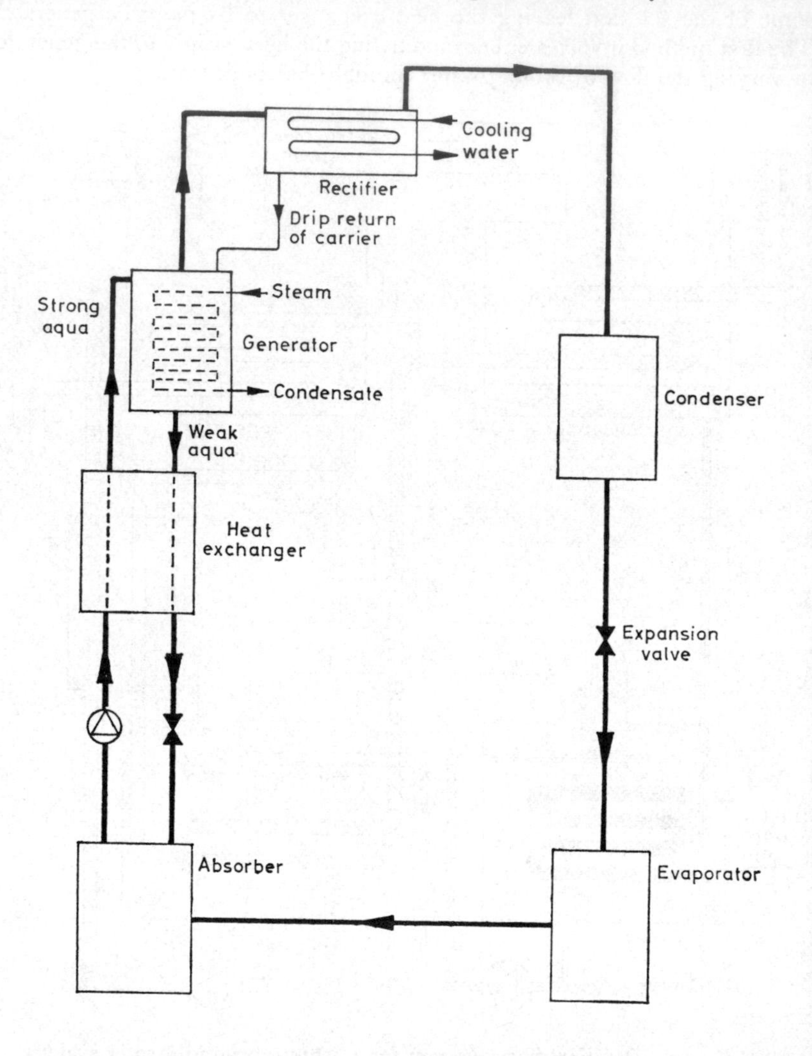

Fig. 14.6 Flow diagram for an aqua-ammonia system

has to be provided with a rectifier to condense and return as much of the carrier as possible. A diagram of the system is shown in Fig. 14.6.

Steam consumptions for water-ammonia systems generally exceed those attainable with lithium bromide-water systems and are in the region of 1·1 to 1·5 g/s kW.

EXERCISES

1. Show on a line diagram the essential components of a steam heated aqua-ammonia absorption refrigerating plant. Briefly outline the cycle of operation and describe the function of each component of the plant.

2. Make a neat sketch showing the layout of a lithium bromide-water absorption refrigeration plant using high-pressure hot water for the supply of energy to the generator. How may its capacity be most economically controlled? Show a suitable control system for this purpose in your diagram.

BIBLIOGRAPHY

1. W. F. STOECKER. *Refrigeration and Air Conditioning.* McGraw-Hill, 1958.
2. W. H. CARRIER, R. E. CHERNE, W. A. GRANT and W. H. ROBERTS. *Modern Air Conditioning, Heating and Ventilating.* Pitman, 1959.
3. N. R. SPARKS. *Theory of Mechanical Refrigeration.* McGraw-Hill, 1938.

15

The Fundamentals of Airflow in Ducts

15.1 Viscous and turbulent flow

When a cylinder of air flows through a duct of circular section its core moves more rapidly than its outer annular shells, these being retarded by the viscous shear stresses set up between them and the rough surface of the duct wall. As flow continues, the energy level of the moving airstream diminishes, the gas expanding as its pressure falls with frictional loss. The energy content of the moving airstream is, macroscopically speaking, in the kinetic and potential forms corresponding to the velocity and static pressures. If the section of the duct remains constant then so does the mean velocity and, hence, the energy transfer is at the expense of the static pressure of the air. The magnitude of the loss depends on the mean velocity of airflow, $\bar{v}$, the duct diameter d and the kinematic viscosity of the air itself, ν. It is expressed as a function of the Reynolds number (Re), which is given by

$$(Re) = \frac{\bar{v}d}{\nu} \qquad (15.1)$$

By means of first principles it is possible to derive the Fanning equation and to state the energy loss more explicitly:

$$H = \frac{4f\,lv^2}{2gd} \qquad (15.2)$$

where H = the head lost, in m of fluid flowing (air),

f = a dimensionless coefficient of friction,

g = the acceleration due to gravity in m/s^2 and $\bar{v}$ and d have the same meaning as in eqn. (15.1).

If pressure loss is required we can write

$$\Delta p = \frac{2f\bar{v}^2\rho l}{d}$$

because pressure equals $\rho g H$, ρ being the density of the fluid.

The Fanning equation provides a rather over-simplified picture of the way in which energy is lost from a moving airstream. Further examination of the problem shows that f assumes different values as the Reynolds number changes with alterations in duct size and the mean velocity of airflow. Figure 15.1 illustrates this.

A further appeal to first principles yields the equation

$$f = 2C(Re)^n \tag{15.3}$$

Again the simplicity is misleading; the determination of C and n requires considerable research effort, as Fig. 15.1 implies. C and n are not true constants, but f can be expressed approximately by the Poiseuille formula:

$$f = \frac{16}{(Re)} \tag{15.4}$$

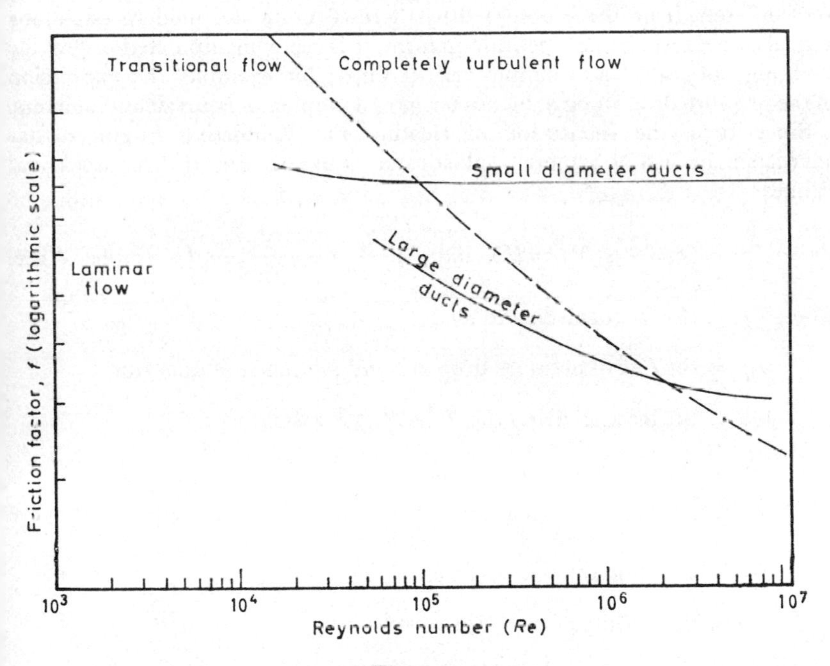

Fig. 15.1

This applies only to streamline flow where (Re) is less than 2000. For turbulent flow, Re, being greater than 3000, eqn. (15.4) no longer holds good and instead the Colebrook White function is used

$$\frac{1}{\sqrt{f}} = -4 \log_{10}\left(\frac{k_s}{3 \cdot 7d} + \frac{1 \cdot 255}{(Re)\sqrt{f}}\right) \tag{15.5}$$

Attempts have been made to rearrange the Fanning equation by making use of eqn. (15.5) and an experimental constant. One such approach, due to Fritzsche, yields the following:

$$\Delta p = \frac{91{\cdot}16 v^{1{\cdot}852} l}{d^{1{\cdot}269}} \tag{15.6}$$

where Δp = the pressure drop in N/m^2 for air at 0°C saturated, 1013·25 mbar barometric pressure, and 0·778 m^3/kg.

v = the mean velocity of airflow in m/s,

l = the length of duct in m,

d = the internal diameter of the duct in mm.

Fritzsche's formula is mentioned because it gives results which are not very different from those obtained by more accurate and modern equations and, also, because, being algebraic in form, it is easily manipulated to provide a number of useful and simple relationships; for example, the expression of the pressure drop along a duct in terms of a number of equivalent diameters.

Since 1959, the Institution of Heating and Ventilating Engineers has advocated the use of a more sophisticated equation, due to Colebrook and White:

$$Q = -4(N_3 \Delta p d^5)^{\frac{1}{2}} \log_{10}\left(\frac{k_s}{3{\cdot}7d} + \frac{N_4 d}{(N_3 \Delta p d^5)^{\frac{1}{2}}}\right) \tag{15·7}$$

where Q = the rate of airflow in m^3/s,

Δp = the rate of pressure drop in N/m^2 per metre of duct run,

d = the internal diameter of the duct in metres.

k_s = the absolute roughness of the duct wall in metres,

$N_3 = \pi^2/32\rho = 0{\cdot}30842\ \rho^{-1}$

$N_4 = 1{\cdot}255\pi\mu/4\rho = 0{\cdot}98567\ \mu\rho^{-1}$,

ρ = the density of the air in kg/m^3,

μ = the absolute viscosity of the air in kg/m s.

The expression is not easily solved in a straightforward manner. The I.H.V.E. publishes a chart (*see* Fig. 15.2), using logarithmic co-ordinates, which gives what are almost linear relationships between air quantity flowing, duct diameter, mean velocity of airflow and rate of pressure drop. The chart is expressed for the flow of dry air at 20°C and 1013·25 mbar in ducts of clean galvanised sheet-steel having joints and seams in accord with good commercial practice. Correction factors are tabulated for ducts of other materials, and an equation is given for the pressure drop of air at other temperatures—

$$\Delta p_2 = \Delta p_1 \left(\frac{293}{273+t_2}\right)^{0.857} = \Delta p_1 E \left(\frac{293}{273+t_2}\right) \quad (15.8)$$

E is a correction factor for the variation in viscosity and is also tabulated, varying from 1·014 at 50°C to 1·054 at 150°C. It is worth noting that, unlike air, the viscosity of water decreases with increase in temperature.

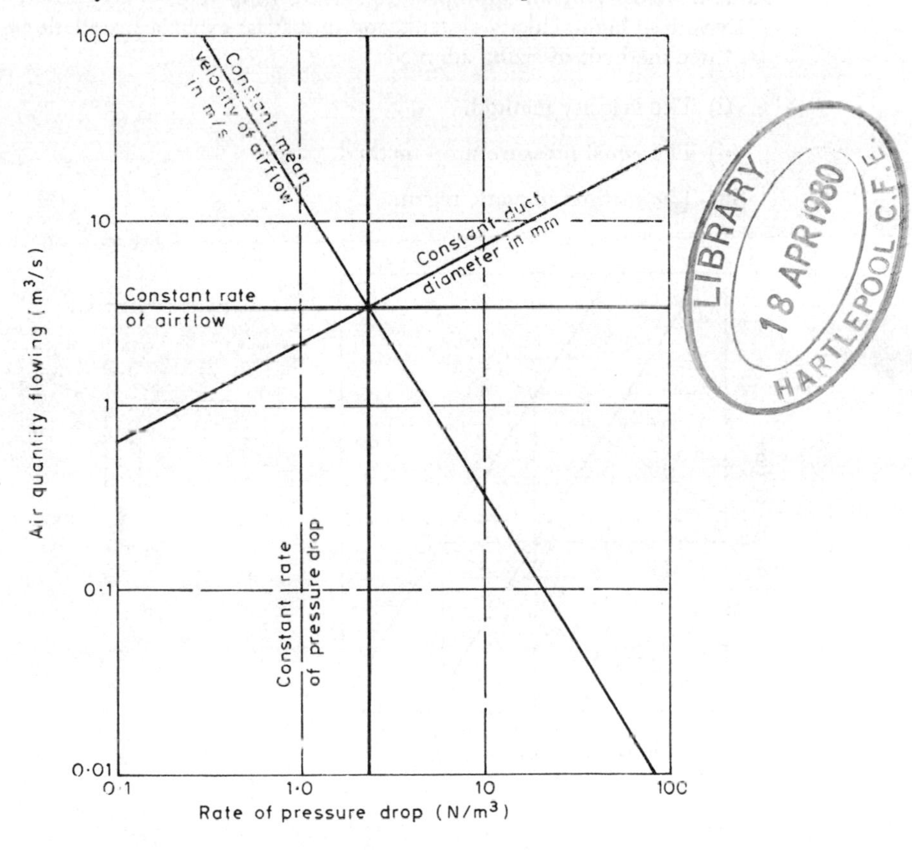

Fig. 15.2

It is also worth noting that a chart based on the Fritzsche formula and constructed with logarithmic co-ordinates is almost indistinguishable from one using the Colebrook-White function.

15.2 Basic sizing

The problem of sizing a duct reduces to the solution of a basic relationship between Q, the quantity of air flowing in m³/s, A, the cross-sectional area of the duct in m^2 and V, the mean velocity of airflow in m/s, given by the simple equation,

$$Q = AV \tag{15.9}$$

Of the three variables in eqn. (15.8), Q is known from other considerations (the required ventilation rate or the amount of air needed to offset a calculated sensible heat gain) and the problem degenerates into one of choosing a suitable velocity or an appropriate pressure drop rate.

Excluding high-velocity systems and industrial exhaust installations, there are three methods of sizing adopted:

(i) The velocity method.

(ii) The equal pressure drop method.

(iii) The method of static regain.

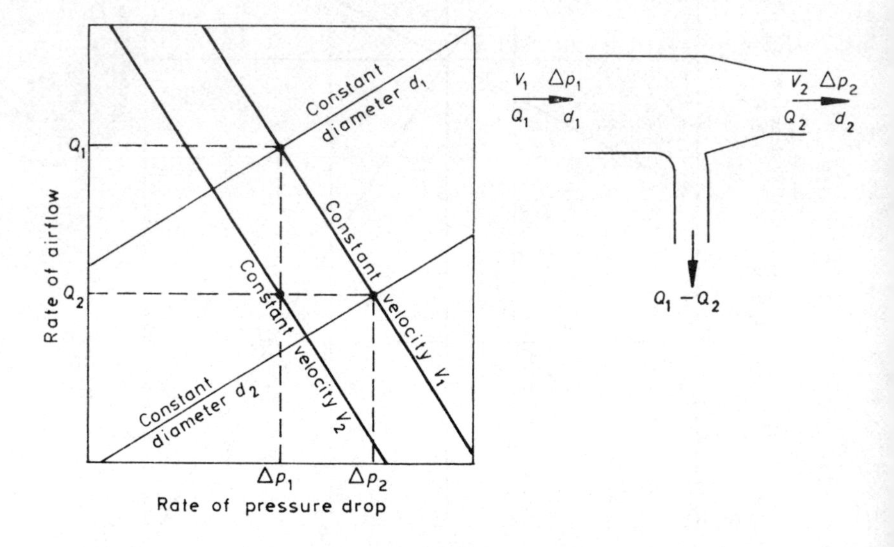

Fig. 15.3

Of these, the first two are in common use but the third is unsuitable for sizing entire systems of ducting. It may, however, be adopted with advantage to size selected parts of systems, provided the initial velocity exceeds about 9 m/s.

In conventional or low-velocity installations, the velocity method perhaps leaves something to be desired. It consists of picking a section of the ducting system thought to be critical (this usually means noisy, and so the section chosen is frequently that immediately after the fan discharge), referring either to personal experience or some authority, and choosing a velocity appropriate to the section of duct or to the premises through which the duct runs. The duct is then sized by making use of eqn. (15.9). The chosen velocity is not kept constant throughout the system but is reduced progressively as the quantity of air handled by the main duct diminishes, air being fed off through branches. Reference to Fig. 15.3 shows the desirability of this reduction: moving along a line of constant velocity on a duct-sizing chart results in an increase of the rate of pressure drop as the quantity of air handled reduces. If a quantity $Q_1 - Q_2$ is fed through a branch duct, the rate of pressure drop in the main duct will increase from Δp_1 to Δp_2, if the velocity of airflow is kept constant in the main by a suitable reduction in the duct diameter. Since noise is of paramount importance in many systems and since the noise generated in a duct through which air is flowing is related to the pressure drop along it, then it is probable that a continued rise in the rate of pressure drop cannot be tolerated.

It is in making the decision on the reduction of velocity that the inadequacy of the method arises. However, provided that some commonsense is used, little trouble should be experienced in low-velocity systems.

EXAMPLE 15.1. Size the ducting shown in Fig. 15.4, given that the velocity in the duct section immediately after the fan outlet must not exceed 7·5 m/s, that this is the critical section of the system, and that the velocity in any branch must not be greater than 3·5 m/s.

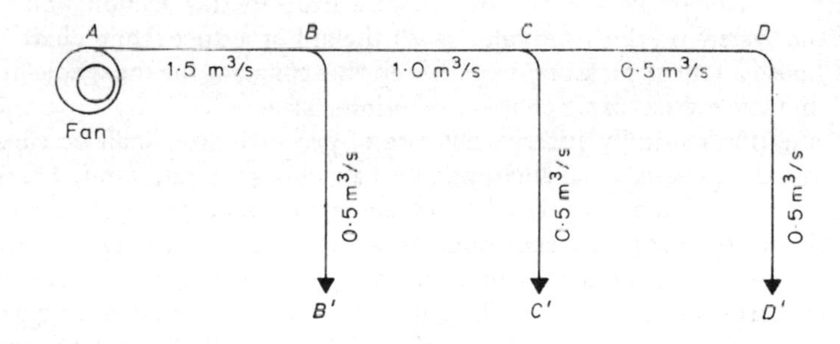

Fig. 15.4

Answer

Make a diagram of the duct system and on it mark the allowable velocities in the sections mentioned, namely 7·5 m/s in section *AB* and 3·5 m/s in sections *BB'*, *CC'* and *DD'*. Common sense suggests that the velocity in *CD* should also be 3·5 m/s and that it is reasonable to choose 5·0 m/s for the velocity in section *BC*.

Since $A = Q/V$, the following table can be compiled:

Section	Q m³/s	V m/s	A m²	d mm	Δp N/m³
AB	1·5	7·5	0·2	505	1·15
BC	1·0	5·0	0·2	505	0·53
CD'	0·5	3·5	0·143	427	0·34
BB'	0·5	3·5	0·143	427	0·34
CC'	0·5	3·5	0·143	427	0·34

The pressure drop rates in the various sections can reasonably be determined only from a duct-sizing chart. Referring to the one published by the I.H.V.E., based on the Colebrook-White function, the values shown in the table are obtained.

The run of duct having the biggest overall energy-loss is termed the *index run*. It is not always possible to leap to the conclusion that the run of ducting greatest in length is that with the largest loss of energy and is, hence, the index run. For example, it cannot be concluded from Fig. 15.4 that the index run is from A to D'. No lengths of ducting are given and no details are provided about the loss of pressure round bends and past branches and other duct fittings.

The second method, that of equal pressure drop, is much favoured in low-velocity systems. There are two approaches:

(*a*) Pick a velocity in what is regarded as the critical section. Size this by eqn. (15.9). Determine the rate of pressure drop in this section. Size the rest of the system on the same rate—with the aid of a duct sizing chart.

(*b*) Choose a rate of pressure drop. Keep this constant for the whole of the system and use a duct-sizing chart to determine sizes.

The question naturally arises, what rate of pressure drop shall be chosen? The naive designer is in a dilemma: he can choose a rate (and, hence, a velocity) so low that the overall loss of energy through the system will be negligible but the duct sizes enormous or, alternatively, he can choose a high rate of pressure drop (and so a high velocity) resulting in very small ducts but a very large loss of energy. The choice lies between a system of gigantic ducts, which are expensive in material and labour and difficult to accommodate in the building, and a system of diminutive ducts which are low in first cost and easily installed. The first arrangement is cheap to run because of the small expenditure of power necessary to overcome the insignificant loss of energy. The second approach to sizing will be expensive to run because the energy losses will be very high if the velocities are high. The question is much more complicated than this, and consideration should be given to the economics of duct-sizing. For the present it suffices to say that a compromise is struck, between the two farcical extremes cited above, which is dictated by experience. For conventional or 'low-velocity' systems, it is recommended that a rate of 1·0 N/m² per m (N/m³) of duct run be chosen.

There is some latitude in this choice but one would expect the chosen rate to be not greater than 1·5 N/m³ nor less than 0·5 N/m³. It is possible, though, that engineers' ideas on this may change in the future.

EXAMPLE 15.2. Size the system of ducting shown in Fig. 15.4 on a constant pressure drop rate of 1·0 N/m³.

Answer

From a duct-sizing chart the diameters and velocities can be read off immediately and then tabulated as follows:

Section	*AB*	*BC*	*CD'*	*BB'*	*CC'*
Diameter (mm)	520	449	340	340	340
Velocity (m/s)	7·1	6·4	5·4	5·4	5·4

15.3 The energy balance in a system of ductwork

Before any attempt is made to determine the losses and exchanges of energy accompanying airflow through a duct system, it must be appreciated that airflow takes place only because energy is supplied to the system by a fan and that this fan receives the energy from a driving motor. The object of calculating the loss of energy along the index run is to obtain enough information to establish the power of the driving motor and the speed at which the fan must run.

Part of the energy drawn from the electricity mains by the driving motor is wasted because the motor itself is not 100 per cent efficient and also because the power output at the driving shaft of the motor is not transmitted to the driven shaft of the fan without some slight loss. Nevertheless, there is a net rate of energy input to the fan, and this is termed the ' fan power '. Some of the fan power is wasted within the fan itself in overcoming bearing loss and in being dissipated as friction and turbulence over the blades of the fan impeller. What remains is termed the ' air power ', and this is expended in overcoming energy losses throughout the ducting system, including those losses occurring across items of air-conditioning plant.

If a certain pressure p is used to propel a small quantity of air δq through a duct in a short time δt, against a frictional resisting force equal and opposite to p, then the rate at which energy must be fed into the system to continue this process is $p(\delta q/\delta t)$. This rate of energy input is what was termed air power.

$$\begin{aligned}\text{air power} &= \text{force} \times \text{distance per unit time}\\ &= (\text{pressure in N/m}^2) \times (\text{area in m}^2) \times (\text{velocity in m/s})\\ &= (\text{fan total pressure in N/m}^2) \times (\text{volumetric airflow in m}^3\text{/s}).\end{aligned}$$

The units are coherent and the product of N/m² and m³/s equals Nm/s or J/s or W.

The total pressure drop through a system of ductwork and its associated items of air-conditioning plant is a measure of the energy expended by the fan; it is defined as the ' fan total pressure ' (*see* section 15.4).

If a symbol η is introduced to represent the total efficiency of the fan, then the expression for fan power is—

$$\text{fan power (W)} = \frac{\text{fan total pressure (N/m}^2) \times \text{volumetric airflow (m}^3/\text{s)} \times 100}{\text{fan efficiency } (\eta)} \tag{15.10}$$

The air power delivered to the airstream by the fan is used to effect the following.

(*a*) Air outside the system is accelerated from rest to the velocity in the air intake.

(*b*) The energy losses resulting from any turbulence and friction as the air enters the system are offset.

(*c*) The frictional resistance offered to airflow by each item of plant is overcome.

(*d*) The resistance to airflow offered by the ducts themselves is offset.

(*e*) The frictional resistances presented by the inlet louvres and by the final discharge grille are nullified.

(*f*) The losses of energy incurred by the maintenance of pockets of turbulence anywhere in the system are made good.

(*g*) The kinetic energy loss from the system, represented by the mass of moving air delivered through the index grille, is offset.

The energy loss as friction and turbulence throughout the system would cause a temperature rise in the airstream, were it is not exactly offset by the temperature fall resulting from the adiabatic expansion accompanying the pressure drop. The only rise in temperature occurs at the fan, where adiabatic compression takes place.

Thus, an energy blance is struck between the electricity drawn from the mains and the frictional and other losses occasioned by the passage of air through the air-conditioning system. To proceed further with a study of the energy losses it is necessary to establish a framework of definitions, axioms and theorems.

15.4 Basic concepts of energy changes

Six fundamental ideas must be grasped before calculations concerning the changes of pressure occurring when air flows through an air-conditioning system can be carried out successfully. These six principles are summarised below, and then discussed in more detail.

(*a*) $TP = SP + VP$ (15.11)

(*b*) Energy loss corresponds to fall of *TP* (15.12)

(*c*) *VP* is always greater than zero, in the direction of airflow (15.13)

(*d*) $FTP = (TP$ at fan outlet$)-(TP$ at fan inlet$)$ (15.14)

(*e*) $FSP = (SP$ at fan outlet$)-(TP$ at fan inlet$)$ (15.15)

(*f*) $FSP = FTP-(VP$ at fan outlet$)$ (15.16)

The symbols used have the following meanings:

TP = total pressure, *SP* = static pressure, *VP* = velocity pressure, *FTP* = fan total pressure, *FSP* = fan static pressure.

One very important and immediate conclusion can be drawn from the above statements: energy loss throughout a system corresponds to the fan total pressure.

Equation (15.16) is an alternative to eqn. (15.15) and is sometimes easier to use; it is derived from eqns. (15.11), (15.14) and (15.15).

(*a*) This is a simplified version of Bernouilli's theorem. It states that, in a moving airstream, the total energy of the moving mass is the sum of its potential energy and its kinetic energy. Since energy is the product of an applied force and the distance over which it is acting and since pressure is the intensity of force, total pressure may be taken as the equivalent of energy per unit mass of air flowing. The potential energy of an airstream is its static pressure. That is to say, its ' collapsing pressure ' if it is less than atmospheric in value, or its ' bursting pressure ' if it is greater than atmospheric. Because it is common to measure pressure with a manometer, balancing the measured pressure against that of the surrounding air, it is convenient to adopt atmospheric pressure as a measurement datum and to assign to it the value of zero. Thus, static pressure can be expressed either as negative or positive, in a relative sense.

(*b*) It is a corollary of Bernouilli's theorem that a fall in energy should correspond to a fall in total pressure and so, in assessing the energy loss through a system, it is the overall change of total pressure which must be computed. For air to flow from one position in a duct to another, the total energy content must be greater upstream than it is downstream if the retarding forces such as friction are to be overcome. It follows that if air is to flow from an energy source at zero (atmospheric) pressure along a ducting system and be delivered into an energy sink which is also at zero pressure, then energy must be supplied to the airstream from an external source. As was pointed out in section 15.3, the fan and motor are this external source of energy.

(*c*) If the velocity of airflow is constant in a duct, then its kinetic energy is also constant, this being a function of velocity. Under these circumstances losses of energy appear as a fall in static pressure. An observation of a change of static pressure is not, in itself, however, a reliable indication of a corresponding change in total energy. For example, if a constant mass of air is flowing through a frictionless duct of expanding section, its velocity of airflow must decrease in accordance with eqn. (15.9). A corresponding drop in velocity pressure must take place and, for Bernouilli's theorem to hold, the static pressure must rise.

(*d*) This is a definition and, therefore, requires no proof. Since the change of total energy across a fan is equal to the amount of energy it is feeding into the system, the definition is merely a way of stating the energy requirement of the fan. It also emphasises the fact that it is fundamentally more correct to work in terms of total pressure changes than to evaluate changes of static pressure—although such assessments are necessary from time to time.

Equation (15.14) states that the energy input required for a ducting system is equal to the algebraic difference between the total pressure at fan outlet and the total pressure at fan inlet. If there is a duct or other piece of energy-consuming equipment on the upstream side of the fan inlet, the total pressure at fan inlet must be negative by an amount equal to the upstream loss. (Energy upstream must always be greater than energy downstream, in order that airflow shall take place.) On the other hand, if there is no such upstream loss, the total pressure at fan inlet will be zero—ignoring any entry loss into the suction eye of the fan. On the discharge side of the fan, the total pressure can never be negative: first it must be large enough to overcome any frictional losses in the downstream ducting and, secondly, it must also be the source of the kinetic energy of the moving airstream issuing forth from the index grille. A consequence of this is that the fan total pressure can never be anything but positive.

(*e*) Fan manufacturers find it convenient to express fan performance in terms of fan static pressure, rather than fan total pressure. *It is to be noted that the static pressure at fan outlet is not the same as the fan static pressure.* The former plays a part in the definition of the latter. The objection need never be raised that one cannot calculate the fan total pressure since one does not know the velocity pressure at fan discharge. The size of a fan does not really depend on the resistance the fan is expected to overcome. It depends instead, for a particular type of fan, on the quantity of air the fan is to handle. Consequently, if the type of fan and its size is known, so also is its discharge velocity pressure.

15.5 Velocity pressure

The units of pressure are synonymous with those of energy per unit volume, as can be seen by multiplying the denominator and the numerator of N/m^2 by metres. This supports the earlier contention (section 15.3) that the product of fan total pressure and volumetric airflow rate is the rate of energy flow, or power. It follows that velocity pressure may be regarded as kinetic energy per unit volume and hence

$$p_v = \frac{1}{2}\frac{mv^2}{\text{vol}} = \frac{1}{2}\rho v^2 \qquad (15.17)$$

Similarly, static pressure may be considered as potential energy per unit volume.

If the density of air is taken as 1·20 kg/m^3, eqn. (15.17) becomes

$$p_v = 0{\cdot}6\,v^2 \tag{15.18}$$

and the converse is

$$v = 1{\cdot}291\sqrt{p_v} \tag{15.19}$$

15.6 The flow of air into a suction opening

Consider a simple length of ducting attached to the inlet side of a fan and another length of ducting attached to its outlet. Air is accelerated as it approaches the suction opening and, in order to produce this increase in kinetic energy, a negative potential energy has to be set up within the opening. Figure 15.5(*a*) illustrates that the air, in negotiating the entry to the duct, is compelled to undergo a change of direction (unless it happens to be on the centre-line of the duct) and that this involves the setting up within the duct of a pocket of turbulence which reduces the area of entry available to the air. The reduced area is termed the ' vena-contracta '.

Three immediate conclusions can be drawn from this.

1. The velocity of airflow through the vena-contracta must be higher than that prevailing in the succeeding downstream length of duct.

2. The curved paths followed by the air in the eddies within the pocket of turbulence involve the expenditure of energy—in accordance with Newton's first law of motion.

3. There is a drop in total pressure as the air flows through the open end of a suction duct, because of conclusion 2 above, regardless of the presence of any grille at the opening. If a grille is present then the loss of total pressure will be greater.

An application of Bernouilli's theorem (eqn. (15.11)) permits the changes of total, static and velocity pressure to be determined as air enters the system. Figure 15.5(*a*) shows a plot of such pressure changes for a suction opening at the end of a duct, and (*b*) shows airflow into a ' no-loss ' entry. The end of the duct is constructed in such a way that solid material occupies the space normally filled with turbulence and prevents the formation of a vena-contracta. Virtually no losses occur, and the static suction set up just within the open end of the duct, where it has attained its proper diameter, is numerically equal to the velocity pressure, but opposite in sign.

(No-loss entry pieces of this kind provide a very reliable and accurate method of measuring the rate of airflow through a duct. It is only necessary to take a careful measurement of the static pressure on the section where the taper has just ceased and to convert this to velocity, by means of eqn. (15.17). Such a method is usually suitable for laboratory uses only.)

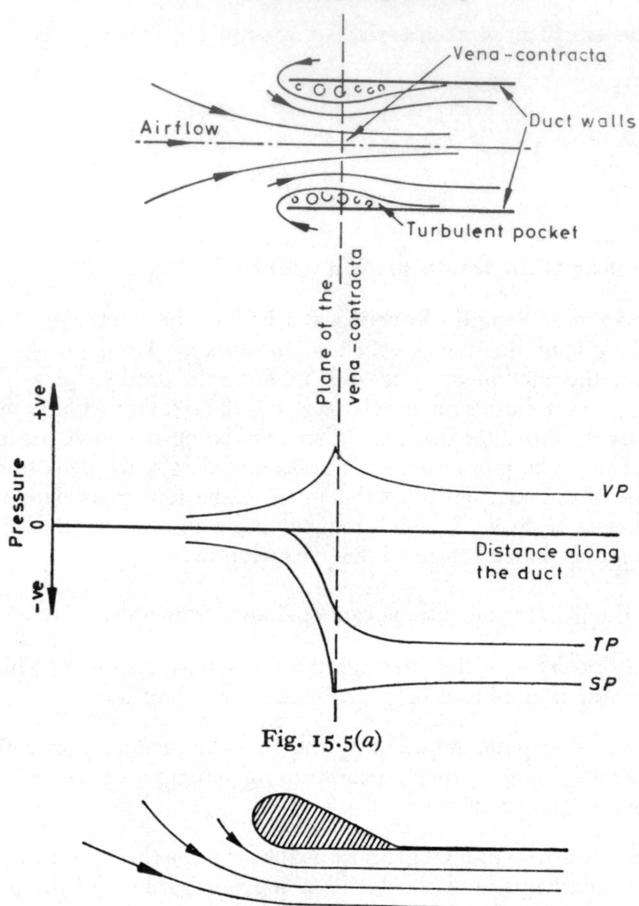

Fig. 15.5(a)

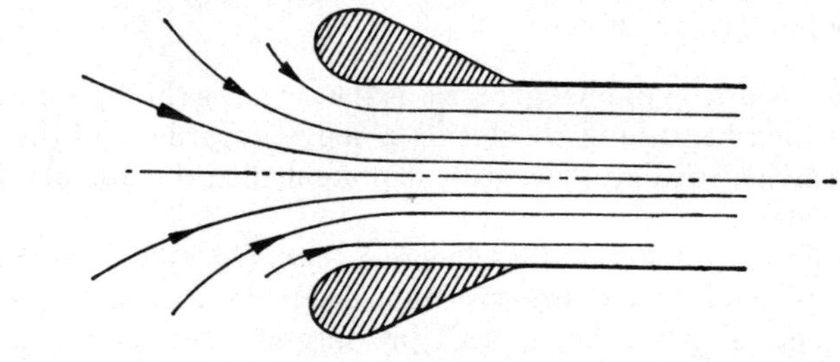

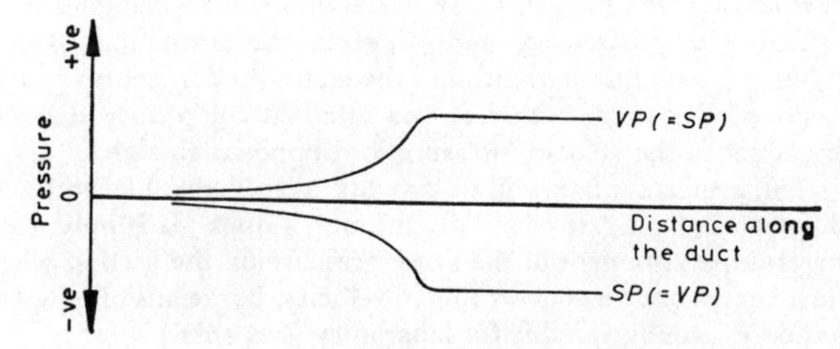

Fig. 15.5(b)

15·7 The coefficient of entry (C_E)

For normal entry-pieces not all the static suction is used to accelerate the air to the velocity prevailing in the downstream duct. Some of the potential energy of the suction set up is wasted in offsetting losses due to turbulence and friction. It is customary to express these losses in terms of the steady velocity pressure in the downstream duct, after any vena-contracta. Thus, the loss due to eddies formed at the vena-contracta can be written as k multiplied by this velocity pressure, where k is obviously less than unity.

The fact that there is a reduced area, A', available for airflow at the vena-contracta gives rise to the concept of a coefficient of area, defined by

$$C_A = \frac{\text{area at the vena-contracta}}{\text{area of the duct}} \tag{15.20}$$

There is also the concept of a coefficient of velocity. This arises from the fact that the velocity at the vena-contracta is less than that which would be attained if all the static pressure were to be converted into velocity. There is some friction between the air flowing through the vena-contracta and the annular pocket of turbulence which surrounds it. This is consequent on the energy transfer needed to maintain the eddies and whorls within the pocket. The coefficient of velocity is defined by

$$C_V = \frac{\text{actual velocity at the vena-contracta}}{\text{velocity which would be attained at the same section in the absence of losses}}$$

Since the quantity flowing is the product of velocity and area it can be seen that a flow coefficient, C_E, may be inferred as the product of the coefficients of velocity and area—

$$C_E = C_V \times C_A$$

Thus, Q, the actual rate of flow of air entering the system, may be expressed by

$$Q = C_E \times Q'$$

where Q' is the theoretical rate which would flow if no loss occurred. It may also be written as

$$Q = C_E(A \times V')$$

But $V' = 1{\cdot}291\sqrt{p_v}$, the theoretical maximum possible velocity, where p_v is equivalent to the static suction set up in the plane of the vena-contracta.

Thus, $$Q = 1{\cdot}291\sqrt{p_v}.$$

15.8 The discharge of air from a duct system

Air in the plane of the open end of the discharge duct must be at virtually atmospheric pressure—since there is no longer any resisting force to prevent the equalisation of pressure. We can therefore say that the potential energy of the air leaving the system through an open end is zero. The kinetic energy is not zero however. There is an outflow of mass from the system. Applying Bernouilli's theorem, it is seen that the total energy of the airstream leaving the ducting must be equal to its kinetic energy or velocity pressure.

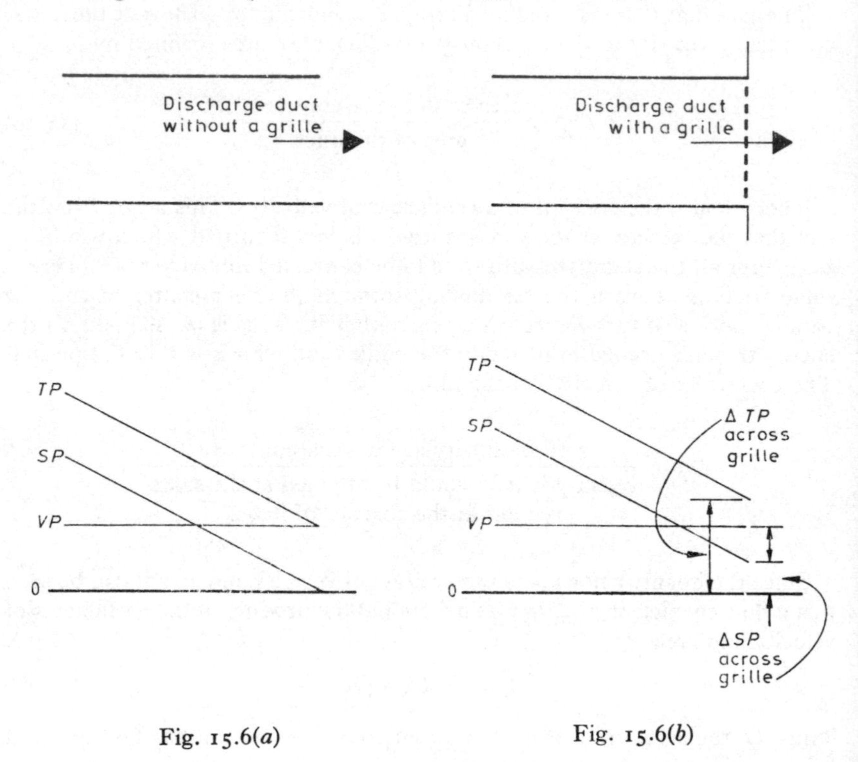

Fig. 15.6(a) Fig. 15.6(b)

If a grille or diffuser is placed over the open end of the duct, the total pressure on its upstream side must be greater than that on its downstream side by an amount equal to the frictional loss incurred by the flow of air through the grille. If any change of velocity that may take place as the air flows between the bars of the grille is ignored, the velocity pressure upstream must equal the velocity pressure downstream. It follows that the static pressure of the upstream air is used to overcome the frictional loss through the grille. This is illustrated in Fig. 15.6.

Denoting the upstream side of the grille by the subscript 1 and the downstream side by subscript 2, we can write

$$TP_1 = TP_2 + \text{frictional loss past the grille.}$$

$$VP_1 + SP_1 = VP_2 + SP_2 + \text{frictional loss past the grille.}$$

But
$$VP_1 = VP_2$$

Hence, $SP_1 - SP_2$ = frictional loss past the grille.

Since SP_2 is virtually the same as atmospheric pressure—the zero datum—this becomes

$$SP_1 = \text{frictional loss past the grille.}$$

To sum up: the fan has to make good the energy loss at the end of the system incurred by the kinetic energy of the airstream leaving the system *and* the friction past the last grille.

15.9 Airflow through a simple duct system

Figure 15.7 shows a simple straight duct on the suction side of a fan, followed by a similar duct of smaller cross-section on its discharge side. There is a grille over both the inlet and outlet openings of the duct.

Air entering the system is accelerated from rest outside the duct to a velocity V_1 prevailing within the duct, once the vena-contracta has been passed and the flow has settled down to normal. If, for convenience of illustration, the physical presence of the vena-contracta is ignored but the loss which it causes is included with the frictional loss past the inlet grille, the entry loss can be shown as occurring in the plane of entry. We can then write

$$\begin{aligned} \text{entry loss} &= 0 - TP_1 \\ &= -TP_1 \\ \text{static suction} &= -TP_1 - VP_1 \\ &= -SP_1 \end{aligned}$$

The static depression immediately within the inlet grille is numerically greater than the velocity pressure by an amount equal to the energy loss past the inlet grille.

The loss due to friction along the duct between plane 1 and plane 2 equals $TP_1 - TP_2$. The static depression at fan inlet is clearly $TP_2 - VP_2$. It is also clear that the total pressure at fan inlet, TP_2, represents the loss of energy through the system up to that point, and that it comprises the frictional loss past the inlet grille, the loss past the vena-contracta and the loss due to friction in the duct itself.

On the discharge side of the system it is easier, for purposes of illustration, to start at the discharge grille. The change of static pressure past this grille equals SP_4—o but the change of total pressure across the grille equals TP_4—o, or the friction plus the loss of kinetic energy. Between the planes 3 and 4, the loss of energy due to friction equals $TP_3 - TP_4$, in the direction of airflow. The total energy upstream must exceed that downstream if airflow is to take place. Hence, TP_3 must be greater than TP_4 and so also must TP_1 be greater than TP_2, in a similar way. But TP_3 is the total energy at

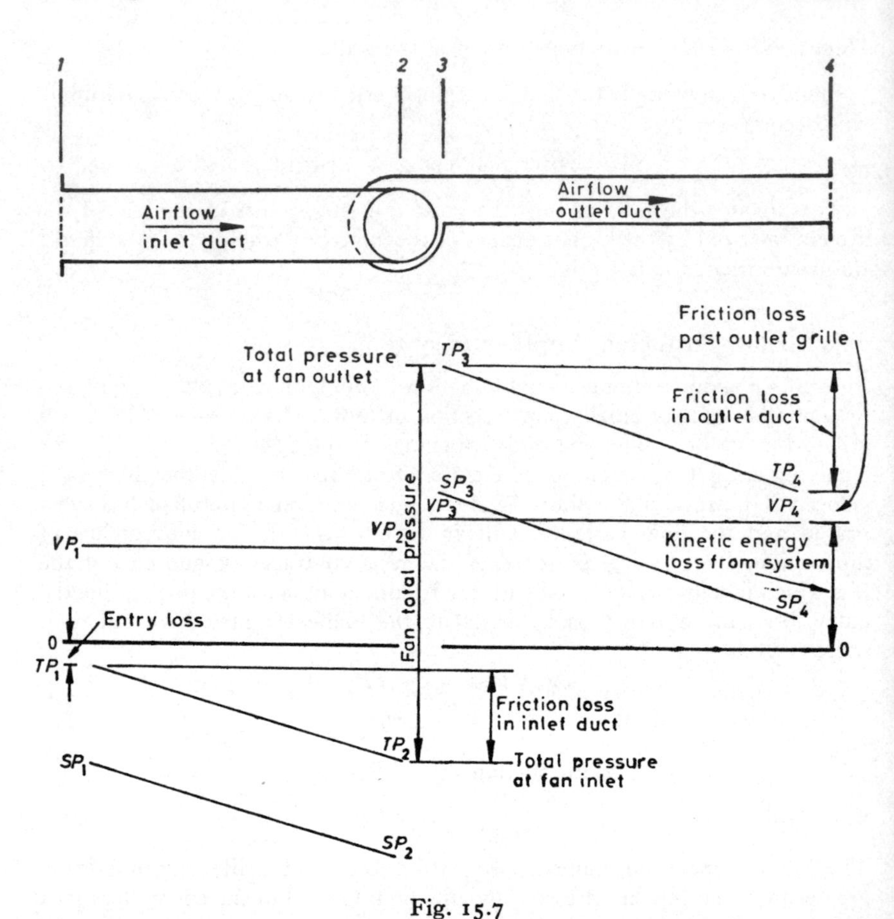

Fig. 15.7

fan outlet and TP_2 is the total energy at fan inlet. Consequently, the difference between these two quantities, TP_3 and TP_2, must be the energy supplied to the system by the fan. This statement fits in with eqn. (15.14) and we can write

$$\begin{aligned} FTP &= TP_3 - TP_2 \\ &= TP_3 + TP_2, \end{aligned}$$

numerically speaking, since TP_2 has a negative sign. Thus, the fan total pressure can never be negative.

To sum up:

FTP = [friction past the inlet grille+energy loss at the vena-contracta] +[friction loss in the inlet duct]+[friction loss past the outlet grille+kinetic energy lost from the system]+[friction loss in the outlet duct]

$$= (0-TP_1)+(TP_1-TP_2)+(TP_4-0)+(TP_3-TP_4)$$

$$= TP_3-TP_2$$

15.10 Airflow through an abruptly expanding duct section

Using first principles, it is possible to derive an expression for the loss of pressure Δp, resulting from the sudden expansion of an airstream—

$$\Delta p = \tfrac{1}{2}\rho(V_1-V_2)^2 \qquad (15.21)$$

where V_1 and V_2 are the mean velocities of airflow upstream and downstream, respectively, as Fig. 15.8. The data for such losses are presented in the current edition of the I.H.V.E. Guide in terms of the difference of the upstream and downstream velocity pressures, multiplied by a factor k, the value of which depends upon the ratio V_2/V_1. Since V_2 equals $V_1(a_1/a_2)$, eqn. (15.21) may be re-written in the form

$$\Delta p = \tfrac{1}{2}\rho V_1{}^2\left(1-\frac{a_1}{a_2}\right)^2$$

If also it is stipulated that Δp equals k multiplied by the difference of the velocity pressures, then

$$\Delta p = k(\tfrac{1}{2}\rho V_1^2-\tfrac{1}{2}\rho V_2^2)$$

$$= \tfrac{1}{2}\rho k V_1^2\left(1-\frac{a_1^2}{a_2^2}\right)$$

Thus

$$k = \frac{\left(1-\dfrac{a_1}{a_2}\right)^2}{\left(1-\dfrac{a_1^2}{a_2^2}\right)} \qquad (15.22)$$

EXAMPLE 15.3. If an airsteam moving at 20 m/s suffers a sudden expansion which reduces its velocity to 10 m/s, calculate the loss of total pressure occurring,

(*a*) by making use of the difference of the upstream and downstream velocity pressures and,

(*b*) by making use of the fundamental expression given in eqn. (15.21).

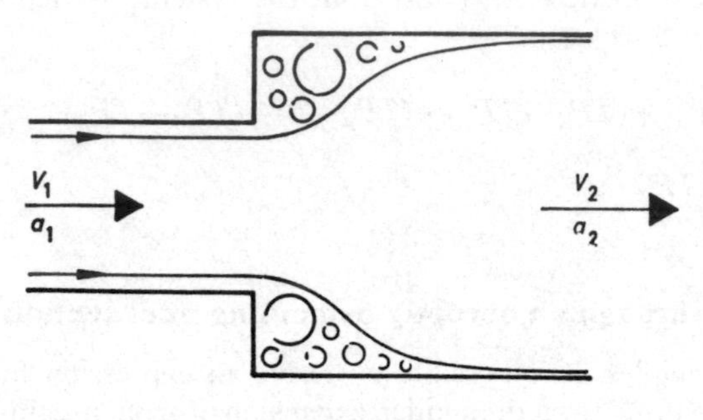

Fig. 15.8

Answer

(a) From eqn. (15.22)

$$k = \frac{\left(1-\frac{10}{20}\right)^2}{\left(1-\frac{10^2}{20^2}\right)}$$

$$= \left(\frac{1-0.5}{1-0.25}\right)^2$$

$$= 0.33.$$

Hence, the loss of total pressure is

$$\Delta TP = 0.33\{0.6\times 20^2 - 0.6\times 10^2\}$$

$$= 0.33\,(240-60)$$

$$= 60\ \mathrm{N/m^2}.$$

(*b*) By eqn. (15.21),

$$\Delta p = \tfrac{1}{2}\rho(V_1 - V_2)^2$$

$$= 0.6\,(20-10)^2$$

$$= 60\ \mathrm{N/m^2}$$

The I.H.V.E. Guide tabulates values of k for velocity ratios from 0.20 to 0.75.

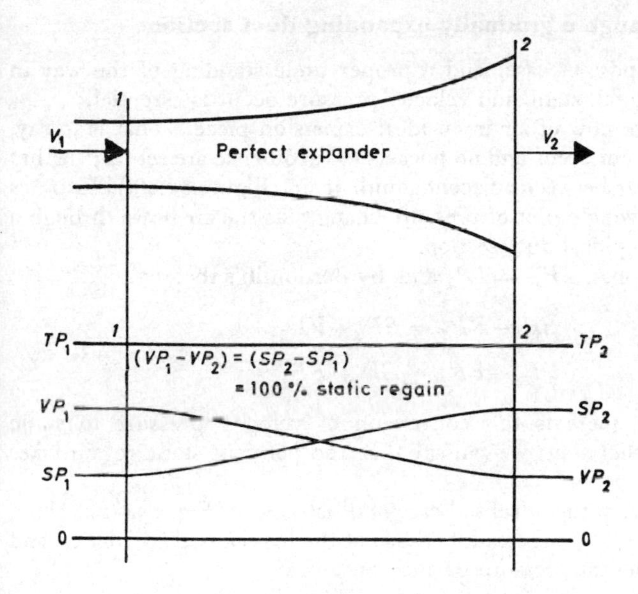

Fig. 15.9(*a*)

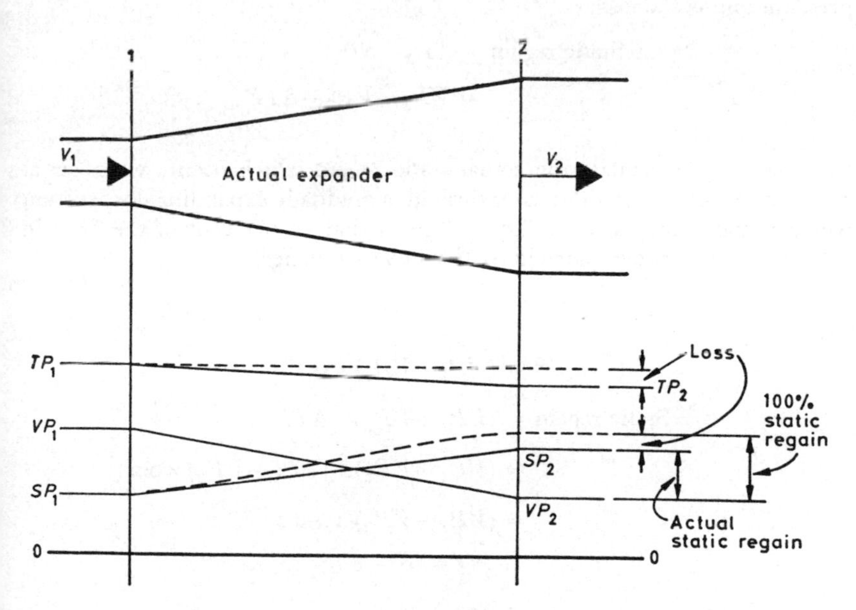

Fig 15.9(*b*)

15.11 Airflow through a gradually expanding duct section

This is a most important case, and a proper understanding of the way in which changes of total, static and velocity pressure occur is essential.

Consider first the flow of air in an ideal expansion piece. That is to say, frictional loss does not occur and no pockets of turbulence are set up; neither is there viscous shear between adjacent annuli of air. Figure 15.9(*a*) illustrates such a case by showing a plot of pressure changes as the air flows through a gradually expanding ideal duct section.

Since there is no loss, $TP_1 = TP_2$ and, by Bernouilli's theorem,

$$SP_1 + VP_1 = SP_2 + VP_2$$

Hence,

$$VP_1 - VP_2 = SP_2 - SP_1$$

This states that there is full conversion of velocity pressure to static pressure. Put another way: we can say that 100 per cent static regain takes place.

The departure from this idealised case is illustrated in Fig. 15.9(*b*). Here, TP_1 does not equal TP_2. Instead, because of the loss caused by friction and turbulence, the relevant pressure balance becomes

$$\Delta TP = TP_1 - TP_2$$

$$= SP_1 + VP_1 - (SP_2 + VP_2)$$

The actual static regain taking place is clearly equal to the fall in velocity pressure minus the loss:

$$\text{Static regain} = SP_2 - SP_1$$

$$= VP_1 - VP_2 - \Delta TP$$

EXAMPLE 15.4. Calculate the actual static regain which occurs when air at an initial velocity of 20 m/s passes through a gradually expanding duct section which reduces its velocity to 10 m/s, given that 20 per cent of the drop in velocity pressure is dissipated in friction and turbulence.

Answer

$$\Delta TP = (VP_1 - VP_2) \times 0{\cdot}2$$

$$\text{Static regain} = (VP_1 - VP_2) - \Delta TP$$

$$= (VP_1 - VP_2) - (VP_1 - VP_2) \times 0{\cdot}2$$

$$= (VP_1 - VP_2)(1 - 0{\cdot}2)$$

$$= (0{\cdot}6 \times 20^2 - 0{\cdot}6 \times 10^2) \times 0{\cdot}8$$

$$= 144\ \text{N/m}^2.$$

Another way of stating the problem would have been to say that 80 per cent static regain takes place. Among the various authorities there is no uniformity of expression of static regain. It is American practice to express the regain in terms of a factor which is used to multiply the difference between the upstream and downstream velocity pressures, the magnitude of the factor depending on the included angle of expanding duct section. The I.H.V.E., on the other hand, choose to express the loss as a fraction of that which would be incurred if the expansion were abrupt. The value of the fraction depends on the included angle of section. However, the various methods yield results which are in reasonably close agreement with one another. If the static regain is required, it is probably safest to regard this as the difference between the velocity pressures minus the loss of total pressure, rather than to work slavishly to a formula.

EXAMPLE 15.5. (*a*) Calculate the static regain which occurs when air at 20 m/s flows through a gradually expanding duct section which reduces the velocity to 5 m/s. The included angle of the section is 20° and the loss is 0·45 times the difference between the upstream and downstream velocity pressures.

(*b*) Repeat the calculation for the same expanding duct section but base it on the assumption that the loss is 80 per cent of that for an abrupt expansion.

Answer

(*a*)

$$\begin{aligned} \Delta TP &= 0{\cdot}45(0{\cdot}6 \times 20^2 - 0{\cdot}6 \times 5^2) \\ &= 0{\cdot}45 \times 225 \\ &= 101{\cdot}25 \text{ N/m}^2. \end{aligned}$$

$$\begin{aligned} \text{Static regain} &= \text{difference of velocity pressures minus the loss} \\ &= 225 - 101{\cdot}25 \\ &= 123{\cdot}75 \text{ N/m}^2. \end{aligned}$$

(*b*)

$$\Delta TP = 0{\cdot}80 \times \text{the loss for an abrupt expansion.}$$

From eqn. (15.22), the loss factor for an abrupt expansion is

$$\begin{aligned} k &= \frac{\left(1 - \frac{5}{20}\right)^2}{\left(1 - \frac{5^2}{20^2}\right)} \\ &= \frac{(1 - 0{\cdot}25)^2}{(1 - 0{\cdot}25^2)} \\ &= 0{\cdot}6. \end{aligned}$$

Hence,

$$\Delta TP = 0{\cdot}80 \times 0{\cdot}60 \times \text{difference of velocity pressures}$$
$$= 0{\cdot}48 \times 225$$
$$= 108\ \text{N/m}^2.$$
$$\text{Static regain} = 225 - 108$$
$$= 117\ \text{N/m}^2.$$

15.12 Airflow through an abruptly reducing duct section

Figure 15.10 shows the flow patterns that tend to develop when an airstream undergoes a sudden contraction. There are two areas where pockets of turbulence occur: at the shoulders of the contraction in the large section, and shortly after the neck. It is this latter pocket of turbulence, set up by the formation of a vena-contracta, which is the major source of energy loss. The velocity of airflow downstream is larger than it is upstream and the energy transfer is correspondingly greater.

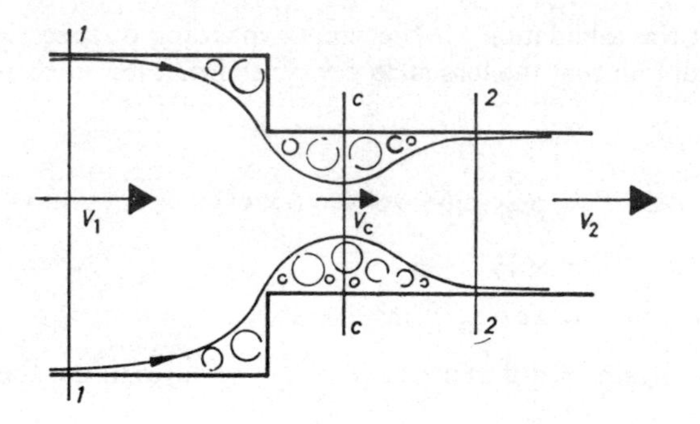

Fig. 15.10

A consideration of basic mechanics of fluids permits a simple derivation to be made of the loss occurring, assuming the loss is substantially due to the downstream pocket of turbulence. Using the notation of Fig. 15.10, we can write

$$\Delta p = \tfrac{1}{2}\rho(V_c - V_2)^2$$

The coefficient of area being defined as

$$C_A = \frac{\text{area}_{cc}}{\text{area}_{22}},$$

it is also defined as

$$C_A = \frac{V_2}{V_c}$$

thus,

$$V_c = \frac{V_2}{C_A}$$

and we can write an expression for the loss as

$$\Delta p = \tfrac{1}{2}\rho V_2^2\left(\frac{1}{C_A} - 1\right)^2 \qquad (15.23)$$

This is the basic equation for the loss due to a sudden contraction. It is slightly in error because the upstream losses are ignored, and it is open to different interpretations because the answer depends on the value assigned to C_A. For circular orifices and waterflow, C_A is often taken as 0·62. This gives the loss as 0·375 times the downstream velocity pressure. Experiment shows that a better answer is 0·5 times the downstream velocity pressure, and this is the value commonly adopted. For the flow of air through sudden contractions, the factor used to multiply the downstream velocity pressure is given different values, depending on the ratio of the upstream and downstream velocities.

15.13 Airflow through a gradually reducing duct section

The loss through a typical section of tapering duct is quite small, as a rule, and certainly much less than that through a similar expanding section. The reason is that, in the reducer, the opportunity for the formation of the vena-contracta downstream is restricted by the smaller cross-section of the down-stream duct. It is customary to express the loss as a fraction of the down-stream velocity pressure, the value of the fraction depending on the included angle of the taper.

The obvious feature of a reducing duct section is that since the velocity increases *there is no possibility of any static regain occurring*. Energy is required to accelerate the air from its lower upstream velocity to its higher downstream velocity. This increase in the kinetic energy of the airstream is recoverable, in part, by the subsequent use of an expanding section. There is a further fall in a static pressure however, occasioned by the need to make good the losses at the vena contracta and the losses resulting from skin friction. This fall in static pressure is not recoverable at a later stage in the duct system.

Figures 15.11(*a*) and 15.11(*b*) illustrate the two-fold drop in static pressure which takes place. In (*a*) an ideal case is considered in which it is assumed that no loss occurs.

Hence $TP_1 = TP_2$, and $(SP_1 + VP_1) = (SP_2 + VP_2)$ and so, the energy loss being zero, we can write

$$\begin{aligned} 0 &= TP_1 - TP_2 \\ &= (SP_1 + VP_1) - (SP_2 + VP_2) \end{aligned}$$

from which it can be seen that the fall in static pressure is given by

$$(SP_1 - SP_2) = (VP_2 - VP_1).$$

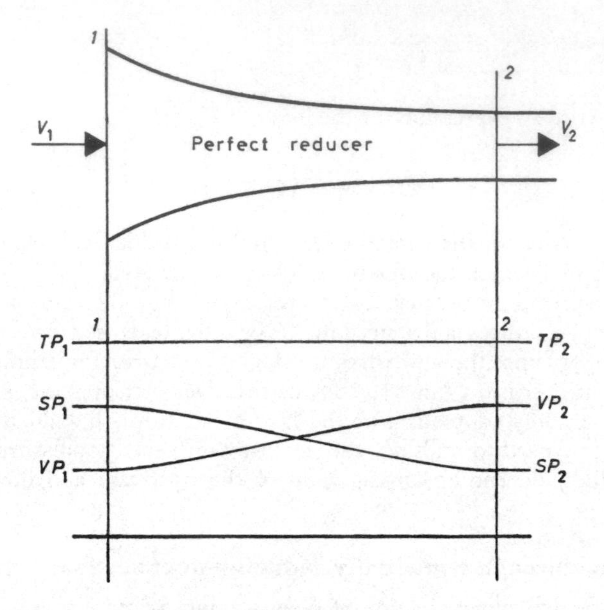

Fig. 15.11(*a*)

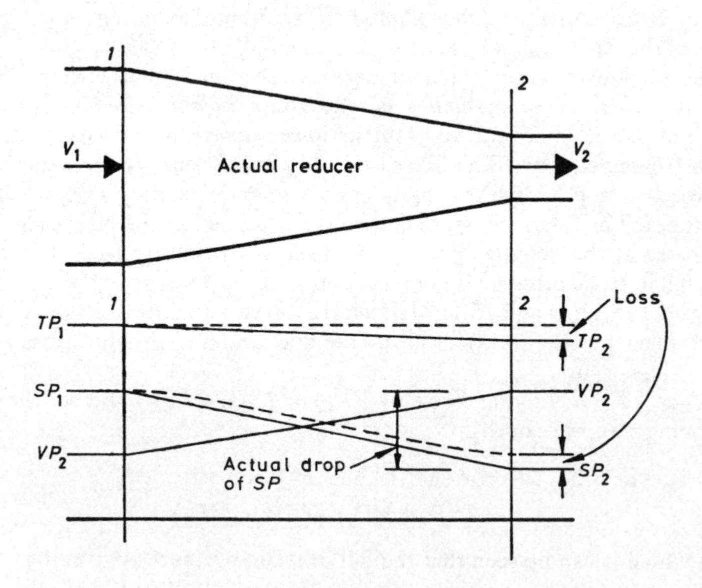

Fig. 15.11(*b*)

In Fig. 15.11(*b*) a more realistic case is illustrated in which a loss of energy, due to turbulence and friction, does occur. This time TP_1 does not equal TP_2 and, hence, the energy loss is given by

$$\begin{aligned}\Delta TP &= TP_1 - TP_2 \\ &= (SP_1 + VP_1) - (SP_2 + VP_2)\end{aligned}$$

and the fall in static pressure is expressed by

$$(SP_1 - SP_2) = (VP_2 - VP_1) + \Delta TP$$

It must be remembered that the value of ΔTP is negative in the direction of airflow, and so the overall drop in static pressure exceeds that due to the rise in velocity pressure.

EXAMPLE 15.6. Calculate (*a*) the loss and (*b*) the change in static pressure which occurs when air flows through a reducing section having an included angle of taper of 45°, if its velocity is increased from 5 to 20 m/s.

Answer

From the I.H.V.E. Guide, $k = 0{\cdot}05$ for an angle of taper of 45°.

(*a*)
$$\begin{aligned}\Delta TP &= 0{\cdot}04 \times 0{\cdot}6 \times 20^2 \\ &= 9{\cdot}6\ \text{N/m}^2\end{aligned}$$

(*b*)
$$\begin{aligned}SP &= -(VP_2 - VP_1 + \Delta TP) \text{ in the direction of airflow.} \\ &= -(0{\cdot}6 \times 20^2 - 0{\cdot}6 \times 5^2 + 9{\cdot}6) \\ &= -(240 - 15 + 9{\cdot}6) \\ &= -234{\cdot}6\ \text{N/m}^2 \text{ in the direction of airflow.}\end{aligned}$$

15.14 Airflow around bends

As with other pieces of ducting, the loss incurred when air flows around a bend is expressed as a fraction of the mean velocity pressure in the bend, provided that the section is constant. If the section varies, precise calculations are not possible but, with a bit of commonsense, a good approximation can frequently be obtained.

For a normal bend of constant cross-sectional area, the loss depends on three structural properties:

(i) The curvature of the throat.
(ii) The shape of the section.
(iii) The angle through which the airstream is turned.

It is customary to express the curvature of the throat either in terms of the ratio of the throat radius R_t to the dimension W, parallel to the radius, or, in terms of the ratio of the centre-line radius R_c to the dimension W. Figure

15.12(*a*) illustrates these two methods. The mode using the centre-line radius is the more common and the one adopted by the I.H.V.E. A very large value of R_c/W implies that the air is only very gradually turned and that, turbulence having little opportunity to form, the loss is small. It is evident, however, that not only will skin friction play an increasing part if the bend is excessively gradual, but that the bend will be expensive to make and be unsightly in appearance, occupying as it does a very large amount of space. A good practical value for R_c/W is 1·0.

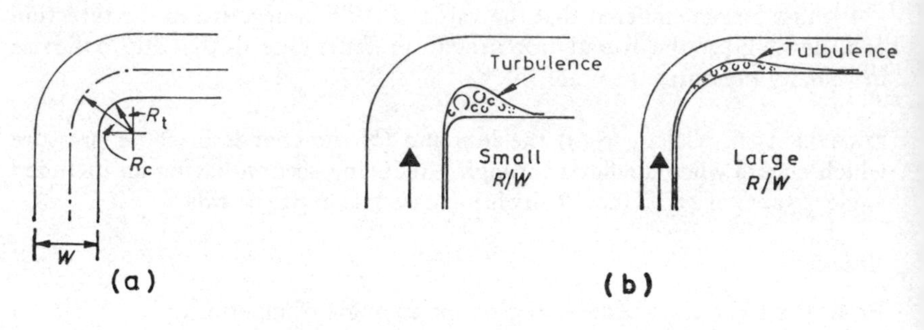

Fig. 15.12

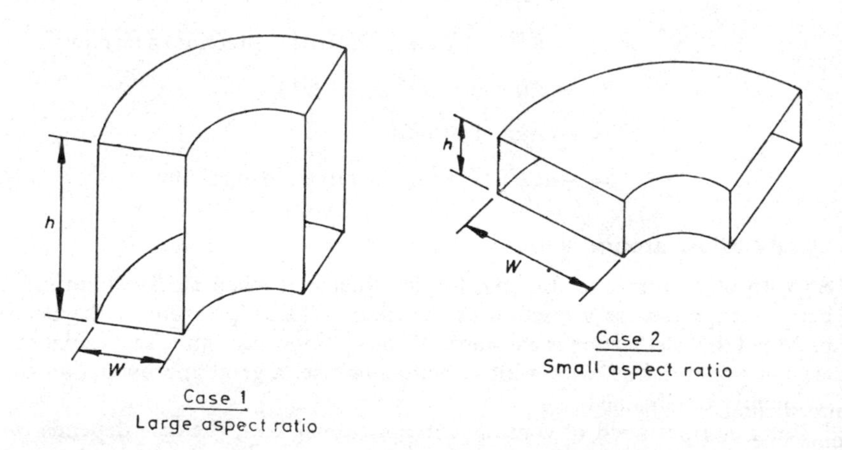

Fig. 15.13

Figure 15.12(*b*) illustrates that if the curvature of the throat is great (and so R_c/W is small), the pocket of turbulence downstream is extensive and the energy losses which result from maintaining the eddies are large.

The shape of the section, or the aspect ratio, has an effect which is shown by Fig. 15.13. Case 1 illustrates a duct where the aspect ratio, h/W, is large and the curvature of the throat is small. Case 2 shows the reverse situation; h/W is small, the curvature of the throat is great, and the loss is comparatively large.

One might expect the loss through a bend to be proportional to the angle turned through, but it appears this is not quite so. The continuation of the bend beyond 90° seems to inhibit the formation of some of the downstream turbulence. Some authorities claim that the loss for a bend of 180° is a little less than that for two bends of 90° but that, on the other hand, the loss for a bend of 45° is a little more than half that for a bend of 90°. If the bend of 180° is not all in one plane, however, then the loss is greater, the k-value being proportional to the angle turned through. When the two bends are not in the same plane, as shown in Fig. 15.14, then use a k-value equal to twice that for a single 90° bend. The multiplying factor becomes 2·4 if the two bends in series in the same plane form an off-set.

There are two methods of minimising the energy loss round a bend: the use of mid-feathers or splitters and the use of turning vanes.

Mid-feathers are of value only when the aspect ratio of the duct section is small; here the effect of inserting splitters is to divide the section of the duct

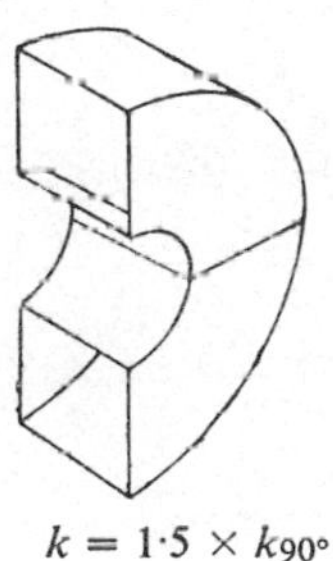

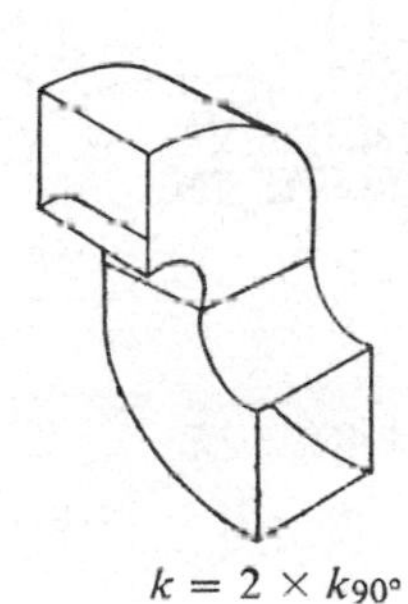

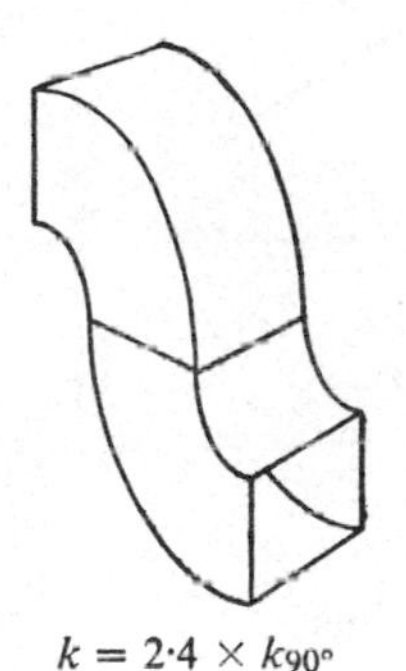

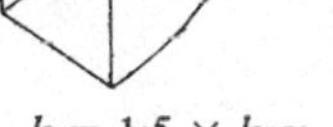

$k = 1{\cdot}5 \times k_{90°}$ $k = 2 \times k_{90°}$ $k = 2{\cdot}4 \times k_{90°}$

Fig. 15.14

into a number of smaller sub-sections. Noise is kept to a minimum and the long side of the duct is adequately supported, but there is little reduction of the energy loss. (This can be appreciated if a comparison is made with a simple bend having a R_c/W value of 1·0, but lacking splitters.) It is usual to arrange the positioning of the splitters so that they cluster near to the throat of the bend. Experiment suggests that it is the curve ratio, defined as throat radius/heel radius, which determines the energy loss, rather than the aspect ratio directly. The consequence is that the splitters should be arranged to produce a number of sub-sections of the duct, each having the same curve ratio.

An expression for the curve ratio, C, in terms of the number of splitters, n, the throat radius, R_0 and the heel radius, R_{n+1} can be obtained. The following statements hold good for the bend shown in Fig. 15.15:

$$\frac{R_0}{R_{n+1}} = K = \frac{\text{throat radius of the bend without splitters}}{\text{heel radius of the bend without splitters}}$$

$$C = \frac{R_0}{R_1} = \frac{R_1}{R_2} = \frac{R_2}{R_3} = \text{etc.} \ \ldots = \frac{R_n}{R_{n+1}} \tag{15.24}$$

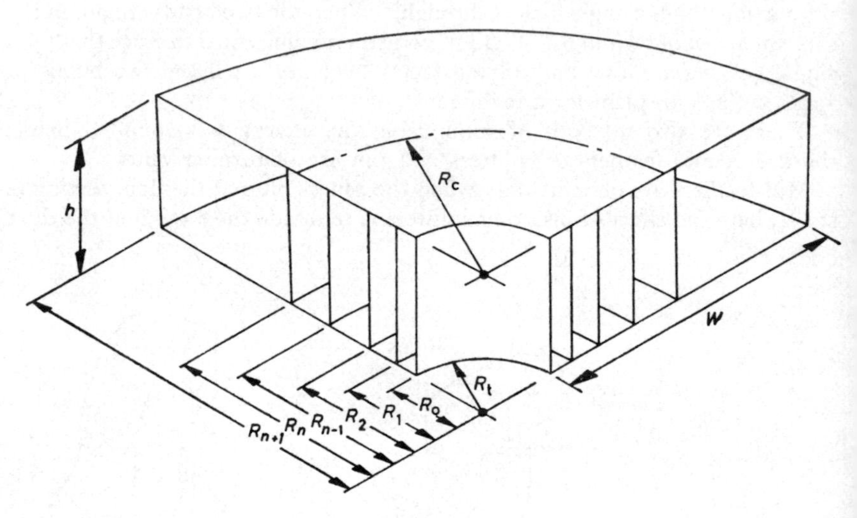

Fig. 15.15

Thus

$$R_1 = \frac{R_0}{C}$$

$$R_2 = \frac{R_1}{C} = \frac{R_0}{C^2}$$

$$R_3 = \frac{R_2}{C} = \frac{R_0}{C^3}$$

etc.

$$R_{n+1} = \frac{R_0}{C^{n+1}}$$

Therefore

$$C = \left(\frac{R_0}{R_{n+1}}\right)^{1/(n+1)} = K^{1/(n+1)} \tag{15.25}$$

Note also that

$$\frac{R_c}{W} = \frac{0{\cdot}5\,(R_0 + R_{n+1})}{(R_{n+1} - R_0)}$$

Dividing above and below by R_{n+1}, this becomes

$$\frac{R_c}{W} = \frac{0{\cdot}5\,(K+1)}{(1-K)} \tag{15.26}$$

EXAMPLE 15.7. Calculate the best position for the insertion of one splitter in a bend of section 1500 mm wide by 250 mm high, if the ratio of R_c to W is unity.

Answer

From eqn. (15.26)

$$1 = \frac{0{\cdot}5\,(K+1)}{(1-K)}$$

therefore

$$K = 0{\cdot}333.$$

By eqn. (15.25),

$$\begin{aligned} C &= (0{\cdot}333)^{1/1+1} \\ &= (0{\cdot}333)^{0.5} \\ &= 0{\cdot}577. \end{aligned}$$

By eqn. (15.24),

$$R_1 = \frac{R_0}{C}$$

but R_0 is 750 mm, since W is 1500 mm and R_c/W is unity. Therefore

$$\begin{aligned} R_1 &= \frac{750}{0{\cdot}577} \\ &= 1300 \text{ mm.} \end{aligned}$$

The radius of the splitter is thus less than the centre-line radius, which has a value of 1500 mm. It is seldom necessary to use more than 3 splitters in one bend.

One of the most effective bends, both from the point of view of accommodating the ductwork neatly and minimising the energy loss, is the mitred bend containing turning vanes. There are two types of turning vanes, the simple kind and the so-called aerofoil sort, as illustrated in Fig. 15.16. Of these, the former have an energy loss which is about double that of a conventional bend with R_c/W equal to unity, and the latter a loss somewhat less than the conventional bend. Figure 15.16(*c*) shows a typical method of construction adopted for aerofoil vanes; it can be seen that the throat radius

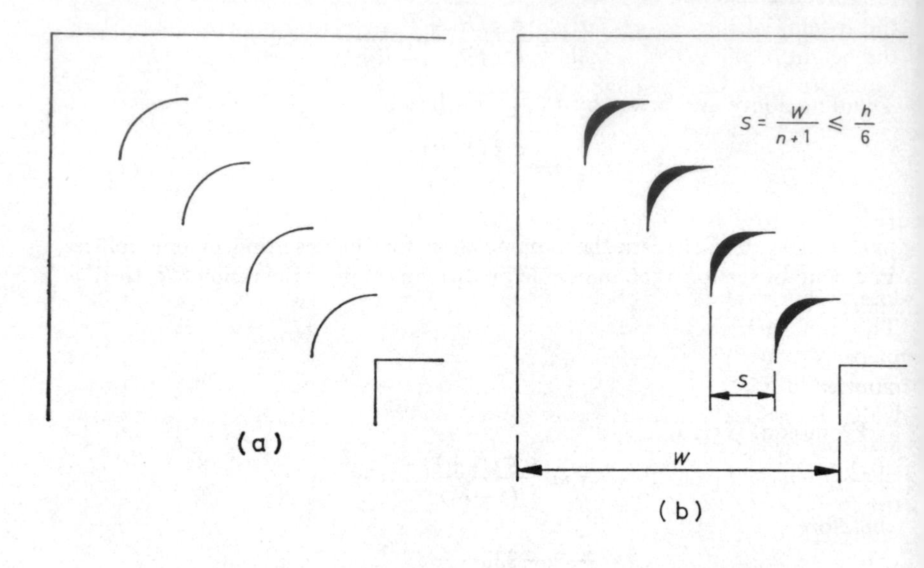

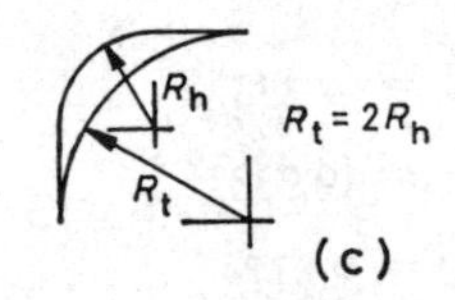

Fig. 15.16

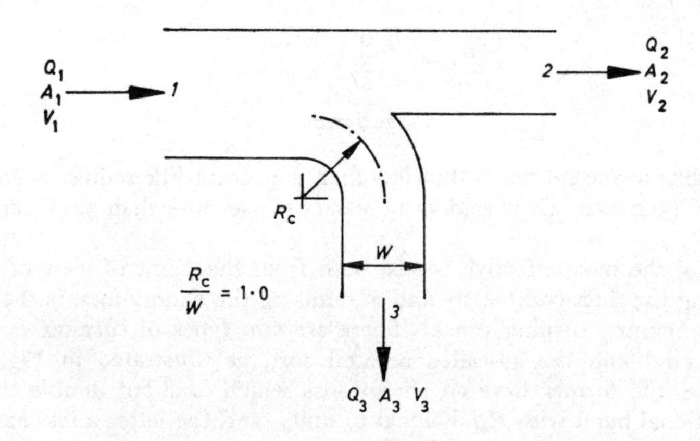

Fig. 15.17

for each vane should be twice the heel radius and that this also determines the spacing of the vanes. To achieve a fairly low pressure drop, the ratio of the height of the vanes to their spacing should be six, or greater. Thus, if n vanes are used, the spacing, s, is given by

$$s = \frac{W}{n+1} \leq \frac{h}{6}$$

15.15 Airflow through supply branches

Loss occurs along two paths when air flows through a branch-piece: there is a loss of energy for flow through the main and also for flow through the branch. The exact magnitude of the loss depends on the way in which the branch-piece is constructed but it is, nevertheless, possible to generalise. The current edition of the I.H.V.E. Guide quotes multiplying factors (k-values) which refer to the velocity pressure in the off-take and which are expressed in terms of both a velocity and an area ratio. Figure 15.17 illustrates this: there is a loss of total pressure between points 1 and 2, when airflow through the main is considered, and also a loss of total pressure (not necessarily the same) when air flows from point 1 to point 3, by way of the branch.

It is of some importance to realise that, since a quantity of air is fed off through the branch, the velocity in the main downstream is less than that upstream, unless a considerable reduction is made in the cross-sectional area of the main duct. Such a drop in velocity will result in a rise in static pressure, and such an increase may occur even though the main duct decreases in section, unless the decrease is sufficient to keep the velocity constant.

EXAMPLE 15.8. (*a*) Given that the loss of total pressure through the main duct from point 1 to point 2, as shown in Fig. 15.18, is 10 per cent of the difference of the upstream and downstream velocity pressures, calculate the energy loss suffered by the airstream flowing through the main duct, and determine the change of static pressure.

(*b*) If the energy loss for airflow through the branch from point 1 to point 3 is 3·7 times the branch velocity pressure, calculate the energy loss through the branch and the change of static pressure.

Answer

(*a*) Energy loss $= \Delta TP$

$$= -0{\cdot}10\,(0{\cdot}6 \times 15^2 - 0{\cdot}6 \times 10^2)$$
$$= -0{\cdot}10\,(135 - 60)$$
$$= -7{\cdot}5 \text{ N/m}^2 \text{ in the direction of airflow.}$$

Static pressure change = velocity pressure change minus the loss.

$$= VP_1 - VP_2 - 7{\cdot}5$$
$$= 135 - 60 - 7{\cdot}5$$
$$= 67{\cdot}5 \text{ N/m}^2.$$

There is thus a regain of static pressure even though the size of the main duct is not increased.

(*b*)

$$\begin{aligned}\text{Energy loss} &= \Delta TP\\ &= 3{\cdot}7\times 0{\cdot}6\times 20^2\\ &= 3{\cdot}7\times 240\\ &= 888\ \text{N/m}^2\end{aligned}$$

$$\begin{aligned}\text{Static pressure change} &= VP_1-VP_3-888\\ &= 135-240-888\\ &= -993\ \text{N/m}^2.\end{aligned}$$

Here it is evident that there is no static regain; in the first place there is a fall of static pressure because the airstream is accelerated from V_1 to V_3, and in the second there is a fall due to the loss.

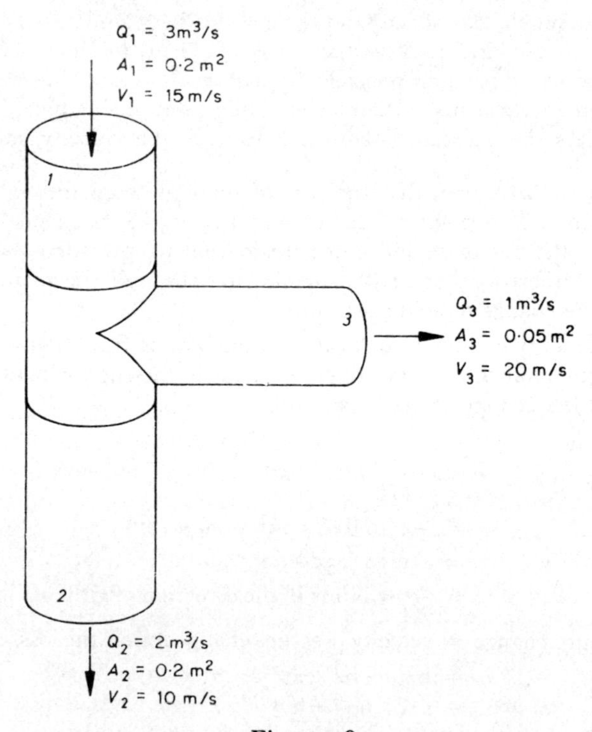

Fig. 15.18

15.16 Flow through suction branches

The pattern of airflow is slightly different in suction branch-pieces as Fig. 15.19 shows.

The distribution of air in comfort conditioning installations is more important on the discharge side of the fan than on its suction side. The positioning and arrangement of extract grilles is not usually very critical, and for this

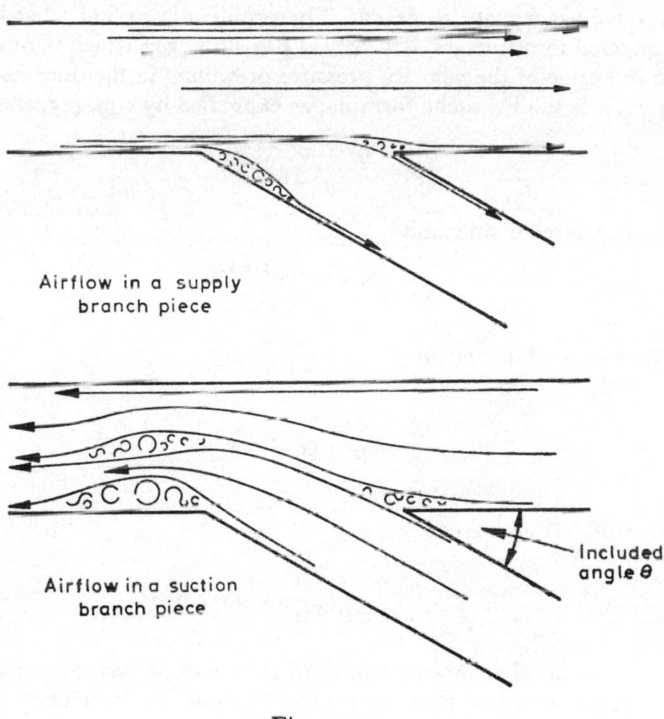

Fig. 15.19

reason supply branch-pieces receive more attention in the literature on the subject. However, in industrial exhaust systems the arrangement of the ducting and the configuration of the suction hoods and openings is of vital importance. For such applications it is thus necessary to know something of the losses incurred when air flows into a suction branch-piece.

The loss through the branch depends largely on the velocity in the branch, but the loss through the main is related to the ratio of the quantity of air flowing in the upstream main to the quantity flowing in the branch. It also depends on the angle θ, which the branch makes with the main. In low-velocity systems the energy losses past suction branch-pieces are usually quite small, and are generally ignored with safety. For industrial exhaust systems, where conveying velocities are high, losses can be substantial, and reference should be made to the appropriate authority for details of the losses.

15.17 The length of duct to absorb one velocity head

It is not uncommon for data on energy losses through ducts and bends to be expressed in terms of an equivalent length of straight duct. (Indeed, in the case of water flow through pipes and fittings, this is almost the only way of expression.) Although this equivalent length may be in metres of duct, it is also often quoted in diameters.

It turns out, on examination, that quite a simple relationship exists which yields results of adequate accuracy. The problem is to find what length of duct, expressed in diameters, will have a pressure drop which is numerically equal to the value of the velocity pressure prevailing in the duct itself. The starting point is the Fritzsche formula, as expressed by eqn. (15.6):

$$\Delta p = \frac{91{\cdot}16v^{1\cdot852}l}{d^{1\cdot269}}$$

where d is expressed in mm, and

$$\Delta p = \frac{0{\cdot}01422v^{1\cdot852}l}{d^{1\cdot269}}$$

where d is expressed in metres.

Hence, if $\Delta p = 0{\cdot}6v^2$

$$\frac{l}{d} = 42{\cdot}19d^{0\cdot269}v^{0\cdot148}$$

But $d = 1{\cdot}128\sqrt{\dfrac{Q}{v}}$ and hence

$$\frac{l}{d} = 57{\cdot}11Q^{0\cdot1345}v^{0\cdot0135} \qquad (15.27)$$

This appears cumbersome to use until it is noticed that 0·0135 is almost zero, in this context, and that consequently variations in velocity will have little effect. At 2·5 m/s, the value of $v^{0.0135}$ is 1·012 and at 25 m/s it is only 1·044. Equation (15.27) can therefore be further simplified by taking an average value of 1·03 for $v^{0.0135}$ over the range of velocity from 2·5 to 25 m/s. The result of this is

$$\frac{l}{d} = 58{\cdot}82Q^{0\cdot1345} \qquad (15.28)$$

The conclusion to be drawn from this is that, for the range of velocities considered, the length of duct that absorbs an amount of pressure numerically equal to one velocity pressure is virtually independent of the size of the duct and is related only to the quantity of air flowing, for all practical purposes.

EXAMPLE 15.9. Calculate the fall of total pressure along 30 m of straight duct, 300 mm in diameter, which carries (*a*) 0·5 m^3/s and (*b*), 2·0 m^3/s.

Answer

(*a*)
$$v = \frac{Q}{A}$$

$$Q = 0{\cdot}5\ \mathrm{m^3/s} \text{ and } A = 0{\cdot}07071\ \mathrm{m^2}$$

Therefore
$$v = 7{\cdot}07\ \mathrm{m/s}.$$

Hence, one velocity pressure is equivalent to

$$0{\cdot}6 \times 7{\cdot}07^2 = 30\ \mathrm{N/m^2}$$

Then, by eqn. (15.28),

$$\begin{aligned}\frac{l}{d} &= 58{\cdot}82 \times (0{\cdot}5)^{0{\cdot}1345}\\ &= 53{\cdot}58 \text{ diameters}\\ &= 16{\cdot}08\ \mathrm{m}, \text{ since the diameter is } 300\ \mathrm{mm}\end{aligned}$$

This means that 16·08 m of duct, 300 mm in diameter and carrying 0·5 $\mathrm{m^3/s}$, has a pressure drop of 30 $\mathrm{N/m^2}$. The answer to the question is that, for a duct length of 30 m, the drop of pressure is

$$\frac{30}{16{\cdot}08} \times 30 = 55{\cdot}99\ \mathrm{N/m^2}.$$

(Incidentally, the duct-sizing chart in the I.H.V.E. Guide gives a pressure drop rate of 1·9 $\mathrm{N/m^3}$ and so the loss in 30 m is 57 $\mathrm{N/m^2}$.)

(*b*) If the quantity of airflow in the same size of duct in 2·0 $\mathrm{m^3/s}$, that is, quadrupled, then the velocity pressure is multiplied by a factor of 16 and becomes 480 $\mathrm{N/m^2}$. Using eqn. (15.28) once more,

$$\begin{aligned}\frac{l}{d} &= 58{\cdot}82 \times (2{\cdot}0)^{0{\cdot}1345}\\ &= 58{\cdot}87 \times 1{\cdot}0977\\ &= 64{\cdot}6\\ &= 19{\cdot}4\ \mathrm{m}, \text{ in this particular case.}\end{aligned}$$

The pressure drop is, therefore,

$$\frac{30}{19{\cdot}4} \times 480 = 743\ \mathrm{N/m^2}$$

or about 24·8 $\mathrm{N/m^3}$.

(The I.H.V.E. Guide quotes a rate of about 27·8 $\mathrm{N/m^3}$).

15.18 Conversion from circular to rectangular section

Ducts of rectangular section are more economical in their usage of building space than are those of circular section. For this reason, ducts are sized on a circular basis and are subsequently converted into an equivalent rectangular section. There are two ways in which the equivalence may be established.

1. The rectangular duct is to carry the same quantity of air as the circular duct and the rate of pressure drop is to be the same, but the velocity may be different.

2. The velocity and the rate of pressure drop in the rectangular duct are to be the same as in the circular one, but the quantity handled is to be different.

The first basis is adopted for ordinary air-conditioning and ventilation applications and is of paramount importance. The second method is little used.

The starting point of both approaches is the Fanning formula (eqn. (15.2)), coupled with the appropriate stipulation on the matter of quantity and velocity.

1. *The Rectangular Equivalent having the same Quantity Flowing and the same Rate of Pressure Drop*

$$H = \frac{4fl\bar{v}^2}{2gd} \tag{15.2}$$

and multiplying by ρg to convert to pressure

$$\Delta p = \frac{2\rho fl\bar{v}^2}{d} \tag{15.29}$$

Noting that m, the hydraulic mean gradient, is given by

$$m = \frac{A}{P},$$

where P is the perimeter of the duct,

$$m = \frac{\pi d^2}{4} \bigg/ \pi d$$

$$= \frac{d}{4}$$

and so d can be expressed in terms of A/P, and a substitution made in eqn. (15.29).

$$\Delta p = \frac{\rho fl\bar{v}^2}{2} \cdot \frac{P}{A}$$

but
$$\bar{v} = \frac{Q}{A}$$

hence
$$Q = \sqrt{\left(\frac{2\Delta p}{fl\rho}\right)} \cdot \sqrt{\frac{A^3}{P}}$$

Consider now two ducts, each carrying the same quantity of air, Q. The ducts are of different sizes, one being circular, the other rectangular. However, since it is stipulated that each must have the same rate of pressure drop, Δp and f must be the same in each duct. Since ρ and l are constants, it follows that an equation must be made between $\sqrt{(A^3/P)}$ for the circular duct and $\sqrt{(A^3/P)}$ for the rectangular duct:

$$\sqrt{\left(\frac{\pi^3 d^6}{4^3 \pi d}\right)} = \sqrt{\left(\frac{(ab)^3}{2(a+b)}\right)}$$

where a and b are the sides of the rectangular duct.

Solving this equation for the diameter of the circular duct, D:

$$d = 1{\cdot}265 \sqrt[5]{\left(\frac{(ab)^3}{(a+b)}\right)} \tag{15.30}$$

2. The Rectangular Equivalent having the same Velocity of Airflow and the same Rate of Pressure Drop

In a similar way, the Fanning equation is used as the starting point but this time the manipulation is arranged to give the velocity of airflow as the dependent variable, instead of the quantity of air flowing. As before, f, Δp and l are constants and an equation is set up involving the A/P terms for the two ducts, one circular and the other rectangular.

$$\Delta p = \frac{\rho f l \bar{v}^2}{2} \cdot \frac{P}{A}$$

hence

$$\bar{v} = \sqrt{\left(\frac{2\Delta p}{\rho f l}\right)} \cdot \sqrt{\left(\frac{A}{P}\right)}$$

Equating $\sqrt{(A/P)}$, we obtain

$$\sqrt{\left(\frac{\pi d^2}{4\pi d}\right)} = \sqrt{\left(\frac{(ab)}{2(a+b)}\right)}$$

$$d = \frac{2ab}{(a+b)} \tag{15.31}$$

Rectangular equivalents are seldom calculated; they are usually quoted in the form of a table or a graph.

It is to be noted that it is inadvisable to round off a diameter to the nearest mm when making the initial sizing. Rounding off at this stage introduces a slight error which is likely to be forgotten, and hence perhaps magnified when further rounding off is done during the process of finding the rectangular equivalent. However, for so-called low-velocity systems, such an error is not likely to be very large.

15.19 Calculation of fan total and fan static pressure

The principles having been explained and the definitions made, the method of calculation is best illustrated by means of an example.

EXAMPLE 15.10. For the system shown in Fig. 15.20 and making use of the data below, size all ducts on a rate of pressure drop of 1 N/m^3 and calculate the fan total and fan static pressures.

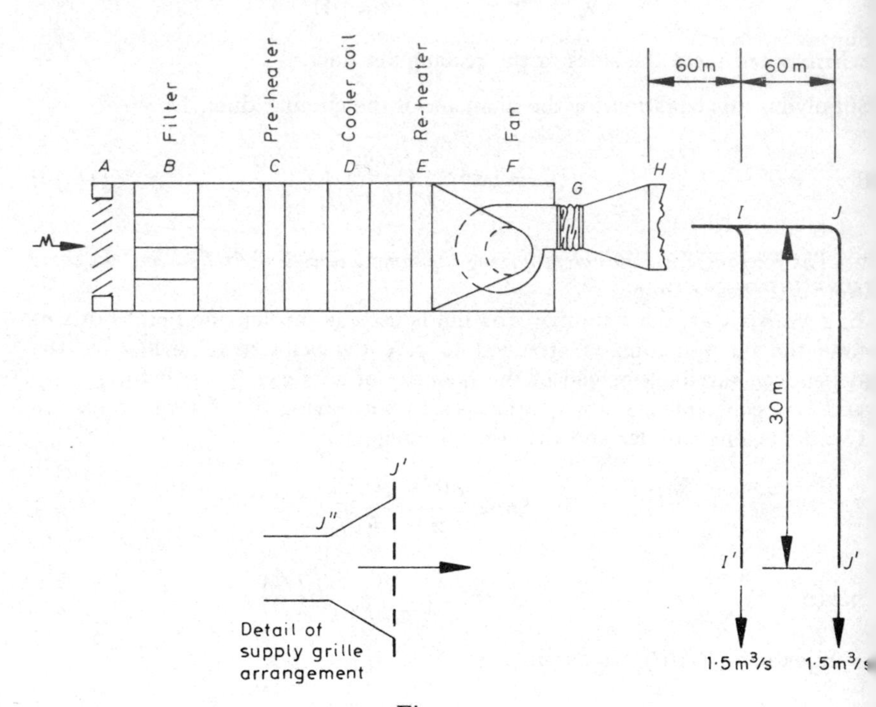

Fig. 15.20

k-values for all expanders are to be taken as applying to the difference between the upstream and downstream velocity pressures.

k-values for all reducers are to be taken as applying to the downstream velocity pressure only.

Air inlet at *A*	square-edged louvres of 80 per cent free area, mean face velocity 4 m/s.
Filter at *F*	$k = 0{\cdot}35$ for the upstream expander, mean face velocity 1·5 m/s, energy loss 90 N/m^2.
Pre-heater at *C*	$k = 0{\cdot}02$ for the upstream reducer, mean face velocity 3 m/s, energy loss 50 N/m^2.

Cooler Coil at D — $k = 0{\cdot}35$ for the upstream expander, mean face velocity 2 m/s, energy loss 150 N/m^2.

Re-heater at E — $k = 0{\cdot}02$ for the upstream reducer, mean face velocity 3 m/s, energy loss 50 N/m^2.

Fan inlet at F — $k = 0{\cdot}02$ for the upstream reducer, diameter 625 mm.

Fan outlet at G — $k = 0{\cdot}30$ for the downstream expander, size 600 mm × 500 mm.

Supply branch piece at I — $k = 0{\cdot}34$ for loss through the branch, zero loss of energy through the main.

Supply grilles at I' and J' — $k = 0{\cdot}5$, mean face velocity 2 m/s, $k = 0{\cdot}35$ for loss through the upstream expander

Bend at J — $k = 0{\cdot}23$.

Use the I.H.V.E. duct-sizing chart and convert all circular sizes to rectangular equivalents which can carry the same amount of air with the same rate of pressure drop.

Answer

From a duct-sizing chart the following diameters are determined for a pressure drop rate of 1 N/m^3:

HI 3 m^3/s, 675 mm dia., 8·4 m/s.

IJ' and II' 1·5 m^3/s, 520 mm dia., 7·1 m/s.

Making use of the tabulated data in the I.H.V.E. guide (based on eqn (15.30)), the following rectangular sizes are established:

HI 3 m^3/s, 500 mm × 750 mm, 8 m/s.

IJ' and II' 1·5 m^3/s, 350 mm × 750 mm, 5·71 m/s.

The following calculations must now be carried out:

Area at fan discharge: $0{\cdot}6 \times 0{\cdot}5 = 0{\cdot}3$ m^2

Velocity at fan discharge: $\dfrac{3{\cdot}0}{0{\cdot}3} = 10$ m/s.

Velocity pressure at fan discharge: $0{\cdot}6 \times 10^2 = 60$ N/m^2.

Area at fan inlet: $\dfrac{\pi \times 0{\cdot}625^2}{4} = 0{\cdot}3068$ m^2.

Velocity at fan inlet: $\dfrac{3}{0{\cdot}3068} = 9{\cdot}778$ m/s.

Velocity pressure at fan inlet: $0{\cdot}6 \times 9{\cdot}778^2 = 57{\cdot}4$ N/m^2.

The dual set of calculations which follows is intended to show that it i usually easier to determine fan total or fan static pressure by working ir changes of total pressure rather than in changes of static pressure. To assis in obtaining clarity it is proposed to adopt the convention that a negativ sign denotes a fall in pressure in the direction of airflow. (It is rare to adop a convention in practice; intuition is generally sufficient to determine th sign of a pressure change or the way in which it should be considered and indeed, it is often to be preferred, since it requires an understanding of wha is happening and this is better than working slavishly to a formula or a convention.)

The calculations are now itemised and duplicated, one set being for loss o total pressure and the other for change of static pressure. The relationshi between the two sets is clarified at the end of the example.

1. Outside Air Intake at *A*

From I.H.V.E. tables, $k = 1{\cdot}40$.

TP loss: $= -1{\cdot}4 \times (0{\cdot}6 \times 4^2) = -13{\cdot}44\ \mathrm{N/m^2}$.

SP change $= -13{\cdot}44 - (0{\cdot}6 \times 4^2) = 23{\cdot}04\ \mathrm{N/m^2}$.

Accumulated downstream pressures: SP $-23{\cdot}04\ \mathrm{N/m^2}$.

VP $+\ 9{\cdot}6\ \mathrm{N/m^2}$.

TP $-13{\cdot}44\ \mathrm{N/m^2}$.

2. Expander, *AB*, to Filter

$$
\begin{aligned}
TP\ \text{loss} &= -0{\cdot}35 \times \{(0{\cdot}6 \times 4^2) - (0{\cdot}6 \times 1{\cdot}5^2)\} \\
&= -0{\cdot}35\,(9{\cdot}6 - 1{\cdot}35) \\
&= -0{\cdot}35 \times 8{\cdot}25 \\
&= -2{\cdot}89\ \mathrm{N/m^2}
\end{aligned}
$$

$$
\begin{aligned}
SP\ \text{change} &= (0{\cdot}6 \times 4^2 - 0{\cdot}6 \times 1{\cdot}5^2) - 2{\cdot}89 \\
&= 8{\cdot}25 - 2{\cdot}89 \\
&= +5{\cdot}36\ \mathrm{N/m^2}
\end{aligned}
$$

Accumulated downstream pressures: SP $-17{\cdot}68\ \mathrm{N/m^2}$

VP $+\ 1{\cdot}35\ \mathrm{N/m^2}$

TP $-16{\cdot}33\ \mathrm{N/m^2}$.

3. Filter at *B*

TP loss = *SP* change = −90 N/m²

Accumulated downstream pressures: *SP* −107·68 N/m²

VP + 1·35 N/m²

TP −106·33 N/m²

4. Reducer, *BC*, to Pre-heater

$$TP \text{ loss} = -0{\cdot}02(0{\cdot}6\times 3^2) = -0{\cdot}11$$

$$SP \text{ change} = \{(0{\cdot}6\times 1{\cdot}5^2)-(0{\cdot}6\times 3^2)\} + \text{the loss.}$$

$$= 1{\cdot}35-5{\cdot}4-0{\cdot}11$$

$$= -4{\cdot}16 \text{ N/m}^2$$

Accumulated downstream pressures: *SP* −111·84 N/m²

VP + 5·40 N/m²

TP −106·44 N/m²

5. Pre-heater at *C*

TP loss = *SP* change = −50 N/m²

Accumulated downstream pressures: *SP* −161·84 N/m²

VP + 5·40 N/m²

TP −151·44 N/m²

6. Expander, *CD*, to Cooler Coil

$$TP \text{ loss} = -0{\cdot}35\{(0{\cdot}6\times 3^2)-(0{\cdot}6\times 2^2)\}$$

$$= -0{\cdot}35\times(5{\cdot}4-2{\cdot}4)$$

$$= -1{\cdot}05 \text{ N/m}^2$$

$$SP \text{ change} = (5{\cdot}4-2{\cdot}4)-1{\cdot}05$$

$$= +1{\cdot}95 \text{ N/m}^2$$

Accumulated downstream pressures: *SP* −159·89 N/m²

VP + 2·40 N/m²

TP −157·49 N/m²

7. Cooler Coil at *D*

TP loss = *SP* change = -150 N/m^2

Accumulated downstream pressures: *SP* $-309 \cdot 89$ N/m^2

VP $+ \; 2 \cdot 40$ N/m^2

TP $-307 \cdot 49$ N/m^2

8. Reducer, *DE*, to Re-heater

$$TP \text{ loss} = -0 \cdot 02(0 \cdot 6 \times 3^2)$$
$$= -0 \cdot 11 \text{ N/m}^2$$

$$SP \text{ change} = (2 \cdot 4 - 5 \cdot 4) - 0 \cdot 11$$
$$= -3 \cdot 11 \text{ N/m}^2$$

Accumulated downstream pressures: *SP* -313 N/m^2

VP $+ \; 5 \cdot 4$ N/m^2

TP $-307 \cdot 6$ N/m^2

9. Re-heater at *E*

TP loss = *SP* change = -50 N/m^2

Accumulated downstream pressures: *SP* -363 N/m^2

VP $+ \; 5 \cdot 4$ N/m^2

TP $-357 \cdot 6$ N/m^2

10. Transformation Piece, *EF*, to Fan Inlet

Assume that this is the same as a reducer, as far as changes of pressure are concerned.

$$TP \text{ loss} = -0 \cdot 02 \times 57 \cdot 4$$
$$= -1 \cdot 15 \text{ N/m}^2$$

$$SP \text{ change} = (5 \cdot 4 - 57 \cdot 4) - 1 \cdot 15$$
$$= -53 \cdot 15 \text{ N/m}^2$$

Accumulated downstream pressures: *SP* $-416 \cdot 15$ N/m^2

VP $+ \; 57 \cdot 40$ N/m^2

TP $-358 \cdot 75$ N/m^2

These are the pressures at the fan inlet.

It is now convenient to start at the index grille *J'*, easily identified in this simple example, and work back to the fan discharge. The same sign convention is adopted and, in this way, a record of actual progressive pressures in the ducting system can be kept, since all changes of pressure are related to the atmospheric datum for zero, which forms the starting point of the calculation, just outside the index grille.

11. Index Discharge Grille at *J'*

As Figure 15.20 shows, the approach to the index grille is regarded as an expanding section. This is a simplification. Every case should be considered on its merits and, if it is considered that the work of the extra calculation will be warranted by the result, then more complicated approaches should be regarded as a combination of simple sections, such as expanders plus bends, and so on.

TP loss $= -0{\cdot}5 \times (0{\cdot}6 \times 2^2)$—(the kinetic energy lost from the system)

$= -1{\cdot}2 - (0{\cdot}6 \times 2^2)$

$= -3{\cdot}6\ \text{N/m}^2$

SP change $= -0{\cdot}5 \times (0{\cdot}6 \times 2^2)$

$= -1{\cdot}2\ \text{N/m}^2$

Accumulated upstream pressures: SP $+1{\cdot}2\ \text{N/m}^2$

VP $+2{\cdot}4\ \text{N/m}^2$

TP $+3{\cdot}6\ \text{N/m}^2$

Observe that the sign of the change of pressure must be reversed if the accumulated upstream pressure is to be determined, except, of course, for the value of velocity pressure, which is always regarded as positive.

12. Expander, *J''J'*, to the Index Grille

TP loss $= -0{\cdot}35 \times (0{\cdot}6 \times 5{\cdot}71^2 - 0{\cdot}6 \times 2^2)$

$= -0{\cdot}35 \times (19{\cdot}56 - 2{\cdot}40)$

$= -0{\cdot}35 \times 17{\cdot}16$

$= -6{\cdot}01\ \text{N/m}^2$

SP change $= (19{\cdot}56 - 2{\cdot}40) - 6{\cdot}01$

$= 17{\cdot}16 - 6{\cdot}01$

$= 11{\cdot}15\ \text{N/m}^2$ (Note that this is a regain.)

Accumulated upstream pressures: SP $-\ 9{\cdot}95\ \text{N/m}^2$

VP $+19{\cdot}56\ \text{N/m}^2$

TP $+\ 9{\cdot}61\ \text{N/m}^2$

Note that the static pressure at *J″* is actually less than atmospheric. This is not an error. The static pressure within the discharge duct is quite independent of the outside air pressure, *except* immediately before the discharge grille, where it must be greater than atmospheric by an amount equal to the frictional loss past the grille. On the other hand, total pressure, which is the index of the energy content of the airstream, must *always* be positive inside the discharge duct. The corollary of this is that the total pressure inside the suction duct must always be negative, but the static pressure therein could never be positive, since even a reduction of the velocity of airflow to zero would only raise the static pressure to atmospheric (ignoring losses).

13. Straight Duct *JJ″*

TP loss = *SP* change = $-1{\cdot}0\times 30 = -30\ \text{N/m}^2$

Accumulated upstream pressures: *SP* $+20{\cdot}05\ \text{N/m}^2$
VP $+19{\cdot}56\ \text{N/m}^2$
TP $+39{\cdot}61\ \text{N/m}^2$

14. Bend at *J*

TP loss = *SP* change $= -0{\cdot}23\times 19{\cdot}56$
$= -4{\cdot}50\ \text{N/m}^2$

Accumulated upstream pressures: *SP* $+24{\cdot}55\ \text{N/m}^2$
VP $+19{\cdot}56\ \text{N/m}^2$
TP $+44{\cdot}11\ \text{N/m}^2$

15. Straight Duct *IJ*

TP loss = *SP* change = $-1{\cdot}0\times 60 = -60\ \text{N/m}^2$
Accumulated upstream pressures: *SP* $+\ 84{\cdot}55\ \text{N/m}^2$
VP $+\ 19{\cdot}56\ \text{N/m}^2$
TP $+104{\cdot}11\ \text{N/m}^2$

16. Supply Branch Piece at *I*

Since the index run is through the main duct to the grille at *J′*, only changes of pressure which occur as the air flows through the main duct past the branch are relevant to this calculation.

TP loss = 0 (this is given)

SP change $= (0{\cdot}6\times 8^2 - 0{\cdot}6\times 5{\cdot}71^2)$
$= 38{\cdot}4 - 19{\cdot}56$
$= 18{\cdot}84\ \text{N/m}^2$ (Note that this is a regain.)

Accumulated upstream pressures: *SP* $+65{\cdot}71$ N/m^2

VP $+38{\cdot}40$ N/m^2

TP $+104{\cdot}11$ N/m^2

17. Straight Duct *HI*

TP loss $= SP$ change $= -1{\cdot}0\times 60 = -60$ N/m^2

Accumulated upstream pressures: *SP* $+125{\cdot}71$ N/m^2

VP $+38{\cdot}40$ N/m^2

TP $+164{\cdot}11$ N/m^2

18. Expander, *GH*, at Fan Discharge

$$
\begin{aligned}
TP\text{ loss} &= -0{\cdot}3(0{\cdot}6\times 10^2-0{\cdot}6\times 8^2)\\
&= -0{\cdot}3(60-38{\cdot}4)\\
&= -0{\cdot}3\times 21{\cdot}6\\
&= -6{\cdot}48\ \text{N/m}^2
\end{aligned}
$$

$$
\begin{aligned}
SP\text{ change} &= (60-38{\cdot}4)-6{\cdot}48\\
&= 21{\cdot}6-6{\cdot}48\\
&= 15{\cdot}12\ \text{N/m}^2
\end{aligned}
$$

Accumulated upstream pressures: *SP* $+110{\cdot}59$ N/m^2

VP $+60{\cdot}00$ N/m^2

TP $+170{\cdot}59$ N/m^2

By means of eqns. (15.14) and (15.15) we can now readily determine the fan total and the fan static pressures:

$$
\begin{aligned}
FTP &= TP\text{ at fan outlet minus }TP\text{ at fan inlet}\\
&= +170{\cdot}59-(-358{\cdot}75)\\
&= +170{\cdot}59+358{\cdot}75\\
&= 529{\cdot}34\ \text{N/m}^2
\end{aligned}
$$

$$
\begin{aligned}
FSP &= SP\text{ at fan outlet minus }TP\text{ at fan inlet}\\
&= +110{\cdot}59-(-358{\cdot}75)\\
&= +110{\cdot}59+358{\cdot}75\\
&= +469{\cdot}34\ \text{N/m}^2
\end{aligned}
$$

The alternative definition of fan static pressure, given by eqn. (15.16), provides a check upon the accuracy of the calculations:

FSP = FTP minus VP at fan outlet

$= +529{\cdot}34 - 60$

$= +469{\cdot}34\ \mathrm{N/m^2}$

15.20 The interaction of fan and system characteristics

Air will flow through a system of ductwork (and its associated conditioning plant) only if the system is connected to a fan. To find out what happens when the system and its fan are coupled it is necessary to consider how each

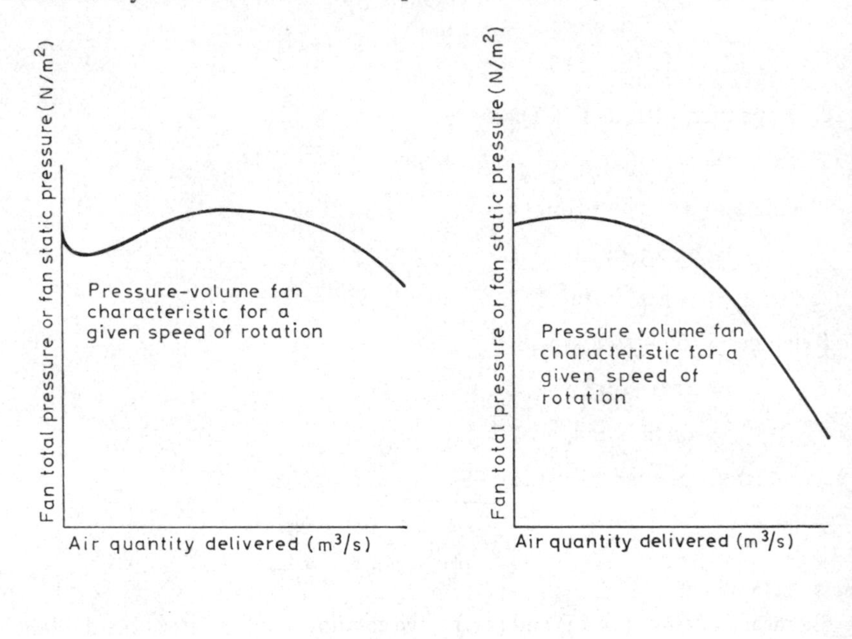

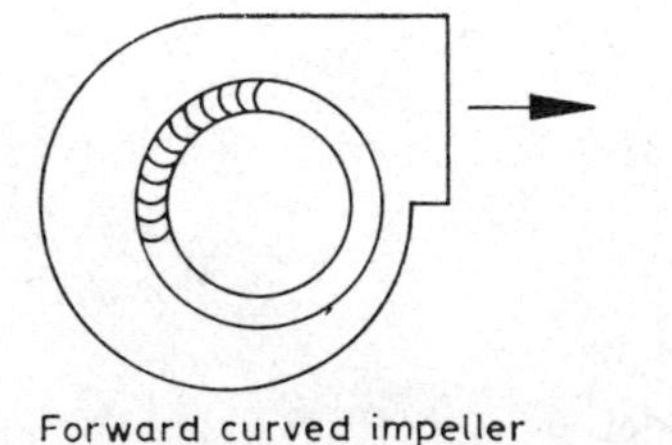

Fig. 15.21

behaves as the air quantity handled varies. Rational considerations and test data provide the basis for the construction of a pair of independent pressure-volume characteristic curves. For a particular fan, running at a constant speed, experimental tests show that the pressure developed (that is, the fan total or the fan static pressure) may change in a manner such as that illustated in Fig. 15.21. It can be seen that the precise form of the curve depends on the type of impeller. Thus, an impeller having many forward-curved blades has a curve with a point of inflection in it, but one that has relatively few blades which are backward-curved does not. On the other hand, the ducting system itself behaves in a manner which is expressed by eqns. (15.29) and (15.6), to an approximate extent. To the standard of accuracy attainable in practice (dictated largely by the accuracy involved in balancing

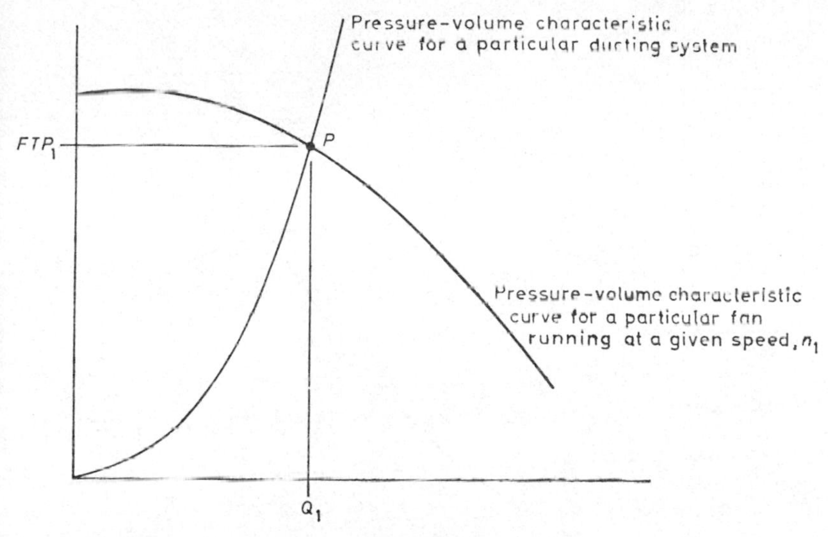

Fig. 15.22

systems of ducting after installation and by the fact that so-called constant-speed motors, depending as they do on the value of the frequency of the electrical supply remaining fixed, may vary by as much as ± 5 per cent), it is sufficient to assume a square law as relating air volume and pressure drop. Thus, we may say that the energy lost through a system is proportional to the square of the volume of air handled. Since energy loss is synonymous with fan total pressure, we can write this relationship as

$$FTP_2 = FTP_1\left(\frac{Q_2}{Q_1}\right)^2 \tag{15.32}$$

When, by means of this equation, system energy loss is plotted against air quantity handled, a simple parabola, passing through the origin, results.

Any fan may be coupled with a system of ducting, but when a *particular* fan, running at a *particular* speed, is installed, then the capacity of the combination can be readily determined by plotting the pressure-volume characteristic curves for the two components on the same co-ordinates, as Fig. 15.22

shows. The intersection of the two curves, at P, the point of rating, gives the volume, Q_1, which the combination will produce, and the fan total pressure, FTP_1, the fan will develop. This, as was made clear in section 15.3, equals the energy loss involved.

EXAMPLE 15.11. Plot the pressure-volume characteristic curve for a system having a total pressure loss of 531 N/m² when handling 3 m³/s.

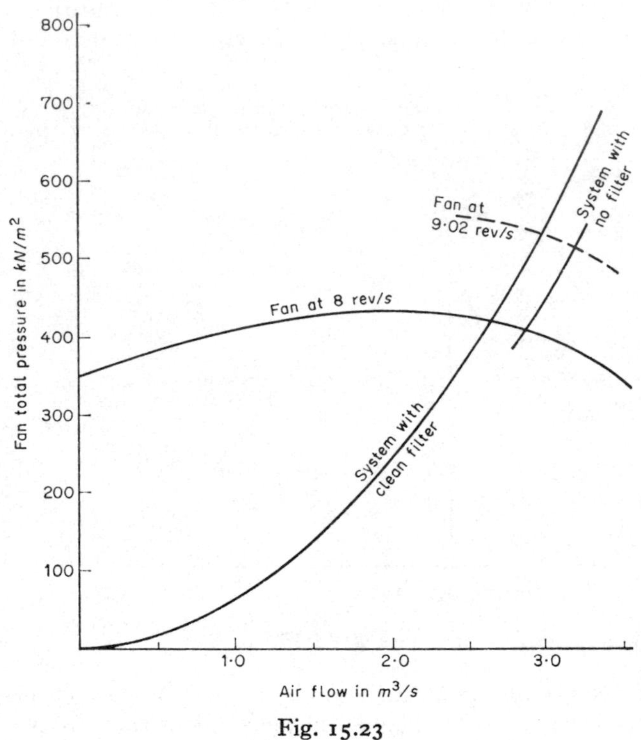

Fig. 15.23

Answer

For the moment, the selection of a fan not yet having been considered, how air is persuaded to flow is entirely irrelevant.

Using eqn. (15.32) we can write—

$$FTP_2 = 531 \times \left(\frac{Q_2}{3}\right)^2$$

A series of suitable assumptions is now made for different values of Q_2 and the corresponding energy loss values of FTP_2 are calculated. It is then convenient to tabulate the results:

Q(m³/s)	0·5	1·0	1·5	2·0	2·5	3·0	3·5
FTP(N/m²)	15	59	133	236	369	531	722

These values are then plotted, as shown in Fig. 15.23.

It is more than likely that the fan first chosen, when running at the speed first chosen, will not result in the delivery of the required air quantity when it is coupled with the system in question. To deal with this situation it is necessary to consider further the behaviour of fans and to consider also what means can be adopted to alter the pressure-volume characteristic curve of the system.

EXAMPLE 15.12. A fan has a pressure-volume characteristic given by the following information, obtained on test; when running at 8 rev/s:

following information, obtained on test; when running at 8 rev/s:

Q (m³/s)	0	0·5	1·0	1·5	2·0	2·5	3·0	3·5
FTP (N/m²)	350	385	410	427	433	424	400	343

Calculate the quantity of air handled and the fan total pressure developed if this fan is installed in the system used in Example 15.11 and continues to run at 8 rev/s.

Answer

The characteristic data for the fan is plotted on the same pressure-volume co-ordinates used for plotting the system data in Example 15.11. Figure 15.23 shows that these two curves intersect at a point having co-ordinates of 2·66 m^3/s and 418 N/m^2.

It is clear that this fan when running at the speed quoted is unsuitable for the design load on the system. It will just not deliver 3 m^3/s. It is impossible to modify the system so that more air will be handled so one must consider what can be done to modify the output of the fan.

15.21 The fan laws

Since this is not a text on fan engineering, only a brief exposition of the laws that express the behaviour of any given fan will be included here. The following is confined to those laws of most common use in air conditioning.

A. For a given Fan Size, Ducting System and Air Density

1. The volume handled varies directly as the fan speed.
2. The pressure developed varies as the square of the fan speed.
3. The power absorbed varies as the cube of the fan speed.

The third law is a consequence of the first two since, as was defined by eqn. (15.10), fan power is proportional to the product of the fan total pressure and the air quantity handled.

B. For a given Fan Size, Ducting System and Speed

The only possible independent relevant variable now is the density of the air handled by the fan. If this changes then the other dependent variables alter as follows:

1. The volume handled remains constant.
2. The fan total pressure, fan static pressure, and the velocity pressure at fan discharge vary directly as the density.
3. The fan power varies directly as the density.
4. The fan efficiency remains constant.

Since density is inversely proportional to absolute temperature and directly proportional to barometric pressure, items 2 and 3 will vary in a similar fashion. As regards the statement that fan total pressure varies directly as the density, reference to eqn. (15.8) gives approximate verification.

EXAMPLE 15.13. A fan running at 8 rev/s handles 2·66 m^3/s at a fan total pressure of 418 N/m^2 when connected to a ducting system, if the air temperature is 0°C and its barometric pressure is 1013·25 mbar. Given that the fan efficiency is 69 per cent, calculate the air quantity delivered, the fan total pressure and the fan power, when the air temperature is increased to 60°C and the barometric pressure falls to 950 mbar.

Answer

(*a*) The air quantity is unaltered at 2·66 m^3/s.

(*b*)
$$FTP_2 = 418 \times \left(\frac{273+0}{273+60}\right) \times \frac{950}{1013 \cdot 25}$$
$$= 321 \text{ N/m}^2$$

(*c*)
$$FP_1 = \frac{2 \cdot 66 \times 418}{0 \cdot 69} \text{ (\textit{see} eqn. (15.10))}$$
$$= 1611 \text{ W.}$$

Hence, from fan laws B.3 and B.4,

$$FP_2 = 1 \cdot 611 \times \left(\frac{273+0}{273+60}\right) \times \frac{950}{1013 \cdot 25}$$
$$= 1 \cdot 238 \text{ kW.}$$

15.22 Power-volume and efficiency-volume characteristics

The power absorbed at a fan shaft (the fan power) and the efficiency of operation, are conveniently expressed in relation to the volume handled by the fan. Thus, in addition to the pressure-volume characteristic curve mentioned in section 15.20, there exist two other characteristic curves. These are also the result of actual tests carried out, and the shape of them depends on the type of impeller and the type of fan. Figure 15.24 shows three different cases. At (*a*) it can be seen that the power increases as the volume handled rises and that the efficiency reaches a peak and then falls away again. If the pressure-volume curve were also considered (shown as a broken line) it would be observed that the point of maximum efficiency was at a volume corresponding to a very flat part of the pressure-volume curve. This means that if the fan is selected to operate at peak efficiency, as is desirable, then any slight changes in system resistance will produce large fluctuations in the volume delivered. This renders a fan having a forward-curved impeller unsuitable for certain applications (for example, as a forced-draught

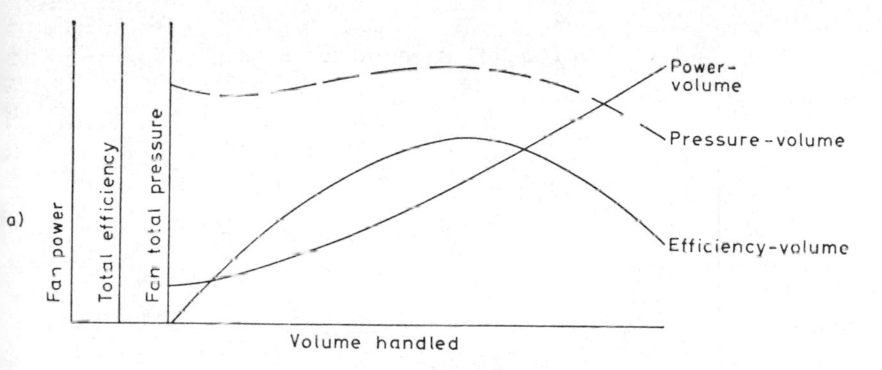

Fig. 15.24(*a*) Forward-curved fan

fan in a boiler installation). The fact that the power rises as the volume increases means that any system using a fan with a forward-curved impeller is liable to overloading. If the main dampers are left wide open when the fan is first started up, too much air will be handled and the excessive power absorbed will overload the driving motor.

In Fig. 15.24(*b*), for the case of a fan having a backward-curved impeller, peak efficiency occurs at a steeper part of the pressure-volume curve, and power as well as efficiency falls off as the volume increases. The backward-curved fan is thus less liable to overload than is the forward-curved. It is also capable of delivering a relatively constant amount of air as the system resistance varies, compared with the forward-curved fan.

To carry the comparison further it must be pointed out that all is not in favour of the backward-curved fan impeller. Because of the shape of the impeller blades and the direction of rotation, a backward-curved fan must run at a higher speed of rotation in order to deliver the same amount of air as a forward-curved fan. The vector diagrams in Fig. 15.25 illustrate this.

The consequences are that forward-curved centrifugal fans are smaller and slower running than are backward-curved fans, for the same air quantity

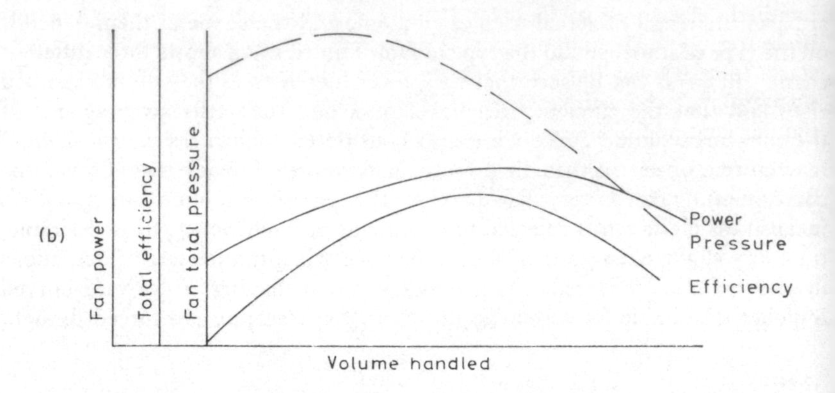

Fig. 15.24(*b*) Backward-curved fan

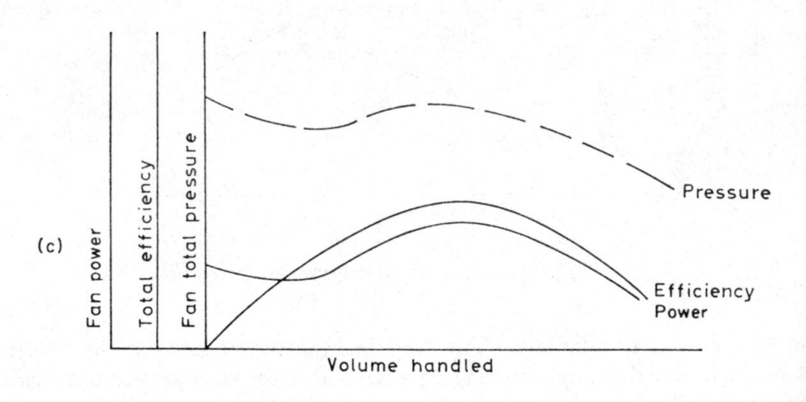

Fig. 15.24(*c*) Axial flow fan

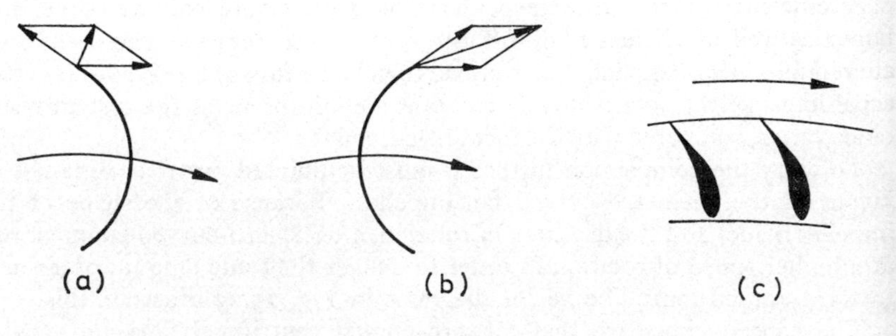

Fig. 15.25

handled. As a result, they tend to be quieter and cheaper for fan total pressures up to about 750 N/m². Above this, the backward-curved impeller starts to take over, particularly when it has aerofoil section blades (*see* Fig. 15.25(*b*)), which result in higher efficiencies than are otherwise obtainable (80 per cent and over are claimed).

Figure 15.24(*c*) shows that there is some similarity between the axial flow fan and the centrifugal fan with backward-curved impeller blades. The axial flow fan is very convenient from an installation point of view, and it is

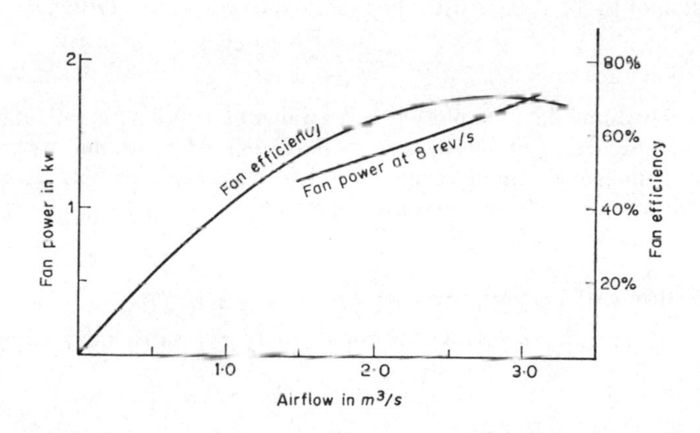

Fig. 15.26

hard to understand why it is not more used. One reason is that, at the time of writing, such fans tend to be noisy if used for fan static pressures much in excess of 70 N/m². This is not always a drawback and, in any case, quite efficient silencers can be installed.

One final point about power-volume characteristic curves:

A particular curve is for the fan when running at a given speed. If the fan speed is altered the position of the curve on the co-ordinate system also alters. If a fan is speeded up, for example, then the absorbed power rises according to fan law A.3.

From the above it can be concluded that, in order to determine the fan power at a given point of rating, the intersection of the pressure-volume curves for the system and the fan must first be found. Armed with this information, namely the air quantity handled when the two are coupled, reference must be made to the power curve for the fan in order to find out the power absorbed at the fan shaft when, running at a given speed, it delivers the known air quantity.

EXAMPLE 15.14. The fan quoted in Example 15.12 has a power-volume relationship as follows, when running at 8 rev/s:

m^3/s:	1·5	2·0	2·5	3·0
kW:	1·185	1·354	1·537	1·715
effcy.:	54%	64%	69%	70%

What power is absorbed at the fan shaft if the fan delivers 2·66 m^3/s? (It is assumed that it has already been established by Example 15.12 that the intersection of the fan curve with the system curve is at this rating.)

Answer

The power-volume data are plotted and a value of 1·58 kW is read off against 2·66 m^3/s. Figure 15.26 shows this. (Reference to Example 15.13 shows that, if the efficiency-volume relationship had been given instead, an answer could have been found by an alternative method.)

15.23 Methods of varying fan capacity in a duct system

There are three basic ways in which the capacity of a fan-duct system can be altered:

1. Changing the fan speed.
2. Partly closing a damper in the duct system.
3. Partly closing variable position guide vanes in the fan inlet eye.

1. The first method can be used to increase or decrease the air quantity handled, provided the driving motor is of adequate size. If the fan speed is increased from n_1 to n_2 rev/s, the pressure-volume curve for the fan takes up a new position above the old one, as shown in Fig. 15.27. It can be seen that the new point of rating is P_2, and the fan delivers a larger air quantity, Q_2, at a higher fan total pressure, FTP_2, through the same duct system. Conversely, a reduction in speed from n_1 to n_3 gives a fan-system intersection at point P_3, the fan curve having been lowered.

EXAMPLE 15.15. Calculate the speed at which the fan mentioned in Example 15.12 must run, if it is to deliver 3 m^3/s.

Answer

In Example 15.12 the point of operation is at 2·66 m^3/s and 418 N/m^2 fan total pressure, when the fan is running at 8 rev/s. The first fan law mentioned in section 15.21 allows the determination of the higher speed at which the fan must run in order to deliver the greater air quantity:

$$n_2 = n_1 \times \frac{Q_2}{Q_1}$$

$$= 8 \times \frac{3}{2\cdot66}$$

$$= 9\cdot02 \text{ rev/s.}$$

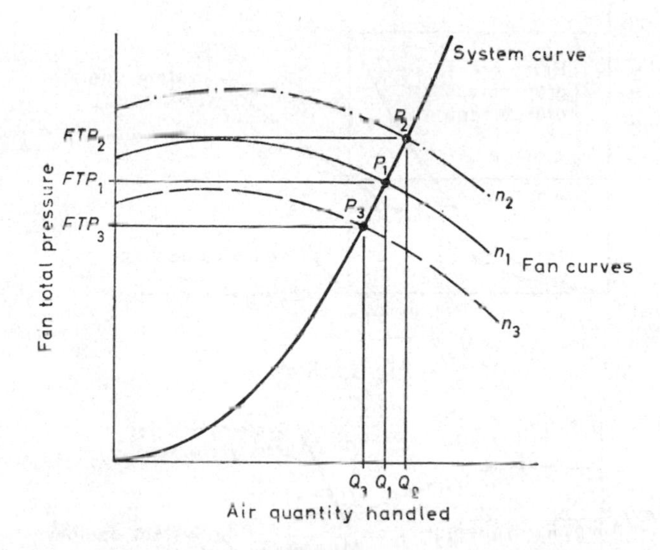

Fig. 15.27

If a new characteristic pressure-volume curve is plotted for the fan when running at this higher speed, it will follow the broken line in Fig. 15.23. Intersection with the system curve must occur, of course, at the required duty of 3 m^3/s and 531 N/m^2 fan total pressure. The second fan law (A.2) verifies this:

$$FTP_2 = FTP_1 \times \left(\frac{n_2}{n_1}\right)^2$$

$$= 418 \times \left(\frac{9\cdot02}{8}\right)^2$$

$$= 531 \text{ N/m}^2.$$

A further verification is provided by the square law assumed for the system pressure-volume characteristic curve. This must be the case since the point of operation is the intersection of the fan and system curves.

2. If a damper in the duct system is closed it changes the position of the system characteristic. The damper is part of the system and, in common with all other system components, there is a certain pressure drop across it for the flow of a certain quantity of air past it. If, by some unspecified means, the flow of air through the damper is kept fixed, regardless of the position of

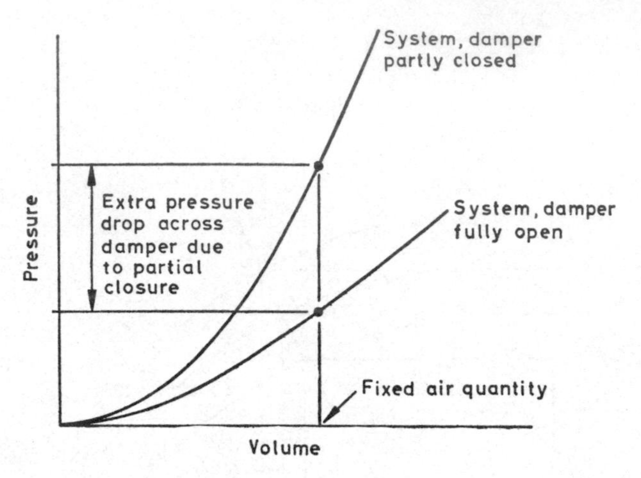

Fig. 15.28(*a*)

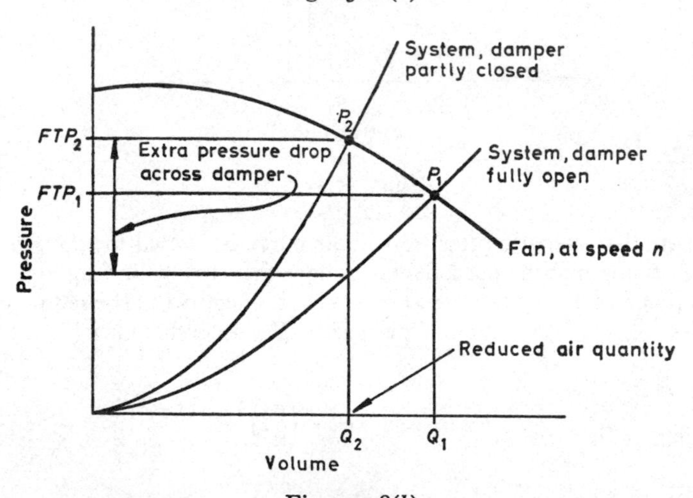

Fig. 15.28(*b*)

the damper, an experimental relationship can be found between the pressure drop across the damper and its position, for a fixed airflow. Section 13.13 discusses this.

The result of partly closing a damper which forms part of a system is shown in Fig. 15.28(*a*). It can be seen that the system curve rotates upwards in an anti-clockwise direction about the zero volume point, as the damper is closed.

Figure 15.28(*b*) indicates what happens when a damper is used to reduce the capacity of a fan-duct system. If the fan speed remains constant at n rev/s and the damper in the duct is fully open, the duty is Q_1 at FTP_1. In order to obtain a reduced airflow Q_2, the damper is partly closed and the point of operation shifts from P_1 to P_2, the required capacity being obtained at a fan total pressure of FTP_2. The failing of the method is that it is wasteful of air power. It is, however, convenient.

EXAMPLE 15.16. If the output of the fan-duct system used in Example 15.12 is reduced to 2 m^3/s by partly closing a damper, calculate the air power wasted across the damper.

Answer

When the fan runs at 8 rev/s the duty of the system is 2·66 m^3/s at 418 N/m^2. Reference to Fig. 15.23 shows that if the air quantity handled is to be 2 m^3/s the system curve must cut the fan curve at a point which has co-ordinates of 2 m^3/s and 433 N/m^2. At 2 m^3/s the total pressure co-ordinate is 236 N/m^2—with the main system damper fully open—on the system characteristic curve. It follows that the energy loss through the system must be increased from 236 to 433 N/m^2 if 2 m^3/s is to be delivered. If this is done the new system characteristic will intersect the fan characteristic at 2 m^3/s and 433 N/m^2. Partly closing the main damper until the pressure drop across it is 197 N/m^2 (the difference of 433 and 236) will achieve this.

$$\text{Wasted air power} = 2 \times 197$$

$$= 394 \text{ W.}$$

In a similar way, as a filter gets dirty the pressure drop across it increases and the air quantity handled by the system reduces.

EXAMPLE 15.17. When the filter in the fan-duct system used for Example 15.12 is clean it has a resistance of 90 N/m^2 and passes 3 m^3/s. When the filter is dirty it has a resistance of 180 N/m^2. Calculate the air quantity handled by the system with a dirty filter, if the fan runs at 9·02 rev/s.

Answer

When the system handles 3 m^3/s the energy loss is 531 N/m^2 and, of this, the clean filter absorbs 90 N/m^2. If the filter were absent, the loss of energy through the system would be the difference of 531 and 90 or 441 N/m^2 for an airflow of 3 m^3/s. A new system characteristic curve can now be plotted for the installation without a filter. Figure 15.29 shows this. A further curve is then plotted for the system such that its intersection with the fan curve takes place at a point P_2. The position of this point is established by measuring an ordinate value of 180 N/m^2 between the newly plotted system curve and the fan curve. Thus, when the filter is dirty, the point of operation is at P_2, with co-ordinates of 550 N/m^2 and 2·75 m^3/s, determined from the figure. At 2·75 m^3/s the system without a filter has a total pressure loss of 370 N/m^2 (= 550 − 180 N/m^2). On the other hand, when the filter is clean, the point of operation is P_1, at 3 m^3/s and 531 N/m^2. Finally, when the filter is absent, the point of operation is P_0, at 3·23 m^3/s and 510 N/m^2.

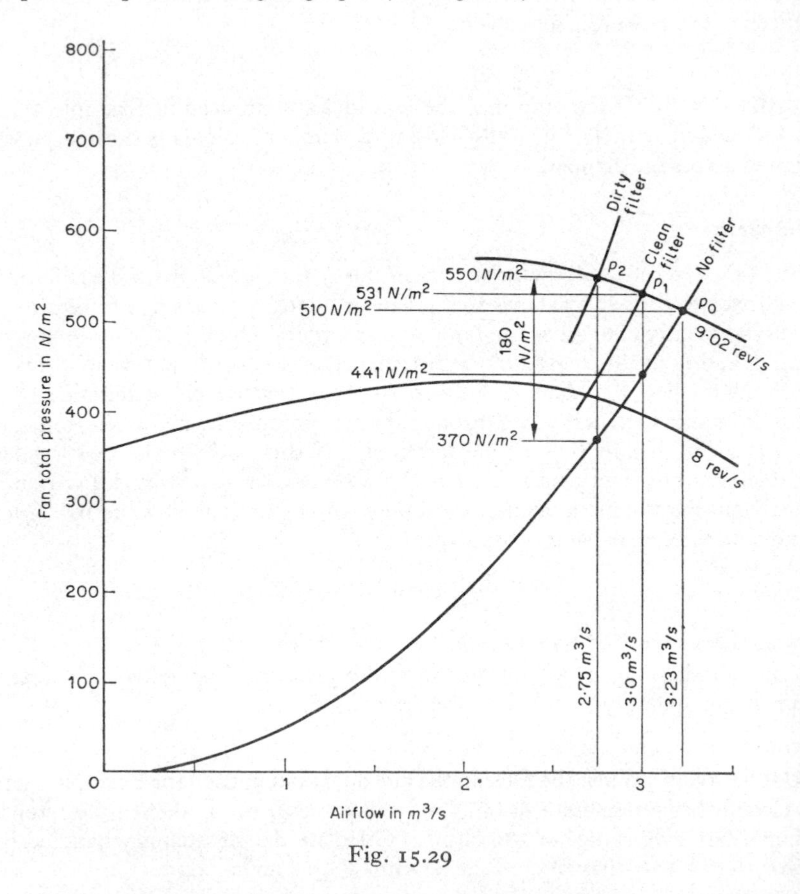

Fig. 15.29

3. Variable inlet guide vanes provide an efficient way of reducing the capacity of a fan-duct system, in a manner analogous to that discussed in section 12.9 for centrifugal compressors in refrigeration systems. Such vanes are positioned in the inlet eye of a centrifugal fan in a way that permits them to assist or retard the airflow through the impeller. Each vane is radially mounted and is hinged along a radial centre-line, being thus able to vary its inclination to the direction of airflow. The inclination may be such that the

swirl imparted to the airstream changes its angle of entry to the impeller blades. This is accomplished without appreciable loss of power over a large range of the fan capacity. The result of partly closing the vanes is not the same as a throttling action. Instead, it rotates the pressure-volume characteristic curve for the fan about the origin, in an clockwise direction. Figure 15.30 shows that a duty Q_2 at FTP_2 can be obtained if the vanes are closed enough to shift the point of operation from P_1 to P_2. If the vanes were capable of complete closure, then, when they were so closed, the pressure-volume curve for the fan would be the ordinate through the origin. This is a practical impossibility since leakage always occurs past the vanes.

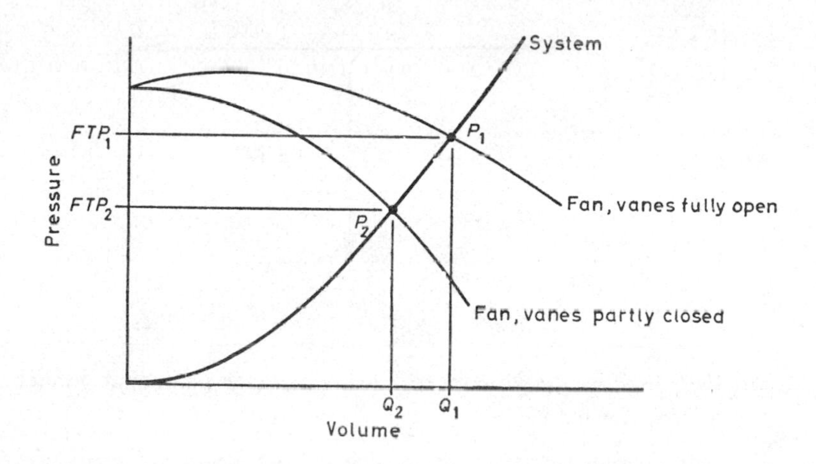

Fig. 15.30

15.24 The effect of opening and closing branch dampers

In any system it is usual to ensure that the correct quantity of air flows along the branches off the index run by partially closing regulating dampers in the various branches. (The alternative to this would be to size down the branch ducts so that they absorbed the correct amounts of energy by virtue of their reduced diameters.) Only the index run has its damper fully open. Thus, in a simple system, as illustrated in Fig. 15.31(*a*), if the index run is *A-B-C*, then the natural energy loss along *BB'* is less than it is along *BC*, for the flow of the design air quantities. This situation is clearly anomalous; only one total pressure can exist at *B*, hence the loss along *BB'* must be artificially increased by the partial closure of a regulating damper. Then

$$\Delta TP_{bc} = \Delta TP_{bb'}$$

Two interesting questions arise. What happens if the damper in *BB*, is fully closed? What happens if it is fully opened? The answers are best provided by means of an example.

EXAMPLE 15.18. Suppose the simple system shown in Fig. 15.31(*a*) is designed to deliver 1 m^3/s with a total pressure loss of 625 N/m^2, the air quantity handled being divided equally between the two duct runs, *BC* and *BB'*. The total pressure loss is made up as follows:

Plant	375 N/m^2 when 1·0 m^3/s is flowing.
Duct *A'B*	60 N/m^2 when 1·0 m^3/s is flowing.
Duct *BC*	190 N/m^2 when 0·5 m^3/s is flowing.
Duct *BB'*	60 N/m^2 when 0·5 m^3/s is flowing and its regulating damper is wide open.

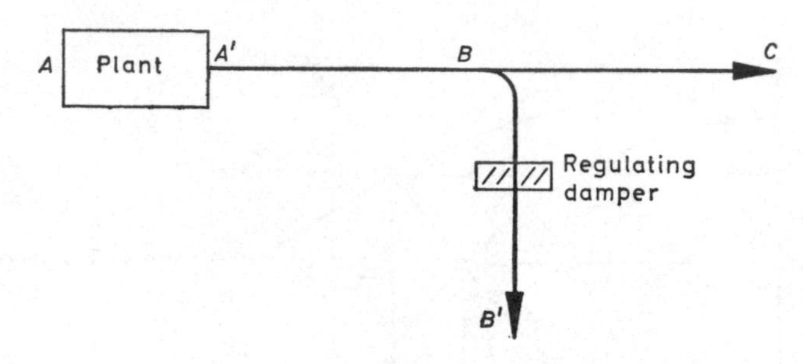

Fig. 15.31(*a*)

Assume that there is negligible energy loss across the damper in branch *BB'* when it is completely open.

Calculate:

(*a*) the pressure drop which must occur across the damper in *BB'* when it is partially closed, in order that 0·5 m^3/s may pass along the branch,

(*b*) the total air quantity handled by the system when the damper in *BB'* is fully closed and,

(*c*) the air quantity handled by the system if the damper in *BB'* is fully opened.

The pressure-volume characteristic for the fan is given in Fig. 15.31(*b*). Assume a square law for the system characteristic.

Answer

(*a*) Pressure to be absorbed by the partially closed damper in *BB'*

$$= \Delta TP_{bc} - \Delta TP_{bb'}$$

$$= 190 - 60$$

$$= 130\ \text{N/m}^2.$$

(*b*) Any assumption can be made about the air quantity flowing along *BC* in order to plot a new system characteristic for the revised arrangement with the damper in *BB'* closed. It is, however, most convenient to assume an air flow rate of 1 m³/s along *BC*, since the loss of energy from *A* to *B* is already known for this rate.

$$\Delta TP_{ab} = 435 \text{ N/m}^2 \text{ when } 1 \text{ m}^3\text{/s is flowing.}$$

$$\Delta TP_{bc} = 190 \times \left(\frac{1\cdot0}{0\cdot5}\right)^2$$

$$= 760 \text{ N/m}^2 \text{ when } 1 \text{ m}^3\text{/s is flowing.}$$

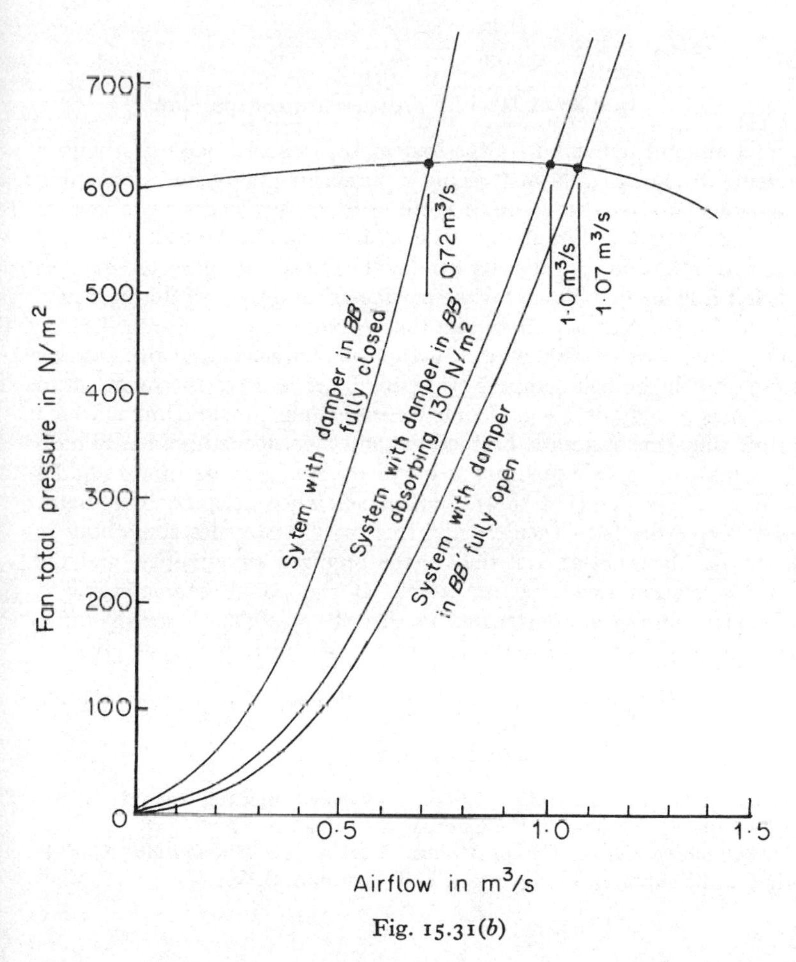

Fig. 15·31(*b*)

Hence

$$TP_{ac} = 1195 \text{ N/m}^2 \text{ when } 1 \text{ m}^3\text{/s is flowing.}$$

From this information a new system characteristic is plotted, as shown in Fig. 15·31(*b*), and the intersection of it with the fan characteristic determined.

It is seen that 0·72 m^3/s is handled by the system when the damper in BB' is fully closed.

(*c*) If the damper in BB' is fully open, then the loss of total pressure from B to C must equal that from B to B'. Any value for this loss of pressure may be assumed and a new system curve established. For ease of computation, assume that the pressure drop along BB' is 190 N/m^2. Hence the air quantity flowing along BB' may be calculated:

$$Q_{bb}' = 0{\cdot}5 \text{ m}^3/\text{s with a pressure drop of 60 N/m}^2.$$

$$Q_{bb}' = 0{\cdot}5 \times \sqrt{\frac{190}{60}}$$

$$= 0{\cdot}89 \text{ m}^3/\text{s with a pressure drop of 190 N/m}^2$$

The total amount delivered by the system thus equals 1·39 m^3/s, with a total pressure drop of 1030 N/m^2*. This is the starting point for establishing a new system characteristic curve, according to an assumed square law, as shown in Fig. 15.31(*b*). The intersection with the fan curve occurs at 1.07 m^3/s, and this is the total air quantity handled by the system when the damper in BB' is left fully open. Thus, for the particular conditions of this example, an overload of 7 per cent is imposed on the system.

It is clear that a risk exists, when a system is first started, of overloading the fan motor if the branch dampers are open. This could cause the motor to burn out. Accordingly, it is sound practice to include a main damper which can be shut when the system is first started and then opened gradually until the approximate design air quantity is handled. It is also essential to add a margin of 25 to 35 per cent to the fan power when selecting the motor required to drive the fan. Incidentally, this margin provides something in hand if the fan does not at first deliver the required air quantity, and the speed of its rotation must be increased. If the power characteristic is known for a fan running at a certain speed, then the power curve for any other speed can be determined and plotted by means of Fan Law A.3, as given in section 15.21.

BIBLIOGRAPHY

1. LUDWIG PRANDTL. *Essentials of Fluid Dynamics*. Blackie, 1953.
2. *Fan Engineering*. Buffalo Forge Company, Buffalo, N.Y.
3. *Air Conditioning System Design Manual, Part 2: Air Distribution*. Carlyle Air Conditioning & Refrigeration Ltd., London, S.W.1.

* $190 + (375 + 60) \times 1{\cdot}39^2 = 1030 \text{ N/m}^2$.

16

High Velocity and Other Systems

16.1 Development

Necessity is the mother of invention, and it has been the need to air condition buildings in spite of restricted space that has provided the stimulus to development of so-called high-velocity air conditioning. Conventional, or low-velocity systems, suffer from the disadvantage that the medium they employ to distribute cooling throughout the building, namely, air, has both a low density and a low specific heat. Large quantities of air are therefore necessary and these demand ducts of large cross-section. Two answers to this difficulty suggested themselves: (i) to distribute the air at as high a velocity as is acceptable without incurring excessive noise or running cost, and (ii), to augment the cooling capacity of the system by using an auxiliary fluid having a higher density and specific heat, namely, water. A further spur in the improvement of air-conditioning systems has been the need to provide individual control over the conditions maintained in individual rooms.

These pressures have resulted in the following systems being in common use to-day:

(i) Perimeter induction.
(ii) Fan-coil.
(iii) Chilled ceiling.
(iv) Single duct.
(v) Double duct.
(vi) Variable air volume.
(vii) Terminal heat recovery units.
(viii) Integrated Environmental Design.

16.2 Perimeter induction systems

These were first introduced in America in about 1935, using a system of low-velocity primary air distribution, but a change to high velocity was soon made.

The essential feature of such systems (in common with fan-coil and chilled-ceiling systems) is that the dehumidifying function is to a large extent

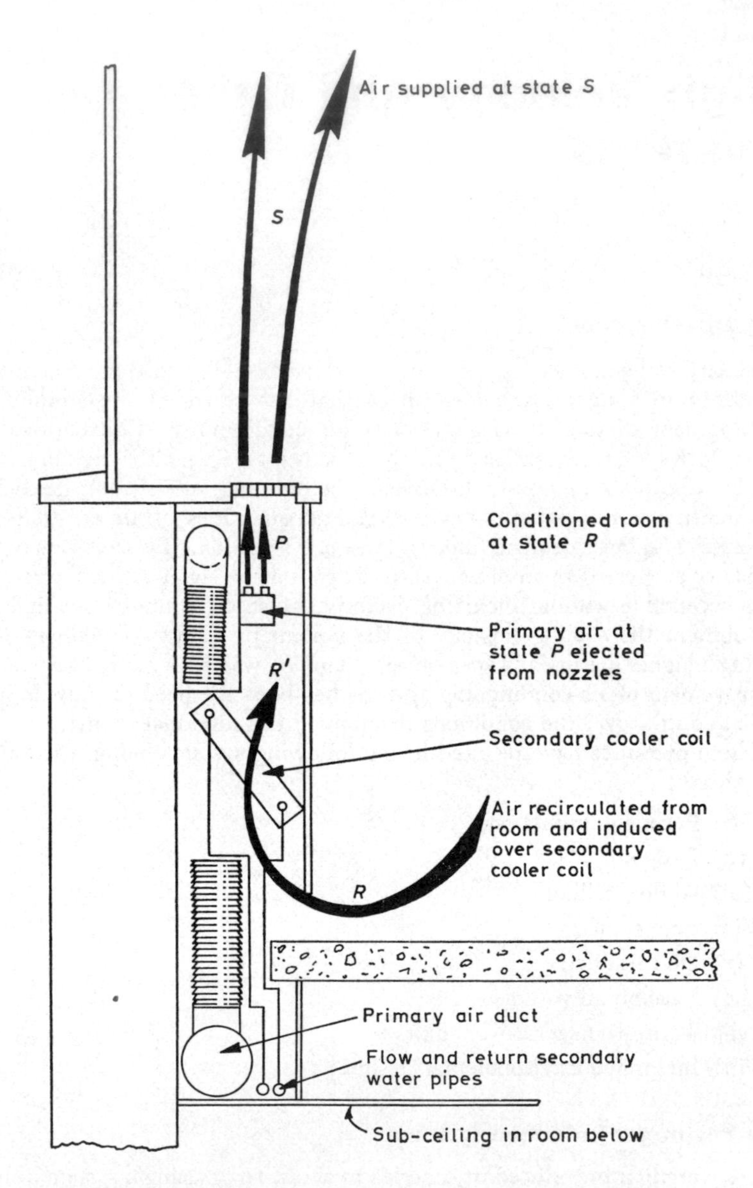

Fig. 16.1(*a*) Section of perimeter induction unit installation

divorced from the sensible cooling function. To achieve this and, at the same time permit a considerable degree of individual control, one or more air-conditioning units are placed in each room. Every unit receives a supply of primary dehumidifying air from a central air-handling plant and a supply of secondary water from a central pumping plant. The arrangement is shown in Fig. 16.1(*a*). Ducted primary air is fed into a small plenum chamber in the unit where its pressure is reduced by means of a suitable damper to the level required at the nozzles. At the same time, any noise that may have been generated in the ducting after leaving the air handling plant and any noise produced in the process of pressure reduction is attenuated in the chamber. The primary air is then delivered through nozzles at a fairly

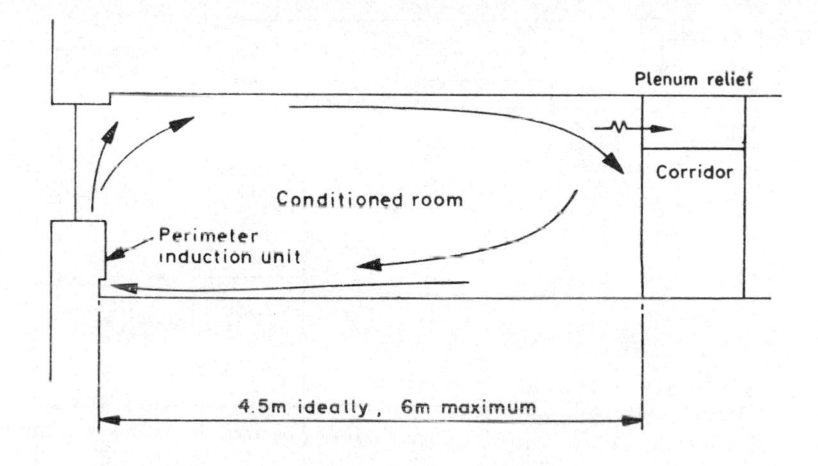

Fig. 16.1(*b*) Air circulation with a perimeter induction system

high pressure (100 to perhaps 700 N/m^2). The jets of high-velocity air issuing from the plenum chamber through the nozzles entrain a good deal of the air which surrounds them, about 3 to 8 volumes of secondary air being induced by each volume of primary air. An induction ratio of 4 to 1 is normal. The secondary air is pulled over a heat exchanger, termed the 'secondary coil', prior to entrainment with the primary jet. A mixture of primary and secondary air then enters the room.

Figure 16.1(*b*) shows the sort of air distribution that results. Airflow is upward past the window, thus offsetting down-draughts in winter, and across the room in depth, returning at low level to the recirculation opening at the bottom of the unit. An effective distribution of this kind is best obtained in a room of about 4·5 mm in depth, from window to wall, although a maximum depth of 6 m is sometimes tolerated. The primary air delivered by the nozzles escapes from the room through a plenum relief grille, which may be mounted as shown in the figure or in some other convenient place. The ideal with a system of this type is achieved if the primary air is solely from outside, no recirculated air being used by the central air-handling plant. However, since the unit capacity is related to the amount of primary air it receives, it is often uneconomic to use 100 per cent of outside air for primary purposes. Outside air in summer has to be cooled and dehumidified, which is expensive in terms of the capital cost of the installed refrigeration plant, to say nothing of the running cost. Because of this the primary air quantity, which is composed of a mixture of fresh air and recirculated air, is often selected to meet the latent load on the unit. The amount of fresh air is dictated by the

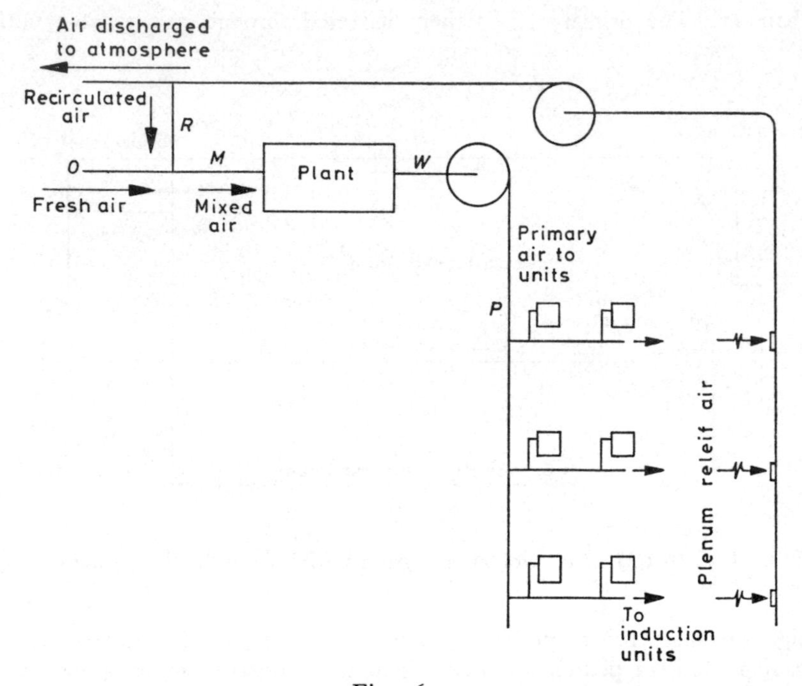

Fig. 16.2

ventilation requirements of the space, consistent with the density of occupation and the activity of the occupants, about 1·3 to 2·0 litres/s m^2 of conditioned floor area being the likely range of fresh air needs with 1·5 litres/s m^2 being typical. A check must be made, of course, that the total sensible cooling capacity of the unit equals or exceeds the total sensible load imposed on it by the heat gains to the room. The total sensible cooling capacity of the unit equals the sensible cooling capacity of the primary air plus that of the secondary coil.

Figure 16.2 is a diagrammatic arrangement of a typical air-distribution system.

16.3 Changeover and non-changeover systems

The induction system was conceived in North America, a country which, over a good deal of its area, has a severely cold winter and a hot summer, with a spring and autumn of negligible duration. This climate has had its impact on the development of air-conditioning in general and on induction systems in particular, fostering the ' changeover ' design.

A changeover system is designed so that chilled secondary water is supplied to the unit coils in summer but warm secondary water in winter. The primary air the units receive in summer is at a temperature which is scheduled in value according to the outside air temperature, the idea being that transmission heat gains and losses can be offset by the primary air alone. In this way, random and cyclic gains due to electric lighting, people and sunlight can be dealt with by the chilled water circulated through the secondary coils. Thus, a room is kept at the correct temperature by the primary air supplied to it, even though there are no heat gains other than those due to the air-to-air temperature difference. When people walk into the room, the lights are switched on and perhaps sunlight shines through the windows, the flow of chilled water through the secondary coil then deals with the extra load, its flow being modulated by a motorised valve controlled from a room thermostat. If several units are mounted in one room their output is controlled from a common thermostat.

This summer schedule of operation is satisfactory provided the sensible heat losses in marginal weather do not become unduly large. In the North American winter, however, outside air temperatures drop to very low values, and offsetting heat losses with primary air alone involves its supply to the units at extremely high temperatures. For this reason a winter schedule of operation was devised. Hot water is supplied to the units and, because of the large heat transfer surface afforded by the finned secondary coils, the severest heat losses can be offset without embarrassment.

Some cooling capacity is still likely to be required at the units. Winter sunshine can be surprisingly strong at times and, in any case, electric lights and people can cause a considerable heat gain. The combined effect of all these may exceed the sensible heat loss. Figure 16.3 is a load diagram which shows the need for cooling in winter. Primary air is supplied to the units under the winter schedule of operation at a constant low temperature (probably obtained naturally without the use of the refrigeration plant). The particular quantity supplied to a unit in a room has, therefore, a certain sensible cooling capacity and, as the figure shows, there is a certain outside air temperature at which it equals the maximum net sensible heat gain. For temperatures below this, the cooling capacity of the primary air exceeds any likely sensible heat gain, but for higher values of the outside air temperature it is less than the sensible gain and the unit will not always be able to maintain the correct conditions in the room. It follows that this balance temperature defines the changeover from the winter to the summer schedule of operation.

In the North American climate the distinction between summer and winter is usually sharp and well defined, so there is little difficulty experienced in making a decision to change over a system. In our climate, winter drifts into summer with a prolonged spring of a variable nature and vice-versa. For many days of the year in spring and autumn the temperature may change considerably about the value at which the system should have its schedule of operation altered. In March, for example, a system dealing with a south-west orientation may have to operate on the winter schedule in the mornings and on the summer schedule in the afternoons. This can be an acute embarrassment, particularly if the system is a large one and the mass of water stored in the secondary and primary pipework is large.

The non-changeover system is simply one which never changes over. It operates on the summer schedule throughout twelve months of the year. For such a system to be effective, it may be used only in a climate that does

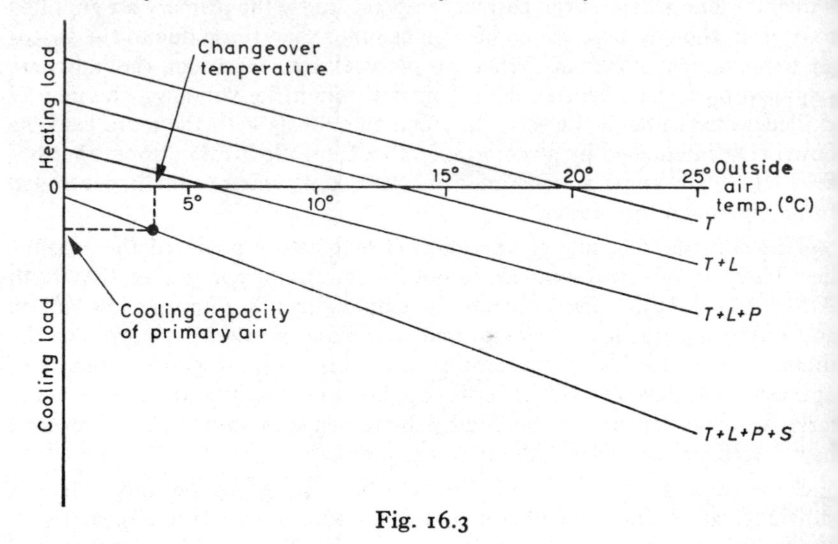

Fig. 16.3

not suffer a very severe winter. Generally speaking, the climate in England meets such a requirement and the non-changeover system is suitable. There are exceptions, of course; large U-values and considerable areas of glass may demand a changeover system.

EXAMPLE 16.1. A perimeter induction system is used to air-condition a room measuring 3 m high × 3 m wide × 6 m deep, for which the loads are as follows:

Summer	Outside air temperature	28°C
	Solar heat gain	1200 W
	Sensible gain from electric lights	600 W
	Sensible gain from people	300 W
	U-value for the walls	1·7 W/m² °C

	Area of the walls	4·5 m^2
	U-value for the windows	5·7 W/m^2 °C
	Area of the windows	4·5 m^2
Winter	Outside air temperature	−1°C
	Solar heat gain	600 W

Room air temperature 21·5°C in summer and in winter.

Assume that the solar heat gains are proportional to the outside air temperature.

(*a*) If a non-changeover system is to be used and if 0·05 m^3/s of primary air is to be supplied at the unit, what is the maximum primary air temperature?

(*b*) If the system is made to operate on a winter schedule and if primary air at 13°C is available under these circumstances, what is the changeover temperature?

(*c*) What primary air temperature is required to offset transmission heat gains under summer design conditions?

Answer

$$\text{Summer transmission gain} = (1{\cdot}7\times4{\cdot}5+5{\cdot}7\times4{\cdot}5)(28-21{\cdot}5)$$
$$= 33{\cdot}3\times6{\cdot}5$$
$$= 216{\cdot}4 \text{ W}$$

Other summer gains = 2100 W

Total summer gain = 2316·4 W

$$\text{Winter transmission loss} = 33{\cdot}3\times(21{\cdot}5+1)$$
$$= 749{\cdot}2 \text{ W}$$

Winter gains = 1500 W

Total net winter gain = 750·8 W

(*a*) Using eqn. (6.13):

$$(t_p-21{\cdot}5) = \frac{0{\cdot}7492}{0{\cdot}05}\times0{\cdot}8$$

$$t_p = 33{\cdot}5°\text{C}.$$

(*b*) The gross sensible heat gain must be equated to the sensible heat loss by transmission plus the cooling capacity of the primary air:

$$1500 = 33 \cdot 3(21 \cdot 5 - t_0) + \frac{0 \cdot 05 \times (21 \cdot 5 - 13) \times 1000}{0 \cdot 8}$$

$$t_0 = -7 \cdot 6^\circ\text{C}.$$

(*c*)
$$0 \cdot 2164 = \frac{(21 \cdot 5 - t_p) \times 0 \cdot 05}{0 \cdot 8}$$

$$t_p = 18^\circ\text{C}.$$

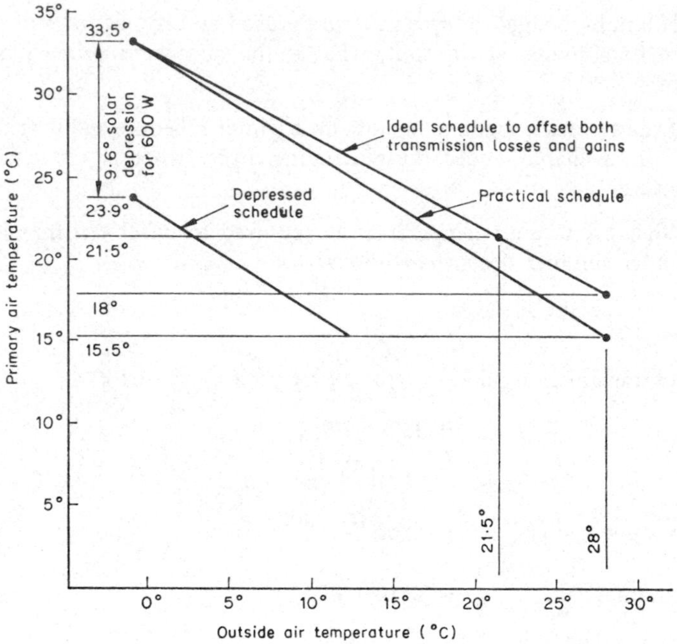

35°
33·5°
30°
9·6° solar depression for 600 W
Ideal schedule to offset both transmission losses and gains
25°
Depressed schedule
Practical schedule
23·9°
21·5°
20°
18°
15°
15·5°
10°
5°
21·5°
28°
0°
5°
10°
15°
20°
25°
30°
Primary air temperature (°C)
Outside air temperature (°C)

Fig. 16.4

This example shows three things:

(*a*) The non-changeover system is suitable, since 33·5°C is quite an acceptable temperature.

(*b*) The changeover system is unsuitable. With a winter design temperature of −1°C, the occurrence of −7·6°C will be very rare in the day-time.

(*c*) A primary air temperature of 18°C is probably too high for summer design operation. Commercial pressures necessitate optimum unit selection. This leaves no room for wasting any sensible cooling potential the primary air may have. A practical temperature for the air leaving the primary air cooler s about 11°C, and if an allowance of, say, 4·5 degrees, is made for the tem-erature rise due to fan power and duct heat gain, a practical primary air emperature of 15·5°C is obtained at the unit. Re-heating this to 18°C merely o get a precise balance with the transmission gains involves wasting about .o per cent of the cooling capacity of the primary air.

This has an impact on the form of the re-heat schedule, as illustrated by 'ig. 16.1. The ideal schedule assigns to the primary air a temperature xactly right for offsetting transmission gains or losses. The practical chedule differs in that the primary air temperature being 2·5 degrees lower han the ideal in summer design conditions all other primary air temperatures re also lower than the ideal, except in winter design conditions.

If the sun shines through the windows it may provide part or all of the eat necessary to offset transmission losses. But this will be of no value if the rimary air has already been re-heated in accordance with the sort of schedule ust discussed. The solar gain will have to be cancelled with cooling by the econdary coil. It is, therefore, sometimes arranged that the re-heat schedule s depressed by an appropriate amount when the sun shines. An element ensitive to solar radiation registers the presence of the radiation and sends signal to the re-heat control system. The schedule takes up a new position, s shown in the figure. There are difficulties in doing this. The intensity of olar radiation is not the same throughout the year, thus the slope of the lepressed schedule ought, strictly speaking, to be greater than that of the ormal schedule. Secondly, instantaneous solar gains through a window do ot immediately become apparent as a load on the air-conditioning system. 'urther, there is a time lag in the response of the solar compensation control ystem; by the time the schedule has been depressed a cloud may have overed the sun, rendering the depression unnecessary. Furthermore, a olar compensating system is difficult to set up.

All this may suggest that solar compensation is not worth the trouble. Iowever, heat gains from people and lights provide a base load on the air-onditioning system which, coupled with the individual thermostatic control t the induction units, may tend to smooth out the variations in response nentioned. In any case, the saving in fuel oil that solar compensation offers y cutting back on re-heat may make its use desirable.

XAMPLE 16.2 Using a value of 600 W for the solar heat gain, determine he corresponding depression of the re-heat schedule if solar compensation s to be used.

Answer

The cooling capacity of 0·05 m³/s of primary air must be equated to 600 W nd the corresponding temperature change calculated.

$$t_p - t_r = \frac{0{\cdot}60 \times 0{\cdot}8}{0{\cdot}05}$$

$$= 9{\cdot}6°C.$$

Thus, the primary air temperature should be 23·9°C when it is −1°C outside. The depressed schedule runs parallel to the normal one and, of course, a temperature of less than the summer design primary air temperature can never be obtained.

Note that all the rooms served by a given primary air schedule must have approximately the same ratio of primary air quantity supplied to fabric heat loss. This is commented on later in section 16.12.

16.4 The psychrometry of induction systems

Figure 16.5 illustrates the operation of an induction system under summer design conditions. Primary air at a state O (it is assumed that 100 per cent fresh air is used) is cooled and dehumidified by the main cooler coil in the central primary air handling plant to a state *W*. On the way to the induction

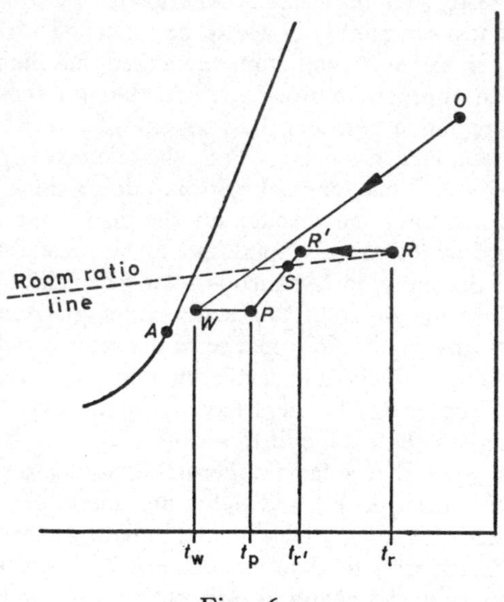

Fig. 16.5

units the temperature of the primary air is undesirably but inevitably increased from t_w to t_p and the air is finally delivered through the nozzles of the units at a state *P*. Secondary air is induced over the cooler coil in the unit and is sensibly cooled from state *R* to state *R'*. (Although it is sometimes arranged that the secondary coils do some dehumidification, this is generally undesirable. First, copper condensate drain lines have to be run from the units, and this is expensive. Secondly, the moisture which forms on the coils eventually picks up and retains some dirt, and this, in due course, results in smells.) The entraining primary air mixes with the induced secondary air, and the mixture state *S* so formed must lie on the room ratio line since it is the total quantity of air, primary plus secondary, which, delivered to the room at this state, offsets the sensible and the latent heat gains.

The total cooling load is the load on the primary coil in the central air-handling plant plus the load on the secondary coil in the induction unit. It may be checked in the usual way.

EXAMPLE 16.3. If 100 per cent of fresh primary air is used with the system mentioned in Example 16.1, determine the primary air cooling load, the relative humidity in the room, the total cooling load, the load on the secondary coil, and the temperature rise of the secondary water, given the following additional information:

Outside air state	28°C dry-bulb, 19·5°C wet-bulb, 55·36 kJ/kg.
State of air leaving primary cooler coil	11°C dry-bulb, 7·539 g/kg, 30·07 kJ/kg.
Humid volume of air delivered to units	0·8272 m³/kg.
Latent heat gain in room	130 W.
Water flow rate through secondary coil	0·12 kg/s.

Answer

$$\text{Primary cooling load} = \frac{0{\cdot}05 \times (55{\cdot}36 - 30{\cdot}07)}{0{\cdot}8272}$$

$$= 1{\cdot}529 \text{ kW}$$

Making use of eqn. (6.14):

$$\text{Moisture pick-up} = \frac{0{\cdot}130}{0{\cdot}05} \times \frac{(273 + 15{\cdot}5)}{856}$$

$$= 0{\cdot}876 \text{ g/kg.}$$

$$\text{Room moisture content} = 7{\cdot}539 + 0{\cdot}876$$

$$= 8{\cdot}415 \text{ g/kg.}$$

At 21·5°C dry-bulb in the room this gives a relative humidity of almost 52 per cent. At this state the enthalpy is 43·01 kJ/kg.

$$\text{Fresh air load} = \frac{0{\cdot}05 \times (55{\cdot}36 - 43{\cdot}01)}{0{\cdot}8272}$$

$$= 0{\cdot}746 \text{ kW.}$$

Load due to fan power and duct heat gain

$$= \frac{0{\cdot}05 \times (15{\cdot}5 - 11{\cdot}0) \times 358}{(273 + 15{\cdot}5)}$$

$$= 0{\cdot}279 \text{ kW.}$$

Total cooling load = Sensible gain + latent gain + fresh-air load + fan power load.

$= 2316 + 130 + 746 + 279$

$= 3471$ W

Secondary coil load $= 3471 - 1529$

$= 1942$ W.

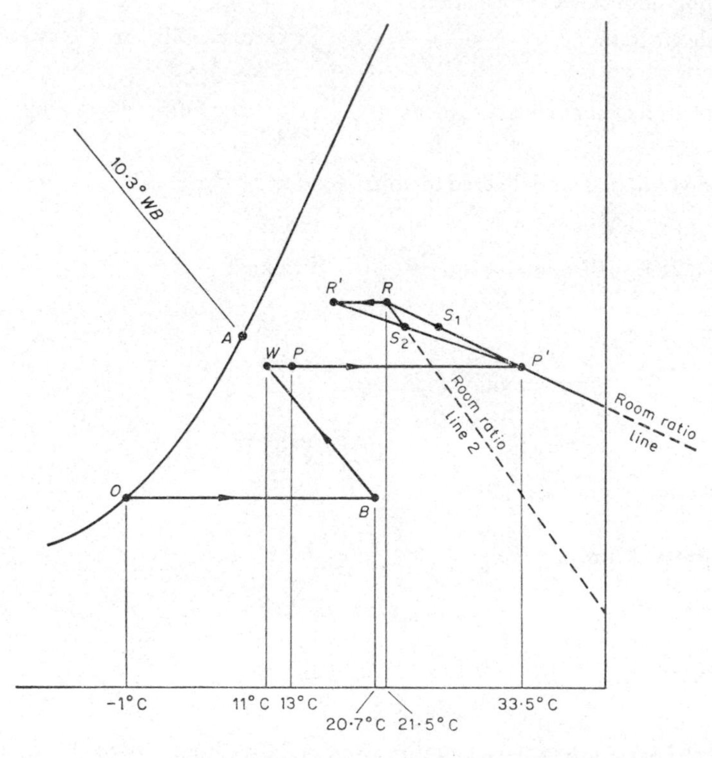

Fig. 16.6

Temperature rise of the secondary water

$$= \frac{1942}{0{\cdot}12 \times 4190}$$

$= 3{\cdot}86$°C.

If this system operates in a similar fashion in winter, without changing over, then the psychrometric behaviour is as shown in Fig. 16.6. Outside air at a state *O* is pre-heated to state *B* and then, by flowing over the sprayed primary air-cooler coil, it is adiabatically humidified to state W. Fortuitous re-heat again occurs to state P and, when it is -1°C outside, the primary air is re-heated, according to the proper schedule, to 33·5°C. At this state, P', it is delivered from the nozzles and induces air over the coil in the unit. If there are no heat gains at all, then the thermostatic control system of the unit will prevent the flow of any chilled water through the secondary coil, and so the mixture state S_1 of the air delivered to the room will be formed by P' mixing directly with R. If some heat gains are occurring then the induced air will be chilled from state R to state R', at the command of the controlling unit thermostat, and the air will be delivered to the room at a state S_2, a mixture of P' with R'. The supply state, S_2, will now lie on a new room ratio line.

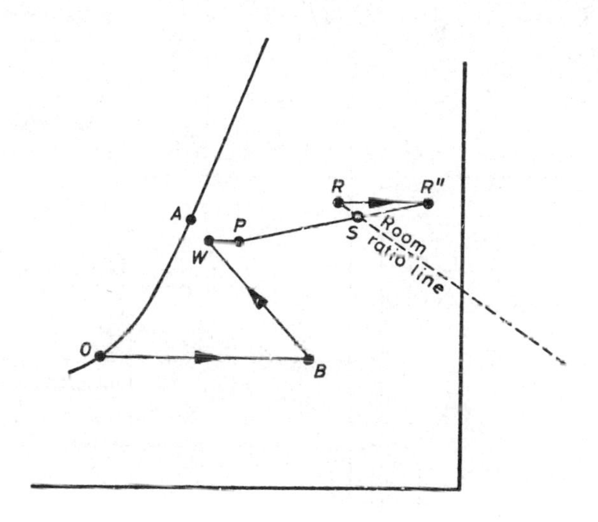

Fig. 16.7

If the system is changed over to operate on the winter schedule then the behaviour will be as shown in Fig. 16.7. Outside air at state O is pre-heated to state B and then adiabatically humidified to state W. The usual undesirable temperature rise to state P takes place, and it is at this state that the primary air is delivered through the nozzles. The control system of the unit modulates the flow of warm water through the secondary coil so that the induced air flowing over it is heated from state R to state R''. R'' mixes with the primary air at P and the mixture state lies on the room ratio line. Thus, air at the mixture state S is supplied from the unit to the room and offsets the heat losses. If heat gains occur, then the automatic control system for the unit throttles the flow of warm water through the secondary coil, allowing the primary air to do the necessary cooling.

16.5 Piping arrangement for an induction system

Naturally enough, some variety exists in the piping circuits adopted for induction systems, but there should be a basic similarity in them all. Figure 16.8 illustrates a basic arrangement for a non-changeover system.

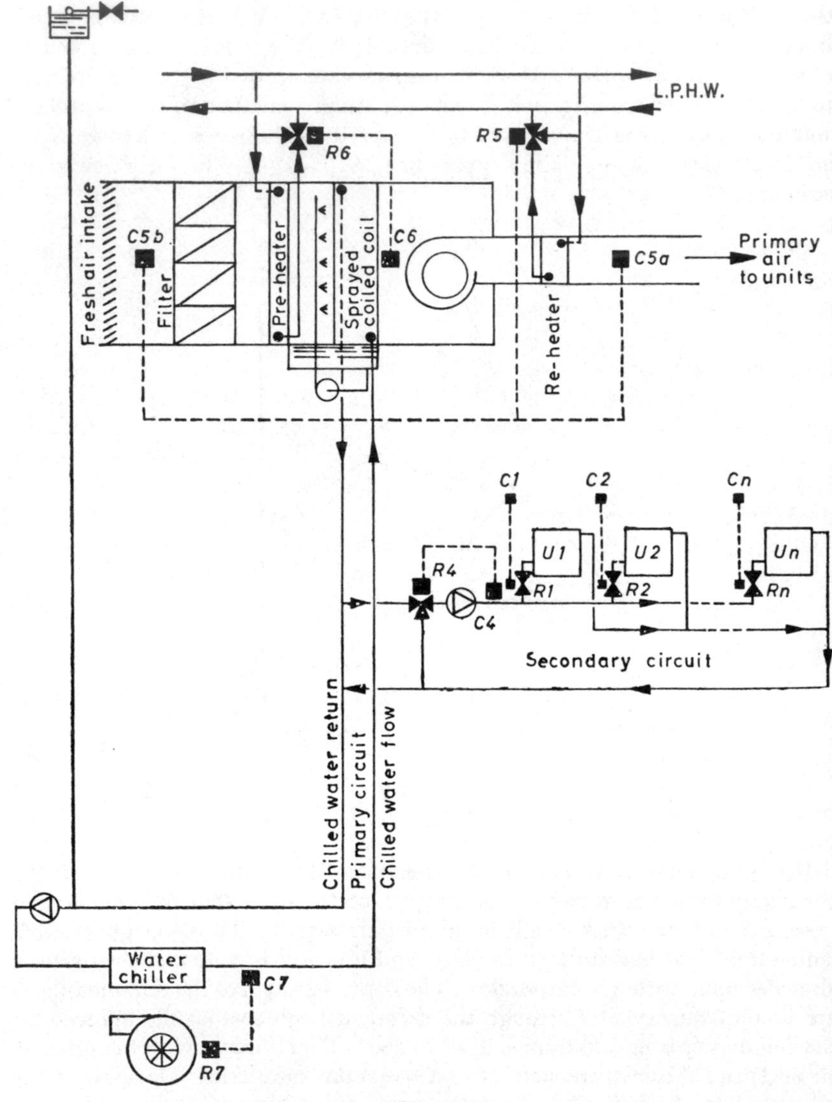

Fig. 16.8

Chilled water is pumped through a chiller to the primary air sprayed cooler coil. Its flow temperature is kept constant at a nominally fixed value by means of an immersion thermostat *C7*, which regulates the capacity of a centrifugal compressor by means of variable inlet guide vanes, *R7*. Suppose the state to be maintained in the room is 22°C dry-bulb with 50 per cent saturation. The dew point corresponding to this is about 11·3°C. Because of the thermal resistance of the water film within the piping which circulates the chilled water to the units, a minimum water temperature of a little less than this can be tolerated without causing sweating on the outside surface of the piping. A value of, say, 10·5°C is suitable then for the flow temperature to the units. If possible, it is arranged that the temperature of the primary water leaving the sprayed cooler coil is at this value. Thus, water flows on to the primary coil at 6°C and leaves it at 10·5°C. This choice is, of course, dictated by the performance required from the primary coil. Complete freedom of choice is not possible.

The connexions to the secondary circuit are taken adjacent to one another on the outflow pipe from the primary cooler coil so that the difference of pressure between them is negligible. Thus, any alteration in the draw-off by the secondary pump, through the automatic control valve *R4*, will have little effect on the flow in the primary piping circuit.

Water is now available at the desired value of 10·5°C for the secondary circuit, under conditions of summer design operation. At other times of the year there will be a reduced load on the primary air-cooler coil and so, the flow temperature of the chilled water being kept constant at 6°C but the temperature rise being less than 4·5 degrees, the draw-off to the secondary circuit will be at less than 10·5°C. To prevent condensation at the units and on the secondary piping, a three-port mixing valve, *R4*, must be used. This is under the control of an immersion thermostat *C4*, having a set point of 10·5°C.

The units themselves may be under the control of individual, motorised, two-port or three-port, valves and room thermostats, denoted by *R1*, *R2*, *Rn* and *C1*, *C2*, *Cn*, in the figure. Suppose the temperature rise through the secondary piping circuit were 3 degrees C under conditions of maximum load, then water would be returned to the primary system at 13·5°C and would be mixed with what primary water had not been drawn off by the secondary pump but had by-passed the flow connexion to the induction units. Thus, depending on the mass flow rates of the water at the different temperatures, the return temperature to the chiller would be somewhat higher than 10·5°C.

Complications may arise. If the rate of water flow through the primary coil is less than that through the induction units, then some revision of the above simple arrangement is necessary. A by-pass pipe may be required across the primary coil so that the primary pump may handle more water than flows through the coil.

The primary air is re-heated to its scheduled temperature (*see* Fig. 16.4) by means of the re-heater battery shown. The output of this is controlled by a sub-master duct thermostat *C5a*, which regulates the position of the three-port valve *R5*. A master thermostat *C5b*, located in the fresh air intake duct, re-sets the set point of *C5a* as the outside air temperature changes.

The arrangement shown in the figure permits the dew point of the primary air to fall as summer changes to winter. This may be desirable if the conditioned space has single glazing, so that condensation on the windows is avoided in the coldest weather. A duct thermostat *C*6 would have a set point of, say, 5°C. This would roughly correspond to a dew point temperature of 5°C, the sprays on the cooler coil not being 100 per cent efficient. Thus, condensation would not occur on the inside of the windows when the outside air temperature was about −1°C. *C*6 achieves its control by regulating the output of the pre-heater through the agency of the three-port motorised valve *R*6, the primary air cooler coil being sprayed. (If it had been desired that the nominal dew point remain constant throughout the year, then it would have been necessary to fit a three-port mixing valve in the return line from the primary cooler coil. This would have to be controlled in sequence with *R*6, from *C*6, and the set point of *C*6 would have to be raised to, say, 9°C.)

One last point: it is usually desirable to adopt a reversed return piping arrangement for the distribution of the secondary water to the induction units. This minimises the amount of regulation necessary when the system is commissioned. The piping to the units in Fig. 16.8 is reversed return.

16.6 Ducting arrangement for an induction system

There is an optimum velocity for sizing ducts (*see* Chapter 15). Apart from this, the only virtue that adopting a high velocity has, for sizing, is that it permits the ducts to be accommodated in the building more easily. Ignoring the question of running costs, there are three criteria according to which the ducts are sized.

(i) Accommodating the ducts in the building.
(ii) Minimising the first cost.
(iii) Achieving inherent self-balance.

The first factor is one that varies from building to building and is one on which it is not possible to generalise.

Minimising the cost of the ducting does not necessarily mean using the smallest size of duct. Ducting lengths have to be joined together, and this involves labour and materials. If a duct main delivers some air to an induction unit (*see* Fig. 16.9), a branch-piece is required. This often takes the form of a tee. It may be that the cost of a tee that does not reduce its size in the main is less than that of the one that does. Consequently, it may be cheaper to retain the duct size at a constant diameter for a while.

This retention of a constant diameter results in a certain amount of static regain occurring as the air flows in the main past each tee. When static pressure increases along the main in this fashion it becomes difficult to identify the index unit.

The static regain that will occur if duct sizes are not reduced may be used to some advantage in achieving an inherent system balance. Although fire

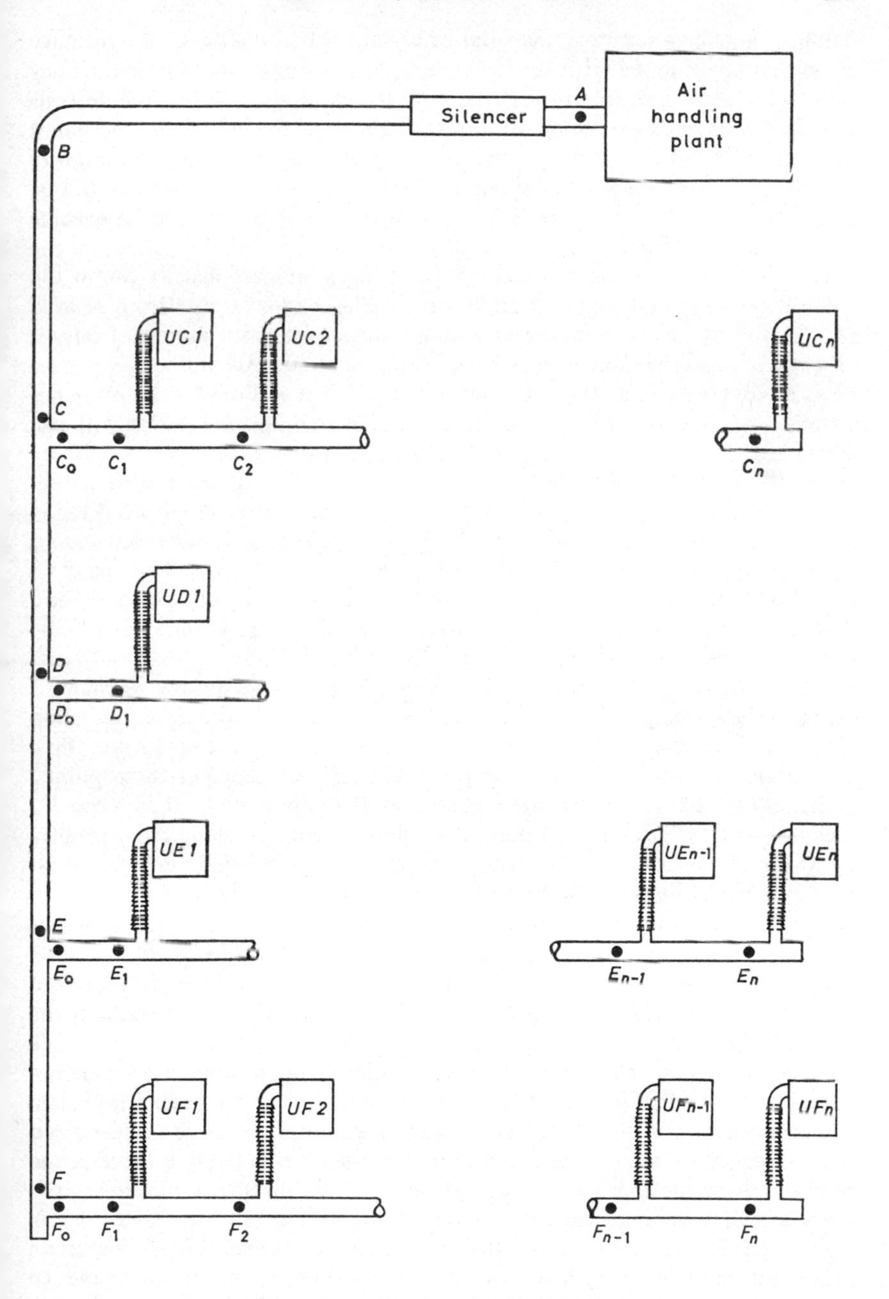

Fig. 16.9 Horizontal distribution

dampers may be a statutory essential in certain urban districts, the presence of any dampers in a high-velocity system of ducting is undesirable. They cause noise, are a source of leakage and are expensive. For these reasons attempts should always be made to size the system so that all the induction units are persuaded to deliver the correct quantities of primary air only by adjusting the dampers in their plenum boxes. That is to say, there should be no branch dampers. The method of achieving this inherent balance may be understood by reference to Fig. 16.9.

It is probable that the primary air-handling plant will usually be on the roof of the building being air conditioned. The length of duct from A to B will, therefore, lie on the roof and will not be subject to the same restrictions of space affecting the ducts elsewhere in the building. Accordingly it should be sized for use with a relatively low velocity. This will tend to keep the fan static pressure, and hence the running costs, as low as possible. It will also mean that the difference of pressure between A and B or between A and any other dropping main duct will not vary greatly. The dropper B to F will almost certainly have to be sized with a higher velocity than was used for A to B, because it will run within the building. However, this velocity should be kept as low as possible. The disadvantages of doing this are that the costs of the duct and lagging will not be as low as they might otherwise be and that the space occupied will be greater. The advantage is that some static regain can be contrived past each tee and that the static pressures at the points C, D, E and F, will be approximately the same, if proper thought is given to the sizing. This assumes that the quantities of primary air required along each horziontal branch are the same. If different quantities flow through each branch then different pressures will be needed at these points.

Strictly speaking, it is the static pressures at the points C_0, D_0, E_0 and F_0, which should be considered if the correct flow is required along the branches. However, since the loss through the branch of a tee is largely dependent on the branch velocity, equalisation of the pressures at C, D, E and F is usually an adequate step. If the branches are identical, of course, then the branch losses are also identical and it does not matter which points are involved.

It will not be possible, in general, to achieve an exact equalisation of pressure. The index branch is the one with the lowest static pressure at its beginning, e.g. at point E_0.

Generally, the space available for the installation of the horizontal branches is even more restricted than it is for the installation of the dropping mains. For this reason the duct at the beginning of any branch run is usually sized with a higher velocity than in the falling main. If this is done there is the further advantage that the larger resistance of the branches minimises the effect of the variations in pressure in the main.

Once the initial duct section in the branch has been sized, it is advantageous to size the succeeding sections so as to achieve enough static regain to reduce to a minimum the variation in static pressure upstream of each unit damper. The unit damper which has the lowest pressure on its upstream side defines the index unit.

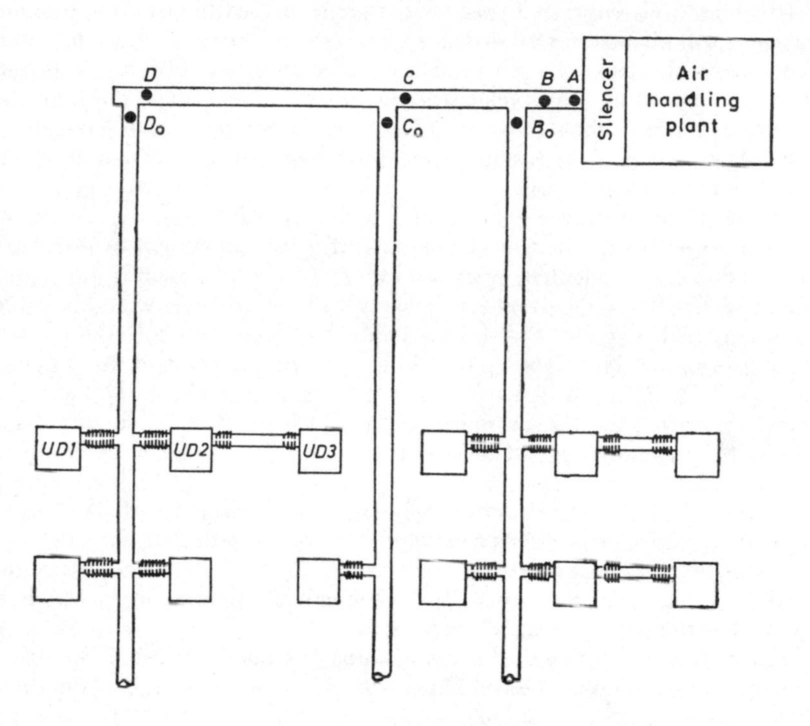

Fig. 16.10 Vertical distribution

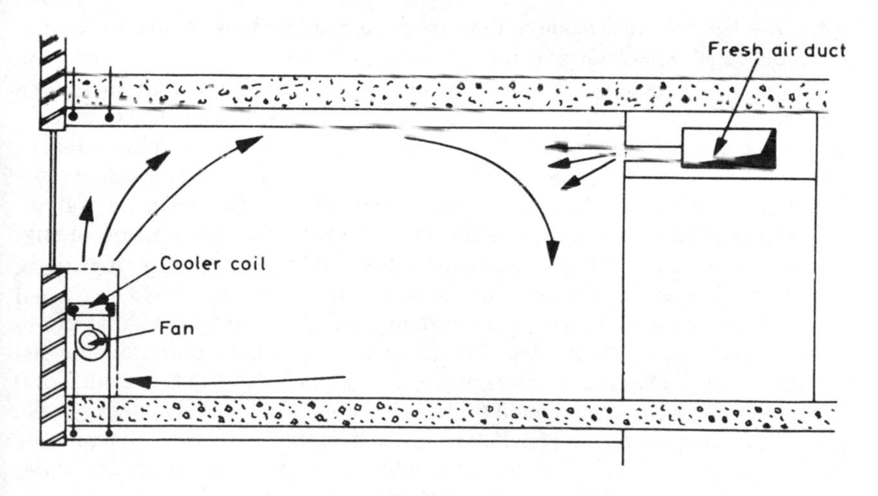

Fig. 16.11

It is tentatively suggested that 10 to 15 m/s be used to size the dropping mains in a horizontal distributions system (as illustrated in Fig. 16.9) and that the initial section of each branch duct be sized on 15 m/s. Although there are indications that velocities very much higher than this may be optimum; when sizing is considered (as it should be) in terms of the owning and operating cost of the system, velocities in excess of 20 m/s tend to result in rather large energy losses.

Figure 16.9 illustrates a horizontal system of distribution. Perimeter induction systems are best accommodated if a vertical system of ductwork distribution can be adopted, as shown in Fig. 16.10. The main from A to D would be sized on a relatively low velocity and the droppers would be sized on a fairly high velocity. Sizing would aim at achieving the correct values of static pressure at the points B_o, C_o, D_o, etc. It will be observed that some of the induction units are blown through; for example, units UD_2 and UD_3 share a branch from the dropping main. Such an installation permits a compact arrangement of the units and minimises the number of dropping mains needed.

In a horizontal system, the distribution duct feeding the units does so from below; it frequently runs beneath the floor slab and delivers the primary air to the units through a flexible duct connexion. With a vertical system, the vertical distribution ducts run at the columns and are encased by builders' work. Flexible connexions are again used.

The piping may be arranged in a horizontal or a vertical fashion. A vertical piping arrangement may be combined with a horizontal ducting system and vice-versa, to suit the situation.

16.7 Fan coil systems

Here, the fan coil unit replaces the induction unit, and an auxiliary ducting system is usually necessary. A fan coil unit, as its name implies, contains one or more centrifugal fans. Thus, a supply of air directly to the unit is unnecessary. The fan draws air from the room and blows it over the cooler coil within the unit. Any auxiliary air delivered to the room is for dehumidification and ventilation purposes. Although it is possible to duct auxiliary air to the units, or even to draw air in from outside through a hole in the wall of the building behind the unit, it is not uncommon to duct the air in a ceiling void in the corridor outside the room and to deliver it to the room at high level through a grille. Figure 16.11 illustrates this.

The operation and design of the system can be very similar to that of an induction system. The unit does sensible cooling only in summer, and a non-changeover or a changeover scheme can be adopted. The fact that air does not necessarily have to be ducted to the units gives the fan coil system a little more flexibility, as far as installation goes, than the induction system has. Two-pipe fan coil units can only provide proper heating if operated on a changeover basis, a method generally undesirable in the U.K. Relying on heating by means of the auxiliary air supply is frequently unsatisfactory, if this is at high level (Fig. 16.11): stratification and the difficulty of balancing airflow in the low velocity duct system so as to deliver the correct-heating capacity to each module combine to give poor results.

16.8 Chilled ceilings

Three forms of chilled ceiling are possible:

(i) Embedded piping coils.

(ii) An acoustic metal pan ceiling, clipped to piping coils and suspended below the ceiling slab.

(iii) A perforated ceiling with a plenum above it, into which chilled air is fed.

Coils embedded in the lower portion of the ceiling slab provide a cooling surface which is at a uniform temperature, provided the pipes are spaced fairly close together. It thus provides a large surface area for the reception of heat. Its disadvantages are that it is sluggish in responding to automatic control, and it may prove expensive.

The second type of chilled ceiling takes the form of perforated metal pans which have slabs of sound-absorbing material above and which are clipped to pipes. The ceiling in this form requires only about three or four inches of space below the slab for its installation. It offers the advantage that if the need for an acoustic ceiling exists, part of the cost of air conditioning can be set against this. In common with the embedded ceiling, it occupies no valuable floor space but it tends to be cheaper than that system. Its response to automatic control is also quicker, since the thermal capacity of the metal pans is a good deal less than that of the vast mass of the concrete in the floor slab.

Both types of ceiling are arranged to do sensible cooling only, a ducted supply of auxiliary air being necessary for dehumidification and ventilation purposes. Whereas the embedded ceiling offers quite a range of cooling capacity, because there is freedom of choice in selecting the distance between pipe centres, the acoustic metal-pan ceiling has a capacity which is related to the sizes of the metal pans available. Commonly, these pans are available in 450 mm, 300 mm and 225 mm widths.

Panels of 450 mm have a cooling capacity of about 4·7 W/m^2 for each degree of difference between mean water temperature and ambient air temperature. Of this total output, about 20 per cent is upward and the remainder downward. For 300 mm panels, the total capacity is about 7·0 W/m^2 °C and the percentage upward is about the same as for the 450 mm panels.

The flexibility of individual thermostatic control is not always possible with embedded piping coils because of their thermal inertia. So, although the principles used for designing induction and fan coil systems are also followed, after a fashion, for the design of chilled ceiling systems, differences of control exist. There is the further major difference that the quantity of

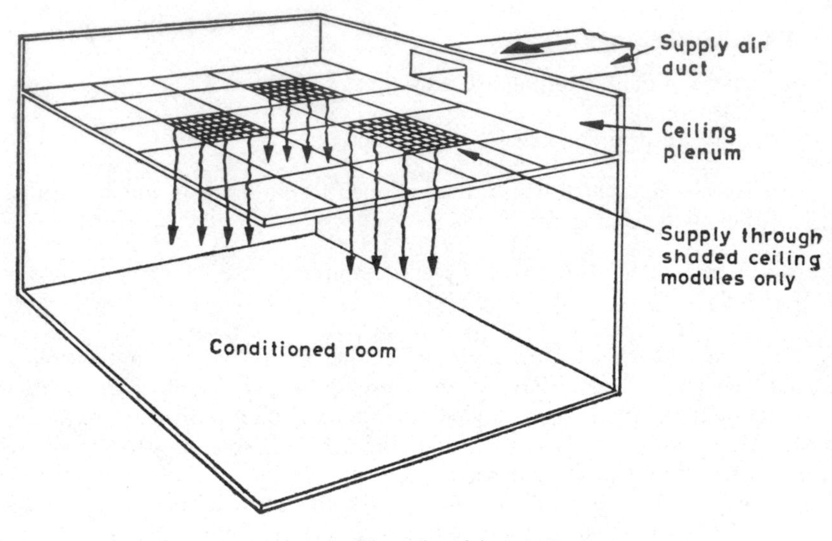

Fig. 16.12(*a*)

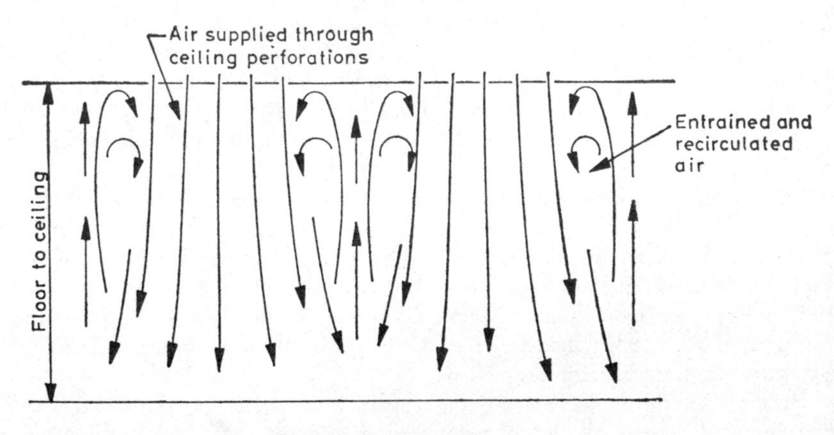

Fig. 16.12(*b*)

auxiliary air needed may possibly exceed the amount of primary air that would be used with an induction system. This is because the cooling capacity of the secondary coils in induction and fan coil units is large compared with a ceiling, for a given treated floor area.

The third form of chilled ceiling is rather different from the other two, both in intention and in performance. An alternative name for it is the 'ventilated ceiling'. It is used in two different applications:

(i) As an alternative to a more conventional method of distribution through ceiling diffusers, for the convenience offered when an acoustic ceiling is combined with an air-distribution system.

(ii) To deliver very large quantities of air to a conditioned space without excessive draught and with uniformity.

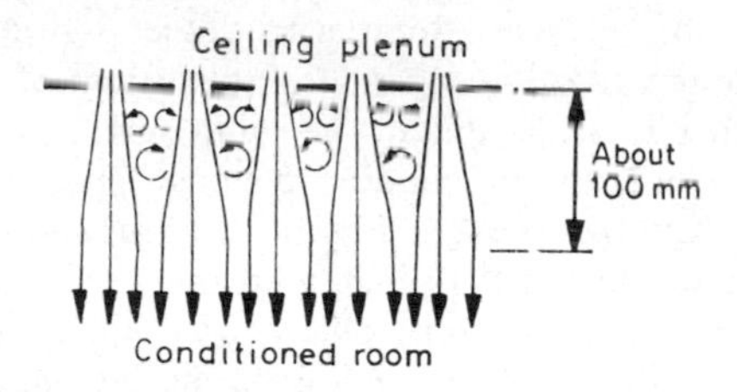

Fig. 16.12(*c*)

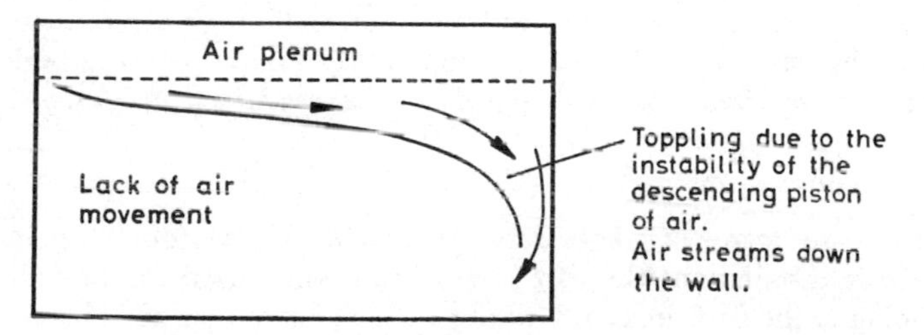

Fig. 16.12(*d*)

In either case, whether the ceiling is truly chilled will depend on the material of the ceiling. If aluminium pans, of the type adopted for the piped chilled ceiling mentioned above, are used, then the high thermal conductivity of the material will give a ceiling which is cooler than the air temperature maintained in the room and is also cooler than the other surfaces in the room. If an acoustic material of a non-metallic type is used then the effect is less marked, and little radiant cooling is obtained from the ceiling.

In the first type of ceiling, its entire surface is not used for the delivery of air. Figure 16.12(*a*) illustrates this case. All the cooling is done by the air supplied to the plenum chamber. The fact that the ceiling becomes chilled by contact with the air in the plenum is incidental; this serves to increase the temperature of the air slightly before it is delivered to the room and, thus, to reduce its convective cooling by the amount that the ceiling provides radiant cooling. Some of the cooling capacity of the air in the plenum chamber is lost to above. This may be of value, if the space above is also conditioned.

In Fig. 16.12(*a*) it can be seen that the air is supplied to the room through three only of the ceiling panels. These panels have perforations which penetrate the full depth of their thickness. The other panels have perforations which penetrate only through part of the thickness, and thus do not provide an air path. The three panels in question could well have been replaced by diffusers. With this arrangement, a good deal of air movement is promoted in the room. Air issuing from the perforations entrains ambient air, and circulation occurs somewhat as illustrated in Fig. 16.12(*b*).

If the whole surface of the ceiling is perforated and used to supply air, very little entrainment with the air in the room occurs (see Figure 16.12(*c*)). This means that a piston of cold air attempts to descend from the ceiling. The piston is balanced momentarily on a rising mushroom of warm air currents produced by the sensible heat gains in the room and, since this is inherently an unstable arrangement, the descending piston topples and the supply air streams down one side of the room, as Figure 16.12 (*d*) shows.

Proper mixing of the cool air supplied, with the warm convection currents, is necessary if sensible gains are to be offset and uniform temperature conditions obtained. Ceiling diffusers achieve this for air change rates of up to, say, 30 per hour. A perforated ceiling, having a large proportion of its area used for supply purposes, after the fashion of Figure 16.12 (*a*), can be used to handle air change rates in excess of 30 per hour. For really high air change rates, say 600 per hour, it is necessary to use the entire surface area of a perforated ceiling, *but in conjunction with a similar floor area used for extract purposes.*

16.9 Single-duct systems

There are many forms of single duct system, but in the context of modern high-velocity installations they are commonest when used for providing air conditioning to the core areas of buildings which have their peripheral spaces treated by perimeter induction or fan coil units. Although high velocity is sometimes used for the distribution of the air all the way to the final diffuser or grille, more often it is used only for part of the distribution, the remainder being at conventional velocity. Taking consideration of these possibilities, we are able to make a loose classification of single duct, 'high-velocity' systems as follows:

(i) Total high velocity.
(ii) Partial high velocity with pressure-reducing valves.
(iii) Partial high velocity with 'octopus' boxes.

The first type of system demands that the grille or diffuser which supplies air to the conditioned space shall incorporate a pressure-reducing and noise-attenuating device. Some diffusers are available which do this but, if grilles are to be used, a self-contained combination is not currently on the market, and the functions of air distribution and pressure reduction with attenuation have to be achieved with distinct appliances. The system then becomes rather like (ii) above. Here, for example, the velocity in the vertical mains may be high and, at each floor where a horizontal duct is taken from the main, a pressure reducing and attenuating valve is used to permit the use of a lower velocity of air flow in the duct. Figure 16.13(*a*) illustrates the arrangement. The pressure-reducing valve itself does some of the attenuation required but

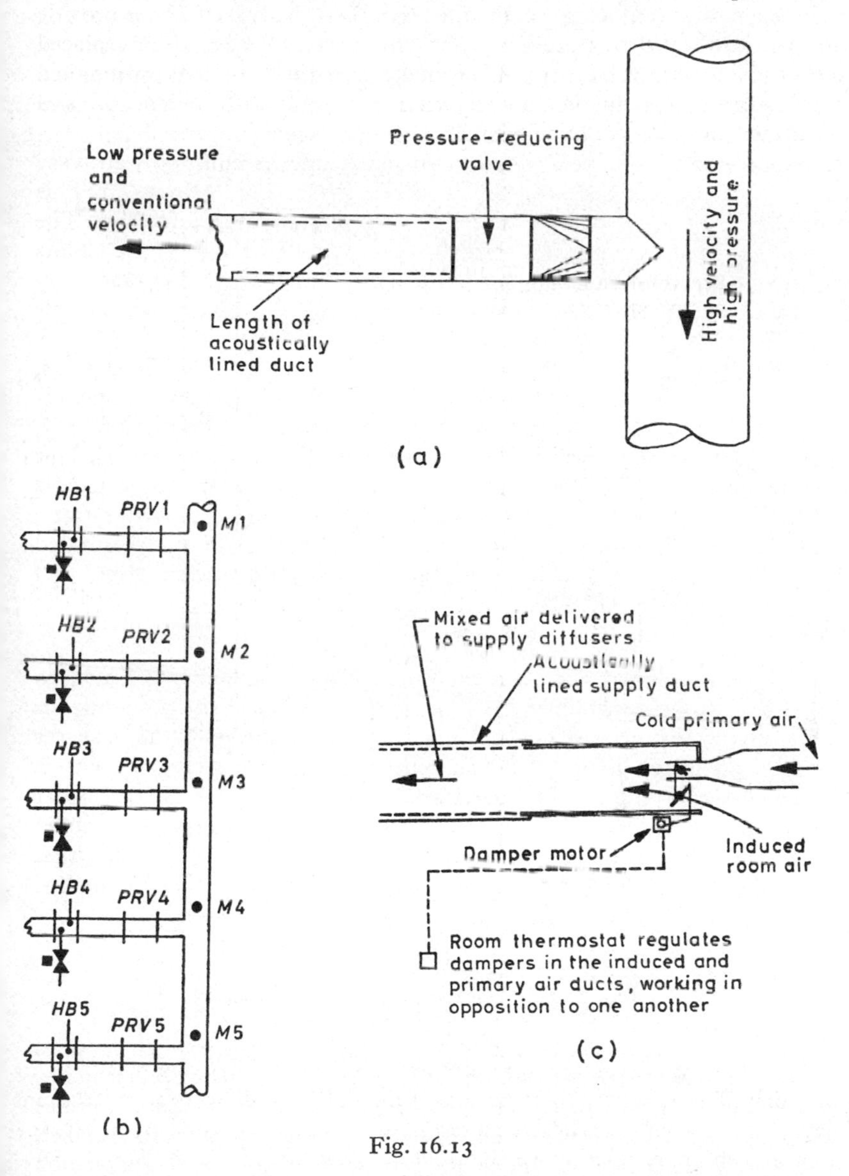

Fig. 16.13

it is usually necessary to provide a length of lined duct downstream of the valve, to ensure quietness at the first grille. Figure 16.13(*b*) shows the sort of pressure distribution the use of these valves permits. The dropping main from $M1$ to $M5$ is sized on a high velocity, say 15 m/s, in order to economise in duct space, and a large pressure drop ensues. There might be a difference of 100 N/m^2 between the static pressures on the upstream side of each successive reducing valve, for example. Thus $PRV1$ might have an upstream static pressure 400 N/m^2 greater than $PRV5$. Supposing that $PRV5$ is the index valve, it would normally be required to do very little in the way of pressure reduction, merely having to deal with any discrepancies resulting from errors of calculation or other contingencies. Ideally, of course, a pressure-reducing valve is not required on the index circuit.

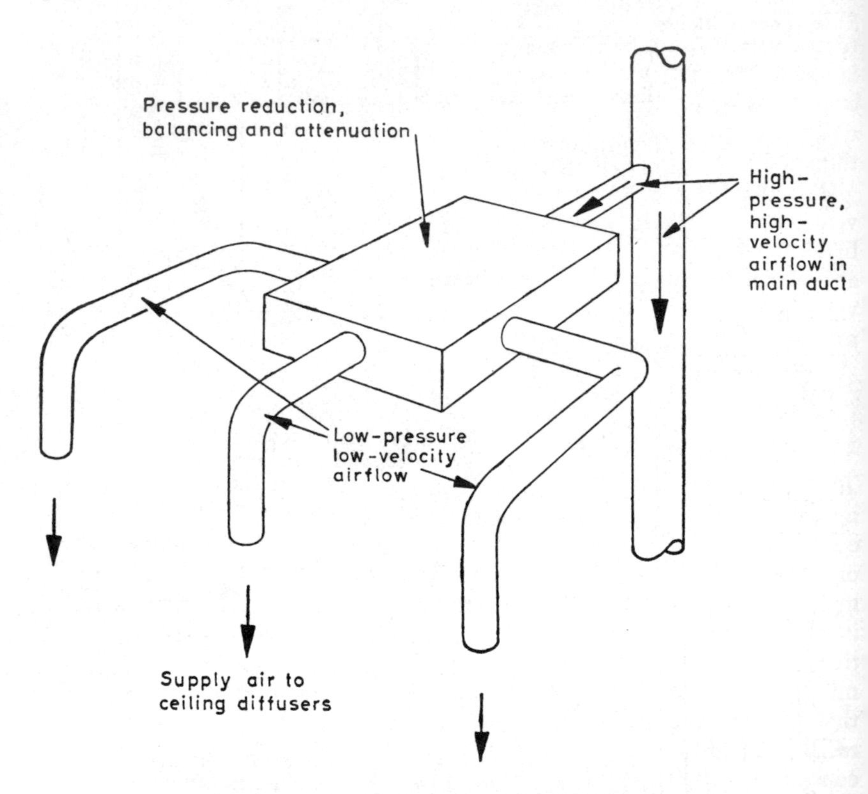

Fig. 16.14 'Octopus' box distribution

Systems of this type are very often re-heat systems. An independent air handling plant, used for the core area of the building, delivers air at constant dew point to a set of re-heater batteries, each being associated with, say, the core area on a particular floor. Control is exercised from a local room thermostat. This is showndiagrammatically in Fig. 16.13(*b*), batteries being denoted by $HB1$, etc. Re-heat systems like this are wasteful of refrigeration, and other means are sought to provide air conditioning for core areas. There are two notable variants. The first makes use of a device which delivers a variable amount of air to the core and, this air being a mixture of primary air and induced secondary air, throttling the quantity of primary air permits a control to be achieved over the local temperature in the core. This is possible because the secondary air is unheated; only the primary air is warmed, and this is done at the main air-handling plant. Figure 16.13(*c*) illustrates this. Some variation in the total amount of air delivered occurs, but this is not very objectionable. More serious is the fact that the primary air quantity is altered. It is true that a limit is set on the reduction of the quantity but, even so, a variable ventilation rate may not be desirable. The other alternative to re-heat is to use sensible coolers in place of the re-heater batteries. Cancellation of the cooling done by the primary air-handling plant does not then occur, but the system may be expensive in capital cost. Cooler coils, particularly sensible ones, are usually more expensive than heater batteries for the same heat exchange.

The third single-duct system is very similar. Air is distributed at high velocity along the vertical mains, also along part of the horizontal ducting. A break-down of pressure takes place at what are termed 'octopus boxes', any associated noise being attenuated, and further distribution is at low velocity to ceiling-mounted diffusers. Figure 16.14 shows such an arrangement.

Other single-duct designs are also used.

16.10 Double-duct systems

The double-duct system is later in concept than the perimeter induction, installations dating from 1945, or thereabouts. Several varieties of design exist but all are based on the same principle. Two airstreams, one cold and one hot, are distributed in separate ducts. In the conditioned space, one or more mixing boxes are positioned, either at high level above a sub-ceiling or, in the manner of induction units, below the window cills. Connexions from the two ducts are made to the boxes. Each box or group of boxes then mixes the two airstreams in varying proportions at the dictate of a room thermostat. The mixing boxes execute the additional functions of pressure reduction and noise attenuation. It has also been found necessary to add a constant-volume regulator, in order to obtain stable operation for the system as a whole.

16.11 Calculating the cold duct air quantity

The double-duct installation is an all-air system, and so the entire design cooling load must be offset by the air delivered from the mixing boxes. The calculation of the cold air quantity is therefore performed in an entirely conventional manner, but allowing 3 to 5 degrees of temperature rise to cover duct heat gains and fan horsepower. When any box is dealing with its design cooling load it receives its maximum amount of cold air. No hot air is used at all. The valve in the box which admits air from the cold duct is fully open, and the one governing the entry of hot air is fully closed (*see* Fig. 16.15).

It is desirable that the rate of air movement in the conditioned space should remain substantially constant and, also, that the amount of cold air used by one box should not affect that handled by another. To this end, self-acting, factory-set, constant-volume regulators are mandatory on all double-duct mixing boxes. Such regulators commonly have a throttling range of $\pm$ 5 per cent.

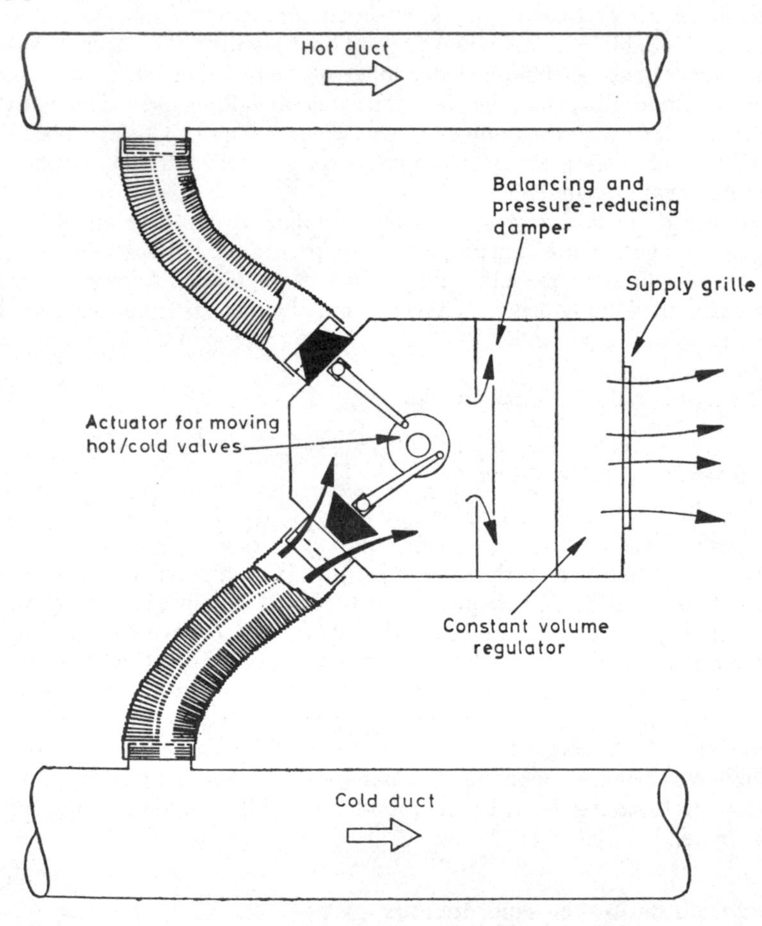

Fig. 16.15

EXAMPLE 16.4. If the room mentioned in Example 16.1 is to be air conditioned by means of a double-duct system, calculate (*a*) the cold air quantity, (*b*) the relative humidity maintained in the room and (*c*) the cooling load. Assume that the latent heat gains are 130 W, the outside air state is 28°C dry-bulb with 55·36 kJ/kg, the state of the air leaving the main cooler coil is 11°C, dry-bulb with 7·539 g/kg and 30·07 kJ/kg, the temperature rise due to fan power etc. is 5 degrees and that the amount of fresh air is 0·05 m³/s at the state delivered to the room.

Answer

(*a*) The total sensible heat gain is 2316 W.

$$\text{Cold air quantity} = \frac{2{\cdot}316}{(21{\cdot}5-16)} \times \frac{(273+16)}{358}$$

$$= 0{\cdot}340 \text{ m}^3/\text{s}.$$

(*b*)

$$\text{Moisture pick-up} = \frac{0{\cdot}130}{0{\cdot}340} \times \frac{(273+16)}{856}$$

$$= 0{\cdot}129 \text{ g/kg}.$$

Hence,

$$g_r = 7{\cdot}539 + 0{\cdot}129$$

$$= 7{\cdot}668.$$

At 21·5°C dry-bulb, this gives a relative humidity of just over 47 per cent.

(*c*) Since the cold air delivered by the box is a mixture of 0·290 m³/s which has been recirculated, and 0·05 m³/s from outside, the mixture state of the air entering the cooler coil in the air handling plant must be calculated.

At 21·5°C dry-bulb and 7·668 g/kg the enthalpy is 41·11 kJ/kg. Hence the enthalpy of the air entering the cooler coil is, approximately,

$$h_m = \frac{0{\cdot}050}{0{\cdot}340} \times 55{\cdot}36 + \frac{0{\cdot}290}{0{\cdot}340} \times 41{\cdot}11$$

$$= 8{\cdot}14 + 35{\cdot}06$$

$$= 43{\cdot}20 \text{ kJ/kg}.$$

At 16°C dry-bulb and 7·539, the state at which the cold air enters the room from the mixing box, the humid volume is 0·8287 m³/kg.

$$\text{Cooling load} = \frac{0{\cdot}340}{0{\cdot}8287} \times (43{\cdot}20 - 30{\cdot}07)$$

$$= 5{\cdot}39 \text{ kW}.$$

16.12 Calculating the hot duct quantity and temperature

The temperature necessary in the hot duct depends on the quantity of air handled, in the following way. The room requiring the largest quantity of hot air is that which has the smallest value of the ratio of supply air quantity to transmission heat loss (sometimes termed the A/T ratio). For this room, the quantity of hot air delivered under winter design conditions is the same as the quantity of cold air delivered under summer design conditions. This means that the full movement of the valves in the hot and cold connexions on the mixing box is used for this room, and the best control achieved. Elsewhere, less than the full movement is used for the hot valve, and the control is a little poorer. The remarks in section 13.12 on water valves are relevant. Other rooms then, which have larger values of A/T, will require smaller quantities of hot air (at the same temperature) in winter. So the smallest value of A/T determines the hot duct temperature.

EXAMPLE 16.5. Two rooms, *X* and *Y*, are to be air conditioned by means of a double-duct system. Each room has a summer design sensible heat gain of 2316 W, but whereas the heat loss from *X* in winter is 749 W, it is only 500 W from room *Y*. Making use, where necessary, of the information given in Examples 16.1 to 16.4, inclusive, determine

(*a*) the fan duty,
(*b*) the temperature of the air in the hot duct,
(*c*) the quantities of hot air to be supplied to rooms *X* and *Y*.

Answer

(*a*) In Example 16.4 it was determined that the cold air quantity needed to deal with a sensible heat gain of 2316 W was 0·340 m³/s at 16°C. Since the sensible heat gains are identical in rooms *X* and *Y*, this amount must be supplied to each room, and so the fan duty is the air quantity at 11°C which corresponds to 0·680 m³/s at 16°C.

$$\text{Fan duty} = 0{\cdot}680 \times \frac{(273+11)}{(273+16)}$$
$$= 0{\cdot}668 \text{ m}^3/\text{s at } 11°\text{C}.$$

For most practical purposes, where constant-volume regulators are fitted on each mixing box, the fan duty can be regarded simply as the sum of the maximum cold air quantities to the rooms.

(*b*)
$$\text{For room } X: \text{A/T} = \frac{340}{749} = 0{\cdot}454$$

$$\text{For room } Y: \text{A/T} = \frac{340}{500} = 0{\cdot}680$$

Room *X* has the smaller value of A/T and so this room receives 0·340 m³/s from the hot duct in winter, and its sensible heat loss is used to determine the hot duct temperature, t_h:

$$t_h - t_r = \frac{0{\cdot}749}{0{\cdot}340} \times \frac{(273+23)}{358}$$
$$= 1{\cdot}8°\text{C}$$
$$t_h = 23{\cdot}3°\text{C}.$$

(See example 6.10.)

(*c*) Room *Y* also receives 0·340 m³/s in winter, but if this were delivered at 23·3°C, a room temperature of more than 21·5°C would be maintained, the sensible heat loss being only 500 W. Assuming that the cold duct temperature is kept constant at 16°C, a heat and mass balance must be struck to establish the quantity of air required from the hot duct. Denoting the air mass flow rates by *M*, with the appropriate subscripts, we can write

$$\frac{M_h \times 23{\cdot}3 + M_c \times 16}{M_h + M_c} = t_s$$

The value of t_s, the temperature of the air supplied from the mixing box to the room, can be easily determined:

$$t_s = t_r + \frac{0{\cdot}500}{0{\cdot}340} \times \frac{(273+23)}{358}$$

$$= 22{\cdot}7^\circ\text{C}.$$

Assuming that the relative humidity in the room is at about 47 per cent, even in winter, then air will be delivered from the mixing boxes at a humid volume of about 0·83 m³/kg.

Approximately, then

$$M_h + M_c = \frac{0{\cdot}340}{0{\cdot}830} = 0{\cdot}410 \text{ kg/s}.$$

Substituting for M_c,

$$M_h \times 23{\cdot}3 + (0{\cdot}41 - M_h) \times 16 = 22{\cdot}7 \times 0{\cdot}41$$

$$M_h = 0{\cdot}376 \text{ kg/s}.$$

The volume of hot air handled is therefore about

$$0{\cdot}376 \times 0{\cdot}83 = 0{\cdot}312 \text{ m}^3\text{/s at } 23{\cdot}3^\circ\text{C}.$$

Thus, 0·340 m³/s of air is delivered from the hot duct at 23·3°C for room *X*, but only 0·312 m³/s at 23·3°C is supplied from the hot duct to the mixing-box serving room *Y*.

16.13 Types of double-duct design

Many diverse types of system have been designed, most with success but some without. Only two types of design are considered here: both have been found to work well, although variations on these themes are possible, some of them yielding improvements of one kind or another.

The double-duct system, like more conventional systems, can be designed so as not to use wasteful re-heat but to take advantage of diversity in the heat gain and so economise in installed refrigeration tonnage. On the other hand, it can be devised to run as a dew-point plant. This entails cancelling refrigeration with re-heat by using air from the hot duct which has previously passed over the main cooler coil. Figure 16.16(*a*) illustrates the first system and 16.16(*b*) the second.

A double-duct dew-point plant is illustrated which, for simplicity, uses 100 per cent of fresh air. In summer, air is filtered, chilled and dehumidified by being passed over a sprayed cooler coil and delivered to the ducting system

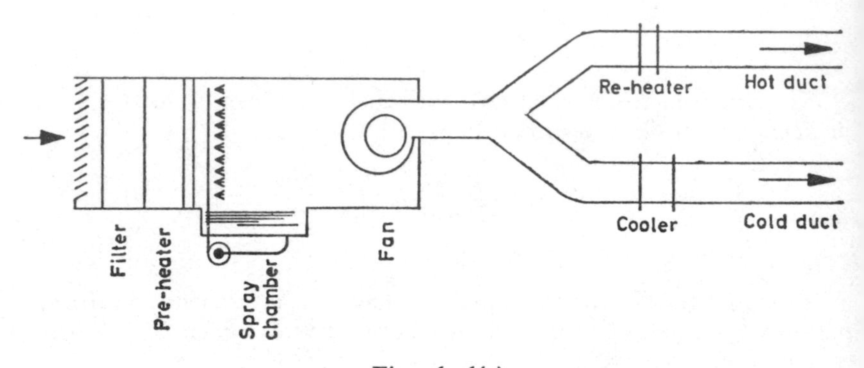

Fig. 16.16(*a*)

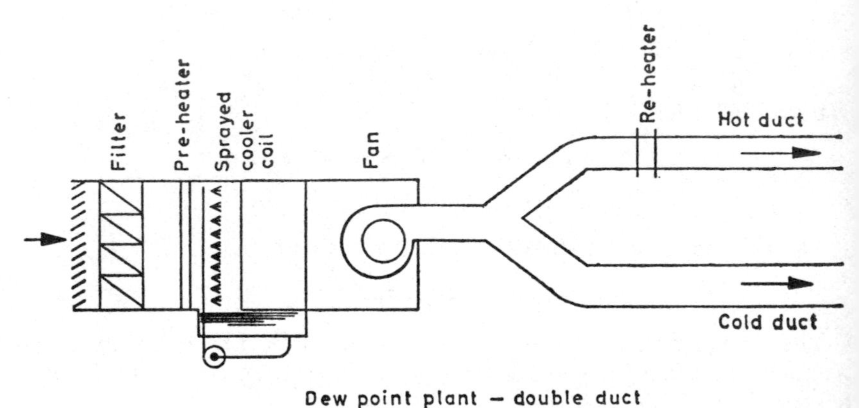

Fig. 16.16(*b*)

by a fan. The air is then at state *C*, as required for the cold duct. The state *H*, needed for the hot duct, is achieved by passing the hot duct quantity through a re-heater battery. The psychrometry of this is shown in Fig. 16.17(*a*). It can be seen that the correct supply air state *S* can be obtained by mixing air at state *C* with air at state *H*. In winter, air is passed over a pre-heater and adiabatically humidified at the sprayed cooler coil. The necessary state *C* is obtained after passing through the fan, and state *H*' for the hot duct, is obtained by re-heat. It can be seen, by examining the

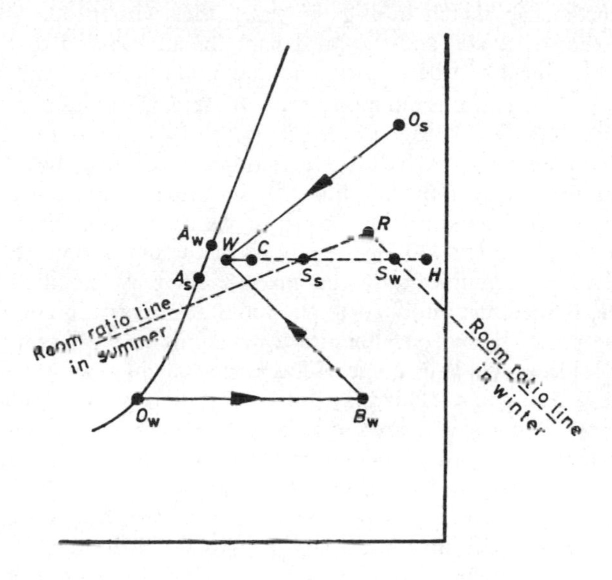

Fig. 16.17(*a*) Dew-point plant—double duct

Subscript s denotes summer operation
Subscript w denotes winter operation

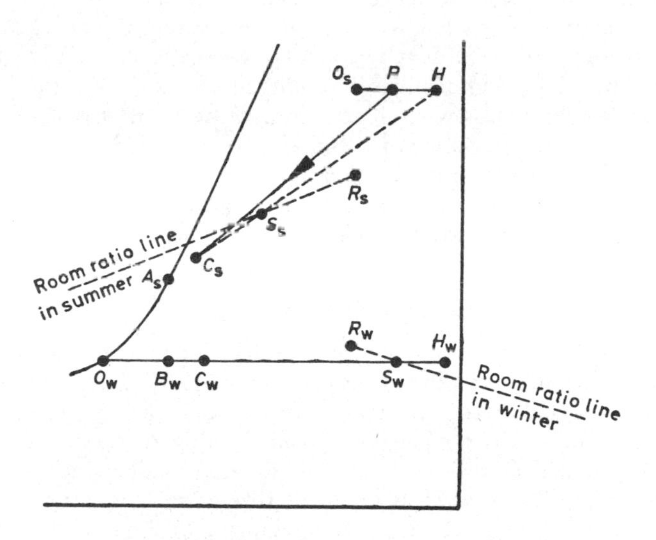

Fig. 16.17(*b*)

Subscript s denotes summer operation
Subscript w denotes winter operation

psychrometric sketch in Fig. 16.17(*a*), that with proper sequence control over the cooler coil and the pre-heater the air handled by both the hot and the cold ducts will be at the same dew point summer and winter.

The other and more popular type of system has the psychrometry shown in Fig. 16.17(*b*). If the sprays of Fig. 16.16(*a*) are omitted, the pre-heater is used merely to give frost protection to the cooler coil in the cold duct and the heater battery in the hot duct. Under these circumstances, in summer, air is filtered and delivered by the fan to the two ducts, where it is heated and cooled to states *H* and *C*. The process of cooling to state *C* for the cold duct involves dehumidification; the mixture state *S*, handled by the mixing boxes, is therefore supplied to the room at a moisture content which varies as the room thermostat alters its instructions to the mixing box. A variety of humidities will thus be provided in rooms which are suffering a variety of sensible heat gains. This may not be objectionable in comfort conditioning. In winter, if the sprays are not used, air is pre-heated from state *O* to state *B*, for frost protection, is fortuitously re-heated by fan horsepower to *C*, the cold duct state, and intentionally re-heated to state *H* for the hot duct. Both *C* and *H* have the same dew point and, therefore, so has *S*. This means that there is a considerable seasonal variation in the relative humidity in the conditioned rooms. Again, this may not be very objectionable in comfort conditioning but the situation can be improved by using the sprays in winter. The psychrometric process is then as shown in Fig. 16.17(*a*) for winter. It would probably be very undesirable for the sprays to remain on in the summer, hence it might be arranged for the spray pump to be switched off when the outside wet-bulb rose above a certain value.

It may be wondered why it is necessary that the air in the hot duct be heated in summer; this is, of course, necessary to meet a reduction in sensible load by mixing warmer air with cooler air. Such could arise in a building that had two long facades, each containing a lot of glazing, and facing east and west. In the morning, in summer, the rooms on the east side would be receiving air solely from the cold duct, while those on the west side received a mixture of cold air and hot air, their sensible gains not being at a maximum because of the absence of sunlight.

16.14 Terminal heat recovery units

Self-contained, water-cooled, air conditioning units operate in a conventional manner to deal with total heat gains in the range 1·8 kW to 3·5 kW but may also cope with heat losses for maxima from 2·7 kW to 5·3 kW by working as heat pumps. The changeover from heating to cooling is effected in the refrigerant circuit, the roles of the evaporator and condensor being mutually reversed. Clean water at a temperature of 27°C is circulated to the units in summer, falling to about 24°C in winter, providing a sink for the rejection of heat from the condenser when units operate in a conventional manner but a source of heat for the evaporator when acting as heat pumps. If most units in a conditioned building are rejecting heat, the water is circulated through the secondary side of a plate heat-exchanger (see Fig. 16.18), the primary side of which handles dirty water from a cooling tower. At other times, when most units are taking heat from the circulated water, the thermal deficiency is rectified by passing the water through the secondary side of another heat-exchanger, which receives hot water from the boilers on its primary side. During intermediate seasons, when units on one face of a building may be dealing with a heat loss while those on the opposite face are offsetting a heat

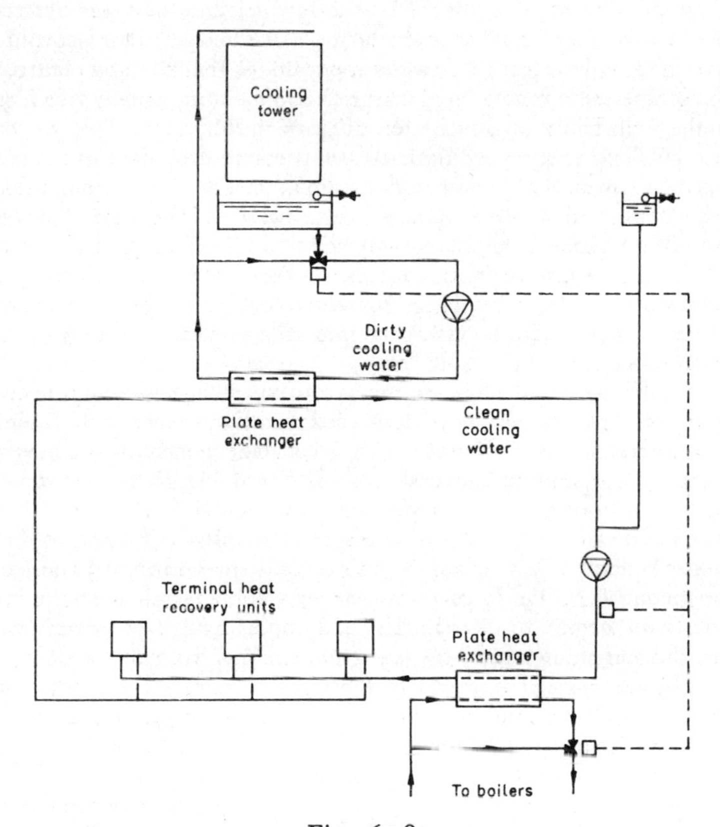

Fig. 16.18

gain, a transfer of heat is effected between the faces, through the cooling water, without the cooling tower or boilers playing any part. The system thus conserves heat.

The overall coefficient of performance is not as high as that of a larger, conventional refrigeration plant and consequently about one third to one quarter of the energy provided when heat-pumping is electrical, being the power drawn from the mains to drive the compressors in the units. To mitigate the cost of this some units offer auxiliary heating coils which may be fed with LPHW in the winter.

16.15 Variable air volume systems

As an alternative to delivering warmer air to meet a reduced sensible heat gain it is logical to deliver less air, at the original supply temperature, avoiding wasteful re-heat and saving running cost by only using the minimum fan power. Variable air volume systems of a practical nature stem from about 1960 and have gained popularity in recent years. Two common designs used are: (i) stabilising static pressure in the low-velocity approach duct (Fig. 16.19(*a*)) by means of a pressure-reducing valve, whilst motorised diffusers regulate the supply of cold air to the room under thermostatic control and, (ii), combining these functions in a single terminal unit, usually also forming part of the air distribution duct system (Fig. 16.19(*b*)).

When air flows along a ceiling frictional pressure drop next to the surface ensures the room air will press the airstream against it. Sometimes called the ' Coanda ' effect, this phenomenon is the basis of many air distribution devices. With variable volume supply the Coanda effect decays as the flow reduces (since the surface frictional loss is a function of the velocity of air-flow) and what is called ' dumping ' sometimes occurs—objectionable cold air falling downwards. This is one of the limitations on the turn-down possible with such systems.

The air diffusers and slots in Fig. 16.19 share a common property of variable geometry: as the airflow is reduced the area available also diminishes and the airstream tends to leave with a constant velocity. Such variable geometry air distribution devices minimise dumping to some extent and permit a turn-down to about 25 per cent of design airflow. Other devices, which modulate the airflow before it reaches the diffuser (Fig. 16.19(*c*)) tend to be less adaptable to load changes and may only turn-down to 50 per cent of full design airflow. One further variable air volume terminal on the market attempts (with success) to avoid the risk of dumping by delivering air in pulses, each at the full design rate, the variation in total volume handled being achieved by alternating the length of the pulses, under thermostatic control (*see* Fig. 16.19(*d*)). When air is not delivered to the conditioned room it is re-cycled to the air handling plant through the ceiling void and a ductwork system. No economy in fan power is achieved.

Another limitation on turn-down is that of inadequate air movement in the conditioned space and it is reckoned that about 3 litres/s m^2 of floor area is a desirable minimum. Otherwise, the need to provide sufficient fresh air may also impose a restriction but this can be dealt with by varying the mixing proportions of fresh and recirculated air at the air handling plant. Variable air volume systems are appropriate within the range of supply air quantities from 15 to 3 litres/s m^2.

Throttling the airflow from individual room terminals will cause static pressure variations in the associated ductwork system. These should be kept to a minimum by the use of motorised, variable inlet guide vanes on the supply fan (plus similar vanes, working in unison, on the extract fan). Pressure tappings should be located in a position on the index duct run where it is

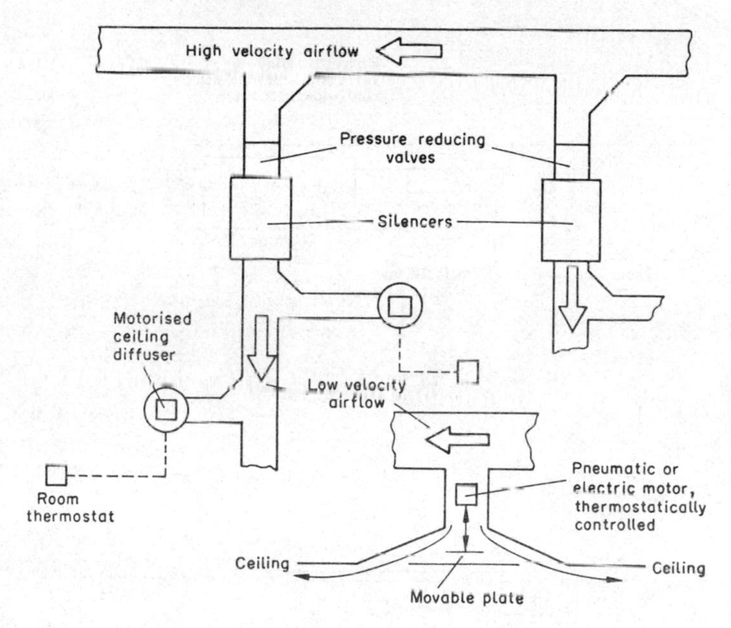

Fig. 16.19(*a*)

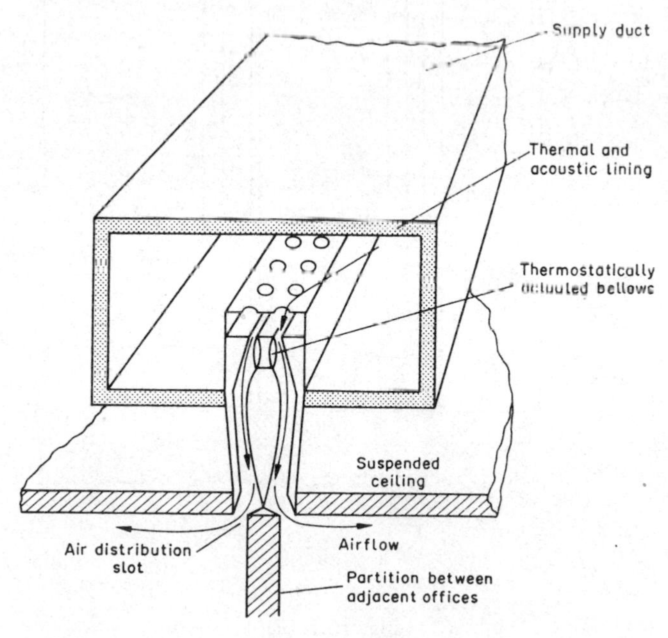

Fig. 16.19(*b*)

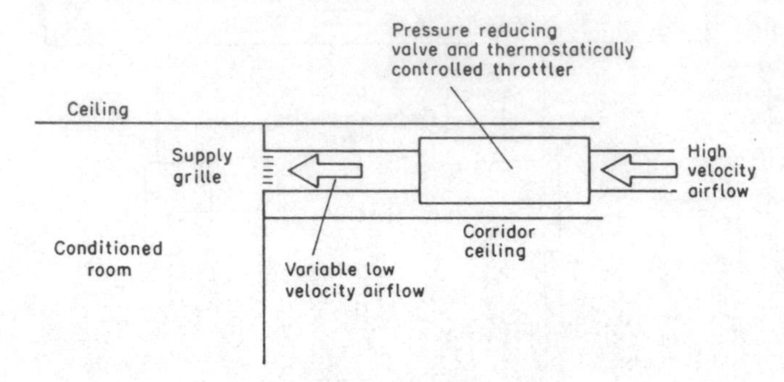

Fig. 16.19(*c*)

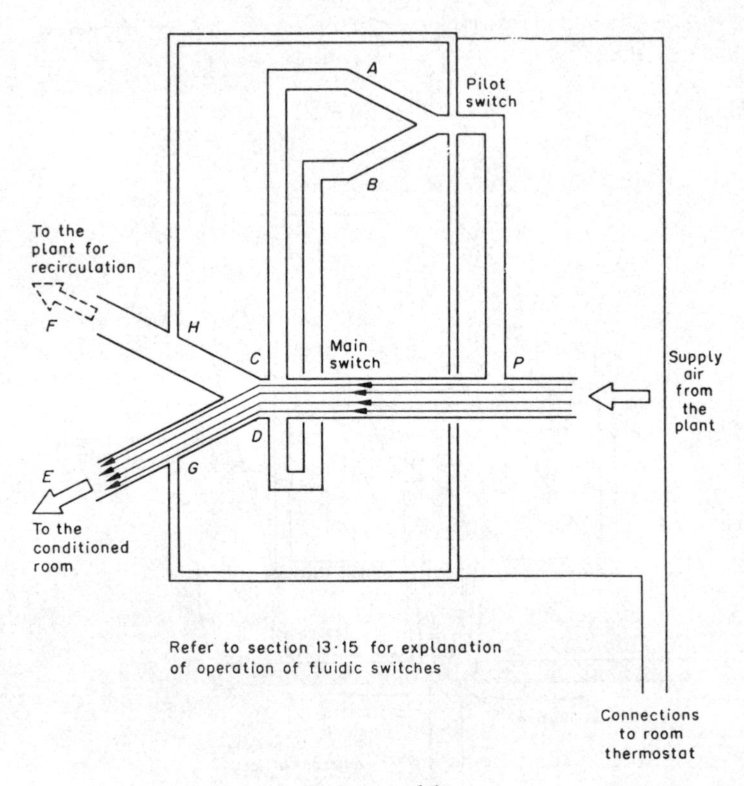

Fig. 16.19(*d*)

calculated that the maximum pressure variation will occur—usually about two-thirds to three-quarters of the distance from the fan outlet. These tappings then sense changes in the duct pressure as the terminals throttle and reduce fan capacity through the inlet guide vanes. A limitation of the amount of turn-down possible for the whole system is the risk of surge (see section 12.9). Choosing a slightly under-sized fan of an appropriate type helps to reduce the risk but, depending on the fan choice, system design and pressures involved, it is unlikely that a whole system can be turned down effectively to much below 50 per cent.

Intrinsically the variable volume system can only deal with reductions in sensible heat gain. It cannot deal with a heat loss unless it is modified in one of three ways by: (*a*) adding a perimeter system of compensated LPHW heating, (*b*) providing terminal re-heaters for each module, or (*c*) adding a hot duct which only provides hot air to its variable volume mixing box when the cold quantity has been reduced to a minimum and heating is called for. The second method is wasteful, no account being taken of diversity, and should only by used to a limited extent. The third is expensive and is wasteful of building space. Stratification and air distribution and heating problems may result from both methods at low air volumes and high air temperatures. The first modification is the best solution to the problem of dealing with heat loss and it permits the variable air volume design to be used to deal with the whole of the floor area, not just the core.

16.16 Integrated environmental design

Abbreviated as I.E.D. this is a design aimed at producing a combination of thermal, acoustic, illumination and aesthetic elements which provides a harmonius, satisfying environment for the occupants of a building whilst, at the same time, conserving the energy required for the purpose. In terms of air conditioning, the emphasis is thus on heat recovery techniques, the design of the building being made subordinate to the design of the mechanical services to achieve economical operation.

Bearing in mind that the six main components in the thermal balance for a building are transmission (T), people (P), electric lighting (L), solar radiation (S), ventilation (V) and humidification (H), the building should be designed to achieve a steady-state equation for which

$$T+V+H = L+P \qquad (16.1)$$

The solar component is ignored as unpredictable and probably absent for much of the time in the winter.

On the left-hand side of eqn. (16.1) the outside air temperature, t_0, is the relevant variable and it is usual to aim at achieving a balance for t_0 equal to the winter design value. The internal heat gains ($L+P$) can then offset the steady-state energy demand and the air conditioning designer's problem is to devise a system capable of distributing the gains to those places where they can make good a need.

Figure 16.20 shows a steady-state balance, achieved when $T+V = L+P$ at the winter design value of outside air temperature. If humidification were added the balance would be achieved at a higher temperature, more energy being needed at the winter design temperature than is available from lights and people.

Since, according to the 1970 I.H.V.E. Guide, volume A,

$$V = 0{\cdot}33nv(t_r - t_0) \tag{16.2}$$

where n is the ventilation rate in air changes per hour, v is the volume of the room, 0·33 is the ventilation allowance factor in W/m^3 °C and t_r is the room temperature, eqn. (16.1) can be modified:

$$t_0 = t_r - \frac{L+P}{\Sigma AU + 0{\cdot}33nv} \tag{16.3}$$

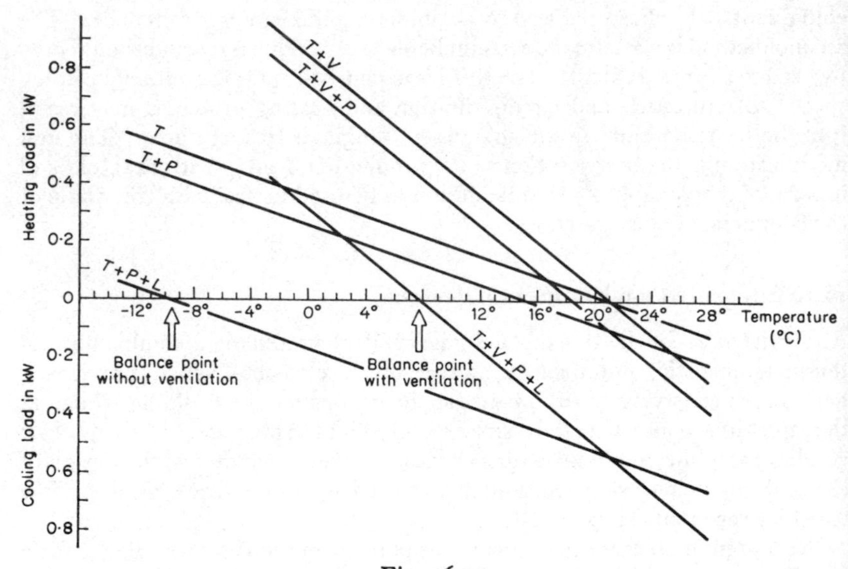

Fig. 16.20

This shows that the variable open to manipulation is ΣAU, where A is an element of area and U its thermal transmittance coefficient, referred to the building envelope. An examination of the influence of ΣAU makes it clear that, to obtain steady-state balance points at the winter design value, window areas must be reduced to less than 20 per cent of the facade, U-values must be at 0·6 W/m^2 °C or less and the ratio of external building surface to usable floor area must be minimised. The outcome is a building of near cubical shape having small windows and low U-values. This has benefits for the summer air conditioning loads, performance and running costs; the efficient insulating value of the envelope filters the effects of the climate and makes the air conditioning system more able to cope with the smaller load variations which result.

Equation (16.1) conceals the facts that the energy produced by the lights and people is in low-grade form and is not liberated where it is wanted. The air conditioning system design deals with these matters. The sensible and latent heat gains removed by the system are released at the condenser, often for ultimate rejection at a cooling tower. This waste heat can be recovered by using a second condenser or bundle of tubes which can provide clean, warm water for the heater batteries of the air conditioning system. Because the water is only at about 40°C, at the most, such heat recovery batteries have to have a large surface area and thus usually comprise several rows. Figure 16.21 illustrates a typical design. A four-pipe design is shown and this is safest—three-pipe versions should never be attempted, they are fraught with danger.

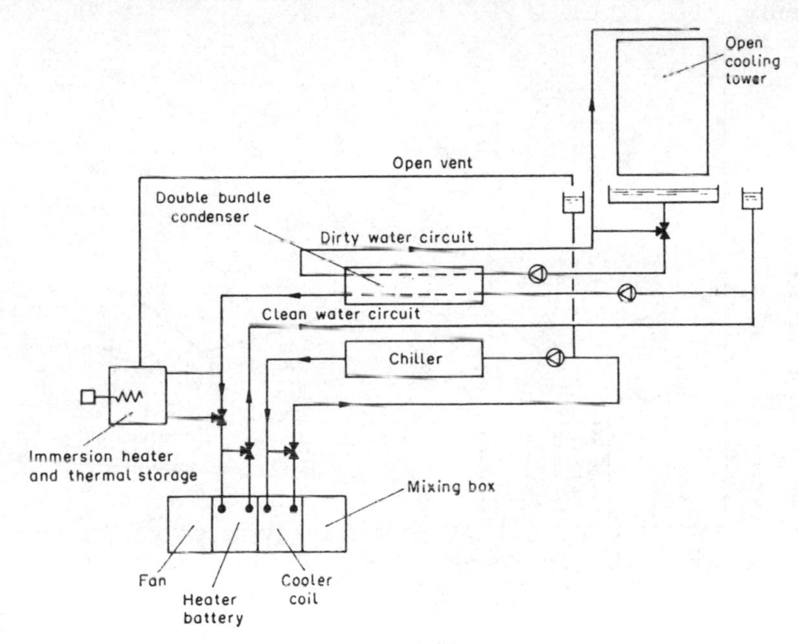

Fig. 16.21

Air conditioning systems designed along these lines must operate with fixed, minimum proportions of fresh air throughout the year if the heat from the condensers is to be always available for offsetting heat losses, thus apparently obviating the need for a boiler and flue.

There is an economic case for running heating systems intermittently, closing them down during periods of non-occupancy; the mass of the building cools at night and the system must make good this loss on start-up, as well as warming the air in the building, if comfort is to be preserved. Figure 16.22 shows that a light-weight building will save more in running cost than a heavy one because little heat is stored in its structure and its mean internal temperature is less over the winter.

Because the condenser cooling water has a relatively low temperature and little heating potential, an auxiliary source of heating, not necessarily electrical, must be provided with an I.E.D. system intended for intermittent operation, to deal with the boost heating load in the mornings in winter. Any source of heating can be used but if it is electrical consideration must be given to the tariff under which it is supplied. During the daytime, with heat-recovery methods, advantage can be taken of the thermodynamics of refrigeration which give about five times as much heat at the condenser as is drawn in electrical energy by the compressor; the consequence of this is that, effectively, the cost of electricity for heat-pumping in this way is only about one fifth of the normal tariff. The cost of electricity is then on a par with the cost of oil or gas for heating.

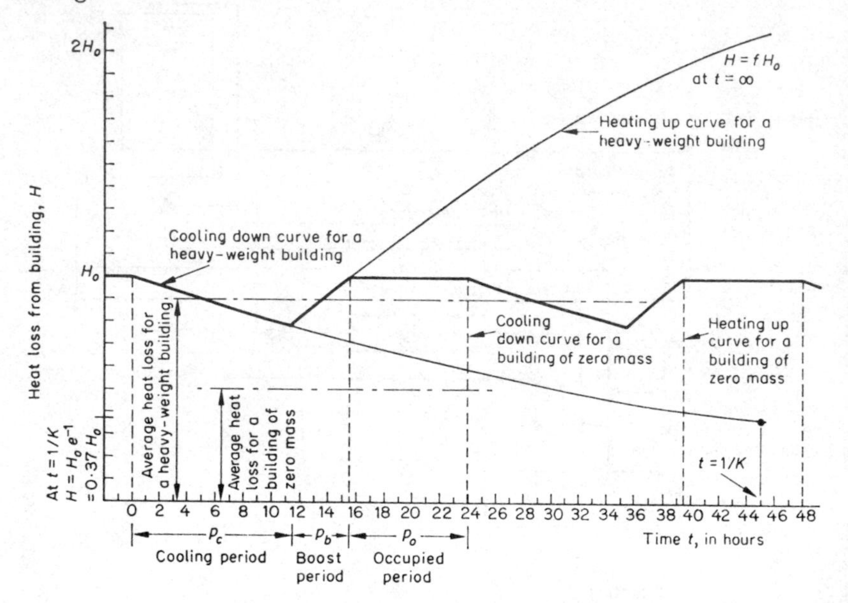

Fig. 16.22

At night time, however, the air conditioning system would be off and there would be no heat available at the condenser. One method of obtaining the high temperature heat needed for morning boost is to store hot water in electrically heated vessels. The height of the building dictates the temperature of storage and the amount stored depends on the mass of the building structure, its heat loss, the boost period and the occupied period, as is seen later. The electrical input to heat the stored water can either be at times when an off-peak tariff (at about half the normal tariff) may be used, or at any time during the day, provided that the maximum demand is not increased thereby. (The cost of electrical units is about half the total cost with a maximum demand tariff and so if the load factor can be improved in this way the cost of the extra electricity is about half the normal rate.)

If the entire mass of a building is assumed to have no thermal resistance and if heat loss from its external surfaces follows Newton's law of cooling, then the equation for the cooling curve in Fig. 16.22 is

$$H = H_0 e^{-Kt} \tag{16.4}$$

The area under the curve represents the total amount of heat in the building mass above a datum temperature, usually the outside design temperature, and can be found by integrating between $t = 0$ and $t = \infty$, yielding the answer H_0/K. The symbol $1/K$ is called the 'time constant' for the building and can thus be easily calculated as the heat stored in the building mass divided by the design rate of heat loss. In evaluating the heat stored in the building material it is suggested that, for the floor slab in contact with the earth, the thermal capacity be taken above a datum of 10°C, rather than above −1°C, the winter design value in the United Kingdom used for the rest of the building.

Harrison has shown that the concept of Newtonian cooling is probably not significantly far from the truth.

If, with intermittant heating, H_0 is the design heat loss, f is the boost factor, p_0 the period of occupancy, p_b the boost period and p_L the running period at full capacity (H_0) to meet the 24 hour needs of the building, then the following relationship holds:

$$p_L = p_0 + \frac{1}{K}[1 - e^{K(p_0 - 24)}] \tag{16.5}$$

This holds true as p_b tends to zero and hence it defines the optimum performance of an intermittent operation and can be used to assess the greatest saving possible.

EXAMPLE 16.6 A building has a time constant of 45 hours and is occupied between 9.00 a.m. and 5.30 p.m. What is the greatest possible saving by using intermittent heating?

Answer

$1/K = 45$, $p_0 = 8{\cdot}5$ hours, and so the period that the plant must run at the design heat capacity is given by eqn. (16.5)

$$\begin{aligned} p_L &= 8{\cdot}5 + 45(1 - e^{-15{\cdot}5/45}) \\ &= 8{\cdot}5 + 13{\cdot}1 \\ &= 21{\cdot}6 \text{ hours.} \end{aligned}$$

$$\text{The saving possible} = \frac{24 - 21{\cdot}6}{24} \times 100 = 10\%$$

The building cools after shut-down before the morning boost begins and the boost factor, f, must be large enough when applied to the design heating capacity, H_0, over the boost period, p_b, to provide for this cooling period, p_c. It can be shown that

$$f = e^{-Kp_c} + \frac{(1 - e^{-Kp_c})}{(1 - e^{-Kp_b})} \tag{16.6}$$

EXAMPLE 16.7 Calculate the boost factor required for the last example, if the boost period is to be 4 hours.

Answer

$p_0 = 8{\cdot}5$ hours, $p_b = 4$ hours, therefore $p_c = 11{\cdot}5$ hours and by eqn. (16.6)

$$f = e^{-11\cdot5/45} + \frac{(1-e^{-11\cdot5/45})}{(1-e^{-4/45})}$$
$$= 0{\cdot}775 + 0{\cdot}225/0{\cdot}0851$$
$$= 3{\cdot}42$$

The larger the boost factor the smaller will be the boost period and the greater the economy in operation, reaching a saving of 10 per cent over continuous operation, in the limit when p_b equals zero, as Example 16.6 shows. The boost factor attainable depends on the performance of the heat transfer surfaces used for heating and on the initial temperatures of the heating medium and the air being heated, compared with those temperatures normally used in the daytime.

The thermal storage provided must be enough to provide for the heating of the building over 24 hours.

EXAMPLE 16.8 For the building considered in the last two examples, determine the necessary storage if the design heat loss is 200 kW.

Answer

From Example 16.6 it is found that the input at the full rate is for 21·6 hours to meet the 24 hour heating requirements. But the condensers are heating the building for 8·5 hours therefore 13·1 hours input is needed, or 26 200 kWh. This is the absolute minimum amount of storage.

For a boost period of more than zero the following applies instead of eqn. (16.5).

$$p_L = p_0 + \left[e^{-Kp_c} + \frac{(1-e^{-Kp_c})}{(1-e^{-Kp_b})}\right]p_b \qquad (16.7)$$

EXAMPLE 16.9 For the building already considered, determine the necessary thermal storage if the boost period is 4 hours.

Answer

$p_0 = 8{\cdot}5$ hours, $p_b = 4$ hours, $p_c = 11{\cdot}5$ hours. Hence by eqn. (16.7)

$$p_L = 8{\cdot}5 + \left[e^{-11\cdot5/45} + \frac{(1-e^{-11\cdot5/45})}{(1-e^{-4/45})}\right]4$$
$$= 8{\cdot}5 + \left[0{\cdot}775 + \frac{0{\cdot}2250}{0{\cdot}0851}\right]4$$
$$= 8{\cdot}5 + 13{\cdot}68$$
$$= 22{\cdot}18 \text{ hours.}$$

The necessary storage is therefore $(22{\cdot}18 - 8{\cdot}5) \times 200$ or 27 360 kWh.

A prudent safety margin is recommended on top of this to cover possible losses and any doubt in the attainable boost factor or in the calculation of the time constant.

BIBLIOGRAPHY

1. *Conduit Inductionaire System Design Manual.* Carlyle Air Conditioning & Refrigeration Ltd., London, S.W.1.
2. C. M. Wilson. Handbook on high velocity air distribution design, *Heating, Piping and Air Conditioning*, November 1954.
3. *Air Conditioning System Design Manual, Part 2: Air Distribution.* Carlyle Air Conditioning & Refrigeration Ltd., London, S.W.1.
4. N. S. Shataloff. Comparative costs of dual duct air conditioning systems, *Air Conditioning, Heating and Ventilating*, June, 1960.
5. N. S. Shataloff. Elements of dual duct design and performance, *Trans. A.S.H.R.A.E.*, 1956.
6. D. Rickelton. Psychrometrics of dual duct systems, I and II, *Heating, Piping and Air Conditioning*, March and April, 1964.
7. T. Mariner. Plenum engineering—a key to success with ventilating ceilings, *Heating, Piping and Air Conditioning*, October, 1962.
8. J. Rydberg. *Introduction of Air to Perforated Ceilings.* H. V. R. A. Translation No. 45, June 1962.
9. H. C. Jamieson. Meteorological data and design temperatures. *J. Instn Heat. Vent. Engrs*, 1954, **22**, 465–495.
10. E. Harrison. The intermittent heating of buildings. *J. Instn Heat. Vent. Engrs*, 1956, **24**, 145–187.
11. E. Harrison. The intermittent heating of buildings by off-peak electricity supplies. *J. Instn Heat. Vent. Engrs*, 1959, **27**, 189–208.
12. W. P. Jones. (1973). Integrated environmental design. *Steam Heat. Eng.* **42**, 496, 6-10; **42**, 497, 14-18, **42**, 498, 10-15; **42**, 499, 58-63; **42**, 500, 18-23.

17

Ventilation and a Decay Equation

17.1 The need for ventilation

The minimum amount of fresh air required for breathing purposes is really quite small; about 0·2 litres/s person (*see* section 4.7). For comfort conditioning, however, it is insufficient to supply this small amount of fresh air; two other factors enter into consideration and so one can say that enough fresh air must be delivered to achieve the following:

1. The satisfaction of the oxygen needs of the occupants.
2. The dilution of the odours present to a socially acceptable level.
3. The dilution of the concentration of carbon dioxide to a satisfactorily low level.

The dilution of odours by the introduction of odourless fresh air is one of the two satisfactory methods of dealing with smells. (The other is the use of an activated carbon filter.) The rate at which air must be introduced will naturally depend on the rate of odour production; this, in turn, depends on the use to which the space is put.

Carbon dioxide is present in fresh air to the extent of 0·03 per cent (*see* section 2.2), whereas concentrations up to 0·1 per cent (section 4.7) are quite acceptable in conditioned or ventilated spaces occupied by human beings. It follows that the introduction of fresh air is a cogent means of reducing the level of concentration in a room. The level can never be brought down to less than 0·03 per cent, of course, and the effectiveness of the chosen air change rate as a dilution agent is complicated by the fact that the people present in the room continually produce carbon dioxide in their normal course of respiration. In fact, the rate of production is $4{\cdot}72 \times 10^{-3}$ litres/s.

The rate of supply of fresh air necessary to meet the three requirements mentioned above depends on the density of occupation in the space and on the usage of it. The rate is expressed in three different ways, as an air change rate (*n* changes of the cubical content of the room in one hour), as a quantity supplied in litres/s person, or as a quantity supplied in litres/s m^2 of floor area. It may be uneconomical to quote air change rates for rooms with large floor-to-ceiling heights, and for this reason the other two methods are used. The use of litres/s person is preferred when the density of occupation is high, as in a ballroom or a banqueting space (as dense as 1 m^2/person).

On the other hand, the use of litres/s m^2 of floor area is satisfactory for offices where the typical densities of occupation are well established, varying from 6 m^2/person to 12 m^2/person, with 10 m^2/person as an average.

The following values for fresh air supply rates are satisfactory:

TABLE 17.1

Application	litres/s person	litres/s m^2
Private dwellings	8–12	—
Board rooms	18–25	—
Cocktail bars	12–18	—
Department stores	5–8	—
Factories	—	0·8
Garages	—	8·0
Operating theatres	—	16
Hospital wards	8–12	—
General offices	5–8	1·3 to
Private offices	8–12	2·0
Restaurants	12–18	—
Theatres	5–8	—

Lavatories, especially those without external openable windows, are a special case (as often are some others of those mentioned above). The recommendation is that they be provided with extract ventilation by mechanical means to give 15 air changes an hour, or 80 litres/s lavatory pan, or 16 litres/s m^2, whichever is the greater. The supply of fresh air may also be necessary, or even mandatory, by mechanical means.

17.2 The decay equation

Although from the point of view of a purist it might be better to derive the decay equation in general terms, it is certainly easier to understand a derivation phrased with reference to a particular application. Accordingly, the equation is derived in terms of the rate at which a contaminant decays in a ventilated room under the influence of a diluting influx of fresh air. The contaminant chosen is carbon dioxide.

Consider a room, as shown in Fig. 17.1, having a volume V m^3, in which the concentration of carbon dioxide is c, expressed as a fraction (e.g. parts of carbon-dioxide per million parts of air). Suppose that during time Δt, a small quantity of air, Δq, entirely free of carbon dioxide (for simplicity), enters the room. A similar small quantity, Δq, of contaminated air is forced out of the room. The concentration within the room is therefore reduced by an amount $(\Delta q/V)\,c$. This reduction in concentration can be expressed by Δc, defined as follows:

$$\Delta c = -\frac{\Delta q}{V}c \qquad (17.1)$$

The negative sign is used because the concentration is decreasing.

It follows that the rate of change of concentration is given by $\Delta c/\Delta t$, defined as follows:

$$\frac{\Delta c}{\Delta t} = -\frac{c}{V}\frac{\Delta q}{\Delta t} \tag{17.2}$$

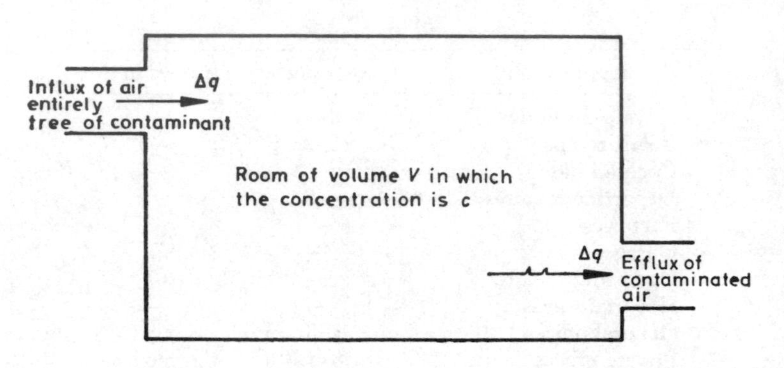

Fig. 17.1

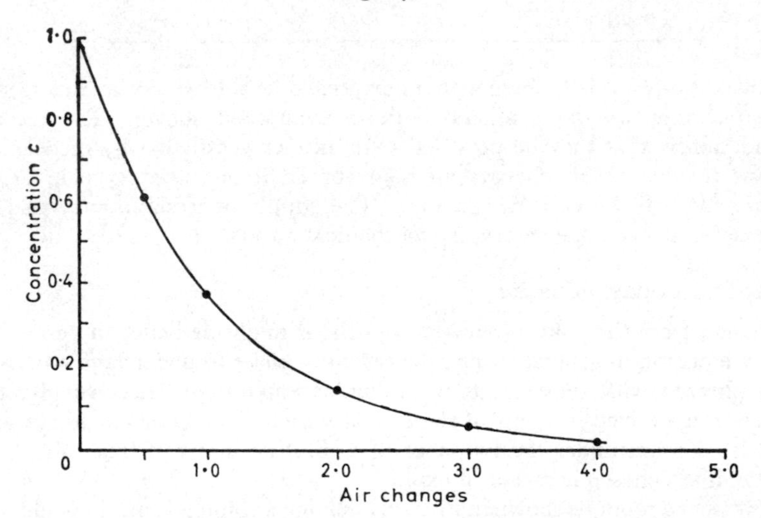

Fig. 17.2

But $\Delta q/\Delta t$ is the rate of influx of ventilating air and is a constant, Q, say. Thus the rate of change of the concentration with respect to time may be written as

$$\frac{\mathrm{d}c}{\mathrm{d}t} = -\frac{cQ}{V} \tag{17.3}$$

The physical problem has now been phrased as a simple differential equation, and a solution to this will be of practical value in determining the answers to real problems.

By integration, the solution to eqn. (17.3) is

$$\int \frac{dc}{c} = -\frac{Q}{V}\int dt$$

$$\log_e c = -\frac{Qt}{V} + \log_e A$$

where $\log_e A$ is a constant of integration.

Hence,

$$\log_e c - \log_e A = -\frac{Qt}{V}$$

and

$$A\,e^{-(Qt/V)} = c$$

The value of the constant A is established by considering the boundary condition $c = c_0$ (the initial concentration in the room) at $t = 0$ (the instant that the ventilation began).

The solution to eqn. (17.3) is therefore

$$c = c_0\,e^{-(Qt/V)} \tag{17.4}$$

If it is observed that in a ventilated room, where V is in m^3, Q is in m^3/s and t is in seconds, Qt/V is the number of air changes, then one can rephrase eqn. (17.4) to

$$c = c_0\,e^{-n} \tag{17.5}$$

where n is the number of times the cubical content of the room is changed.

The graph of eqn. (17.5) is an exponential curve, as shown in Fig. 17.2. It can be seen that concentration of the contaminant decays rapidly if the ventilating air is entirely free of the contaminant. After one air change it is 36·8 per cent of its initial value and after three air changes it is only 5 per cent.

EXAMPLE 17.1. If the air in a room has an initial concentration of 1000 ppm (parts per million) of hydrogen, how many air changes are required to reduce the value of this to 50 ppm?

Answer

From eqn. (17.5) we can write

$$n = \log_e (c_0/c) = \log_e 20 = 3$$

Suppose a more practical case where, with carbon-dioxide as a contaminant, the fresh air used for ventilating purposes also contains some carbon dioxide. Suppose further, that there are people in the room who top up the level of carbon dioxide continually by respiratory activity.

Let c = the concentration of carbon-dioxide in the room at any instant, expressed as parts per million of air,

Q' = the rate of fresh air supply in m^3/s person,

V' = the volume of the room in m^3/person,

t = the time in seconds after the beginning of occupancy and ventilation,

c_a = the concentration of carbon dioxide present in the ventilating air, expressed in parts per million of air,

and V_c = the volume of carbon dioxide produced by breathing on the part of the occupants, in m^3/s person.

The volume of air entering the room in time Δt is $Q'\Delta t$, expressed in m^3/person. Hence, the increase of carbon dioxide in time Δt, due to the contamination of the ventilating air, is $(Q'\Delta t)(c_a/10^6)$, in m^3/person. (This is because even if the air in the room were initially free of the contaminant, the fraction of the entering airstream which is carbon dioxide is $c_a/10^6$ and, therefore, this fraction of the entering volume is an addition to the contamination in the room in m^3.)

The volume of air forced out of the room by the entering airstream is also $Q'\Delta t$ and so, in a similar way, the amount of carbon dioxide leaving the room in m^3/person, is $(Q'\Delta t)(c/10^6)$.

Thus, a balance can be drawn up for the net change of carbon dioxide, in time Δt, expressed in m^3/person:

$$\text{Net change of } CO_2 = V_c\Delta t + (Q'\Delta t c_a/10^6) - (Q'\Delta t c/10^6)$$

In this equation, $V_c\Delta t$ is the volume of CO_2 produce in time Δt by one person breathing, and is expressed in m^3/person.

Since the concentration is the volume of CO_2 divided by the volume of the room, we can write the change in concentration, per person, as

$$\frac{\text{net change of } CO_2 \text{ in } m^3/\text{person}}{\text{volume of the room in } m^3/\text{person}}$$

and this can be defined by

$$\Delta c = \{V_c\Delta t + (Q'\Delta t c_a/10^6) - (Q'\Delta t c/10^6)\}/V'$$

expressed as a fraction.

Hence

$$\frac{\Delta c}{\Delta t} = \{V_c + (Q'c_a/10^6) - (Q'c/10^6)\}/V' \tag{17.6}$$

also expressed as a fractional change in concentration per unit time. Equation (17.6) may be rearranged in a form which is recognisable as amenable to solution:

$$\frac{dc}{dt} + \frac{Q'c}{V'} = \frac{10^6 V_c + Q'c_a}{V'} \tag{17.7}$$

This is the differential equation which expresses the physical problem in mathematical terms, just as was eqn. (17.3) for the first simple case dealt with. Equation (17.7) is obtained from eqn. (17.6) by multiplying throughout by 10^6, thus giving the rate of change of concentration in parts of CO_2 per 10^6 parts of air, in unit time. It may be solved by multiplying throughout by an integrating factor, $e^{(Q't/V')}$.

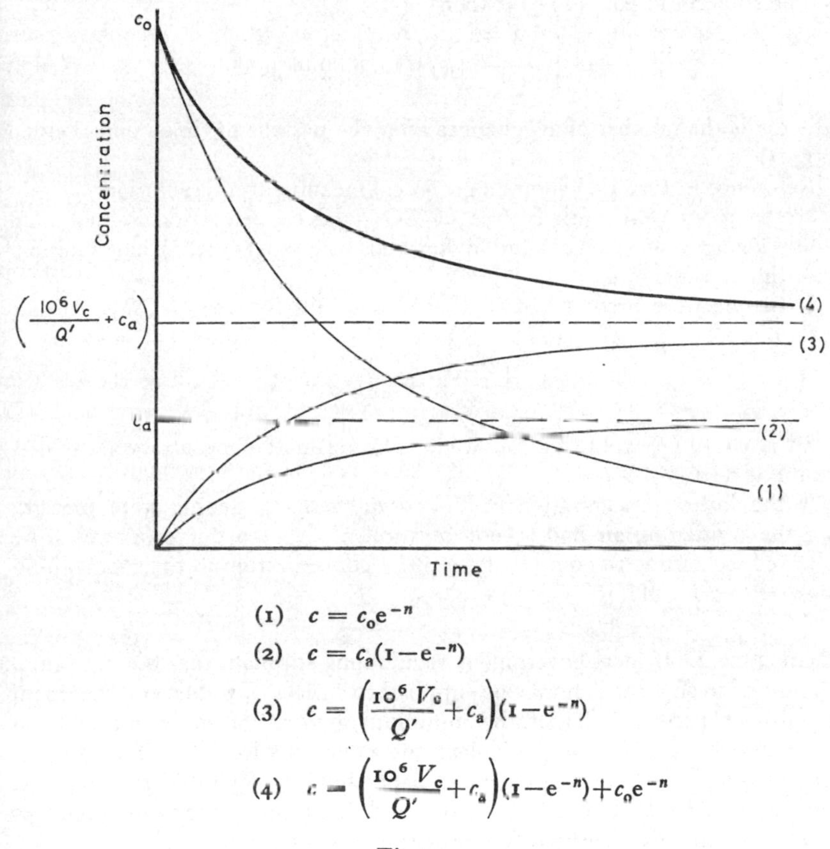

(1) $c = c_o e^{-n}$

(2) $c = c_a(1-e^{-n})$

(3) $c = \left(\dfrac{10^6 V_c}{Q'}+c_a\right)(1-e^{-n})$

(4) $c = \left(\dfrac{10^6 V_c}{Q'}+c_a\right)(1-e^{-n})+c_o e^{-n}$

Fig. 17.3

Then

$$\frac{dc}{dt}e^{(Q't/V')}+\frac{Q'c}{V'}e^{(Q't/V')} = \left(\frac{10^6 V_c+Q'c_a}{V'}\right)e^{(Q't/V')}$$

The left-hand side of this is the derivative of a product and so, by integrating, we get

$$A+c\,e^{(Q't/V')} = \frac{10^6 V_c+Q'c_a}{V'}\,\frac{V'}{Q'}\,e^{(Q't/V')}$$

where A is a constant of integration.

Write $B = AQ'$, then

$$B\,e^{(-Q't/V')} = (10^6 V_c + Q'c_a - Q'c)$$

The boundary condition is $c = c_o$ when $t = 0$, and so

$$B = 10^6\,V_c + Q'c_a - Q'c_0$$

The solution to eqn. (17.7) is then

$$c = \left[\frac{10^6\,V_c}{Q'} + c_a\right](1 - e^{-n}) + c_0 e^{-n} \qquad (17.8)$$

where n is the number of air changes after the passage of time t and is equal to $Q't/V'$.

Reference to Fig. 17.3 shows a graphical meaning of this solution.

If the room were initially free of CO_2, the concentration in the room would change along curve (2), following the law $c = c_a(1 - e^{-n})$, and attaining an ultimate value of c_a.

If people were present but the room was initially free of CO_2, the law would be

$$c = \left(\frac{10^6\,V_c}{Q'} + c_a\right)(1 - e^{-n})$$

and the curve (3) would be followed. The ultimate concentration attained would be $\{(10^6\,V_c/Q') + c_a\}$.

If the initial concentration in the room were c_o, people were present, and the ventilating air had a concentration of c_a, then curve (4) would be followed, according to eqn. (17.8), and the concentration in the room would approach a value of $\{(10^6\,V_c/Q') + c_a\}$.

EXAMPLE 17.2. If local government regulations stipulate that the minimum amount of fresh air which may be supplied to a place of public entertainment is 8 litres/s person and that the minimum amount of space allowable in the room is 12 m^3/person, calculate the concentration of carbon dioxide present after one hour, expressed as a percentage. Assume that fresh air contains 0·03 per cent of carbon dioxide and that human respiration produces $4{\cdot}72 \times 10^{-3}$ litres/s of carbon dioxide per person.

Answer

$Q' = 0{\cdot}008$ m^3/s person,

$c_a = 300$ parts per million,

$V_c = 4{\cdot}72 \times 10^{-6}$ m^3/s person,

$n = Qt/V'$ (for minimum ventilation),

$V' = 12$ m^3/person.

Hence,

$$n = 0{\cdot}008 \times 3600/12 = 2{\cdot}4 \text{ air changes.}$$

From eqn. (17.8)

$$c = \left(\frac{10^6 \times 4.72 \times 10^{-6}}{0.008} + 300\right)(1 - e^{-2.4}) + 300e^{-2.4}$$
$$= 890 \times (1 - 0.907) + 300 \times 0.0907$$
$$= 809.3 + 27.2$$
$$= 836.5 \text{ ppm}$$
$$= 0.084 \text{ per cent.}$$

EXAMPLE 17.3 A garage measures 60 m × 30 m × 3 m high and contains a number of motor cars which produces a total of 0·0024 m^3/s of carbon-monoxide.

(*a*) If the maximum permissible concentration is to be 0·01 per cent of carbon monoxide, what number of air changes per hour are required if the garage is in continual use?

(*b*) If the garage is in use for periods of 8 hours only, and if at the start of any such period the concentration of carbon monoxide is zero, what number of air changes per hour is needed if the concentration is to reach 0·01 per cent only by the end of the 8-hour period?

(*c*) What is the concentration after the first 20 minutes of an 8-hour period?

(*d*) If at the end of an 8-hour period the concentration is 0·01 per cent, for how long should the ventilation plant be run in order to reduce the concentration to 0·001 per cent?

Answer

(*a*) $c_{max} = 100$ ppm

Since the garage is in continual use, c_0 also equals one part per million.

The number of persons present is irrelevant but, for ease of understanding, it may be convenient to assume an occupancy of one person.

$Q' = V'n/3600$, since t is one hour.

$V' = 5400 \text{ m}^3$.

$c_a = 0$ and $V_c = 0.0024 \text{ m}^3/\text{s}$.

Hence, from eqn. (17.8),

$$100 = \left(\frac{10^6 \times 0.0024 \times 3600}{5400n} + 0\right)(1 - e^{-n}) + 100e^{-n}$$

$$1 = \frac{16}{n}(1 - e^{-n}) + e^{-n}$$

$$(1 - e^{-n}) = \frac{16}{n}(1 - e^{-n})$$

Hence $\quad n = 16.$

(*b*) From eqn. (17.8)

$$100 = \left(\frac{10^6 \times 0{\cdot}0024 \times 3600}{5400 \times n} + 0\right)(1 - e^{-n}) + 0e^{-n}$$

In this equation, n is the total number of air changes that has occurred after 1 hour, if the other terms of the equation are for one hour. It follows that a factor of 8 must be introduced in the appropriate places:

$$100 = \left(\frac{10^6 \times 0{\cdot}0024 \times 3600 \times 8}{5400 \times n}\right)(1 - e^{-n})$$

where n is the total number of air changes that has occurred after 8 hours.

$$1 = \frac{128}{n}(1 - e^{-n})$$

Therefore,

$$\frac{n}{128} = (1 - e^{-n})$$

If n equals 128, e^{-n} is almost zero and so we can conclude that the required air change rate is 16/hour.

(*c*) From eqn. (17.8)

$$c = \left(\frac{10^6 \times 0{\cdot}0008 \times 3600}{5400 \times 5{\cdot}33}\right)(1 - e^{-5{\cdot}33}) + 0 \times e^{-5{\cdot}33}$$

$$= 99 \text{ ppm.}$$

(In the above solution, 0·0008 is the production of CO in m^3 in twenty minutes.)

(*d*) Equation (17.4) is used because there is no production of carbon monoxide after the eight-hour period and the fresh air used to ventilate contains no contaminant of this sort.

Observing that 0·001 per cent is a concentration of 10 ppm, we can write

$$10 = 100e^{-n}$$

because c_0 is 100 ppm.

Hence, $$n = 2{\cdot}3.$$

At a rate of 16 air changes an hour, it takes 8·6 minutes to effect 2·3 air changes.

17.3 An application of the decay equation to changes of enthalpy

Throughout this book it has been usual to consider the steady-state case. Air has been supplied to the room being conditioned at a nominally constant temperature and, this temperature being lower than the design temperature in the room, the sensible heat gains to the room have been offset. Under these conditions the cooling capacity of the airstream exactly matches the sensible heat gain to the room, and if the sensible gain stays unaltered, the temperature of the air in the room also remains unaltered.

The picture is different, however, when the air-conditioning plant is first started up. Suppose that with the plant off the conditions within the room are equal to those outside and that no heat gains are occurring. When the air-conditioning system is started it delivers air to the room at a temperature very much below the value of the temperature prevailing there. Thus, the initial cooling capacity of the airstream is very large and the temperature of the air in the room is rapidly reduced. As this reduction is effected, the difference between supply air temperature and room air temperature decreases, and so the cooling capacity of the supply airstream diminishes.

This is a somewhat simplified picture of what is occurring. For example, heat gains are not solely due to transmission; solar radiation, electric lighting and occupants provide additional sensible gains and there are, of course, latent gains as well. There are, thus, complicated changes of load occurring.

By a process similar to that used to derive eqn. (17.8) it is possible to formulate a differential equation which represents the physical situation and to obtain a solution to it in terms of the relevant heat exchanges, enthalpies and masses.

The solution is

$$H = M[\{h_a + H(t)\}(1 - e^{-n})] + H_o e^{-n} \tag{17.9}$$

In this equation the following notation has been used:

M = mass of air contained in the room, in kg.

H_0 = initial enthalpy of the air in the room, in kJ. (Hence $H_0 = Mh_0$, where h_0 is the specific enthalpy in kJ/kg.)

h_a = specific enthalpy of the air supplied to the room in kJ/kg.

$H(t)$ = the enthalpy gain to the room, expressed in kJ/s per kg/s of air supplied, at any time t. (Hence the units of $H(t)$ are kJ/kg. In general, $H(t)$ is a function of time and the ease with which eqn. (17.9) may be made to yield a useful answer depends on how complicated this function is.)

G_a = rate of mass flow of the air supplied to the room, in kg/s.

The use of this equation is not limited to problems relating to changes of temperature or moisture content in an air-conditioned room. It may be used to solve a variety of problems involving unsteady state operation. For instance, it could be used to determine the amount of storage of chilled water necessary to prevent a refrigeration compressor from having too many starts an hour.

EXAMPLE 17.4. A room measures 3 m × 6 m × 3 m high and is air conditioned. Making use of the information given below, determine the dry-bulb temperature and relative humidity in the room three minutes after the air-conditioning plant has been started, assuming that the initial state in the room is the same as the state outside.

Sensible heat gain	2 kW
Latent heat gain	0·2 kW
Outside state	28°C dry-bulb, 20°C wet-bulb (screen).
Design inside state	22°C dry-bulb, 50 per cent saturation.
Constant supply air state	13°C dry-bulb, 8·055 g/kg.
Constant supply air quantity	0·217 kg/s.

Answer

The humid volume at a state midway between the inside and the outside states is about 0·8572 m^3/kg of dry air and this establishes the fact that the mass of air contained in the room is about 63 kg.

Consider first the change of temperature that occurs when the air-conditioning plant is started. Since temperature alone is the concern, the terms in eqn. (17.9) involving enthalpy may be conveniently modified so as to express only the sensible components of the enthalpy.

The information available may now be summarised—

$$G_a = 0{\cdot}217 \text{ kg/s}$$

$$M = 63 \text{ kg}$$

$$H_0 = 63 \times 1{\cdot}025 \times (28 - 0) = 1807 \text{ kJ}$$

$$H(t) = \frac{2{\cdot}0 \text{ kW}}{0{\cdot}217 \text{ kg/s}} = 9{\cdot}22 \text{ kJ/kg}$$

$$h_a = 1 \times 1{\cdot}025 \times (13 - 0) = 13{\cdot}32 \text{ kJ/kg}$$

$$n = G_a t/M = 0{\cdot}217t/63 = 0{\cdot}00344t$$

Then, from eqn. (17.9),

$$H = 63\{(13{\cdot}32 + 9{\cdot}22)(1 - e^{-0{\cdot}00344t})\} + 1807e^{-0{\cdot}00344t}$$

As t approaches a value of infinity, H approaches a value of 63 × 22·54 kJ, and so the specific sensible enthalpy of the air in the room tends towards a value of 22·54 kJ/kg. This means that the dry-bulb temperature of the air in the room eventually reaches a value of 22°C, as designed for. This is its ‘ potential value ’ (*see* section 13.10).

This may not seem very satisfactory. It must be remembered though, than when the system is started up it does not usually face its full design load. There is, thus, an opportunity for the system to pull down the room air temperature to 22°C under conditions of partial load.

After three minutes, for the case under consideration—

$$t = 3 \text{ minutes} = 180 \text{ seconds}$$

$$e^{-0.00344 \times 180} = e^{-0.62} = 0{\cdot}538.$$

Hence,

$$H = (63 \times 22{\cdot}54)(1 - 0{\cdot}538) + 1807 \times 0{\cdot}538$$

$$= 1627 \text{ kJ}.$$

Thus

$$\text{Dry-bulb temperature} = \frac{1627}{63 \times 1{\cdot}025}$$

$$= 25{\cdot}2\text{°C}.$$

A similiar approach, with appropriate modifications to the values of enthalpy so that latent heat is taken into account, but not the sensible component, yields a figure for the change in moisture content. From tables the specific enthalpy of dry air at 28°C is 28·17 kJ/kg and at the outside design state it is 55·36 kJ/kg, hence

G_a = 0·217 kg/s

M = 63 kg

H_0 = $63 \times (55{\cdot}36 - 28{\cdot}17) = 1712$ kJ

$H(t) = 0{\cdot}2/0{\cdot}217 = 0{\cdot}9215$ kJ/kg

h_a = $33{\cdot}41 - 13{\cdot}08 = 20{\cdot}33$ kJ/kg where 33·41 kJ/kg is the enthalpy of the supply air and 20·33 kJ/kg is the enthalpy of dry air at 13°C.

Then, from eqn. (17.9), as before,

$$H = 63(20{\cdot}33 + 0{\cdot}9215)(1 - e^{-0.00344t}) + 1712e^{-0.00344t}$$

As t approaches infinity, H tends to $63 \times 21{\cdot}25$, and so the specific enthalpy in the room tends to $21{\cdot}25 + 22{\cdot}13 = 43{\cdot}38$ kJ/kg, the design value, where 22·13 kJ/kg is the enthalpy of dry air at 22°C, after an infinitely long period of time. After three minutes, as before, we get a solution as follows—

$$H = 63 \times 21{\cdot}25(1 - e^{-0.62}) + 1712e^{-0.62}$$

$$= 1541 \text{ kJ or } 1541/63 = 24{\cdot}5 \text{ kJ/kg}.$$

At 25·2°C the sensible component of enthalpy, from tables, is 25·25 kJ/kg and so the total enthalpy is 49·75 kJ/kg.

At a dry-bulb temperature of 25·2°C and an enthalpy of 49·75 kJ/kg, the relative humidity is about 48 per cent.

EXERCISES

1. If the heat gain to a conditioned space is suddenly changed from a steady value q_1 to a new steady value q_2 and the supply air temperature is changed from a corresponding steady value t_{s1} to t_{s2}, the space temperature t_r changes according to the equation:

$$-mc\frac{dt_r}{dT} = Mc(t_r - t_{s2}) - q_2$$

where m is the mass of air in the room, c is the specific heat of air, M is the mass flow rate of the supply air, and T is the time. Hence, prove that the instantaneous space temperature t_{r2} is given by

$$E = \frac{t_{r2} - t_{r1}}{t_{r\,max} - t_{r1}} = 1 - e^{-NT}$$

where N is the number of air changes and t_{r1} is the initial space temperature. Show that for 10 air changes t_{r2} attains 90 per cent of its maximum value in 14 minutes.

2. A classroom having a cube of 283 m^3 undergoes $1\frac{1}{2}$ air changes per hour from natural ventilation sources. The concentration of CO_2 in the outside air is 0·03 per cent and the production of CO_2 per person is $4{\cdot}72 \times 10^{-6} m^3/s$

(*a*) What is the maximum occupancy, if the CO_2 concentration is to be less than 0·1 per cent at the end of the first hour, assuming that the initial concentration is 0·03 per cent?

(*b*) What is the maximum occupancy if the classroom is continuously occupied and the concentration must never exceed 0·1 per cent?

Answers: (*a*) 225, (*b*) 175.

BIBLIOGRAPHY

1. T. E. Bedford. *Basic Principles of Ventilation.* H. K. Lewis, 1948.
2. W. P. Jones. Theoretical aspects of air-conditioning systems upon start-up. *J. Instn Heat. Vent. Engrs*, 1963, **31**, 218–223.

18

Filtration

18.1 Contaminants

Atmospheric contaminants fall into four classes: solid, liquid, gaseous and organic. Distinction between some members of these classes is not clear cut and not particularly important, but recognition of their existence is relevant. For this reason the following broad statements are made; they should not be regarded as a set of definitions.

In speaking of small particles—which all atmospheric contaminants are—it is customary to use the micrometre as a unit of measurement of their size.

(*a*) *Dusts* (< 100 μm)
These are solid particles produced by natural or man-made processes of erosion, crushing, grinding or other abrasive wear. Dusts do not agglomerate, except under the influence of electrostatic forces, but settle on the ground by the force of gravity.

(*b*) *Fumes* (< 1 μm)
These also are solid particles but formed in a different way from dusts. Fumes are produced by the sublimation, or by the condensation and subsequent fusion, of gases which are solids at normal temperature and pressure. Metals can be made to produce fumes. The term is often misused to indicate merely a gaseous substance which has a pungent smell.

(*c*) *Smokes* (< 1 μm)
Smokes may be regarded as small solid particles which are the product of incomplete combustion or, more truly, as a mixture of solid, liquid and gaseous particles resulting from partial combustion. Excluding the gaseous particles, which are molecular in size, industrial and domestic furnaces produce particles which vary in size from 0·1 μm to 1·0 μm, but tobacco smokes are much smaller, existing in the range 0·05 to 0·1 μm. Hence the difficulty in their effective removal from air streams.

(*d*) *Mists and Fogs* (< 100 μm)
The distinction between mists and fogs is somewhat blurred. However, they are both airborne droplets which are liquid at normal temperature and pressure. Their normal range of size is 15 to 35 μm.

(*e*) *Vapours and Gases*

A distinction between the two has already been drawn in section 2.2. From a filtration point of view, they are substances which are in the gaseous phase at normal temperature and pressure, but whereas a vapour may be removed by cooling to below the dew point, a gas cannot. Both gases and vapours diffuse uniformly throughout an enclosing space. Separation by inertia is not possible.

(*f*) *Organic Particles*

The commonest of these are: bacteria (0·2–5·0 μm), pollen (5·0–150 μm), the spores of fungi (1·0–20 μm) and viruses (much less than 1·0 μm in size). Bacteria are generally larger in size than 1·0 μm and rely on dust particles as a mode of transport. Hence the importance of dust filtration in controlling the spread of infection by bacteria. Viruses are very small indeed, and for this reason many have never been identified. Some are transported by airborne liquid droplets.

Individual particles of size less than 0·1 μm are not thought to be of great importance. They may, however, become permanent atmospheric impurities, particularly as smells. Other very much larger particles, such as insects and even birds, must be kept out of air-conditioning systems but, except in the case of insects and electric filters, they do not usually constitute a special filtration problem.

18.2 Efficiency

There are two criteria for establishing the effectiveness of an air filter: the gravimetric test and the discoloration test. While it must be acknowledged that the first of these is of value in expressing the dust-holding capacity of a filter, it must also be recognised that it is not a true indication of the cleansing ability of the filter. The staining of walls and fabrics resulting from the presence of the dust in the air is not a function solely of the mass of the dust and its size; small particles of small mass are just as efficacious in soiling as are larger and heavier particles. The discoloration test is therefore a better way of determining the cleansing ability of a filter, because it actually measures the staining capacity of the air before and after it passes through the filter.

(*a*) *The Gravimetric Test*

Gravimetric efficiency is defined by the ratio

$$\frac{\text{weight of dust retained by the filter}}{\text{weight of dust fed through the filter}} \times 100.$$

The test and the equipment and standards to be used in carrying it out are described in B.S. 2831:1971, entitled 'Methods of Test for Air Filters used in Air-conditioning and General Ventilation'.

A specified dust is fed into the airstream, before the filter being tested, in measured quantity. The weight of the filter is taken before and after the test, the two figures enabling the efficiency to be calculated by the ratio given above. It is specified that the test be carried out in three steps, each for the same approximate increment of air-pressure drop across the filter. (As a filter gets dirty, the air pressure drop increases.) This is so that the dust-holding capacity of the filter may be established at the same time as its gravimetric efficiency.

Dust-holding capacity is defined as the weight of dust that is collected by the filter while its air pressure drop increases from an initial value by 125 N/m^2, or to a value which is twice the initial value, whichever is the greater.

There are two kinds of test dust specified:

(i) Aloxite optical powder No. 50, consisting of a fused aluminium dust with a distribution of particle size as follows:
60 to 80 per cent by weight, 3·5 to 7·0μm,
99·5 per cent by weight, less than 13μm,
2·0 per cent by weight, less than 2·5μm.

It is also termed Test Dust No. 2.

(ii) Aloxite optical powder No. 225, consisting of fused aluminium dust with a distribution of particle size as follows:
60 to 80 per cent by weight, 15 to 25μm,
99·5 per cent by weight, less than 35μm,
2·0 per cent by weight, less than 10μm.
This is also termed Test Dust No. 3.

Either one of these or a mixture of both is recommended for use. It is important in comparing the relative efficiencies of two different filters to know the type of dust used in the efficiency test for each.

(b) The Methylene-blue Test

This is the discoloration test recommended in B.S. 2831:1971. The standard describes in detail the equipment and method to be used. In brief, the test is carried out as follows:

A solution of methylene-blue dye in water is injected into the airstream before the filter is tested. During its passage through the duct leading to the filter the water is evaporated and the blue dye remains in the airstream as small solid particles. The dust is known as Test Dust No. 1. These particles are within the range of 0·2 to 2·0 μm in diameter and about 50 per cent by weight of them are over 0·5 μm in size. By means of a vacuum pump a measured sample of this airstream is drawn off and passed through a white filter paper. Another measured sample is taken from the airstream after the filtering, and also passed through a white filter paper.

The extent to which the airstream has stained the filter paper in each case is established by passing a beam of light through the filter paper, and measuring its emergent intensity by means of a photo-sensitive cell.

Let the following symbols be used.

V = the measured volume of the air-dust mixture drawn through a filter = paper, in m^3.

I_o = the reading of the photo-sensitive cell (termed a 'densitometer') for a clean filter paper.

I = the reading of the densitometer for a stained paper.

L = the optical density of the stain on the paper.

= $\log_e (I_o/I)$

L_0 = the optical density of the clean paper (having a value of about 0·1 for esparto paper).

Then the relationship between the optical density and the volume sampled is given by

$$L - L_0 = CV$$

where C is a constant, the value of which depends on the concentration of the dust and the type of the filter paper, provided that L lies between about 0·2 and 0·5.

Methylene-blue efficiency is defined by the equation

$$\text{efficiency} = \left\{1 - \frac{V_1(L_2 - L_0)}{V_2(L_1 - L_0)}\right\} \times 100 \text{ per cent.}$$

In this equation the subscripts 1 and 2 denote unfiltered and filtered air, respectively.

Details are given in the British Standard for eliminating error and for obtaining a fair average result from several readings; it is stipulated that agreement to the extent of ±5 per cent must be obtained between three tests. It is also recommended that the methylene-blue tests be carried out in conjunction with the gravimetric test.

The methylene-blue test is a stringent one and gives very low values for many filters. This is not to say that these filters are of no use; a filter should be chosen to have an efficiency appropriate to the application. However, since it is a test based on a well-established standard, both as regards dust size and the method of test, it forms a yardstick by which the efficiencies of different filters can be compared, and it is only by comparing like with like that relative merits are seen.

(*c*) *Other Tests of Efficiency*

Discoloration tests are also carried out with other dusts. These may have advantages for their use on site but comparison of results achieved by different methods is not valid.

One other test dust is coming into use which offers a most uniform and accurate way of determining efficiency. This is the use of small crystals of sodium chloride which are left suspended in the airstream in much the same way that the particles of methylene-blue dye were. It is termed the

'Sodium Flame Test' and it gives an instantaneous reading of the extent to which the dust particles (crystals) are penetrating a filter. Like the methylene-blue test, it is so stringent that it is not worth using unless the filter is an efficient one, such as a so-called 'absolute' filter. The test involves withdrawing a specimen of air from the downstream side of the filter and passing it through a flame of burning hydrogen gas. The flame is coloured bright yellow when sodium chloride crystals pass through it, and the intensity of the colour is a direct measurement of the weight of the sodium chloride.

One feature of the test which renders it more accurate than the methylene-blue test is the uniformity of shape of the particles; all crystals of sodium chloride are cubical. The average diagonal measurement of a cubical crystal is about 0·6 μm and the largest size is 1·7 μm. Of the particles, 99 per cent are less than 0·6 μm in size and 58 per cent are below 0·1 μm. Hence, the exacting nature of the test. The test is specified in detail in B.S. 3928 : 1969, and it is finding favour over the methylene-blue test because it is so rapidly performed.

18.3 Classification according to efficiency

In a broad way it is possible to distinguish between filters according to the efficiencies they show on the methylene-blue basis, special types of filters being excluded.

On this basis we can say the following:

Automatic roll	10–20%
Viscous	5–15%
Dry fabric (cotton-wool pads obliquely arranged)	25%
Dry fabric (cotton-wool wadding in V-arrangement)	35%
Dry fabric (cotton-wool wadding plus fine glass wool)	75%
Electrical	50–90%
Brush type	10–20%
Absolute	65–99·9%
Cleanable cells	10%

Efficiencies are related to the velocity of airflow through the filtering medium, and the manufacturer's recommendation should not be exceeded. It is often possible, particularly in absolute filters, to improve efficiency by using a larger filter having a reduced velocity of airflow.

Pressure drops are also related to efficiencies, except in electrical filters which achieve a high efficiency without the need for a large pressure drop.

18.4 Viscous filters

Essentially, these filters have a large dust-holding capacity but a low efficiency, and this defines their sphere of application; for example, they are more suitable for use in industrial areas where a high degree of atmospheric pollution prevails. Their drawback is usually expense, particularly in automatic versions.

The principle of the viscous filter is that if the mixture of dust and air is forced to follow a tortuous path in negotiating a passage through the filtering medium, inertial separation of the more massive dust from the lighter air will occur. If the filtering medium is coated with a suitable oily fluid, the particles of dust will be trapped by the oil and retained, the air passing on. To effect this, the oil must have a surface tension low enough to permit easy entry for the dust. Once within the oil, it is desirable that the dust should disperse, not remaining in one place but being retained in depth throughout the filter (hence its large dust-holding capacity). It is thus necessary that the oil used should have a high enough capillarity to encourage dust to flow away from its point of entry to the oil. It is also essential that the oil should be non-inflammable, non-toxic, germicidal and comparatively non-evaporative. The oil should not deteriorate during its life. There are two types of viscous filter: the cell-type and the automatic type.

The cell type consists of a cheap retaining box, open front and rear, which contains the filtering medium. The medium (often some form of industrial waste such as swarf, brass turnings, etc.) is coated with the oil, and the cells are assembled into a battery of convenient shape and placed across the air-stream for filtration purposes. After use, when dirty, the cells are either thrown away and replaced with new cells or, if they have been selected with this in mind, they may be treated with a cold wash, drained, and re-coated with fresh oil for further use.

The automatic viscous filter takes the form of a continuous roll of material coated with the oil and is motor-driven across the airstream. It is arranged so that the roll is drawn through a trough of oil at the bottom of the assembly. The trough serves the dual purpose of washing off the dirt and re-coating the fabric of the roll with relatively clean oil. The material of the fabric takes several forms: one is an array of hinged metal plates which, dangling downwards, force the dirty airstream to change direction several times, thus achieving the inertial separation of the dust into the oil, as desired; other versions of automatic viscous filter are elaborated by the use of oil sprays and pumps. Figure 18.1 illustrates the cell and automatic types.

Pressure drops across these filters very from 40 to 60 N/m^2, and recommended air velocities over the face are from 1·7 to 2·5 m/s. Viscous filters are not very effective in removing particles of size less than 0·5 μm.

18.5 Dry filters

Cotton-wool, glass-fibre fabric, asbestos-woven fabric, pleated paper of various types, foamed polyurethane, cellular polythene and other materials are used for the construction of dry filters. As with viscous filters, there are cell-type and automatic roll-type filters available. The construction of these, in broad outline, is very similar to that of viscous filters. No oil is used, of course, and the way in which efficiency is improved is by increasing the surface area of the fabric offered to the airstream for filtration purposes. In cell-types this is achieved by using pads of material placed obliquely across the airstream (Fig. 18.2). An alternative to this, which can be made to yield very high efficiencies, is to use a system of pleating. This is best achieved with paper as a material, although asbestos is also used and gives high efficiencies. Figure 18.3 shows an arrangement of pleating.

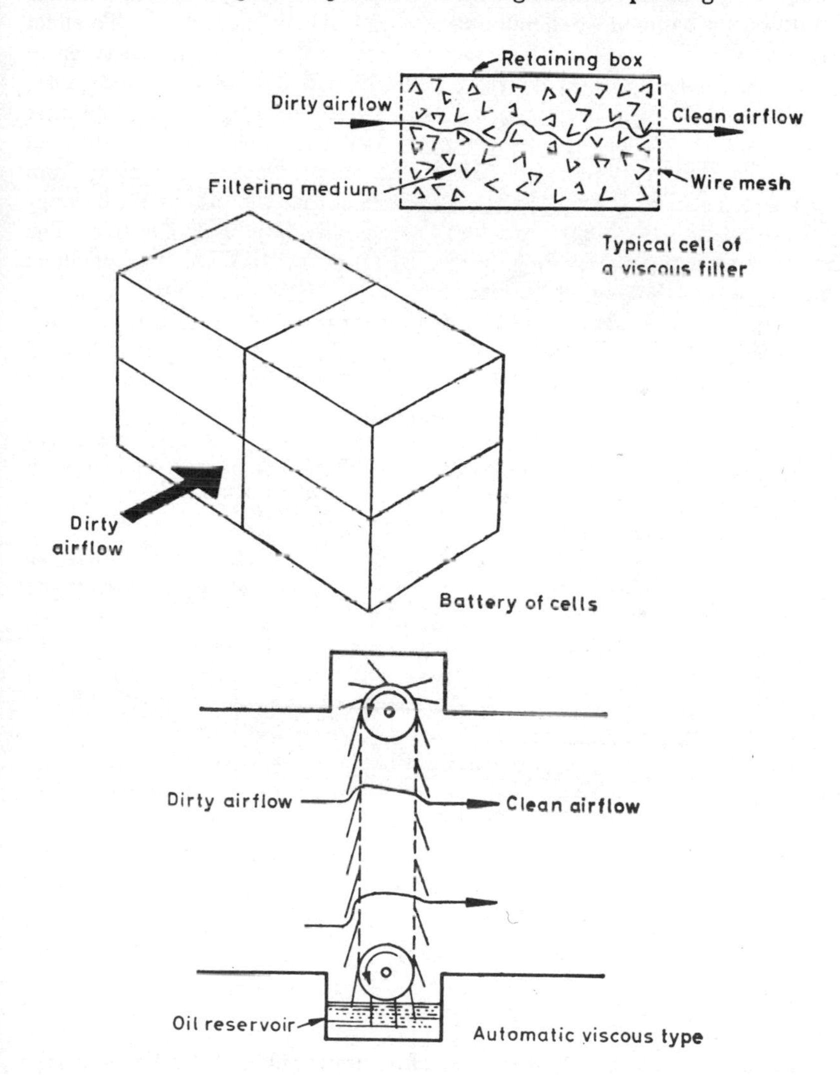

Fig. 18.1

When a very large amount of material is used in a filter, the efficiency becomes very high and the filter is termed an 'absolute' filter. No filter is truly absolute but almost any desired efficiency could be achieved if sufficient filtering fabric were used; this would be associated with a high pressure drop, but one way of achieving a high efficiency without the penalty of wasted energy is to use a very large filter, that is, one with a very low face velocity. Naturally, the capital cost is increased.

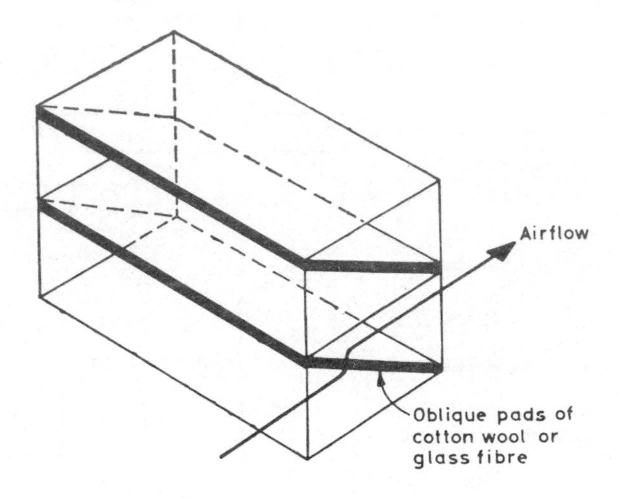

Fig. 18.2

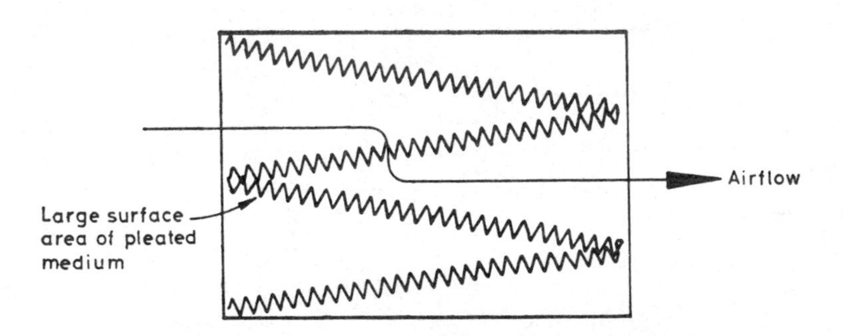

Fig. 18.3 Pleated high-efficiency filter

Automatic dry-fabric filters consist of an upper roll of clean fabric wound downwards across the airstream. The dirtied material is then re-wound into a roll at the bottom of the unit. Figure 18.4 shows this arrangement. The advantage of this type of filter is the low maintenance cost when compared with the cell-type of dry filter. Removing a dirty roll and replacing it with a clean one at relatively infrequent intervals is a cleaner task and is, therefore, easier and cheaper than changing dirty cells. The disadvantage is that the efficiency is not so high. This is overcome to some extent by using a denser material and by introducing V-form or S-shaped changes of direction in the fabric as it crosses the airstream, thus presenting more surface area. The rolls, as in automatic viscous filters, are motorised, and the geared-down electric driving motors are operated either from a time or pressure-differential switch.

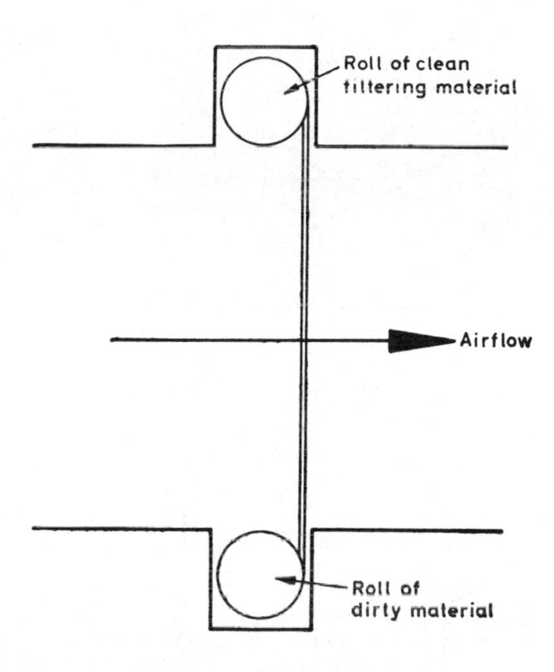

Fig. 18.4

18.6 Electric filters

An electric filter is illustrated in Fig. 18.5. The principle of operation is that when air is passed between a pair of oppositely charged conductors it becomes ionised if the voltage difference between the conductors is sufficiently large. Both negative ions and positive ions are formed, the latter being in the larger quantity. By contact with the dust particles mixed with the airstream, the charge of the ions is shared with the dust. In this way, about 80 per cent of the dust particles passing through the ionising field are given a positive charge and the other 20 per cent a negative charge. The ionising voltage used varies somewhat, and to achieve a given efficiency of ionisation smaller air velocities can be used with smaller voltages. However, typical ionising voltages are from 7800 to 13000.

The electrodes which form the poles of the ionising unit mentioned above consist of alternate small diameter (of the order of 25 μm) tungsten wires and flat metal plates, spaced an inch or so apart. The intensity of an electric field is a function of the curvature of the charged conductor producing it, hence the small diameter of alternate poles. The tungsten wires are positively charged, the plates representing the negative electrodes being earthed.

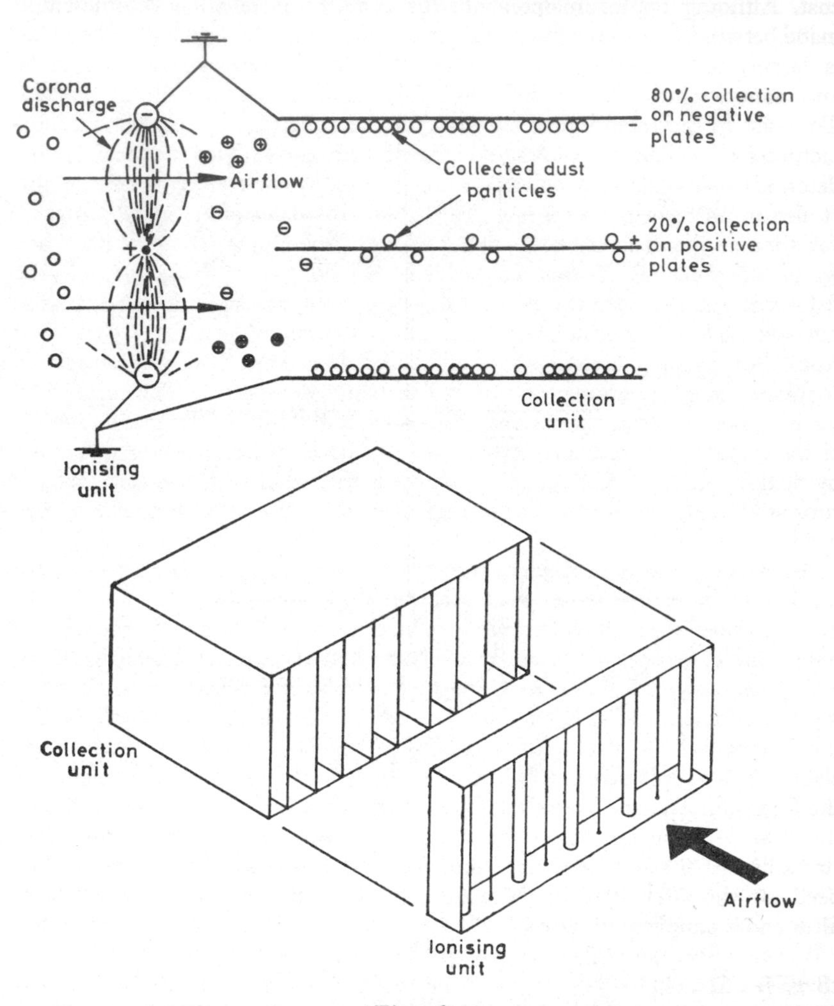

Fig. 18.5

The charged particles of dust which leave the ionising unit then pass through a collecting unit. This consists of a set of vertical metal plates, spaced about half-an-inch or so apart. Alternate plates are positively charged and earthed and attract the negatively and positively charged dust particles, respectively. The voltage difference across the collecting plates is about 6000 or 7000, although at least one manufacturer is currently offering an electrical filter which uses about 8000 volts across the electrodes of both the ionising and collecting sections.

One of the main advantages claimed for electric filters is low maintenance cost, when compared with other filters of similar efficiency but lower capital cost. Although the low maintenance cost is debatable and if a comparison is made between filters on a basis of their owning and operating costs, the result is largely at the choice of the analyst, it is very necessary there should be automatic washing of electric filters if maintenance cost is to be minimised. To assist in the retention of the dust on the collecting plates, some manufacturers arrange for the plates to be coated with an oil. This may mean that a detergent is necessary as well as a water supply when washing is carried out. It also means that fresh oil must be put on the plates after the washing. A variation on this is to use a mixture of oil and detergent together so that when washing is done, usually with cold water, respraying with oil also provides the detergent necessary for the next wash. One manufacturer uses no detergent but asks that hot water be used for washing. Oil is not used in this case. One feature of washing common to all filters is that water at high pressure is required, gauge pressures of 2 to 3 bar being necessary. The method of washing is either from fixed stand-pipes attached to the appropriate reservoirs of water and oil or from a traversing nozzle system, attached to the reservoirs by flexible piping. The pump which feeds the water to the nozzles usually runs at 48 rev/s and is sometimes noisy, because of the requirement of a large head.

An insect screen is necessary upstream of the filter, and it is also necessary to provide either a pre-filter or an after-filter.

A pre-filter is preferred because it relieves the filtration load on the electric filter which follows. However, if it is used it must be washed automatically when the rest of the filter is cleaned, otherwise there is an increased maintenance problem. If an after-filter is used, its duty is merely to ensure that the filter fails safe. Neither the pre-filter nor the after-filter need be very elaborate, a simple screen of closely woven nylon cloth is often enough. If the filter is not washed regularly at suitably short intervals, the thickness of the dust-covering on the collection plates increases and, after a while, dust starts flaking off and is carried into the conditioned space. Hence another need for the after-filter. The ideal then is an automatically washed pre-filter and a simple after-filter.

18.7 Wet filters

Washers and scrubbers of various sorts are used extensively throughout industry, largely for the absorption of soluble gases. They are not very commonly used for cleansing the air of solid dust particles. The effectiveness of a washer in removing a dust depends on the 'wettability' of the dust by water. This is a function of the surface tension of the water when it is in contact with the solid involved. Different forces of surface tension are the rule for different materials. Thus, water finds it very difficult to wet greasy materials but fairly easy to wet non-greasy materials. Examples of greasy materials are atmospheric pollutants present in urban and industrial atmospheres; it follows that washing the air is not a very effective way of cleaning it of the dust normally present. Spray-type air washers are thus not to be regarded as filters. However, capillary air washers are, it is claimed, partial filters, owing to the intimacy of contact between the airstream and the water. It is unsafe, nevertheless, to use an air washer of any sort, without a filter.

As was remarked above, air washers are an effective way of dealing with undesirable atmospheric gases such as sulphur dioxide, which may be taken into solution by the water.

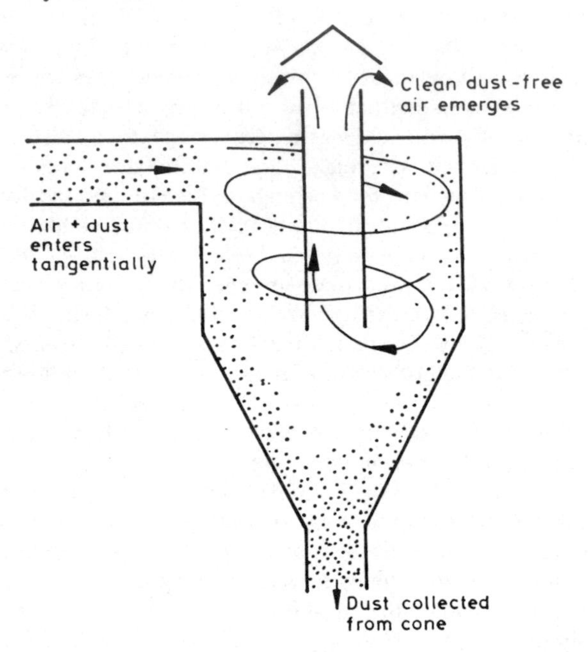

Fig. 18.6

18.8 Centrifugal collectors

If air is made to travel in a circular path, centrifugal force acts both on the molecules of the air and on the associated dust particles. The dust particles being the heavier, the force on them is the greater and so they are forced to the outer boundary of the curved airstream. This is the principle of the cyclone, illustrated in Fig. 18.6. This method of dust removal is not in use in air-conditioning systems. Its application is confined to industrial exhaust installations where the dust mixed with the airstream is relatively massive, such as wood shavings and sawdust.

18.9 Adsorption filters

The process of absorption, which is a chemical process, is to be distinguished from that of adsorption, which is a purely physical process.

Just as there is an attraction between a liquid and a solid at a surface, so there is also an attraction between a gas and a solid at a surface. An explanation of the mechanism of this cohesive force is beyond the scope of this text but the consequences of it are briefly discussed in the following.

If the existence of the attractive force between a gas and a solid is acknowledged, then it is clear that if a very large surface area can be offered to a gas a large amount of the gas will be attracted to the surface of the solid. The process of surface attraction is not limitless, in due course the surface is saturated, as it were, and no further gas will be taken up.

The principal adsorption filter used in air conditioning is the activated carbon filter. It is most effective in removing smells from the atmosphere, and in removing poisonous gases such as sulphur dioxide. The capacity of an activated carbon cell is expressed as an efficiency, but for sulphur dioxide, as an example, a typical cell with an efficiency of 95 per cent cannot adsorb 95 per cent of its own weight in sulphur dioxide. In fact, for the product of one particular manufacturer, a cell with a quoted efficiency of 95 per cent can adsorb 10 per cent of its own weight in sulphur dioxide. The cell contains 20 kg of activated carbon and can adsorb 2 kg of SO_2. (Activated carbon, incidentally, is prepared from the shells of coco-nuts and has a structure which offers an enormous surface area to any stream of gas passing over it.)

The extent of adsorption seems to be related to the molecular weight and boiling point of the substance to be removed. The higher the boiling point the greater the adsorption. Thus, the adsorption capacity of activated carbon for such substances as ammonia, ethylene, formaldehyde, hydrogen chloride and hydrogen sulphide, which boil at between $-104°$ and $-21°C$ is inadequate for practical purposes. On the other hand, substances such as butyric acid (' body odour '), petrol, putrescine and the common mercaptans etc., which boil at the higher temperatures of 8° to 158°C, are very effectively adsorbed. The activated carbon used during the 1939–45 war for adsorbing the poisonous low boiling point substances such as arsine, hydrocyanic acid and other war gases was specially impregnated with other chemicals to increase the adsorption.

When the activated carbon has reached saturation it is removed from the filter for re-activation. This is accomplished by heating the carbon to a high temperature, of the order of 600°C or more. It is customary to return saturated cells to the supplier for re-activation, using spare replacement cells in the meantime.

BIBLIOGRAPHY

1. B.S. 2831: 1971. *Methods of test for air filters used in air conditioning and general ventilation.* British Standards Institution.
2. B.S. 3928: 1969. *Method for sodium flame test for air filters (other than for air supply to I.C. engines and compressors).* British Standards Institution.
3. C. P. McCord and W. N. Witheridge. *Odours, Physiology and Control* McGraw-Hill, 1949.

Index